PESTICIDE INDEX

AN INDEX OF CHEMICAL, COMMON AND TRADE NAMES OF PESTICIDES AND RELATED CROP-PROTECTION PRODUCTS

Up to 30 April 1989
The Royal Society of Chemistry
The University
Nottingham NG7 2RD
England

From 1 May 1989
The Royal Society of Chemistry
Thomas Graham House
Science Park
Milton Road
Cambridge CB4 4WF
England

EDITORS: Hamish Kidd and Douglas Hartley

Cover Design: John Tanton

ISBN 0-85186-733-2

Printed in the United Kingdom by
Unwin Brothers Limited, Old Woking, Surrey.

Note: Governmental and educational establishments, research institutions and non-profit-making organisations in countries eligible for British Government Aid can obtain a copy of this index free-of-charge by applying directly to the Overseas Development and Natural Resources Institute (ODNRI), Central Avenue, Chatham Maritime, Chatham, Kent ME4 4TB, UK.

INTRODUCTION

This index is intended as a QUICK REFERENCE GUIDE to pesticide names and manufacturers, as there are many other good handbooks and manuals dealing with the application, toxicology and use of pesticides; a list of such reference works is given at the back of this book.

Entries in this index are of two kinds – ACTIVE-INGREDIENT entries and TRADE NAME entries.

ACTIVE-INGREDIENT entries consist of an active-ingredient name (BSI, ISO etc.) (in **BOLD**), followed by a letter or series of letters indicating activity (e.g. H for herbicide, I for insecticide, F for fungicide); a full list of these activities is given below. Then follows the Chemical Abstracts Registry Number, which is a unique number for a chemical used internationally in many printed publications and in most chemical databases. This is followed by the fully systematic chemical name (conforming in most cases to the IUPAC English rules). Finally there is a list of the many trade names and code numbers which can be used for that active-ingredient chemical.

SAMPLE ACTIVE-INGREDIENT ENTRY

fentin acetate FLM [900-95-8] triphenyltin acetate, fentine acetate, TPTA, asetat fenolovo, Batasan, Brestan, Chimate, Fenilene, Fentol, Hoe 2824, Hokko Suzu, Liro-Tin, Pandar, Phenostan-A, Stanex, Suzu, Tinestan, Ucetin

TRADE NAME ENTRIES consist of the trade name followed by a letter or series of letters indicating activity (e.g. H for herbicide, etc.), then the active-ingredient name or names (in **BOLD**), and finally the company(ies) who market pesticides of that trade name.

SAMPLE TRADE NAME ENTRIES

Fervinal H **sethoxydim** Schering
Fibenzol F **benomyl** Agrocros
Ficam I **bendiocarb** Schering, NOR-AM
Filariol IA **bromophos-ethyl**
Filex F **propamocarb hydrochloride** Schering
Filitox IA **methamidophos** Bitterfeld
Final H **glufosinate-ammonium** Hoechst

To save space SELF-EVIDENT names, i.e. those which contain the same name as the active-ingredient, have been omitted, together wih those which the formulation variants.

A number of obsolete active-ingredient names have, however, been included for completeness, as some of these do still crop up in literature from time to time. Alternative common names are cross-referred back to the preferred BSI or ISO name.

ACTIVITY CODES USED IN THIS BOOK

A acaricide
B bactericide
E earthworm killer
F fungicide
G plant growth regulator
H herbicide
I insecticide
L algicide
M molluscicide
N nematicide
R rodenticide
S synergist and other additives
T attractant
V virucide
W wound protectant
X repellent

NOTE

The PESTICIDE INDEX was first published in 1976 by the Centre for Overseas Pest Research (COPR) (now the Overseas Development and Natural Resources Institute (ODNRI)) as a supplement for PANS (as *Tropical Pest Management* was then known), and, from 1977 to 1984 was revised every two years as a supplement to *Tropical Pest Management*. In 1984 the Index was extensively enlarged and updated, and separately published as a compact reference book. The format of the 1984 edition has been continued in this new revised and extended edition.

---

1 + 1 H **atrazine + cyanazine** Shell
666 I **lindane**
1081 R **fluoroacetamide**
58 12 315 I **propoxur**
A7 Vapam FIHN **metham-sodium**
A 11 F **mancozeb + zineb** Ciba-Geigy
A 36 F **decafentin**
A 150 F **zineb** Du Pont
A 363 I **aminocarb** Bayer
AA 800 R **piperonyl butoxide + pyrethrins** Schering AAgrunol
AA 10460 H H **amitrole + atrazine** Schering AAgrunol
AA 10478 Z F **carbendazim + imazalil** Schering AAgrunol
AAbantyl Combi H **dicamba + MCPA + mecoprop** Schering AAgrunol
Aabi-Rattentod-Streupulver R **warfarin** Jannausch
AAcaptan F **captan** Schering AAgrunol
AAcaptan M F **captan + maneb** Schering AAgrunol
AAcaptan S F **captan + sulphur** Schering AAgrunol
AAchlorine G **chlormequat chloride** Schering AAgrunol
AAcidon H **bromacil + diuron** Schering AAgrunol
AAcombin H **MCPA + mecoprop** Staehler
AAcuram F **cymoxanil + fentin acetate + mancozeb** Schering AAgrunol
AAcuram-C FW **carbendazim** Schering AAgrunol
AAdifec I **chlorpyrifos** Schering AAgrunol
AAdimethoat IA **dimethoate** Staehler
AAdimitrol H **amitrole + diuron** Staehler
AAdimitrol fluessig H **amitrole + simazine** Staehler
AAdimitrol-SP H **amitrole + atrazine + 2,4-D** Staehler
AAdinol H **DNOC** Schering AAgrunol
AAdipon H **dalapon-sodium** Schering AAgrunol, Staehler
AAfeniton I **fenitrothion** Schering AAgrunol
AAfertis F **ferbam** Schering AAgrunol
AA-Fleur IA **diazinon** Schering AAgrunol
AA-Fleur spuitbus IA **diazinon + dicofol** Schering AAgrunol
AAfuma F **quintozene** Schering AAgrunol
AAglotyl-4 H **MCPA** Schering AAgrunol
AAgrano-droog-77 F **carbendazim + imazalil + quintozene** Schering AAgrunol
AAgrano GF 2000 F **imazalil** Staehler
AAgrano Spezial Schlaemmbeize F **furmecyclox + thiabendazole** Staehler
AAgrano 2000 UF F **furmecyclox + imazalil** Staehler
AAgrano Universal Schlaemmbeize F **furmecyclox + imazalil + thiabendazole** Staehler
AAgrano 2000 UT F **carbendazim + imazalil** Staehler
AAgrazon H **2,4-D + dicamba + MCPA + mecoprop** Schering AAgrunol
AAgrunol Anti-stuifmiddel S **poly(vinyl acetate)** Schering AAgrunol
AAherba-CIPC H **chlorpropham** Staehler
AAherba-Combi-Fluid H **2,4-D + MCPA** Staehler
AAherba-2,4-D aminezout H **2,4-D** Schering AAgrunol
AAherba-DP H **dichlorprop** Staehler
AAherba-KV-Combi-Fluid H **2,4-D + mecoprop** Staehler
AAherba-M-Fluid H **MCPA** Staehler
AAherba-Oxy-M H **bromoxynil + ioxynil + mecoprop** Staehler
AAherba-Super-Fluid H **mecoprop** Staehler
AAhydraz nieuw G **maleic hydrazide** Schering AAgrunol
AA-Karmex HL **diuron** Schering AAgrunol
AAkarzol H **amitrole + diuron** Schering AAgrunol
AA Kasaerosol I **dichlorvos** Schering AAgrunol
AA-Kelthane A **dicofol** Schering AAgrunol
AAko-CCC 750 G **chlormequat chloride** Aako
AAlindan IA **lindane** Schering AAgrunol
AAlindan-Inkrusta-S FI **lindane + thiram** Staehler
AAmagan F **maneb** Schering AAgrunol
AAmasul N F **maneb + sulphur** Schering AAgrunol
AAmeltex IA **lindane** Schering AAgrunol
AAmitrol H **amitrole + ammonium thiocyanate** Schering AAgrunol
AA Mix H **2,4-D + dicamba + MCPA** Schering AAgrunol
AAmonam S **metham-sodium** Schering, Schering AAgrunol, Staehler
AA Mos-ex H **iron sulphate** Schering AAgrunol

AAnetos-M H **mecoprop** Schering AAgrunol
AApermin I **permethrin** Schering AAgrunol
AAphythora-M F **maneb** Staehler
AAphytora F **zineb** Schering AAgrunol, Staehler
AApirol F **thiram** Schering AAgrunol, Staehler
AApirsan F **ferbam + manam + thiram** Schering AAgrunol
AApropion-4 H **mecoprop** Schering AAgrunol
AAprotect X **ziram** Cillus, Rhodiagri-Littorale, Schering, Schering AAgrunol, Staehler, UCP
AArado FB **copper oxychloride** Schering AAgrunol
AArdisan F **ethylmercury bromide** Schering AAgrunol
AAritna I **lindane** Schering AAgrunol
AArocint Nieuw F **1,3-dichloropropene + etridiazole** Schering AAgrunol
AArosan F **dodemorph acetate** Schering AAgrunol
AArupex I **lindane + piperonyl butoxide + pyrethrins** Schering AAgrunol
AAscabol IA **lindane** Schering AAgrunol
AAservo G **chlorpropham + propham** Schering AAgrunol
AAsprayol VIAS **petroleum oils** Schering AAgrunol
AAstaman F **fentin acetate + maneb** Schering AAgrunol
AAstaneb F **fentin acetate + maneb** Schering AAgrunol
AAstar I **flucythrinate + phorate** Cyanamid
AAstickol W S **polyglycolic ethers** Schering AAgrunol
AAstimasul F **maneb + sulphur + zineb** Staehler
AAstrepto B **streptomycin** Schering AAgrunol
AAsulfa-Supra F **sulphur** Schering AAgrunol
AAsyfun F **carbendazim** Schering AAgrunol
AAtack FX **thiram** Schering AAgrunol
AAtarsan N F **carbendazim + maneb + sulphur** Schering AAgrunol
Aateck-Hoko F **tecoram + thiram** Hokochemie
AAterra F **etridiazole** ICI, Knutson, Schering, Schering AAgrunol, Uniroyal
AAtifon FI **dichlofenthion + thiram** Schering AAgrunol
AAtiram FL **thiram** Schering AAgrunol, Staehler
AAtiram-Combi-Neu FI **dichlofenthion + thiram** Staehler
AA Tisin H **amitrole + 2,4-D + simazine** Schering AAgrunol
AA Tisin spuitbus H **amitrole + atrazine + 2,4-D** Schering AAgrunol
AAtopam N F **carbendazim + thiram** Schering AAgrunol
AAtrex H **atrazine** Ciba-Geigy
AAtulsan F **maneb + zineb** Schering AAgrunol
AAvarol Vloeibaar H **amitrole + bromacil** Schering AAgrunol
AAvolex FX **ziram** Staehler
AAwiedex H **glyphosate** Monsanto
AAwieral IA **quaternary ammonium** Schering AAgrunol
AAzimag F **maneb + zineb** Schering AAgrunol
AAzolin H **benazolin + mecoprop** Schering AAgrunol
AAzomate A **benzoximate** Nippon Soda, Schering AAgrunol
2-AB F **2-aminobutane**
AB 10 I **parathion** Sariaf
abamectin IA [71751-41-2] avermectin B_1, Affirm, Avid, Dynamec, MK-936, Vertimec
Abar I **leptophos**
Abat I **temephos** Cyanamid
Abate I **temephos** Cyanamid, Lapapharm, Ligtermoet, Procida, Zerpa
Abathion I **temephos** Cyanamid
Abavit Universal Trockenbeize F **methoxyethylmercury silicate** Kwizda
Abavit UT mit Beizhaftmittel F **carboxin + prochloraz** Schering
Abavit-Gamma-Beize FI **lindane + methoxyethylmercury silicate** Kwizda
A.B.H. Coumarine R **warfarin** Frere
Abol I **pirimicarb** ICI
Abol Derris Dust F **rotenone** ICI
Abolin IA **tar oils** Bitumen Arnstadt
Abotryl F **captafol + folpet** Enotria
Abru R **warfarin** Haco, Zorka Sabac
AC 293 H **imazamethabenz-methyl** Cyanamid
AC 3422 IA **parathion** Cyanamid
AC 3911 IAN **phorate** Cyanamid
AC 5223 F **dodine** Cyanamid
AC 8911 I **phorate** Cyanamid

AC 12880 I **dimethoate** Cyanamid
AC 18737 I **endothion** Cyanamid
AC 26691 I **cythioate** Cyanamid
AC 38555 G **chlormequat chloride** Cyanamid
AC 47300 I **fenitrothion** Cyanamid
AC 52160 I **temephos** Cyanamid
AC 64475 IN **fosthietan** Cyanamid
AC 82258 I **nitrilacarb** Cyanamid
AC 84777 H **difenzoquat methyl sulphate** Cyanamid
AC 92100 IN **terbufos** Cyanamid
AC 92553 H **pendimethalin** Cyanamid
AC 206,784 H **xylachlor** Cyanamid
AC 217,300 I **hydramethylnon** Cyanamid
AC 222,293 H **imazamethabenz-methyl** Cyanamid
AC 222,705 I **flucythrinate** Cyanamid
AC 252,214 H **imazaquin** Cyanamid
AC 252,925 H **imazapyr** Cyanamid
AC 263,499 H **imazethapyr** Cyanamid
Acaban I **CGA 29170** Ciba-Geigy
Acacide A **chlorbenside**
Acadrex A **amitraz** Shell
Acafor A **dicofol + ethion** Pennwalt
Acaphyd I **disulfoton** La Quinoleine
Acaraben A **chlorobenzilate** Ciba-Geigy
Acarac IA **amitraz** Maag
Acarcid A **dicofol + tetradifon** Scam
Acared 40 S A **dicofol** Terranalisi
Acared F.O. A **fenson + propargite** Terranalisi
Acared O.M. A **propargite** Terranalisi
Acarelte 4-1 AF **dinobuton + tetradifon** Probelte
Acarelte 4-50 AF **dinobuton + sulphur** Probelte
Acarelte 40 A **dinobuton** Probelte
Acarelte Forte AF **dinobuton + tetradifon** Probelte
Acarfen A **dicofol** Inagra
Acarflor A **hexythiazox** Sipcam
Acargil A **propargite** Chimiberg, Sepran
Acaricida A **dicofol + tetradifon** EniChem, Hoechst, Permutadora, Sandoz, Tecniterra
Acaricida 8 A **cyhexatin** Sariaf
Acaricida Oro A **tetradifon** Quimicas Oro
Acaricida Oro doble A **dicofol + tetradifon** Quimicas Oro
Acarie M F **mancozeb** Siapa
Acarin A **dicofol** Edefi, Makhteshim-Agan
Acarion A **propargite** Du Pont
Acaristop A **clofentezine** Epro, Schering
Acarit A **dicofol + tetradifon** Schering
Acarkey A **dicofol + tetradifon** Key
Acarkil A **dicofol + tetradifon** ICI
Acarmate A **benzoximate** Nippon Soda, Sipcam
Acarnet A **dicofol** Sepran
Acaroil A **dicofol + tetradifon** Afrasa
Acaroil K A **dicofol** Afrasa
Acaroil TD A **tetradifon** Afrasa
Acarol A **bromopropylate** Ciba-Geigy
Acarol A **fenson** Sariaf
Acarox A **cyhexatin** Quimigal
Acarozil A **chlorobenzilate** Ellagret
Acarpec A **cyhexatin** Sapec
Acarpolv 1-3 A **dicofol + tetradifon** Inagra
Acarsan A **cyhexatin** Sivam
Acarsivam 34 Bis IA **dicofol + TEPP + tetradifon** Sivam
Acarsivam KT A **dicofol + tetradifon** Sivam
Acarstin A **cyhexatin** Inagra, Lapapharm, Radonja, Sandoz, Universal Crop Protection
Acartal-T A **dicofol + tetradifon** Sapec
Acarthane AF **dicofol + dinocap** Rohm & Haas
Acartop A **cyhexatin + dicofol + tetradifon** Inagra
Acartot A **dicofol + tetradifon** Quimigal
Acartotal A **dicofol + tetradifon** Inagra
Acarvin A **tetradifon** Inagra
Acaryl PL A **cyhexatin + tetradifon** Prochimagro
Acatox KT A **dicofol + tetradifon** Sandoz
Acavers IA **dicofol + methomyl** R.S.R.
Accel G **cytokinin** Shell, Abbott
Accelerate HLG **endothal** Pennwalt
Acclaim H **fenoxaprop-ethyl** Hoechst
Acconem IN **fosthietan**
Accotab H **pendimethalin** Cyanamid
Accothion I **fenitrothion** Cyanamid, Zorka
Accotril I **nitrilacarb**
Ace-Kill New H **bromacil + 2,4-D + diuron** Chimac-Agriphar
Ace-Thios IA **dimethoate** Chimac-Agriphar
Acecap I **acephate** Agricola Ucana, Floris
Aceite Amarillo IA **DNOC + petroleum oils** Argos, Ficoop
Aceite IA **petroleum oils** Argos, CAG, Ficoop, Inagra
Acemeco H **mecoprop** Chimac-Agriphar
Acenit H **acetochlor** Chemolimpex, Nitrokemia
Acephat I **acephate** Ortho
acephate I [30560-19-1] *O,S*-dimethyl

acetylphosphoramidothioate, ENT 27822, Kitron, Orthene, Ortho 12420, Ortran, Ortril, Tornado
acephate-met IA **methamidophos**
Acertrol Trippel H **dichlorprop + ioxynil + MCPA** Bjoernrud
Acesan New F **mercury compounds (unspecified)** Chimac-Agriphar
acetate de dodemorphe F **dodemorph acetate**
acetate de phenylmercure F **phenylmercury acetate**
Acethios 40 IA **dimethoate** Chimac-Agriphar
acetochlor H [34256-82-1] 2-chloro-*N*-ethoxymethyl-6'-ethylacet-*o*-toluidide, acetochlore, Acenit, Harness, Mon-097
acetochlore H **acetochlor**
Acetormone H **2,4-D** Interphyto
Acibel G **gibberellic acid** Bayer
acide cyanhydrique IR **hydrogen cyanide**
acide gibberellique G **gibberellic acid**
acide naphtylacetique G **1-naphthylacetic acid**
acide naphtyloxyacetique G **(2-naphthyloxy)acetic acid**
acide trichlorobenzoique H **2,3,6-TBA**
acifluorfen-sodium H [62476-59-9] sodium 5-(2-chloro-α,α,α-trifluoro-*p*-tolyloxy)-2-nitrobenzoate, acifluorfene-sodium, Blazer, MC 10109, RH-6201, Tackle
acifluorfene-sodium H **acifluorfen-sodium**
Acifon IA **azinphos-methyl** Quiminor
Acifon D IA **azinphos-methyl + dimethoate** Quiminor
aclofen H **aclonifen**
aclonifen H [74070-46-5] 2-chloro-6-nitro-3-phenoxyaniline, aclofen, Bandren, Bandur, Challenge, CME 127
ACN HL [2797-51-5] 2-amino-3-chloro-1,4-naphthoquinone, ACNQ, Mogeton
ACNQ HL **ACN**
Acodol G **etacelasil** Siapa
Acorit A **hexythiazox** Kwizda
Acorit H **butralin + neburon** C.F.P.I., La Littorale
ACP 322 H **naptalam** Union Carbide
ACP 63-303 H **ioxynil**
ACP M-629 H **chloramben** Union Carbide
Acquinite NIFRH **chloropicrin**
ACR-2807 B I **fenoxycarb** Maag
ACR-3651 A F **pyrifenox** Maag
Acrex AF **dinobuton** Efthymiadis, KenoGard, Kwizda
Acrex Super A **dinobuton + tetradifon** Efthymiadis
Acriben AI **chlorfenson + dicofol + tetradifon** ERT
Acricid AF **binapacryl** Chromos, Hoechst, Sopepor
Acrisol A **cyhexatin** Pluess-Staufer
Acritan F **captafol** Siapa
Acritene F **folpet** Chemia
Acritet I **acrylonitrile** Stauffer
acrolein H [107-02-8] prop-2-enal, Aqualin
Acrylofume I **acrylonitrile**
Acrylon I **acrylonitrile** Cyanamid
acrylonitrile I [107-13-1] 2-propenenitrile, Acritet, Acrylofume, Acrylon, Carbacryl, Ventox
Acryptane F **folpet** Bayer
Actellic IA **pirimiphos-methyl** AVG, Compo, Delitia, ICI, Organika-Fregata, Pinus, Prometheus, Siapa, Sopra, Spolana
Actellic Oil IA **petroleum oils + pirimiphos-methyl** ICI-Zeltia
Actellifog I **pirimiphos-methyl** ICI, Spolana
Acticupro F **copper oxychloride** Ciba-Geigy
Actidione FG **cycloheximide** NOR-AM
Actidione RZ F **cycloheximide + quintozene** NOR-AM
Actidione Thiram F **cycloheximide + thiram** NOR-AM
Actifix H **2,4-D** Ciba-Geigy
Actiol F **sulphur** Sipcam-Phyteurop
Actiplus S **petroleum oils** Sopra
Actipron S **petroleum oils** Bayer, BP, Hoechst, Megafarm, Pennwalt
Actispray FG **cycloheximide** FMC
activated 7-dehydrocholesterol R **cholecalciferol**
Activate S **wetter** Leffingswell, Uniroyal
Activol G **gibberellic acid** ICI
Actor H **diquat dibromide + paraquat dichloride** Agroplant, Ciba-Geigy, ICI, Siegfried
Actosin R **pyranocoumarin** Schering
Actosin Kombi R **pindone + warfarin** Schering
Actosin P R **warfarin** Schering
Actril H **ioxynil** Agriben, Rhone-Poulenc
Actril 3 H **dichlorprop + ioxynil + MCPA** Plantevern-Kjemi
Actril AC H **ioxynil + MCPA** Agriben
Actril DP H **dichlorprop + ioxynil** Agrotec

Actril DS H **2,4-D + ioxynil** Rhodiagri-Littorale
Actril M H **ioxynil + mecoprop** Agriben, Chromos, Rhodiagri-Littorale
Actril S H **bromoxynil + dichlorprop + ioxynil + MCPA** Ewos, Kemira
Actrilawn H **ioxynil** Rhone-Poulenc
Acumen H **bentazone + MCPA + MCPB** BASF
Acuprex F **zineb** Eli Lilly
Acuprico F **ziram** Terranalisi
Acuprilene F **sulphur + zineb** Mormino
Acylon P F **maneb + metalaxyl** Ciba-Geigy
Acylon super Flo F **folpet + metalaxyl** Ciba-Geigy
Acylon tabac F **maneb + metalaxyl** Ciba-Geigy
AD 67 S [77430-44-5] 4-(dichloroacetyl)-1-oxa-4-azaspiro[4.5]decane
Adaran S **polyglycolic ethers** Aragonesas
Adder S **petroleum oils** Embetec
Additivo AS S **glycerol** Agrimont
Adesvin S **alkyl phenol ethoxylate** Chemia
Adhaesit S **alkyl phenol ethoxylate** Spiess, Urania
Adhesol S **polyglycolic ethers** Hoechst
Adhesol S **ethylene oxide + terpenes** Procida
Adiabron H **neburon** Tradi-agri
Adiafix F F **copper oxychloride + copper sulphate + folpet** Tradi-agri
Adiafix M F **copper sulphate + maneb** Tradi-agri
Adiakar A **dicofol** Tradi-agri
Adialine I **lindane** Tradi-agri
Adiamix P F **thiram + zineb** Tradi-agri
Adianebe F **maneb** Tradi-agri
Adiapam FIHN **metham-sodium** Tradi-agri
Adiapel F **folpet** Tradi-agri
Adiapon H **dalapon-sodium** Tradi-agri
Adiatra H **atrazine** Tradi-agri
Adiatra Super H **atrazine + simazine** Tradi-agri
Adieu Limaces M **metaldehyde** Fabel-Maxenri
Adine F **dodine** Quiminor
Adiplant S **polyglycolic ethers** Bayer
Adivax S **polyglycolic ethers** Biolchim
adjuvant oils AIHS **petroleum oils**
Adoc Campagnols/Mulots R **crimidine** Desinfection Integrale
Adoc Helicide M **metaldehyde** Desinfection Integrale
Adodin F **dodine** Du Pont
Adol H **lenacil** Chemolimpex, Kobanyai
Adrex S **polyglycolic ethers** Agriplan
Adriv Rumianca S **hydroxyethyl cellulose + lauryl alcoholethoxylate** EniChem
Adrop Polvere G **1-naphthylacetamide + 1-naphthylacetic acid + (2-naphthyloxy)acetic acid** Chimiberg
ADS H **alloxydim-sodium** Nippon Soda
Ad-Spray S **wetter** Helena Chemical
Advance H **bromoxynil + fluroxypyr + ioxynil** ICI
Advantage I **carbosulfan** FMC
Advizor H **chloridazon + lenacil** Farm Protection
Aero Cyanate H **potassium cyanate**
Aerobrom FIHN **chloropicrin + methyl bromide** Lapapharm
Aerobrom AX S **methyl bromide** Lapapharm
Aerosol-15 I **DDT** Zorka Sabac
Aerosol-50 MF I **lindane + malathion** Zorka Sabac
Aerovan I **dichlorvos** Chimac-Agriphar
Aerozol do szklarni I **malathion** Organika-Azot
Aescab IA **lindane** Aesculaap
Aesse L I **carbaryl** Sipcam
Aetos R **zinc phosphide** Ellagret
Aetzmittel H **DNOC** Marktredwitz
AF 96 G **1-naphthylacetamide + 1-naphthylacetic acid** Gobbi
Afalon H **linuron** Argos, Bjoernrud, Hoechst, ICI-Valagro, Nitrokemia, Pluess-Staufer, Procida, Roussel Hoechst, Sandoz, Siegfried, Sopepor, VCH, Zupa
Afalon S H **linuron + monolinuron** Hoechst, Reis
Afalon Special/Spezial H **linuron + monolinuron** Hoechst, Procida
Afalon-kombi H **alachlor + linuron** Zupa
Afarin H **linuron + monolinuron** Hoechst
Afesin H **monolinuron** Hoechst
Affirm IA **abamectin** MSD AGVET
Aficida I **pirimicarb** ICI
Afid I **phosphamidon** Sariaf
Afidamon I **phosphamidon** Du Pont
Afidan IA **endosulfan** Tecniterra
Afidan M IA **endosulfan + parathion-methyl** Tecniterra
Afidex IA **prothoate** Tecniterra
Afidina M I **fenitrothion** Agrimont
Afidol I **pirimicarb** Visplant

Afidon I **dimethoate + endosulfan** Chemia
Afidon IA **dimethoate** Safor
Afidrex IA **dimethoate** Shell
Afilene I **butocarboxim** Inagra, Sipcam, Wacker
Afitan IA **diazinon** Bimex
Afithion IA **dimethoate** ERT
Aflin F **sulphur + zineb** Mylonas
Aflix I **dimethoate + endosulfan** Hoechst
Aflix IA **formothion** Sandoz
Afos IA **mecarbam**
Afourmi I **sodium dimethylarsinate** Sovilo
Afracid I **carbaryl** Afrasa
Afrasulco F **copper oxychloride + sulphur** Afrasa
Afrathion IA **malathion** Afrasa
Afrisect I **cypermethrin** Schering
Afrocap F **captan** Afrasa
Afrocobre F **copper oxychloride** Afrasa
Afrocobre MZ F **copper oxychloride + maneb + zineb** Afrasa
Afrodane FA **dinocap** Afrasa
Afroil IA **petroleum oils** Afrasa
Afroil amarillo IA **DNOC + petroleum oils** Afrasa
Afroland S **metham-sodium** Afrasa
Afrolinde I **lindane** Afrasa
Afrometa M **metaldehyde** Afrasa
Afrosan F **zineb** Afrasa
Afrosan MZ F **mancozeb** Afrasa
Afrosan supercuprico F **copper oxychloride + zineb** Afrasa
Afugan FI **pyrazophos** Argos, Condor, Hoechst, Pluess-Staufer, Procida, Promark, Roussel Hoechst, Zupa
Afugan combi F **captafol + pyrazophos** Hoechst
Afungil F **benomyl** Chemia
AG 100 G **gibberellic acid** Quimicas Oro
Agaclor I **endosulfan** Ital-Agro
Agaril H **alachlor** Visplant
Ager Akar KT A **dicofol** Ager Chemical
Ager Atra H **atrazine** Ager Chemical
Ager Atra S H **atrazine + simazine** Ager Chemical
Ager Azimetil IA **azinphos-methyl** Ager Chemical
Ager Bemil F **benomyl** Ager Chemical
Ager Dan IA **endosulfan** Ager Chemical
Ager Dimetoato IA **dimethoate** Ager Chemical
Ager Diserbo MD H **2,4-D + MCPA** Ager Chemical
Ager Dod F **dodine** Ager Chemical
Ager Esca Limacida M **metaldehyde** Ager Chemical
Ager Etion IA **ethion** Ager Chemical
Ager Faltan F **folpet** Ager Chemical
Ager Flural H **trifluralin** Ager Chemical
Ager Gran H **terbutryn** Ager Chemical
Ager Mal I **malathion** Ager Chemical
Ager Mancoram F **copper oxychloride + mancozeb** Ager Chemical
Ager Mol H **molinate** Ager Chemical
Ager Oil IA **petroleum oils** Ager Chemical
Ager Par I **parathion** Ager Chemical
Ager Par M I **parathion-methyl** Ager Chemical
Ager Paraxone H **paraquat dichloride** Ager Chemical
Ager Pax H **ametryn** Ager Chemical
Ager Prometrine H **prometryn** Ager Chemical
Ager Risina I **lindane** Ager Chemical
Ager Sim H **simazine** Ager Chemical
Ager Solfan F **folpet + sulphur** Ager Chemical
Ager Sulf AF **sulphur** Ager Chemical
Ageram F **copper oxychloride** Ager Chemical
Agermin G **propham** Ciba-Geigy, Fahlberg-List, Stolte & Charlier
Agerzol H **2,4-D** Key, Roussel Hoechst
Agherud Dicamba H **dicamba + MCPA** Du Pont
Aglukon Hydro-Insecticide I **butocarboxim** Aglukon
Aglukon Onkruidverdelger H **diquat dibromide + paraquat dichloride** Duphar
Aglukon plantenspray extra HI **deltamethrin** Aglukon
Aglukon plantenspray tegen insekten en spint IA **piperonyl butoxide + pyrethrins + tetrasul** Aglukon
Aglukon tegen algen L **quaternary ammonium** Aglukon
Aglukon tegen bodeminsekten I **temephos** Aglukon
Aglukon tegen insekten in de moestuin I **diazinon** Aglukon
Aglukon tegen meeldauw vloeibaar F **dodemorph acetate** Aglukon
Aglukon tegen mos H **chloroxuron** Aglukon
Aglukon tegen onkruid onder bomen en heesters H **simazine** Aglukon
Aglukon tegen onkruid op paden, terrassen en onder beplantingenH **amitrole + simazine** Aglukon

Aglukon tegen onkruid op paden, terrassen en in het grindH **dalapon-sodium + dichlobenil** Aglukon
Aglukon tegen rupsen I **carbaryl + malathion** Aglukon
Aglukon tegen slakken M **metaldehyde** Aglukon
Aglukon vloeibaar tegen schimmels F **imazalil** Aglukon
Aglukon voor stekbewoorteling G **4-indol-3-ylbutyric acid** Aglukon
Aglukon Wondbalsem FW **thiophanate-methyl** Aglukon
Agra-Schabi IA **lindane** Agra-Pharm
Agral S **alkyl phenol ethoxylate** ICI
Agravia S **petroleum oils** Avia, Wacker
Agrecina A H **atrazine** Sadisa
Agrecina doble H **atrazine + simazine** Sadisa
Agrecina S H **simazine** Sadisa
Agrecocel G **chlormequat chloride** Sadisa
Agrecozin F **copper oxychloride + zineb** Sadisa
Agrecozin doble F **copper oxychloride + maneb + zineb** Sadisa
Agren 3614 H **MCPA + simazine + terbutryn** Ciba-Geigy
Agren spezial H **mecoprop + simazine + terbutryn** Ciba-Geigy
Agrenocap FA **dinocap** Sadisa
Agrept B **streptomycin** Lapapharm, Meiji Seika
Agrex 40-2 IF **lindane + maneb** Sadisa
Agrex 48-2 FI **lindane + maneb** Sadisa
Agrex 60-2 IAF **parathion-methyl + sulphur** Sadisa
Agrex K A **dicofol** Sadisa
Agrex KT A **dicofol + tetradifon** Sadisa
Agrex L I **lindane** Sadisa
Agrex R IA **dimethoate** Sadisa
Agrex S I **carbaryl** Sadisa
Agrex T A **tetradifon** Sadisa
Agrezufre F **sulphur** Sadisa
Agrezufre Cuprico F **copper oxychloride + sulphur** Sadisa
Agri-mycin B **streptomycin** Pfizer
Agri-Strep B **streptomycin** MSD AGVET
Agrian IA **malathion** Inagra
Agricarb Esca IM **methiocarb** Visplant
Agrichem Bentazon-DP H **bentazone + dichlorprop** Agrichem
Agrichem DB Plus H **2,4-DB + MCPA** Agrichem
Agrichim Antigermes G **chlorpropham** Agricultural Chemicals
Agrichim Antilimace M **metaldehyde** Agricultural Chemicals
Agricol S **sodium alginate** ICI
Agricorn H **MCPA** FCC
Agricorn D H **2,4-D** FCC
Agricorn M H **MCPA** FCC
Agridan I **phosmet** Agriplan
Agridex S **petroleum oils** Bayer
Agriflan H **trifluralin** Agria
Agrimet IAN **phorate** Cyanamid
Agrimul S **wetter** Diamond Shamrock
Agrimycin F **oxytetracycline + streptomycin** Pfizer
Agrisan H **AD 67 + EPTC** Radonja
Agriscab IA **lindane** Aesculaap
Agrisil I **trichloronat** Bayer
Agrisol IAN **phorate** Eli Lilly
Agrisorb S **tallow amine ethoxylate** ABM Chemicals
Agritox I **trichloronat** Bayer
Agritox H **MCPA** Rhone-Poulenc
Agritox I **chlorpyrifos** Kwizda
Agritox plus FI **thiram + trichloronat** Bayer
Agriwet S **alkyl phenol ethoxylate** ABM Chemicals
Agro 45 I **carbaryl + diazinon** Agrochemiki
Agro Stammschutz-neu (Stammschutzmittel E) I **lindane** Agro
Agro straaforstaerker G **chlormequat chloride** Agro-Kemi
Agro-Mix FI **dinocap + malathion + zineb** Agro
Agro-Ravin I **carbaryl** Agro
Agro-strooipoeder I **bromophos** Bayer
Agrobar Super FI **barium polysulphide** Agronova
Agrocape FA **dinocap** Quimigal
Agrocapt F **captan** Key
Agrocarb F **mancozeb + oxycarboxin** Ital-Agro
Agrocel G **chlormequat chloride** Agriben
Agrocide H **MCPA** Quimigal
Agrocide 26 I **lindane** Prometheus
Agrocillina F **8-hydroxyquinoline sulphate** Tecniterra
Agrocit F **benomyl** Chemolimpex
Agrocobre F **copper oxychloride** Shell
Agrocros Sevin I **carbaryl** Agrocros
Agrodin I **diazinon** Ital-Agro
Agrodina F **dodine** Agronova
Agrofac IA **prothoate + tetradifon** Agronova
Agrofid P I **lindane** Ital-Agro

Agrofitol IA **petroleum oils** Agronova
Agrofitol Giallo IA **DNOC + petroleum oils** Agronova
Agrofix S **poly(vinyl propionate)** BASF
Agrofon I **trichlorfon** Hispagro
Agrofos I **parathion + petroleum oils** Geopharm
Agrofos SL-20 I **monocrotophos** Radonja
Agrogas I **carbon tetrachloride + ethylene dichloride** Agronova
Agrol AIHS **petroleum oils** Chemolimpex, Szovetkezet
Agrolan H **trifluralin** Agrocros
Agrolimace M **metaldehyde** Sovilo
Agroluq I **lindane** Luqsa
Agrometil I **parathion-methyl** Agronova
Agronaa G **1-naphthylacetic acid** Aries Agro-Vet
Agronal F **phenylmercury chloride** Spolana
Agronex I **lindane** Shell
Agronex-Spezial FI **lindane + thiram** Shell
Agronexa I **lindane** Celamerck, Permutadora, Sandoz
Agronexa M IA **azinphos-methyl** Sandoz
Agronexa R I **lindane** Sandoz
Agronexa SL I **carbaryl + lindane** Sandoz
Agronexit I **lindane** Celamerck
Agroptan F **captan** Agronova
Agror I **dimethoate** Quimigal
Agrosan H **sodium chlorate** Bitterfeld
Agrosan F **phenylmercury acetate** ICI
Agrosol Dry F **captan + thiabendazole** Chipman
Agrosol Pour-on F **thiabendazole + maneb** Chipman
Agrosyl H **amitrole + atrazine** Agrofarm
Agrosyl SM H **amitrole + simazine** Agrofarm
Agrotect H **2,4-D** Miller
Agroten H **ametryn + prometryn** Agrotechnica
Agrotepp IA **TEPP** Agronova
Agroterr-Rasenduenger mit Unkrautvernichter H **2,4-D + dicamba** Agro
Agrothion I **fenitrothion** ICI
Agrotox I **parathion** Agronova
Agrotox 50 Siapa I **azinphos-ethyl + malathion + parathion-methyl** Agrochemiki
Agroved H **2,4-D** Agronova
Agroxone H **2,4-D** ICI
Agroxone H **MCPA** ICI, Sandoz
Agroxyl H **MCPA** B.B.
Agrozolfo AF **sulphur** Agronova
Agrozon I **diazinon** Radonja
Agrumin IA **petroleum oils** Chimiberg
Agrumol IA **petroleum oils** Sariaf
Agstock Addwet S **tallow amine ethoxylate** Agstock
AIA G **indol-3-ylacetic acid**
AIB G **4-indol-3-ylbutyric acid**
Aim I **chlorfluazuron** Ciba-Geigy
Aimsan IA **phenthoate** All-India Medical
Airone F **propineb** Montedison
Ajutol S **petroleum oils** Siegfried
Akan R **warfarin** Kotzabasis
Akar A **chlorobenzilate** Ciba-Geigy, Viopharm
Akareks A **cyhexatin** Zorka Sabac
Akariaio H **paraquat dichloride** Bredologos
Akarstin A **cyhexatin** KVK
Akatox A **dicofol** Sandoz
Akavit R **warfarin** B.B.
Akotion I **fenitrothion** Zorka Sabac
Akraqueck FX **anthraquinone + methoxyethylmercury silicate** Liro
Aksol H **naptalam** Du Pont
Aktellik-Savupanos I **pirimiphos-methyl** Berner
Aktikon H **atrazine** Chemolimpex
Aktuan F **cymoxanil + dithianon** Shell
Al-Phos IR **aluminium phosphide** All-India Medical
alachlor H [15972-60-8] 2-chloro-2',6'-diethyl-*N*-methoxymethylacetanilide, alachlore, Alanex, Alanox, CP 50144, Lasso, Lazo, Pillarzo
alachlore H **alachlor**
Alagrex H **alachlor** Sadisa
Alagrex Extra H **alachlor + atrazine** Sadisa
Alahlor H **alachlor** Zorka Subotica, Zupa
Alaklor H **alachlor** Pliva, Radonja
Alaklor/Atrazin-T H **alachlor + atrazine** Pliva
Alaklor/Radazin H **alachlor + atrazine** Radonja
Alanap H **naptalam** Argos, Hellafarm, Uniroyal
Alanex H **alachlor** Alpha, Aragonesas, Makhteshim-Agan
Alanox H **alachlor** Crystal
Alapin H **alachlor** Pinus
Alar G **daminozide** Argos, BASF, Bjoernrud, Cillus, Compo, Dow, Hellafarm, Ligtermoet, Liro, Pepro, Spiess, Uniroyal, Urania, Zorka Sabac

Alarsol H **alachlor** Afrasa
Alaxon D IA **diazinon** Maag
Alazin H **alachlor + simazine** Radonja
Alazine H **alachlor + atrazine** Argos, Makhteshim-Agan
Albene IA **petroleum oils** Scam
Albisal FB **8-hydroxyquinoline sulphate** Schering
Albolineum IAH **petroleum oils** ICI, ICI-Valagro, Prometheus
Albrass H **propachlor** ICI
Alcap F **captan**
Alcatraz H **amitrole + simazine** Killgerm
Alco H **sodium chlorate** Donck
Alcosorb S **polyacrylamide** Riba
Aldecid IA **gelatin** KenoGard
aldehyde formique FB **formaldehyde**
Alden G **piproctanyl bromide** Maag, Urania
aldicarb IAN [116-06-3] 2-methyl-2-(methylthio)propionaldehyde *O*-methylcarbamoyloxime, aldicarbe, OMS 771, ENT 27093, Temik, UC 21149
aldicarb sulfone IAN **aldoxycarb**
aldicarbe IAN **aldicarb**
aldimorph F *N-n*-alkyldimethylmorpholine, Falimorph
aldoxycarb IAN [1646-88-4] 2-methyl-2-methylsulphonylpropionaldehyde *O*-methylcarbamoyloxime, aldoxycarbe, aldicarb sulfone, ENT 29261, Standak, UC 21865
aldoxycarbe IAN **aldoxycarb**
Aldrex I **aldrin** Shell
aldrin I [309-00-2] (1*R*,4*S*,5*S*,8*R*)-1,2,3,4,10,10-hexachloro-1,4,4a ,5,8,8a-hexahydro-1,4:5,8-dimethanonaphthalene, HHDN, aldrine, OMS 194, ENT15949, Aldrex, Aldrite, Aldron, Aldrosol, Algran, Compound 118, Octalene
aldrine I **aldrin**
Aldrite I **aldrin** Shell
Aldron I **aldrin**
Aldrosol I **aldrin**
Alegro H **phenmedipham** Agriben, Rhone-Poulenc, Pepro
Alenthion I **parathion-methyl** Agrolinz
Alentisan F **methoxyethylmercury silicate + phenylmercuryacetate** Agrolinz
Alentisan F **methoxyethylmercury silicate** Agrolinz
Alfacron I **azamethiphos** Arkovet, Ciba-Geigy, Ligertermoet
Alfadex F **pyrethrins** Ciba-Geigy
Alfamat RX **chloralose** Jewnin-Joffe
alfamethrin I **alphacypermethrin**
Alfanje H **alachlor** Inagra
Alficron I **azamethiphos** Ciba-Geigy
Algadex H **cyanazine** Shell
Algamine H **benzalkonium chloride** Killgerm
Algenreiniger L **quaternary ammonium** Rowi
Algenstrikker L **quaternary ammonium** Reyer
Alginex LH **quinonamid** Hoechst
Algisol L **benzalkonium chloride** KVK
Algitox L **benzalkonium chloride** KVK
Algopin F **sulphur** Algovit
Algopur-Staub/poudre I **piperonyl butoxide + pyrethrins** Algovit
Algosufrex I **trichlorfon** Platecsa
Algran I **aldrin**
Algrol H **linuron** Agrolac
Alguinex HL **quinonamid** Argos
Alian Unkrauttod H **allyl alcohol** Kwizda
Alibi H **bifenox + linuron** Farm Protection
Alicep H **chlorbufam + chloridazon** BASF
Aliette F **fosetyl-aluminium** Agriben, Agrotec, Condor, Hortichem, Pepro, Ravit, Rhone-Poulenc, Zorka Sabac
Aliette Extra F **captan + fosetyl-aluminium + thiabendazole** Shell
Aliette SD F **captan + fosetyl-aluminium** Pepro
Alimet F **copper oxychloride + folpet** Margesin
Alipal S **alkyl phenol ethoxylate** GAF
Alipur H **chlorbufam + cycluron** Siegfried
Alirox H **EPTC** Chemolimpex, Sajobabony
Alistell H **2,4-DB + linuron + MCPA** Farm Protection
Alizor H **AD 67 + EPTC** Zorka Subotica
Aljaden H **sethoxydim**
Alkron IA **parathion**
All 22 G **1-naphthylacetic acid** Fivat
All Muis Kill R **chloralose** De Rauw
All-TA F **fthalide + kasugamycin**
Alleron IA **parathion**
allethrin I [584-79-2] (*RS*)-3-allyl-2-methyl-4-oxocyclopent-2-enyl (1*RS*)-*cis*,*trans*-chrysanthemate, allethrine, pallethrine, OMS 468, ENT 17510, Alleviate, Esbiol, Pynamin, Pynamin Forte, Pyresin, Pyrexcel, Pyrocide
allethrine I **allethrin**
Alleviate I **allethrin** Fairfield

Alliance F **fenarimol + folpet + fosetyl-aluminium** Pepro
allidochlor H [93-71-0] *N,N*-diallyl-2-chloroacetamide, CDAA, Randox
Allie H **metsulfuron-methyl** Du Pont
Allisan F **dicloran** Agrolinz, Asepta, Schering
Allivin IN **chlorfenvinphos + oxamyl** Shell
Alloxol H **alloxydim-sodium**
Alloxol-S H **sethoxydim**
alloxydim-sodium H [55635-13-7] methyl 3-[1-(allyloxyimino)butyl]-4-hydroxy-6,6-dimethyl-2-oxocyclohex-3-enecarboxylate, sodium salt, ADS, Alloxol, BAS 90210H, Clout, Fervin, Gramit, Grasip, Grasipan, Grasmat, Graspaz, Kusagard, Monalox, NP-48 Na
Alltex Flow H **amitrole + atrazine + 2,4-D + 2,4,5-T** Protex
Alltex Special Flow H **amitrole + atrazine + 2,4-D + dicamba + mecoprop + 2,4,5-T +2,3,6-TBA** Protex
Alltex Super Flow H **amitrole + atrazine + 2,4-D + dicamba + mecoprop + 2,4,5-T** Protex
Ally H **metsulfuron-methyl** Du Pont
allyl alcohol H [107-18-6] 2-propen-1-ol
allyxycarb I [6392-46-7] 4-diallylamino-3,5-xylyl methylcarbamate, OMS 773, Bay 50282, Hydrol
Alodan I **chlorbicyclen**
Alon H **isoproturon** Hoechst, Zupa
Alopex H **clofop-isobutyl**
Alotox I **diazinon + parathion** Mylonas
Alper F **cymoxanil + maneb** Zootechniki
Alpha Simazol H **amitrole + simazine** Alpha
Alpha Simazol T H **amitrole + ammonium thiocyanate + simazine** Alpha
alpha-chloralose RX **chloralose**
alphacypermethrin I [67375-30-8] (1*R* cis *S*) and (1*S* cis *R*) enantiomeric isomer pair of α-cyano-3-phenoxybenzyl-3-(2,2-dichlorovinyl)-2,2-dimethylcyclopropanecarboxylate, alfamethrin, alphamethrin, Bestox, Concord, Dominex, Fastac, Fendona, WL 85871
Alphakil RX **chloralose** Rentokil
alphamethrin I **alphacypermethrin**
Alphate-M H **isoprocarb + monocrotophos**
Alphos-Nebeldose IA **dichlorvos + tetrasul** Schering
Alrato R **antu**
Alsol G **etacelasil** Ciba-Geigy
Alsystin I **triflumuron** Bayer
Altercid N **AITK + D-D** Spolana
Alternil F **chlorothalonil + cuprous oxide** Masso
Alto F **cyproconazole** Sandoz
Alto elite F **chlorothalonil + cyproconazole** Sandoz
Altone F **propineb**
Altorick I **triprene**
Altosid I **methoprene** KenoGard, Sandoz, Zootechniki
Altox I **aldrin** All-India Medical
Altozar I **hydroprene** Sandoz
Altrin H **alachlor + prometryn** Radonja
Alugan I **bromocyclen** Hoechst
aluminium phosethyl F **fosetyl-aluminium**
Alutal IR **phosphine** Wuelfel
Alved F **copper oxychloride + folpet + zineb** Tecniterra
Alzodef HF **calcium cyanamide** Hoechst, Pluess-Stauffer, SKW Trostberg
AM G1 S **dimethylpolysiloxane** Du Pont
Amadene T **hydrolysed proteins** Chimiberg
Amak Extra F **Bordeaux mixture + maneb** Hellafarm
Amarel F **cymoxanil + mancozeb** Siegfried
Amarel cuivre F **copper oxychloride + cymoxanil + folpet** Siegfried
Amarel Folpet F **cymoxanil + folpet** Siegfried
Amarel Kupfer F **copper oxychloride + cymoxanil + folpet** Siegfried
Amarillo H **tebutam + trifluralin** La Quinoleine
Amatin M F **mancozeb** Bayer
Amaze I **isofenphos** Mobay
Ambox AF **binapacryl** Hoechst, Procida
Ambush IA **permethrin** Berner, Compo, ICI, ICI-Valagro, ICI-Zeltia, La Quinoleine, Maag, Plantevern-Kjemi
Ambush C I **cypermethrin** ICI
Ambusz I **permethrin** ICI
Amchem 64-50 G **3-CPA** Union Carbide
Amchem 70-25 HG **butralin**
Amchem A-280 HG **butralin**
Amcide H **ammonium sulphamate** Battle, Hayward & Bower
Amdro I **hydramethylnon** Cyanamid
Amecur I **permethrin** Maag
Ameisen-Ex I **lindane** Staehler
Ameisen-Frei I **lindane** Flora-Frey
Ameisen-Stop I **bendiocarb** Staehler

Ameisen-Streu- und Giessmittel I **lindane** Schacht
Ameisen-Streunex L I **lindane** Shell
Ameisenfrei S I **bromophos** Flora-Frey
Ameisenmittel I **lindane** Schering, Shell
Ameisenmittel I **trichlorfon** Bayer, Spiess, Urania
Ameisenmittel I **phoxim** Bayer
Ameisenmittel Hortex I **lindane** Shell
Ameisenmittel Hortex neu I **bromophos** Shell
Ameisenmittel Tugon I **trichlorfon** Bayer
Ameisenspray I **bendiocarb** Urania
Ameisentod BINAU I **lindane** Propfe
Ameisenvernichter I **lindane** Compo
Ameisenweg I **trichlorfon** Dom-Samen-Fehlemann
Amephyt H **ametryn** Phyteurop
Amepin H **ametryn + atrazine** Pinus
Amerol H **amitrole**
Amethopterin I **methotrexate**
Ametran H **ametryn** Diana
Ametrex H **ametryn** Makhteshim-Agan
ametryn H [834-12-8] 2-ethylamino-4-isopropylamino-6-methylthio-1,3,5-triazine, ametryne, Amephyt, Ametrex, Doruplant, Evik, G 34162, Gesapax, Mebatryne
ametryne H **ametryn**
Amex H **butralin** Amchem, Chimac-Agriphar, Etisa, Sopra, Union Carbide
Amexine H **atrazine + butralin**
Amezin H **ametryn + atrazine** Zorka Sabac
Amiben H **chloramben** Amchem, Bjoernrud, Chimiberg, Gullviks, Union Carbide
Amid-Thin G **1-naphthylacetamide** C.F.P.I., Chimac-Agriphar, EniChem, Etisa, Luxan, Union Carbide, Urania
amidinohydrazone I **hydramethylnon**
amidiphos IA **amidithion**
amidithion IA [919-76-6] *S*-2-methoxyethylcarbamoylmethyl *O,O*-dimethyl phosphorodithioate, amidiphos, ENT 27160, Ciba 2446, Thiocron
Amidron H **amitrole + diuron** SPE
Amigram H **2,4-DB** Visplant
Amilon M **metaldehyde** Leu & Gygax
Aminalon Ultra D H **amitrole + diuron + linuron + monolinuron** Hoechst
Aminatrix I **dichlorvos** Agrochemiki, Siapa
Aminer I **dichlorvos** Agronova
Aminex H **2,4-D** Protex
Aminex pur H **MCPA** Dimitrova
2-aminobutane F [13952-84-6] *sec*-butylamine, 2-AB, secondary butylamine, Butafume, Deccotane, Frucote, Tutane
Amino DHAI H **amitrole** Procida
aminocarb I [2031-59-9] 4-dimethylamino-*m*-tolyl methylcarbamate, aminocarbe, OMS 170, ENT 25784, A 363, Bay 44646, Matacil
aminocarbe I **aminocarb**
Aminopielik D H **2,4-D + dicamba** Azot, Organika-Rokita
Aminopielik M H **2,4-D + mecoprop** Azot, Organika-Rokita
Aminopielik P H **2,4-D + dichlorprop** Azot, Organika-Rokita
4-aminopyridine X [504-24-5] 4-Ap, Avitrol
Aminosal H **amitrole + simazine + sodium thiocyanate** Geophyt
Aminosin Super H **amitrole + atrazine + simazine** SEGE
aminotriazole H **amitrole**
Aminugec H **2,4-D** Sipcam-Phyteurop
Amiral F **triadimefon** Bayer
Amitan Super F **zineb** Platecsa
amiton IA [78-53-5] *S*-2-diethylaminoethyl *O,O*-diethyl phosphorothioate, ENT 24980-X, R 5158
Amitra H **amitrole + atrazine** Procida
amitraz AI [33089-61-1] *N*-methylbis(2,4-xylyliminomethyl)amine, amitraze, OMS 1820, ENT 27967, Acadrex, Acarac, Azadieno, BAAM, BTS 27419, Bumetran, Danicut, Edrizar, Maitac, Mitac, Taktic, Triatix, Triatox, Tudy
amitraze AI **amitraz**
Amitrex H **amitrole** Protex
Amitrol TL H **amitrole + ammonium thiocyanate** Chimac-Agriphar, Pepro, Prost
amitrole H [61-82-5] 1*H*-1,2,4-triazol-3-ylamine, ATA, aminotriazole, ENT 25445, Amerol, Amitrol-T, Amizol, Azolan, Azole, Cytrol, Herbizole, Weedazol
Amizina H **simazine** Sipcam
Amizine H **amitrole + simazine** Union Carbide
Amizol H **amitrole** Union Carbide
Amizol DP neu H **amitrole + bromacil + diuron** Avenarius
Amizol F H **amitrole + diuron** Avenarius
Amizol H H **amitrole + 2,4-D + TCA** Avenarius
Amlure I **propyl 1,4-benzodioxan-2-carboxylate**

Ammate H **ammonium sulphamate** Du Pont
Ammo I **cypermethrin** FMC
ammonium sulphamate H [7773-06-0] ammonium sulphamidate, AMS, sulfamate d'ammonium, Amcide, Ammate, Ikurin, Sepimate, Silvicide, Sulfamate
amobam F **ammonium ethylenebis(dithiocarbamate)**
Amosscabis IA **lindane** Kommer
Amoxone H **2,4-D** ICI
Amozol H **amitrole + atrazine + simazine + sodium thiocyanate** Agrotechnica
Ampelosan R F **copper oxychloride + zineb** Eli Lilly
AMS H **ammonium sulphamate**
Amylone H **MCPA** Chimac-Agriphar, Schwebda
Amylone-Kombi-Fluid H **2,4-D + MCPA** Chimac-Agriphar, Schwebda
AN 69 I **carbaryl + lindane** Epro
ANA Ficoop G **1-naphthylacetic acid** Ficoop
Anamone T **pheromones** Sipcam
Anastop G **1-naphthylacetic acid** Aifar
Anchor F **oxadixyl** Gustafson
Ancrack H **dinoseb + naptalam** Drexel
ancymidol G [12771-68-5] α-cyclopropyl-4-methoxy-α-(pyrimidin-5-yl)benzyl alcohol, ancymidole, A-Rest, EL-531, Reducymol
ancymidole G **ancymidol**
Anelda H **butylate** Chemolimpex, Sajobabony
Anema N **D-D** Procida
Anforstan H **potassium chlorate** Bitterfeld
anilazine F [101-05-3] 2,4-dichloro-6-(2-chloroanilino)-1,3,5-triazine, triazine, ENT 26058, B-622, Botrysan, Direx, Direz, Dyrene, Kemate, Triasyn, Zinochlor
Anilid H **propanil** Eli Lilly
anilofos H [64249-01-0] *S*-4-chloro-*N*-isopropylcarbaniloylmethyl *O,O*-dimethyl phosphorodithioate, Arozin, Hoe 30374, Rico
Animert V 101 A **tetrasul** Duphar, Galenika, KVK, Sodafabrik
anisomycin F **antibiotic from Streptomyces griseolus**
anisylacetone I **4-(*p*-methoxyphenyl)-2-butanone**
Aniten H **flurecol-butyl + MCPA** Shell
Aniten arroz H **flurecol-butyl + MCPA + propanil** Sandoz
Aniten D I **2,4-D + flurecol-butyl** Celamerck, Shell
Aniten DS I **2,4-D + flurecol-butyl** Celamerck, Nitrokemia
Aniten I H **dicamba + flurecol-butyl + MCPA** Dimitrova
Aniten Kombi H **dicamba + flurecol-butyl + MCPA** Shell
Aniten M H **flurecol-butyl + MCPA** Celamerck, ICI, Margesin, Shell
Aniten Mix H **flurecol-butyl + MCPA + pyridate** Margesin
Aniten P H **flurecol-butyl + MCPA + mecoprop** Shell
Aniten-Combi H **dicamba + flurecol-butyl + MCPA** Siegfried
Aniten-DS H **2,4-D + flurecol-butyl** Pinus
Aniten-MPD H **2,4-D + flurecol-butyl + mecoprop** Pinus
Aniten-S/Anitene-S H **flurecol-butyl + MCPA** Siegfried
Anitop H **dichlorprop + flurecol-butyl + ioxynil + MCPA** Epro, Margesin, Shell
Anofex H **chlorotoluron + terbutryn** Ciba-Geigy
Anox DB-Streumittel H **atrazine + diuron** Schering
Anox L H **atrazine + 2,4-D + dichlorprop + diuron + MCPA** Schering
Anox M H **amitrole + atrazine + 2,4-D + diuron** Schering
Anox M Granulat H **amitrole + bromacil + diuron** Schering
Anox WF H **dalapon-sodium + diuron + MCPA** Schering
Ansar H **MSMA** Diamond Shamrock, Fermenta
Ansar-8100 H **DSMA** Fermenta
Ant Killer I **lindane + pyrethrins** Battle, Hayward & Bower
Antak G **decan-1-ol** Drexel, Sipcam
Antene F **ziram** EniChem
Anteor F **cymoxanil + folpet** Procida, Shell
Anteor C F **copper sulphate + cymoxanil + folpet** Procida
Antergon HG **maleic hydrazide** KVK
Anthio I **formothion** Geopharm, Nitrokemia, Organika-Azot, Sandoz, Schering, Zupa
Anthiomix IA **fenitrothion + formothion** Sandoz
Anthonox IA **oxydemeton-methyl** Pepro
anthraquinone X [84-65-1] anthraquinone, Corbit, Corborid, Morkit

Anti-Ant Duster F **pyrethrins** Pan Britannica
Anti-Bladluis N IA **piperonyl butoxide + pyrethrins** Riemvis
Anticarie F **hexachlorobenzene**
Anticascola G **1-naphthylacetic acid** Biolchim, Chemia, CIFO, Sariaf
Anticercospora F **fentin hydroxide** Siapa
Antichiendent H **TCA** CTA
Anticimex 12 IA **piperonyl butoxide + pyrethrins** Anticimex
Anticimex Brumoxon R **warfarin** Anticimex
Anticimex Pyrenonpuder I **piperonyl butoxide + pyrethrins** Anticimex
Anticloque sovilo F **thiram** Sovilo
Anticoagulante P R **sulphaquinoxaline + warfarin** Colkim
Anticoagulante SS R **sulphaquinoxaline + warfarin** Colkim
Anticoagulante W R **warfarin** Colkim
Anticourtilieres Jardisol/Werrenkoerner I **lindane** Meoc
Anticrittogamico MC F **folpet + mancozeb** Chemia
Antideriva Caffaro S **hydroxyethyl cellulose + lauryl alcoholethoxylate** Caffaro
Antidrop G **1-naphthylacetic acid** Efthymiadis
Antifly Combi I **dimethoate + fenitrothion** Burri
Anti-fourmi Umupro I **sodium dimethylarsinate** Umupro
Antigerm-DN G **1-naphthylacetic acid + propham** Agrocros
Antigerme Umupro G **chlorpropham** Umupro
Antigermina G **propham** Sopepor
Antigerminante G **chlorpropham + propham** Probelte
Antigermogliante per patate G **chlorpropham + propham** Sariaf
Antigermoglio G **chlorpropham + propham** Phytoprotect
Anti-gnav X **thiram** Nordisk Alkali
Antigram H **metolachlor** Shell
Antigram H **TCA** Bourgeois
Antigramigna H **dalapon-sodium** Siapa
Anti-Gras H **sodium chlorate** Beulens
Antigril I **malathion** Bimex
Antigrill Forte GI **piperonyl butoxide + propham + pyrethrins** Agriplan
Antigrill P G **chlorpropham + propham** Agriplan
Anti-Herbe H **sodium chlorate** Dewandre, Engrais-Rosier, Lievin, Pauly-Andrianne
Antiherbe Lapaille H **sodium chlorate** Lapaille
Anti-Herbes Radical H **sodium chlorate** Radermecker
Antikiek H **2,4-D + MCPA** Nuyens
Antikolin-mamak R **warfarin** Galenika
Antikvalster A **fenitrothion + tetradifon** Anticimex
Antilesma Eureka M **metaldehyde** Quimigal
Antilimace M **metaldehyde** C.D.P., Rhodic, Rhone-Poulenc, Tradi-agri, Umupro
Anti-Limaces M **metaldehyde** Chimac-Agriphar
Antilimacos M **metaldehyde** Probelte
Antiliserons H **2,4-D** Umupro
Antilumuca M **metaldehyde** Permutadora, Siapa
Antimoos-U-Kombi H **2,4-D + iron sulphate + MCPA** GABI-Biochemie
Antimos H **iron sulphate** Chimac-Agriphar, Van Wesemael
Anti-mousse Gazon Umupro H **iron sulphate** Umupro
Antimousse Sovilo S **dimethylpolysiloxane** Sovilo
Anti-Mus roede musekorn R **crimidine** Laurits Jensen
Antinonnin IAHF **DNOC** Bayer
antiphen LFB **dichlorophen**
Antipilz-Pulver Mioplant F **captan + dinocap + zineb** Migros
Antipilz Stop F **quintozene** Wyss Samen & Pflanzen
Antir F **copper oxychloride + zineb** EniChem
Anti-Scabiosum IA **lindane** Van Wesemael
Anti-Scheut Olie "Olifan" G **vegetable oils** Tabakssyndikaat
Antiscalda Per Mele G **ethoxyquin** Scam
Antischiuma Cifo S **silicones** CIFO
Antischneck M **metaldehyde** Leu & Gygax, Neudorff
Antisork R **calcium phosphide** Anticimex
Anti Spruit G **chlorpropham** Cehave
Anti-stain F **pentachlorophenol** Van Swaay
Anti-Stipp Tradiagri F **calcium chloride** Tradi-agri
Antitavelure F **zineb** Meoc

Anti-Vermine I **piperonyl butoxide + pyrethrins** Riem
Antivermin Svavelkalkvaetska F **lime sulphur** Ohlssons
Antkiller I **pirimiphos-methyl** ICI
Antor H **diethatyl-ethyl** Schering, NOR-AM
Antracol F **propineb** Agroplant, Bayer, Pinus
Antracol BT H **metalaxyl + zineb** Bayer
Antracol-BT F **propineb + triadimefon** Pinus
Antracol Cobre F **copper oxychloride + propineb** Bayer
Antracol cobre especial F **calcium copper oxychloride + propineb** Bayer
Antracol combi F **cymoxanil + propineb** Bayer, Pinus
Antracol M F **propineb + quinomethionate** Bayer
Antracol MN F **propineb** Agro-Kemi, Bayer
Antracol Ramato Micro F **copper oxychloride + propineb** Bayer
Antracol Triple F **calcium copper oxychloride + cymoxanil + propineb** Bayer
Antracol-Kupfer/Cuivre F **copper oxychloride + propineb** Agroplant
Antral H **atrazine + cyanazine** Visplant
antu R [86-88-4] 1-(1-naphthyl)-2-thiourea, krysid, Alrato, Anturat, Bantu, Dirax, Kripid, Rattrack, Smeesana
Anturat R **antu**
Antyperz plynny H **TCA** Organika-Azot
Anvil F **hexaconazole** ICI, Sopra
4-Ap X **4-aminopyridine**
Apache NI **S,S-di-sec-butyl O-ethyl phosphorodithioate** FMC
Apachlor IA **chlorfenvinphos** KenoGard
Apadrin IA **monocrotophos** KenoGard
Apamidon IA **phosphamidon** KenoGard
Aparex F **dodine** Tecniterra
Apaton F **zineb** Tecniterra
Apaton Ramato F **copper oxychloride + zineb** Tecniterra
Apavap IA **dichlorvos** KenoGard
Apavinphos IA **mevinphos** KenoGard
Apell H **fluroxypyr + MCPA** Maag
Aperdex G **1-naphthylacetic acid** ACF, Liro
Apex I **methoprene** Sandoz
Aphamite IA **parathion**
Aphidan I **IPSP** Hokko
Aphox I **pirimicarb** ICI, Quimigal
Apiren G **4-CPA** + (2-naphthyloxy)acetic **acid** Sariaf
Apirenico Cifo G **(2-naphthyloxy)acetic acid** CIFO
Apitium F **captan + etridiazole** Siapa
Apl-Luster F **thiabendazole** Pennwalt
Aplotin F **dinocap** Schering
Apollo A **clofentezine** Elanco, ICI-Valagro, Maag, Schering, Schering AAgrunol, Zupa
Apolo A **clofentezine** Schering
Appa IA **phosmet**
Appat souris super actif Mortis R **crimidine** Sovilo
Appex IA **tetrachlorvinphos** Shionogi
Applaud IA **buprofezin** Nihon Nohyaku, ICI
Apponon G **1-naphthylacetic acid** Bayer
Aprex H **diuron + terbacil** Siegfried
aprocarb I **propoxur**
Aprodas F **ziram** Procida
Apron F **metalaxyl** Ciba-Geigy, Ligtermoet
Apron 70 SD F **captan + metalaxyl** Ligtermoet
Apron T 69 WS F **metalaxyl + thiabendazole** Seedcote Systems
Apropin H **atrazine + prometryn** Pinus
Aquacide H **diquat dibromide** Chipman
Aqua-Kleen H **2,4-D** Union Carbide
Aqualin H **acrolein** Shell
Aquaprop H **dichlobenil** La Quinoleine
Aquaspray Pem I **fenitrothion** Rentokil
Aquathol HLG **endothal** Pennwalt
Aquatox Orange I **malathion** Rentokil
AquaVex HG **fenoprop**
Aquazine H **simazine** Ciba-Geigy
Aquilite FA **sulphur** Craven
Aquinol H **atrazine + cyanazine** Shell
AR Boul'Herb H **lenacil + neburon** Boul'Herb
Arabelex G **gibberellic acid** Aragonesas
Aracet E S **poly(vinyl acetate)** Risnov
Aracloro Super H **alachlor + atrazine** Aragonesas
Aracnol E IA **carbophenothion + dicofol + tetradifon** Chimiberg
Aracnol F A **cyhexatin** Chimiberg
Aracnol K A **dicofol + tetradifon** Chimiberg
Aracusan F **copper oxychloride + zineb** Aragonesas
Aradon H **isoproturon + pendimethalin** Cyanamid
Araflurex H **trifluralin** Aragonesas

Aragol IA **dimethoate** Sipcam
Aragon R **coumatetralyl** Valmi
Aragran IN **terbufos** Maag
Arakol I **petroleum oils** Shell
Aralo IA **parathion + tetradifon** Maag
Aramin A **dicofol + tetradifon** Margesin
Aran H **amitrole + atrazine** Andreopoulos
Aranol A **dicofol + tetradifon** Luqsa
Arapam FNHI **metham-sodium** Aragonesas
Arasem ML IF **lindane + maneb** Aragonesas
Arasulfan IA **endosulfan** Aragonesas
Aration I **parathion-methyl** Aragonesas
Arbax H **linuron** Sariaf
Arbex X **bitumen** Pan Britannica
Arbin X **natural fats + petroleum oils** Staehler
Arbinol X **acridine bases + dicyclopentadiene** Staehler
Arbitex-Staub I **lindane** Berlin Chemie
Arbodrin (Supra) IA **petroleum oils** Kwizda
Arbogal H **triclopyr** Galenika
Arboricida-Z H **2,4,5-T** ICI-Zeltia
Arborol I **DNOC + petroleum oils** Diana
Arborol M IA **DNOC + fenson + petroleum oils** Spolana
Arborol Special IA **chlorfenson + DNOC + petroleum oils** Diana
Arboron H **dichlobenil** Sipcam-Phyteurop
Arbosan F **methfuroxam**
Arbosan spezial F **methfuroxam + thiabendazole** BASF, Ciba-Geigy, Kwizda
Arbosan Universal F **imazalil + methfuroxam + thiabendazole** BASF, Ciba-Geigy, Kwizda
Arbotec F **thiabendazole** MSD AGVET
Arbotect F **thiabendazole** MSD AGVET
Arbre sain FI **DNOC + petroleum oils** SCAC-Fisons
Arbrex W **bitumen + oxine-copper** Pan Britannica
Archer F **fenpropimorph + propiconazole** Ciba-Geigy
Arco Wuehlmaustod R **phosphine** Kwizda
Arcocid Wuehlmauspille R **phosphine** Kwizda
Arcotal S X **acridine bases + dicyclopentadiene** Staehler
Arcotan F **dinocap** Kwizda
Arcotin X **lindane** Staehler
Areb F **copper oxychloride + zineb** Tecniterra
Arelon H **isoproturon** Argos, Hoechst, Nitrokemia, Pleuss-Staufer, Procida, Roussel Hoechst, Siegfried
Arelon Combi/Kombi H **isoproturon + mecoprop** Hoechst, Roussel Hoechst
Arelon P H **isoproturon + mecoprop** Hoechst
Arelon Turbo H **ioxynil + isoproturon + mecoprop** Pluess-Staufer
Arena FG **tecnazene** Tripart
Aresin H **monolinuron** Chromos, CTA, Hoechst, Leu & Gygax, Nitrokemia, Pluess-Staufer, Procida, Roussel Hoechst
Aresin Combi/Kombi H **dinoseb acetate + monolinuron** Hoechst
A-Rest G **ancymidol** Elanco
Arethion I **parathion-methyl** Afrasa
Aretit H **dinoseb acetate** BASF, Hoechst, Leu & Gygax, Nitrokemia, Pluess-Staufer, Procida, RousselHoechst, Siegfried, Zupa
Aretol I **lindane** Kwizda
Argan I **sodium fluosilicate** Ellagret
Argosvin I **carbaryl** Argos
Argylene G **sodium silver thiosulphate** Ingvars
Ariane H **clopyralid + fluroxypyr + MCPA** Prochimagro
Aricosan H **monolinuron + napropamide** Siegfried
Aridal H **diquat dibromide** Enotria
Arikal 67 X **dicyclopentadiene** Staehler
Arione bleu M **metaldehyde** Truffaut
Arionex M **metaldehyde** Protex
Ariosol Schneckenkorn Microgranule anti-limace M **metaldehyde + methiocarb** CTA
Ariotox M **metaldehyde** Sandoz
Arlane I **chlorfenvinphos** CTA
Armex H **diuron** Caffaro
Armoblen S **tallow amine ethoxylate** Radonja
Armogard anti alg L **quaternary ammonium** Akzo
Armor I **cyromazine** Ciba-Geigy
Arotex Extra G **chlormequat chloride + choline chloride** ICI
Aroutox bait R **chlorophacinone** Zarkinou
Arozin H **anilofos** Hoechst
Arpan I **cypermethrin** Agrolinz
Arpan extra I **alphacypermethrin** Agrolinz
arprocarb I **propoxur**
Arquad S **quaternary ammonium** Akzo, Armak
Arrat H **chloridazon + phenmedipham** Schering
Arresin H **monolinuron** Hoechst

Arrex R **aluminium powder + calcium phosphate** Sovilo
Arrex E R **zinc phosphide** Epro, Shell
Arrex M R **zinc phosphide** Shell
Arrex patrone R **phosphine** Chromos, Shell
Arrex Toupeira R **calcium phosphide** Permutadora
Arrhenal H **DSMA**
Arrivo I **cypermethrin** FMC
Arsenagrex sodico F **sodium arsenite** Sadisa
Arsenal H **imazapyr** Chipman, Cyanamid
Arsenat F **sodium arsenite** Agrotechnica
Arsenicros F **sodium arsenite** Agrocros
Arsenito ERT F **sodium arsenite** ERT
Arsenito Sodico F **sodium arsenite** Probelte
Arsinyl H **DSMA**
Arsonate H **MSMA** Fermenta
Artaban A **benzoximate** Procida
Arvicolex R **bromadiolone**
Arvicolon R **bromadiolone** Siegfried
Arvicostop R **bromadiolone** Siegfried
Arvitox R **thallium sulphate** Chimac-Agriphar
Arylam IG **carbaryl**
AS P I **carbaryl** Sipcam
AS-50 B **streptomycin** Pfizer
Asana I **esfenvalerate** Du Pont
Asar F **mancozeb** EniChem
Asarinin S **sesamin**
ASB-Ameisenvernichter I **lindane** ASB-Gruenland
ASB-Schneckenkorn M **metaldehyde** ASB-Gruenland
ASD 2288 H **cliodinate**
Asef Antimousse H **iron sulphate** Moreels-Guano
Asef Greenkeeper H **2,4-D + dicamba** Moreels-Guano
Asef Mosdoder H **iron sulphate** Asef Fison, Moreels-Guano
Asef Plantenspray IA **piperonyl butoxide + pyrethrins** Asef Fison
Asef Rasenduenger mit Moosvernichter H **iron sulphate** Asef Fison
Asef Rasenduenger mit Unkrautvernichter H **2,4-D + dicamba** Asef Fison
Asef Slakkendood M **metaldehyde** Asef Fison
Asef Wieder H **2,4-D + dicamba** Asef Fison
Asepta Anitop H **dichlorprop + flurecol-butyl + ioxynil + MCPA** Asepta
Asepta Antirost F **borax + sulphur** Asepta
Asepta Antrolin H **diethatyl-ethyl + linuron** Asepta
Asepta Bandur H **aclonifen** Asepta
Asepta Ben-Cornox H **benazolin + dicamba + MCPA** Asepta
Asepta benzan H **benazolin + ioxynil + mecoprop** Asepta
Asepta Boomwondbalsem FW **oils** Asepta
Asepta brazolin H **benazolin** Asepta
Asepta Carbapan Speciaal F **ferbam + maneb + zineb** Asepta
Asepta Curbiset G **chlorflurecol-methyl** Felix
Asepta CX 99 I **bromophos** Asepta
Asepta DCP H **dalapon-sodium** Asepta
Asepta Dylopheen H **dinoseb** Asepta
Asepta Entwas Warm FW **natural resins** Asepta
Asepta Fungaflor vloeibaar F **imazalil** Asepta
Asepta Fungazil F **imazalil** Asepta
Asepta Fungicarb F **carbendazim + triforine** Asepta
Asepta Funginex F **triforine** Asepta
Asepta Fytol VIAS **petroleum oils** Asepta
Asepta Marvel K G **maleic hydrazide** Asepta
Asepta Mecoben H **benazolin + mecoprop** Asepta
Asepta Molyso Vier IA **DNOC** Asepta
Asepta Monam FIHN **metham-sodium** Asepta
Asepta Nexagan I **bromophos-ethyl** Asepta
Asepta Nexion I **bromophos** Asepta
Asepta Nexionol I **bromophos** Asepta
Asepta nomolt H **teflubenzuron** Asepta
Asepta Oborex K.B. FW **copper naphthenate** Asepta
Asepta Prebetox H **endothal** Asepta
Asepta Produr I **trichlofenidine** Asepta
Asepta Roxanex I **bromophos-ethyl + dimethoate** Asepta
Asepta Roxion IA **dimethoate** Asepta
Asepta Sintal S **polyglycolic ethers** Asepta
Asepta Slakken korrels M **metaldehyde** Asepta
Asepta Soveurode I **acrylate-based gum** Asepta
Asepta Sovitaup R **crimidine** Asepta
Asepta sporin CT I **Bacillus thuringiensis** Asepta
Asepta Sulfa Micron Vier F **sulphur** Asepta

Asepta Tetranyx IA **amitraz** Asepta
Asepta Tin-Maneb F **fentin acetate + maneb** Asepta
Aseptacarpon IAG **carbaryl** Asepta
Aseptadenol IA **DNOC** Asepta
Aseptaludin H **benazolin + 2,4-D + dicamba + MCPA** Asepta
Aseptamazin F **maneb + zineb** Asepta
Aseptameton IA **thiometon** Asepta
Aseptasolan F **dicloran + maneb** Asepta
Aseptazalin F **captan + dicloran + maneb** Asepta
Asepthion I **parathion** Asepta
Aseptol L **cresol** Asepta
Aseptox IA **piperonyl butoxide + pyrethrins** Asepta
Aservo G **propham** Agrolinz
asetat fenolovo FLM **fentin acetate**
Asilan H **asulam** Shionogi
Asitel H **bromoxynil + 2,4-D** Etisa
ASP-47 IA **sulfotep**
ASP-51 I **O,O,O',O'-tetrapropyl dithiopyrophosphate** Stauffer
A-Specian F **captan + zineb** Du Pont
Asperol FA **copper compounds (unspecified) + cyhexatin + folpet + sulphur** Pluess-Staufer
Aspon I **O,O,O',O'-tetrapropyl dithiopyrophosphate** Stauffer
Aspor F **zineb** Agrimont, ICI, Montedison, Quimigal
A-spray I **malathion + piperonyl butoxide + pyrethrins** Ansvar
Assassin H **ioxynil + isoproturon + mecoprop** ICI
Assault H **imazapyr** Cyanamid
Assert H **imazamethabenz-methyl** Cyanamid, Schering
Assert M H **imazamethabenz-methyl + mecoprop** Cyanamid
Asset H **benazolin + bromoxynil + ioxynil** Schering
Assure H **quizalofop-ethyl** Du Pont
Astaman F **fentin acetate + maneb** UCP
Astonex I **diflubenzuron** Shell
Astrol S **alkyl phenol ethoxylate** Eli-Lilly
Astrol H **bromoxynil + ioxynil + isoproturon** Embetec
asulam H [3337-71-1] methyl 4-aminophenylsulphonylcarbamate, asulame, Asilan, Asulox, M&B 9057
asulame H **asulam**
Asulox H **asulam** Agriben, Agronorden, Agrotec, Burts & Harvey, Condor, Embetec, Imex-Hulst, Maag, Ravit, Rhodiagri-Littorale, Rhone-Poulenc
Asuntol I **coumaphos** Bayer
Asystin Z IA **vamidothion** Kwizda
AT 90 H **amitrole** Custom Chemicides
AT 3020 H **amitrole + atrazine + 2,4-D + 2,4,5-T** Liro
ATA H **amitrole**
Atabron I **chlorfluazuron** Ishihara Sangyo
Atauron H **amitrole + diuron** Agrocros
Atazel H **amitrole** ICI
Atazinax H **atrazine** Agriben, Pepro
Atempo-Unkrautsalz H **sodium chlorate** Agro
Aterox A **cyhexatin** Terranalisi
Atgard IA **dichlorvos** Shell
athidathion I [19691-80-6] *O,O*-diethyl *S*-2,3-dihydro-5-methoxy-2-oxo-1,3,4-thiadiazol-3-ylmethylphosphorodithioate, G 13006
Athlet H **bifenox + chlortoluron** Pepro
Atizon F **maneb** Afrasa
Atlacide H **sodium chlorate** Chipman
Atlacide Extra H **atrazine + sodium chlorate** Chipman
Atladox HI H **2,4-D + picloram** Chipman
Atlas Adherbe S **petroleum oils** Atlas-Interlates
Atlas Adjuvant Oil S **petroleum oils** Atlas-Interlates
Atlas Brown H **chlorpropham + pentanochlor** Atlas-Interlates
Atlas Electrum H **chloridazon + chlorpropham + fenuron + propham** Atlas-Interlates
Atlas Gold H **chlorpropham + fenuron + propham** Atlas-Interlates
Atlas Herbon Yellow H **chlorpropham + fenuron** Atlas-Interlates
Atlas Hermes H **metoxuron + simazine** Atlas-Interlates
Atlas Indigo G **chlorpropham + propham** Atlas-Interlates
Atlas Janus H **linuron + trifluralin** Atlas-Interlates
Atlas Lignum Granules H **atrazine + dalapon-sodium** Atlas-Interlates
Atlas Minerva H **bromoxynil + dichlorprop + ioxynil + MCPA** Atlas-Interlates
Atlas Orange H **propachlor** Atlas-Interlates
Atlas Pink C H **chlorpropham + cresylic acid + diuron + propham** Atlas-Interlates
Atlas Protrum K H **phenmedipham** Atlas-Interlates

Atlas Red H **chlorpropham + fenuron** Atlas-Interlates
Atlas Silver H **chloridazon** Atlas-Interlates
Atlas Solon H **pentanochlor** Atlas-Interlates
Atlas Somon H **sodium monochloroacetate** Atlas-Interlates
Atlas Steward I **lindane** Atlas-Interlates
Atlas Tecgran FG **tecnazene** Atlas-Interlates
Atlas Thuricide HP I **Bacillus thuringiensis** Atlas-Interlates
Atlavar H **atrazine + 2,4-D + sodium chlorate** Chipman
Atlazin H **amitrole + atrazine** Chipman
Atol H **amitrole + atrazine + 2,4-D + dicamba +mecoprop + 2,3,6-TBA** Liro
Atom Zolfo AF **sulphur** Baslini
Atonik G **sodium 5-nitroguaiacolate + sodium1-nitrophenolate + sodium 4-nitrophenolate** Asahi
Atout H **chlorflurecol-methyl + MCPA** O.P.
Atoxan I **carbaryl** Siapa
Atprom H **atrazine + prometryn** Zorka Sabac
Atra H **atrazine** Peppas
Atrab H **atrazine** Bayer
Atrabute Plus H **atrazine + butylate** Griffin
Atracor H **atrazine** Gen. Rep.
Atracor H **atrazine** Gen. Rep.
Atracure H **atrazine** Proval
Atraflow H **atrazine** Burts & Harvey
Atragan H **atrazine** COPCI
Atrakor H **atrazine** Prochimagro
Atralenox H **atrazine** Schering
Atralin H **atrazine + linuron** Schering
Atrall H **amitrole + atrazine + 2,4-D + dicamba +mecoprop + 2,3,6-TBA** Liro
Atralon H **amitrole + atrazine** Hoechst
Atramat H **ametryn + amitrole + atrazine** Zorka
Atramet-Combi H **ametryn + atrazine** Makhteshim-Agan
Atramin H **amitrole + atrazine** Agrochemiki
Atran A **chlorfenethol + chlorfenson** Zupa
Atranex H **atrazine** Aako, Anorgachim, Edefi, Maataloupalvelu, Makhteshim-Agan
Atranozi H **atrazine** Diana
Atraphyt H **atrazine** Sipcam-Phyteurop
Atrapin H **atrazine** Pinus
Atraplus H **benazolin + bromoxynil + mecoprop** Ligtermoet
Atraser H **atrazine** Serpiol
Atrataf H **atrazine** Rallis India
Atrater H **atrazine** Agrishell
Atratex H **atrazine** Protex
Atratol H **atrazine** Ciba-Geigy
atraton H [1610-17-9] 2-ethylamino-4-isopropylamino-6-methoxy-1,3,5-triazine, Gesatamin
Atratylone H **atrazine** Agriphyt, Chimac-Agriphar
Atrazane H **atrazine + simazine** Aseptan
Atrazerba H **atrazine** Sapec
Atrazerbe H **atrazine + 2,4-D + dalapon-sodium** LHN
Atrazinax H **atrazine** Condor
atrazine H [1912-24-9] 2-chloro-4-ethylamino-6-isopropylamino-1,3,5-triazine, AAtrex, Aktikon, Atazinax, Atranex, Atratol, Cekuzina-T, Fenamin, Gesaprim, Griffex, Hungazin, Inakor, Maizina, Mebazine, Primatol A, Radazin, Vectal, Zeazin
Atrazip H **atrazine** Interphyto
Atrazol H **atrazine** Lapapharm
Atrazoni H **atrazine** Diana
Atred H **atrazine** Agrimont, Montedison
Atrimmec G **dikegulac-sodium** PBI Gordon
Atrin H **atrazine** Agrochemiki
Atrinal I **dikegulac-sodium** Bendien, Burts & Harvey, C.F.P.I., Collett, ICI, LaQuinoleine, Maag, Nordisk Alkali, Schering, Siapa, Spiess, Urania, Visplant
Atrizan S H **atrazine + simazine** Visplant
Atroban I **permethrin**
Atromet H **ametryn + atrazine** Radonja
Atrozan H **atrazine** Hellenic Chemical
Attila H **2,4-D + dalapon-sodium + diuron** CARAL
Attivar 90 F **ziram** Eli Lilly
Attraco 7-E IA **petroleum oils** Dow, Du Pont
Attract'n Kill I **pheromones** Scentry
AU Mosdoder H **iron sulphate** Agrarische Unie
Augur H **isoproturon** Pepro
Aurigal H **clopyralid + ioxynil + MCPA + mecoprop** Ciba-Geigy
Auriplak-veterinair I **permethrin** Animed
Auroch H **ioxynil + mecoprop** Ciba-Geigy
Aurocol AF **sulphur** Ciba-Geigy
Auroil R-3 IA **dimethoate** Afrasa
Aus-Maus R **coumafuryl** Agro

Austriebs- und Sitkafichtenlaus-Spritzmittel I **bromophos** Flora-Frey
Austriebsspritzmittel A **petroleum oils** Agrolinz, Kwizda
Autumn Kite H **isoproturon + trifluralin** Schering
Autumn Toplawn I **carbaryl + quintozene** Pan Britannica
Auxene G **gibberellic acid** Terranalisi
Auxin A.A. G **1-naphthylacetamide + 1-naphthylacetic acid + (2-naphthyloxy)acetic acid** Sivam
Auxuran H **2,4-D + dalapon-sodium + diuron + simazine** Urania
Avadex H **di-allate** Monsanto
Avadex BW H **tri-allate** BASF, Bjoernrud, Condor, Hellafarm, Monsanto, Shell, Spolana
Avasint H **linuron + trifluralin** Leu & Gygax
Avenarius-Giftweizen R **zinc phosphide** Avenarius
Avenge H **difenzoquat methyl sulphate** Collett, Cyanamid, Kemira, Lapapharm, Ligtermoet, NordiskAlkali, Quimigal, Schering, Spiess, Urania, VCH, Zorka Sabac
Aventox H **benzoylprop-ethyl** Visplant
Aventox SC H **simazine + trietazine** Dow
Avermil I **carbaryl** Aziende Agrarie Trento
Avermina I **carbaryl + rotenone** Aziende Agrarie Trento
Avicol F **quintozene** Kemira
Avid IA **abamectin** MSD AGVET
Aviocaffaro F **copper oxychloride + petroleum oils** Caffaro
Aviocaffaro PF F **copper oxychloride** Caffaro
Aviosivam S **hydroxyethyl cellulose + lauryl alcoholethoxylate** Sivam
Avirosan H **dimethametryn + piperophos** Ciba-Geigy
Aviso F **cymoxanil + metiram** BASF
Aviso cup F **copper oxychloride + cymoxanil + metiram** BASF
Aviso Super F **cymoxanil + fentin acetate + metiram** BASF
Avisol-DM I **malathion + trichlorfon** Radonja
Avitrol X **4-aminopyridine** Avitrol
Award F **penconazole** Ciba-Geigy
Axall H **bromoxynil + ioxynil + mecoprop** Rhone-Poulenc
AZ 500 H **isoxaben** Eli Lilly
Azac H **terbucarb**
Azadieno AI **amitraz** Quimica Estrella
Azak H **terbucarb**
azamethiphos I [35575-96-3] *S*-[(6-chloro-2-oxooxazolo[4,5-*b*]pyridin-3(2*H*)-ylmethyl] *O,O*-dimethyl phosphorothioate, OMS 1825, Alfacron, Alficron, CGA18809, Snip
Azaplant H **amitrole** Fahlberg-List
Azaplant-Kombi H **amitrole + simazine** Fahlberg-List
Azar H **terbucarb**
Azidem I **azinphos-methyl + demeton-S-methyl sulphone** Leu & Gygax
azidithion I **menazon**
Azifene IA **azinphos-methyl** Ravit
Azimet I **azinphos-methyl** Elanco, Eli Lilly
Azimil IA **azinphos-methyl** Afrasa
Azin I **azinphos-methyl** Siapa
Azinex H **atrazine** Aragonesas
Azinfene IA **azinphos-methyl** Condor
Azinfol IA **azinphos-methyl** Luqsa
Azinfos I **azinphos-ethyl** Ciba-Geigy
Azinfos-metil IA **azinphos-methyl** Zorka Subotica
Azinos IA **azinphos-ethyl**
Azinphos M I **azinphos-methyl** Leu & Gygax
azinphos-ethyl IA [2642-71-9] *S*-(3,4-dihydro-4-oxobenzo[*d*]-[1,2,3]-triazin-3-ylmethyl) *O,O*-diethyl phosphorodithioate, azinphosethyl, triazotion, ENT 22014, Azinos, Azinugec E, Bay 16259, Bionex, Cotnion-Ethyl, Ethyl Guthion, GusathionA, Gusathion K forte, Gutex, R 1513, Sepizin L, Triazotion
azinphos-methyl IA [86-50-0] *S*-(3,4-dihydro-4-oxobenzo[*d*]-[1,2,3]-triazin-3-ylmethyl) *O,O*-dimethyl phosphorodithioate, azinphosmethyl, metiltriazotion, OMS186, ENT 23233, Azimil, Azinugec, Bay 17147, Carfene, Cotnion-Methyl, Crysthyon, Gusagrex, Gusathion M, Guthion, Metazintox, Pancide, R 1582, SepizinM, Toxation
Azinprop I **azinphos-methyl + propoxur** Protex
Azintox-E I **azinphos-ethyl** Phytorgan
Azinugec IA **azinphos-methyl** Phyteurop, Sipcam-Phyteurop
Azinugec E IA **azinphos-ethyl** Phyteurop
aziprotryn H **aziprotryne**

aziprotryne H [4658-28-0] 2-azido-4-isopropylamino-6-methylthio-1,3,5-triazine, aziprotryn, Brasoran, C 7019, Mesoranil
Aziser IA **azinphos-methyl** Sepran
Azithion IA **azinphos-methyl** Scam
azithiram F [5834-94-6] bis(3,3-dimethylthiocarbazoyl) disulphide, azithirame, PP 447
azithirame F **azithiram**
Azition IA **azinphos-methyl** Terranalisi
Azitox I **azinphos-ethyl + azinphos-methyl** T&D Mideast
Azobane IA **monocrotophos + parathion-methyl** Agrishell
Azocord IA **cypermethrin + monocrotophos** BASF
azocyclotin A [41083-11-8] tri(cyclohexyl)-1*H*-1,2,4-triazol-1-yltin, Bay BUE 1452, Peropal
Azodrin I **monocrotophos** Agrishell, Galenika, Shell, Szovetkezet
Azodrin double IA **mevinphos + monocrotophos** Agrishell
Azodrin Extra IA **monocrotophos** Shell
Azodrin soufre FIA **monocrotophos + parathion-methyl + sulphur** Agrishell
Azofene IA **phosalone** Rhone-Poulenc
Azogard H **nometryne** Organika-Azot
Azolan H **amitrole** Makhteshim-Agan
Azole H **amitrole**
azoluron G [4058-90-6] *N*-(2-ethyl-2*H*-pyrazol-3-yl)-*N'*-phenylurea
Azoprim H **atrazine** Organika-Azot
Azosan IA **endosulfan** Eli Lilly
azothoate IA [5834-96-8] *O*-4-(4-chlorophenylazo)phenyl *O,O*-dimethyl phosphorothioate, OMS 1089, L 1058, Slam C
Azotop H **simazine** Organika-Azot
Azufor Pelouse Plus Herbicide H **2,4-D + dicamba + mecoprop** Chimac-Agriphar
Azufre FA **sulphur**
Azufre Cuprico F **copper oxychloride + sulphur** Agrocros, Quimicas Oro
Azufre Oxidante FA **potassium permanganate + sulphur** Serpiol
Azufrex FA **sulphur** Agriplan
Azugreen Herbicide H **2,4-D + dicamba + mecoprop** Chimac-Agriphar
Azumo FA **sulphur** Pallares
Azural H **glyphosate** Monsanto
Azuram F **copper oxychloride** Du Pont
Azurin H **diuron** CEALIN
B-9 G **daminozide** Ligtermoet
B-22 G **chlorpropham + propham** Sandoz
B/77 IA **ethoate-methyl**
B-401 I **Bacillus thuringiensis** Sandoz
B 404 IA **parathion** Agroplant
B-622 F **anilazine** Bayer
B-995 G **daminozide** Uniroyal
B-3015 H **thiobencarb** Kumiai
B1-5452 A **fenothiocarb** Kumiai
BAAM AI **amitraz** Upjohn
Babosil M **metaldehyde** Agrocros
Bacidal Saniflor I **parathion-methyl** Fivat
Bacilan I **Bacillus thuringiensis** PGR-Walcz
Bacillus thuringiensis I Bt, Btk, B 401, Bactimos, Bactospeine, Biobit, Certan, Dipel, Foray, Javelin, Skeetal, Sok-Bt, Teknar, Thuricide, Vectobac
Bacterol-Super F **oxytetracycline + streptomycin** Agrofarm
Bactimos I **Bacillus thuringiensis** Biochem Products, Covagri
Bactospeine I **Bacillus thuringiensis** Agrocros, Applied Horticulture, Biochem Product, Covagri, Duphar, Hellafarm, Koppert, Organika-Fregata, Sapec, Siapa
Bactospeine Jardin I **Bacillus thuringiensis + piperonyl butoxide +pyrethrins** Biochem Product, Covagri, Sovilo
Bactucide I **Bacillus thuringiensis** Sivam
Badacsonyi rezkenpor F **copper hydroxide + sulphur** Osszefogas
Badacsonyi rezmeszpor F **copper hydroxide** Osszefogas
Badilin F **dodemorph acetate + dodine** BASF
Badim IA **dimethoate** Baslini
Bafort S **alkyl phenol ethoxylate + polyglycolic ethers** Farmer
Bafos IA **azinphos-methyl** Baslini
Bagalol F **2-methoxyethylmercury chloride** United Phosphorus
Bagir S **alkyl phenol ethoxylate** EniChem
Bagnante S **polyglycolic ethers** Du Pont, Roussel Hoechst, Sariaf, Scam, Sipcam
Bago de Ouro FA **sulphur** Sapec
Baigon I **propoxur** Berner
Baition I **phoxim** Berner
Bakrocid F **copper oxychloride** Zupa
Bakthane F **mancozeb + myclobutanil** Rohm & Haas
Bakthane F **mancozeb** Condor
Balan I **benfluralin** Elanco, Eli Lilly, Sinteza, VCH

Balanin FI **carbaryl + copper oxychloride** Tecniterra
Balfin H **benfluralin** Elanco
Bamist F **copper oxychloride + zineb** Baslini
Banafine H **benfluralin** Shionogi
Banair H **dicamba** ICI
Banaril H **chlorotoluron + pendimethalin** Ciba-Geigy
Bancol I **bensultap** Takeda, Zorka Subotica
Bandren H **aclonifen** Rhone-Poulenc, Shell, Sovilo
Bandrift S **polyacrylamide** Allied Colloids
Bandur H **aclonifen** BASF, Rhone-Poulenc, Shell
Bangald I **aldrin** Bangalore Pesticides
Ban-Hoe H **lenacil + propham** Shell
Bani I **bioallethrin + permethrin + piperonyl butoxide** Halldor Jonsson
Banitum X **naphthalene + tar oils** Schering AAgrunol
Banlate F **milneb**
Banlene Plus H **dicamba + MCPA + mecoprop** Schering
Banlene Solo H **dicamba + dichlorprop + ioxynil** Schering
Banner F **propiconazole** Ciba-Geigy
Banol F **propamocarb hydrochloride** NOR-AM
Banrot F **etridiazole + thiophanate-methyl** Mallinckrodt
Bantu R **antu**
Banvel H **dicamba** Agrolinz, Plantevern-Kjemi, Sandoz, Serpiol, Siapa, Zorka Subotica
Banvel CMPP H **dicamba + MCPA** Sandoz
Banvel D H **dicamba** Serpiol
Banvel DP H **dicamba + dichlorprop** Agrolinz, Bayer, Zorka Subotica
Banvel K H **2,4-D + dicamba** Agrolinz
Banvel M H **dicamba + MCPA** Agrolinz, Bayer, Budapesti Vegyimuvek, Burri, CTA, Epro, Hokochemie, Kwizda, Leu & Gygax, Mitsotakis, OHIS, Pluess-Staufer, Schering, Shell
Banvel MCPA H **dicamba + MCPA** Sandoz
Banvel MP H **dicamba + MCPA + mecoprop** Agrolinz, Kwizda
Banvel P H **dicamba + mecoprop** Epro, OHIS, Schering, Shell
Banvel PP H **dicamba + MCPA + mecoprop** Leu & Gygax
Banvel T H **tricamba** Velsicol
Bapar I **parathion** Baslini
Bar it gher H **propyzamide + simazine** National Chemsearch
Baram F **copper oxychloride** Baslini
Baran R **fluoroacetamide** Tamogan
barbamate H **barban**
barban H [101-27-9] 4-chlorobut-2-ynyl 3-chlorocarbanilate, barbane, CBN, barbamate, chlorinat, Carbyne, Caryne, Neobyne, Wypout
barbane H **barban**
barbasco IA **rotenone**
Barbetol H **chloridazon** Eli Lilly
Barcap F **captafol + captan + folpet** Protex
Barcotex H **2,4-DB + dicamba + MCPA + mecoprop** Protex
Bardac B **quaternary ammonium** Lonza
Bardew F **tridemorph** Schering
Barenbrug-Combi H **2,4-D + MCPA** KVK
Bargum Rattenkiller R **warfarin** Ebsen
Barin H **atrazine** Baslini
Barioter FI **barium polysulphide** Terranalisi
Barium FI **barium polysulphide** Baslini
Barleyquat G **chlormequat chloride** Mandops
Barnon H **flamprop-M-isopropyl** Shell
Baron H **erbon**
Barquat FB **benzalkonium** Lonza
Barral I **chlorpyrifos** Shell
Barricade I **cypermethrin** Agrishell, Shell
Barrier H **dichlobenil** PBI Gordon
Barrier H **chloridazon + fenuron + propham** Truchem
Barrix H **ethofumesate + phenmedipham** Schering
Barsprout G **chlorpropham + propham** L.A.P.A., Protex
Barthane F **mancozeb** Schering AAgrunol
Barthels Baumteer X **anthracene oil** Asser
Bartol H **MCPA** Shell
Barweed H **2,4-D** Protex
BAS 235I I **fenethacarb** BASF
BAS 238F F **dodemorph acetate** BASF
BAS 255H H **dimidazon** BASF
BAS 263I IN **cloethocarb** BASF
BAS 268I I **zolaprofos** BASF
BAS 276F F **benzamorf** BASF
BAS 290H H **prynachlor** BASF
BAS 305F F **mebenil** BASF
BAS 319F F **furcarbanil** BASF
BAS 327F F **cyclafuramid** BASF
BAS 346F F **carbendazim** BASF
BAS 348H H **flumezin** BASF

BAS 350H H **oxapyrazon** BASF
BAS 351H H **bentazone** BASF
BAS 352F F **vinclozolin** BASF
BAS 379H H **chlorprocarb** BASF
BAS 389F F **furmecyclox** BASF
BAS 392H H **fluchloralin** BASF
BAS 421F F **fenpropimorph** BASF
BAS 436F F **myclozolin** BASF
BAS 461H H **sulglycapin** BASF
BAS 517H H **cycloxydim** BASF
BAS 08306W G **mepiquat chloride** BASF
BAS 2430H H **brompyrazon** BASF
BAS 3050F F **mebenil** BASF
BAS 3053F F **mebenil** BASF
BAS 3170F F **benodanil** BASF
BAS 3201F F **mecarbinzid** BASF
BAS 3920 H **fluchloralin** BASF
BAS 3921 H **fluchloralin** BASF
BAS 3922 H **fluchloralin** BASF
BAS 30000F F **nitrothal-isopropyl** BASF
BAS 46100H H **sulglycapin** BASF
BAS 46204H H **bentazone + dichlorprop + isoproturon** BASF
BAS 47900H H **metazachlor** BASF
BAS 90210H H **alloxydim-sodium** BASF
BAS 90520H H **sethoxydim** BASF
Basagran H **bentazone** BASF, Budapesti Vegyimuvek, Collett, Intrachem, Lucebni zavody, Maag, OHIS, Sapec
Basagran DP H **bentazone + dichlorprop** Agrolinz, BASF, Ciba-Geigy, Intrachem, Zupa
Basagran KV H **bentazone + mecoprop** BASF
Basagran M H **bentazone + MCPA** BASF
Basagran MCPA H **bentazone + MCPA** BASF, Collett
Basagran MP H **bentazone + mecoprop** BASF
Basagran P H **bentazone + mecoprop** BASF
Basagran Ultra H **bentazone + dichlorprop + ioxynil** BASF
Basagran-Top H **bentazone + dichlorprop + MCPA** BASF
Basalin H **fluchloralin** BASF
Basamaize H **prynachlor**
Basamid FIHN **dazomet** Agrolinz, BASF, Chromos, Collett, Compo, Maag, Sandoz, Sapec, Schering
Basanite HI **dinoseb** BASF
Basanor H **brompyrazon** BASF
Basavam S **metham-sodium** Nordisk Alkali
BASF 220 F **tridemorph** BASF
BASF Meeldauwmiddel Meltatox F **dodemorph acetate** BASF
BASF Mehltaumittel F **dodemorph acetate** BASF
BASF Rosenspritzmittel universal FI **endosulfan + metiram + sulphur** Agrolinz
BASF-Gruenkupfer F **copper oxychloride** BASF
BASF-Kupferkalk F **copper oxychloride** BASF
BASF-Rosenspritzmittel F **metiram + sulphur** BASF
Basfapon H **dalapon-sodium** BASF, Chromos
Basforin F **triforine** Compo
Basfungin FA **methylmetiram** BASF, Galenika
Basidazin F **carbendazim + mepronil**
Basilex F **tolclofos-methyl** Fisons
Basiment 450-Extra N I **lindane** Bayer, Desowag-Bayer
Basinex P H **dalapon-sodium** BASF, Bayer
Basirose F **dodemorph acetate + dodine** Agrolinz
Basitac F **mepronil** Kumiai, Schering
Basmetil I **parathion-methyl** Baslini
Bassa I **fenobucarb** Mitsubishi, Kumiai
Basta H **glufosinate-ammonium** Hoechst, Pluess-Staufer, Procida
Bastor H **glufosinate-ammonium** Hoechst
Basudin IA **diazinon** Ciba-Geigy, Kemira, Kirk, KVK, Pinus, Plantevern-Kjemi, Viopharm
Basudin special IA **chloropropylate + diazinon** Ciba-Geigy
Basudine I **diazinon** Ciba-Geigy, Ligtermoet, Liro, Zerpa
Basultra F **thiram** BASF
Batalex G **propham** Sapec
Batapec I **malathion** Sapec
Batasan FLM **fentin acetate** Hoechst
Batazil H **atrazine + simazine** Pepro
Batazina I **azinphos-ethyl** Rhone-Poulenc
Batazina flo H **simazine** Condor
Batazine Flo H **simazine** Pepro
Bathurin I **Bacillus thuringiensis** JZD Slusovice
Battal F **carbendazim** Farm Protection
Baum-Kanker-Balsam FW **captafol** Scheidler
Baumteer schwarz X **anthracene oil** Weyl
Baur's Giftweizen R **zinc phosphide** Baur
Bavical F **carbendazim + maneb + tridemorph** BASF

Bavin IG **carbaryl** Baslini
Bavisfor F **carbendazim** Safor
Bavistin F **carbendazim** BASF
Bavistin M F **carbendazim + maneb** BASF
Bavistin MS F **carbendazim + maneb + sulphur** BASF
Bavistine F **carbendazim** BASF
Bavistine M F **carbendazim + maneb** BASF
Bavizilil F **carbendazim** BASF
Bay 5072 F **fenaminosulf** Bayer
Bay 5621 I **phoxim** Bayer
Bay 5712 F **tolylfluanid** Bayer
Bay 6159H H **metribuzin** Bayer
Bay 6443H H **metribuzin** Bayer
Bay 10756 IA **demeton** Bayer
Bay 15080 F **benquinox** Bayer
Bay 15203 IA **demeton-S-methyl** Bayer
Bay 15922 I **trichlorfon** Bayer
Bay 16259 IA **azinphos-ethyl** Bayer
Bay 17147 IA **azinphos-methyl** Bayer
Bay 18436 IA **demeton-S-methyl** Bayer
Bay 19149 IA **dichlorvos** Bayer
Bay 19639 IA **disulfoton** Bayer
Bay 20315 IA **demeton-S-methyl sulphone** Bayer
Bay 21097 IA **oxydemeton-methyl** Bayer
Bay 22555 F **fenaminosulf** Bayer
Bay 23129 IA **thiometon** Bayer
Bay 23323 IA **oxydisulfoton** Bayer
Bay 23655 IA **oxydeprofos** Bayer
Bay 25141 NI **fensulfothion** Bayer
Bay 25/154 IA **demeton-S-methyl** Bayer
Bay 25634 R **coumatetralyl** Bayer
Bay 25648 M **niclosamide** Bayer
Bay 29493 I **fenthion** Bayer
Bay 30130 H **propanil** Bayer
Bay 30686 FA **thioquinox** Bayer
Bay 33172 F **fuberidazole** Bayer
Bay 36205 FAI **quinomethionate** Bayer
Bay 37289 I **trichloronat** Bayer
Bay 37344 MIAX **methiocarb** Bayer
Bay 38819 R **phosacetim** Bayer
Bay 39007 I **propoxur** Bayer
Bay 41367c I **fenobucarb** Bayer
Bay 41831 I **fenitrothion** Bayer
Bay 44646 I **aminocarb** Bayer
Bay 45432 IA **omethoate** Bayer
Bay 46131 F **propineb** Bayer
Bay 47531 F **dichlofluanid** Bayer
Bay 49854 FA **tolylfluanid** Bayer
Bay 50282 I **allyxycarb** Bayer
Bay 60618 H **benzthiazuron** Bayer
Bay 68138 N **fenamiphos** Bayer
Bay 70143 IAN **carbofuran** Bayer
Bay 70533 H **chlorfenprop-methyl** Bayer
Bay 71628 IA **methamidophos** Bayer
Bay 74283 H **methabenzthiazuron** Bayer
Bay 77049 IA **quinalphos** Bayer
Bay 77488 I **phoxim** Bayer
Bay 78418 F **edifenphos** Bayer
Bay 79770 F **chloraniformethane** Bayer
Bay 92114 I **isofenphos** Bayer
Bay 94337 H **metribuzin** Bayer
Bay 105807 I **isoprocarb** Bayer
Bay BUE 0620 F **fluotrimazole** Bayer
Bay BUE 1452 A **azocyclotin** Bayer
Bay DIC 1468 H **metribuzin** Bayer
Bay ENE 11183B R **coumatetralyl** Bayer
Bay FCR 1272 I **cyfluthrin** Bayer
Bay KUE 2079A H **fluothiuron** Bayer
Bay KWG 0519 F **triadimenol** Bayer
Bay KWG 0599 F **bitertanol** Bayer
BAY MEB 6046 I **plifenate** Bayer
Bay MEB 6447 F **triadimefon** Bayer
Bay NTN 19701 F **pencycuron** Bayer
Bay NTN 8629 I **prothiofos** Bayer
Bay NTN 9306 I **sulfopros** Bayer
Bay S 276 IA **disulfoton** Bayer
Bay S 5660 IA **fenitrothion** Bayer
Bay SLJ 0312 AF **flubenzimine** Bayer
Bay SMY 1500 H **ethiozin** Bayer
Bay SRA 12869 I **isofenphos** Bayer
Bay SRA 7747 I **chlorphoxim** Bayer
Baycap F **bitertanol + captan** Bayer
Baycarb I **fenobucarb** Bayer
Baycid I **fenthion** Bayer
Baycol F **propineb + triadimefon** Bayer
Baycor F **bitertanol** Agro-Kemi, Agroplant, Bayer, Nitrokemia, Pinus
Baycor C F **bitertanol + captan** Agroplant
Baycor Zf F **bitertanol + ziram** Bayer
Baycor-Antracol F **bitertanol + propineb** Bayer
Baycor-Captan F **bitertanol + captan** Bayer
Bayer E 393 IA **sulfotep** Bayer
Bayer Hedonal H **2,4-D + dichlorprop** Bayer
Bayer Svovel F **sulphur** Bayer
Bayfidan F **triadimenol** Agro-Kemi, Bayer, Pinus
Bayfidan BCM F **carbendazim + triadimenol** Bayer
Bayfidan Combi F **sulphur + triadimenol** Bayer
Bayfidan D F **anilazine + triadimenol** Bayer
Bayfidan M-D F **anilazine + triadimenol + tridemorph** Bayer

Bayfidan Triple F **captafol + carbendazim + triadimenol** Bayer
Baygon I **propoxur** Agrokemi, Bayer
Baygon (Spruehdose) I **dichlorvos + propoxur** Bayer
Baygon MEB I **plifenate** Bayer
Baykor F **bitertanol** Berner
Bayleton F **triadimefon** Agro-Kemi, Bayer, Berner, Fahlberg-List, Imex-Hulst, Lucebni zavody, Pinus, Turda
Bayleton A 74 F **propineb + triadimefon** Bayer
Bayleton AN F **propineb + triadimefon** Bayer
Bayleton BCM F **carbendazim + triadimefon** Bayer
Bayleton BM F **carbendazim + triadimefon** Bayer
Bayleton CA F **captan + triadimefon** Bayer
Bayleton CF F **captafol + triadimefon** Bayer
Bayleton CM F **carbendazim + triadimefon** Agro-Kemi
Bayleton Combi F **sulphur + triadimefon** Bayer
Bayleton ME F **tolylfluanid + triadimefon** Bayer
Bayleton Total F **carbendazim + triadimefon** Bayer
Bayleton triple F **captafol + carbendazim + triadimefon** Bayer
Bayleton-Eupareen FA **tolylfluanid + triadimefon** Bayer
Bayluscid M **niclosamide** Bayer
Bayluscide M **niclosamide** Bayer
Baymat F **bitertanol** Bayer
Bayrusil IA **quinalphos** Bayer
Baysan F **climbazol**
Baysol blomspray IA **methiocarb + propoxur** Bayer
Baysol desherbant total 60 H **amitrole + diuron** Bayer
Baysol desherbant total liquide H **amitrole + bromacil + diuron** Bayer
Baytan F **fuberidazole + triadimenol** Bayer, ICI
Baytan 15 F **triadimenol** Bayer, Pinus
Baytan 17.5 F **imazalil + triadimenol** Bayer
Baytan 5 F **triadimenol** Bayer
Baytan Combi F **imazalil + triadimenol** Bayer
Baytan IM F **fuberidazole + imazalil + triadimenol** Bayer
Baytan MO F **quinomethionate + triadimenol** Bayer
Baytan MZ F **mancozeb + triadimenol** Bayer
Baytan Spezial F **fuberidazole + triadimenol** Bayer
Baytan universal F **fuberidazole + imazalil + triadimenol** Bayer
Baytex I **fenthion** Bayer
Baythion I **phoxim** Agro-Kemi, Bayer
Baythion C I **chlorphoxim** Bayer
Baythion fourmis I **phoxim** Bayer
Baythion spray I **phoxim + piperonyl butoxide + tetramethrin** Bayer
Baythion/contra formigas I **phoxim** Bayer
Baythroid I **cyfluthrin** Bayer
Baytop FI **bitertanol + cyfluthrin** Bayer
Baytroid I **cyfluthrin** Bayer
Baza Plantenspray I **allethrin + piperonyl butoxide** Bakker
Baza Plantenspray N op waterbasis I **phenothrin + tetramethrin** Baza
Baza Plantenspray op waterbasis IA **piperonyl butoxide + pyrethrins** Baza
Bazin F **zineb** Baslini
Bazol C AF **sulphur** Baslini
BBS F **Bordeaux mixture** Hoechst, Roussel Hoechst
BCM F **carbendazim**
BCPE A **chlorfenethol**
Beam F **tricyclazole** Elanco
Beaphar slakkenkorrels M **metaldehyde** Beaphar
Bed 526 S **petroleum oils** Kuwait Petroleum
Beeline T **hydrolysed proteins + sugar** Siapa
Beet-Kleen H **chlorpropham + fenuron + propham** Shell
Beetomax H **phenmedipham** Fine Agrochemicals
Beetup H **phenmedipham** Campbell
Befran F **iminoctadine** Dainippon
Bejsin F F **TCMTB** KVK
Bel-Cap F **benomyl + captan** Zootechniki
Bel-Gold G **borax + sulphur** Siapa
Bel-vin F **copper oxychloride + sulphur + zineb** Diana
Belcuram F **cymoxanil + mancozeb** Shell
Belgran H **ioxynil + isoproturon + mecoprop** Shell
Belgrano HBC I **lindane** Sariaf
Bellasol IA **butoxycarboxim** Epro
Bellater H **atrazine + cyanazine** Agrishell, Shell

Bellclo H **2,4-DB + mecoprop** Campbell
Bellmac Plus H **MCPA + MCPB** Campbell
Bellmac Straight H **MCPB** Campbell
Belmark IA **fenvalerate** Shell
Belproil A IA **petroleum oils** Probelte
Belproil MP IA **parathion-methyl + petroleum oils** Probelte
Belpron F **mancozeb** Probelte
Belpron 10 F **zineb** Probelte
Belpron 15 F **copper oxychloride + zineb** Probelte
Belpron 16 F **copper oxychloride** Probelte
Belpron 80 F **zineb** Probelte
Belpron C F **captan** Probelte
Belpron F F **folpet** Probelte
Belpron M F **maneb** Probelte
Belpron T FX **thiram** Probelte
Belt I **chlordane** Sandoz
Beltanol F **8-hydroxyquinoline sulphate** Probelte
Beltasur 30-16 F **copper oxychloride + folpet** Probelte
Beltasur Extra B F **copper oxychloride + maneb + zineb** Probelte
Beltasur M F **copper oxychloride + mancozeb** Probelte
Beltracina H **atrazine** Probelte
Benagro F **benomyl** Safor
benalaxyl F [71626-11-4] methyl *N*-phenylacetyl-*N*-2,6-xylyl-DL-alaninate, Galben, M 9834
Benasalox H **benazolin + clopyralid** Kemi-Intressen, Schering
Benathion IA **malathion** ERT
Benazalox H **benazolin + clopyralid** Schering
Benazim F **carbendazim** Du Pont
benazolin H [3813-05-6] 4-chloro-2-oxobenzothiazolin-3-ylacetic acid, benazoline, Cornox CWK, Cresopur, Galipan, Galtak, Keropur, RD 7693
benazoline H **benazolin**
Benazolinester H **benazolin** Schering
Bencaptan F **captan** ERT
bencarbate I **bendiocarb**
Bencocel G **chlormequat chloride** ERT
Bendazim F **carbendazim** ERT, Hellenic Chemical
Bendazol F **benomyl** Visplant
bendiocarb I [22781-23-3] 2,3-isopropylidenedioxyphenyl methylcarbamate, bendiocarbe, bencarbate, OMS 1394, Dycarb, Ficam, Garvox, Multamat, NC 6897, Niomil, Seedox, Tattoo, Turcam
bendiocarbe I **bendiocarb**
bendioxide H **bentazone**
Benefex H **benfluralin** Makhteshim-Agan, Sinteza
benefin H **benfluralin**
Benefit F **benodanil**
Benelux I **thiofanox**
Benex F **benomyl** Crystal
benfluralin H [1861-40-1] *N*-butyl-*N*-ethyl-α,α,α-trifluoro-2,6-dinitro-*p*-toluidine, benfluraline, benefin, bethrodine, Balan, Balfin, Banafine, Benefex, Bonalan, EL-110, Flubalex, Quilan
benfluraline H **benfluralin**
Benfos IA **dichlorvos** Quimica Estrella
benfuracarb I [82560-54-1] ethyl *N*-[2,3-dihydro-2,2-dimethylbenzofuran-7-yloxycarbonyl(methyl)aminothio]-*N*-isopropyl-β-alaninate, benfuracarbe, OK-174, Oncol
benfuracarbe I **benfuracarb**
benfuresate H [68505-69-1] 2,3-dihydro-3,3-dimethylbenzofuran-5-yl ethanesulphonate
Benil H **dichlobenil** Burri
Benit F **etaconazole** Ciba-Geigy
Benlate F **benomyl** Dillen, Du Pont, Farmos, ICI, Imex-Hulst, NordiskAlkali, Permutadora, Plantevern-Kjemi, Ravit, Rhone-Poulenc, Sapec, Zorka Subotica
Benlate T F **benomyl + thiram** Du Pont
Benoagrex F **benomyl** Sadisa
Benocar F **benomyl + mancozeb** Visplant
benodanil F [15310-01-7] 2-iodobenzanilide, BAS 3170F, Calirus
Benoil IA **petroleum oils** ERT
Benomilo F **benomyl** Aragonesas, Ficoop, Kern
benomyl F [17804-35-2] methyl 1-(butylcarbamoyl)benzimidazol-2-ylcarbamate, Agrocit, Benex, Benlate, Benosan, Du Pont 1991, Fibenzol, Fundazol, Tersan 1991
Benosan F **benomyl** Agrocros
Benox F **benomyl** Enotria, Key
benquinox F [495-73-8] 1,4-benzoquinone 1-benzoylhydrazone 4-oxime, tserenox, Ceredon, Cereline, Tillantox
Bensecal H **benazolin + dicamba + MCPA** Shell
bensulfuron-methyl H methyl 2-[[[[(4,6-dimethoxypyrimidin-2-yl)

amino]carbonyl]amino] sulfonyl]methyl] benzoate, DPX-F 5384, Londax
bensulide H [741-58-2] *O,O*-diisopropyl *S*-2-phenylsulphonylaminoethyl phosphorodithioate, SAP, Betamec, Betasan, Disan, Exporsan, Pre-San, Prefar, R-4461
bensultap I [17606-31-4] *S,S'*-2-dimethylaminotrimethylene di(benzenethiosulphonate), Bancol, Ruban, TI-1671, TI-78, Victenon, ZZ-Doricida
Bentafluid H **bentazone + mefluidide** 3M
Bentazene H **bentazone** Visplant
bentazon H **bentazone**
bentazone H [25057-89-0] 3-isopropyl-(1*H*)-benzo-2,1,3-thiadiazin-4-one 2,2-dioxide, bentazon, bendioxide, BAS 351H, Basagran
benthiocarb H **thiobencarb**
Bentrol H **ioxynil** Union Carbide
Bentrol H **bromoxynil** Chimac-Agriphar
Bentrol HB H **bromoxynil + mecoprop** Union Carbide
Benzacar A **benzoximate + propargite** Sipcam
benzadox H [5251-93-4] benzamido-oxyacetic acid, MC 0035, S 6173, Topcide
Benzafos IA **azinphos-methyl** Du Pont
Benzahex I **lindane**
benzamorf F [12068-08-5] morpholinium 4-dodecylbenzenesulphonate, benzamorphe, BAS 276F
benzamorphe F **benzamorf**
Benzan A **chlorobenzilate** Agrotechnica
benzethazet I **plifenate**
Benzex I **lindane** Woolfolk Chemical
Benzilan A **chlorobenzilate** Alpha, Makhteshim-Agan
Benzinc F **zineb** ERT
benzipram H [35256-86-1] *N*-benzyl-*N*-isopropyl-3,5-dimethylbenzamide, benzipram, S-18510
benziprame H **benzipram**
benzoepin IA **endosulfan**
Benzomarc H **phenobenzuron**
benzomate A **benzoximate**
benzoximate A [29104-30-1] 3-chloro-α-ethoxyimino-2,6-dimethoxybenzyl benzoate, benzomate, Aazomate, Acarmate, Artaban, Citrazon, NA-53M
benzoylprop-ethyl H [22212-55-1] ethyl *N*-benzoyl-*N*-(3,4-dichlorophenyl)-DL-alaninate, Endaven, Suffix, WL 17731
benzthiazuron H [1929-88-0] 1-(benzothiazol-2-yl)-3-methylurea, Bay 60618, Ganon, Gatnon
Beosit IA **endosulfan** Staehler
bercema-Aero-Super I **DDT + lindane** Berlin Chemie
bercema-Akafunin FIA **carbaryl + dicofol + zineb** Berlin Chemie
bercema-Anoxol H **iron sulphate** Berlin Chemie
bercema-Becosal I **DDT** Berlin Chemie
bercema-Bitosen F **carbendazim** Berlin Chemie
bercema-Boragel I **borax** Berlin Chemie
bercema-Ditox I **lindane + methoxychlor** Berlin Chemie
bercema-Haptarex I **chlorfenvinphos** Berlin Chemie
bercema-Haptasol I **chlorfenvinphos** Berlin Chemie
bercema-NMC-Staub I **carbaryl** Berlin Chemie
bercema-Olamin F **carbendazim** Berlin Chemie
bercema-Raps-Inkrustiermittel I **lindane** Berlin Chemie
bercema-Rhagolex I **methoxychlor** Berlin Chemie
bercema-Ruscalin I **lindane** Berlin Chemie
bercema-Ruscalin C I **chlorfenvinphos** Berlin Chemie
bercema-Soltax I **lindane + methoxychlor** Berlin Chemie
bercema-Spritzaktiv-Emulsion I **DDT + lindane** Berlin Chemie
bercema-Spritzpulver NMC 50 IG **carbaryl** Berlin Chemie
Berelex G **gibberellic acid** ICI, ICI-Zeltia, Maag, Sopra
Berghoff 2,4-D Combi H **2,4-D + MCPA** Berghoff
Berghoff MP-Combi H **2,4-D + mecoprop** Berghoff
Bermat AI **chlordimeform** Quimica Estrella
Berta I **lindane + malathion** ERT
Bertram Cumarin R **warfarin** Bertram
Bertram Schneckenfrei M **metaldehyde** Bertram
Bertram-Cumarin-Fertigkoeder R **coumatetralyl** Bertram
Bestol H **chlorotoluron** Sandoz
Bestox I **alphacypermethrin** FMC

Betador H **petroleum oils + phenmedipham** Schering
Betafen F **fentin acetate** Zupa
Betafen TS F **fentin hydroxide** Zupa
Betaflow H **phenmedipham** KVK
Betagri H **phenmedipham** Tradi-agri
Betam H **phenmedipham** Leu & Gygax
Betamat H **ethofumesate + phenmedipham** Schering
Betamec H **bensulide** PBI Gordon
Betamin H **chloridazon** Radonja
Betamix H **desmedipham + phenmedipham** NOR-AM
Betamur H **chloridazon** Caffaro
Betamyn H **chloridazon** Terranalisi
Betanal H **phenmedipham** Ciba-Geigy, Fahlberg-List, Gullviks, Hoechst, Huhtamaki, ICI, Kwizda, Nitrokemia, Organika-Azot, Pinus, Plantevern-Kjemi, Rhone-Poulenc, Schering, Schering AAgrunol, Shell, VCH
Betanal AM H **desmedipham** Schering
Betanal AM 11 H **desmedipham + phenmedipham** Ciba-Geigy, Diana, Kwizda, Nitrokemia, Organika-Sarzyna, Pinus, Schering, VCH
Betanal AM 21 H **desmedipham + phenmedipham** Schering
Betanal Compact H **desmedipham + phenmedipham** Schering
Betanal Tandem H **ethofumesate + phenmedipham** Ciba-Geigy, Schering, Schering AAgrunol
Betanex H **desmedipham** NOR-AM, Schering
Betanil H **lenacil + propham + proximpham** Fahlberg-List
Betapal G **(2-naphthyloxy)acetic acid** Synchemicals
Betaprop H **phenmedipham** Eurofyto
Betarex H **chloridazon** Interphyto
Betaron H **ethofumesate + phenmedipham** Schering
Betasan H **bensulide** Stauffer
Betasana H **phenmedipham** Esbjerg
Betasana combi H **clopyralid + phenmedipham** Esbjerg
Betazen I **bendiocarb** Ital-Agro
bethrodine H **benfluralin**
Betokson G **(2-naphthyloxy)acetic acid** Organika-Fregata
Betoran H **chloridazon + metolachlor** BASF, Ciba-Geigy
Betosip H **phenmedipham** Chiltern, Cillus, Radonja, Sipcam, Staehler
Betoxon H **chloridazon** Oxon
Betozon H **chloridazon** Inagra, Sipcam
Betozon Combi H **chloridazon + lenacil** Sipcam
Bettaquat G **chlormequat chloride** Mandops
Better H **chloridazon** Sipcam-Phyteurop
Bewurzelungsfluid 58e G **1-naphthylacetic acid** Staehler
Bexon Plus H **MCPA + MCPB** Agroplant
Bexton H **propachlor** Dow
Bezin F **zineb** Sofital
BGR Repellent I **putrid egg** Collett
BHZ Tox I **dichlorvos** Zande
Bi 3411 H **chloral hydrate**
Bi 3411-Neu H **chloral hydrate** Bitterfeld
Bi 58 IA **dimethoate** Bitterfeld, OHIS
Bi-Agroxyl H **2,4-D + MCPA** B.B.
Bi-Hedonal H **2,4-D + MCPA** Bayer
Bi-Hormone H **2,4-D + MCPA** Bourgeois
Bi-Monal H **2,4-D + MCPA** Hoechst
bialaphos H [35597-43-4] L-2-amino-4-[(hydroxy)(methyl)phosphinoyl]butyryl-L-alanyl-L-alanine, Herbiace, SF-1293
Biancaneve IA **petroleum oils** Aziende Agrarie Trento
Biancolio IA **petroleum oils** Siapa
Bicep H **atrazine + metolachlor** Ciba-Geigy
Bideron I **prothiofos** Bayer
Bidisin H **chlorfenprop-methyl**
Bidrin IA **dicrotophos** Shell
Bietapost H **phenmedipham** Siapa
Bietazol H **chloridazon** Hermoo
Bietosan F **fentin acetate** Chemia
Bietosti H **chloridazon** Solfotecnica
Bieuroxone H **2,4-D + MCPA** Eurofyto
Bifenix H **bifenox + isoproturon** Agriben, Pepro, Pliva
bifenox H [42576-02-3] methyl 5-(2,4-dichlorophenoxy)-2-nitrobenzoate, MC-4379, Modown
Bifenox-P H **bifenox + mecoprop** Imex-Hulst
bifenthrin IA [82657-04-3] 2-methylbiphenyl-3-ylmethyl (*Z*)-(1*RS*,3*RS*)-3-(2-chloro-3,3,3-trifluoroprop-1-enyl)-2,2-dimethylcyclopropanecarboxylate, bifenthrine, Brigade, FMC 54800, Talstar
bifenthrine IA **bifenthrin**
Bifos IA **monocrotophos** Caffaro
Bifox H **bifenox + mecoprop** Eli Lilly
Big Dipper I **diphenylamine**
Bigramix H **2,4-D + MCPA** Colvoo
Bigrow H **2,4-D + mecoprop** Colvoo

Bihormonex H **2,4-D + MCPA** L.A.P.A.
Bijelo Ulje IA **petroleum oils** Radonja
Bikartol-Neu G **propham** Schering
Bilan I **lindane** OHIS
Bilcan I **carbaryl + diazinon** Mylonas
Bilka algefjerner L **benzalkonium chloride + tributyltin naphthenate** Bilka
Bilobran IA **monocrotophos** Ciba-Geigy
Biloxazol F **bitertanol** Bayer
Bim F **tricyclazole** Elanco, Grima
Bimate H **diuron + tebuthiuron** Elanco
Binab T F **Trichoderma viride** Harrof
binapacryl AF [485-31-4] 2-*sec*-butyl-4,6-dinitrophenyl 3-methylcrotonate, dinoseb-methacrylate, OMS 571, ENT 25793, Acricid, Ambox, Dapacryl, Endosan, Hoe 2784, Morocide, Morrocid
Bio Blatt Mehltaumittel F **lecithin** Meyer, Neudorff
Bio Blatt Mehltauspray F **lecithin** Neudorff
Bio Divercid I **permethrin + pyrethrins** Diversa
Bio DOM Universal-Schutzspray fuer Rosen und ZierpflanzenFIA **piperonyl butoxide + pyrethrins + sulphur** Dom-Samen-Fehlemann
Bio Flydown IA **permethrin** Pan Britannica
Bio Insekt-middel I **permethrin + piperonyl butoxide + pyrethrins** Trinol
Bio Lawn Weedkiller H **2,4-D + dicamba** Pan Britannica
Bio Long-Last IA **dimethoate + permethrin** Pan Britannica
Bio Melduggmiddel F **lecithin** Collett
Bio Moss Killer H **dichlorophen** Pan Britannica
Bio Multirose F **dinocap + permethrin + sulphur + triforine** Pan Britannica
Bio Multiveg F **carbendazim + copper oxychloride + permethrin+ sulphur** Pan Britannica
Bio Myctan Gartenspray FIA **lecithin + piperonyl butoxide + pyrethrins** Neudorff
Bio Myctan Pflanzenspray FIA **piperonyl butoxide + pyrethrins + sulphur** Neudorff
Bio permaforte I **permethrin + pyrethrins** Trinol
Bio Pflanzenspray Hoefter FIA **piperonyl butoxide + pyrethrins + sulphur** Hoefter
Bio Roota H **dichlorophen + 1-naphthylacetic acid** Pan Britannica
Bio Sprayday I **piperonyl butoxide + pyrethrins + resmethrin** Pan Britannica
Bio-Clean L **sodium hypochlorite** Sadolin
Bio-Insektenfrei I **piperonyl butoxide + pyrethrins** Spiess
Bio-Kill I **permethrin** Sturzenegger
Bio-Strip I **dichlorvos** Szovetkezet
Biobit I **Bacillus thuringiensis** Microbial Resources
Biocap F **captan + dinocap + zineb** Sipcam-Phyteurop
Biocid-Spritzmittel LG/Liquide I **piperonyl butoxide + pyrethrins** Leu & Gygax
Biodac H **chlorthal-dimethyl** Inagra
Biodrin I **resmethrin** Agro-Kemi
Biofol FA **dinocap + folpet** Gaillard
Biogard M **ethanol** Sipuro
Biogold F **captan + zineb** Gaillard
404 Bio Insecticide I **bioallethrin + piperonyl butoxide** Beaphar
Bionex IA **azinphos-ethyl**
Biophytoz I **piperonyl butoxide + pyrethrins + rotenone** Agronature
Bioquin F **oxine-copper**
Biorep HR X **androsterone + dehydroepiandrosterone** Nordtend
Biorex I **bioallethrin + bioresmethrin + piperonyl butoxide** T&D Mideast
Bioruiskute IA **piperonyl butoxide + pyrethrins** Kemira
Biothion I **temephos** Cyanamid
Biothrin I **bioallethrin + permethrin + piperonyl butoxide** T&D Mideast
Biotrol R **warfarin** Rentokil
BIPC H **chlorbufam**
BiPC H **chlorbufam** BASF
Biphagittol S **petroleum oils** Bitterfeld
biphenyl F [92-52-4] biphenyl, Lemonene, Phenador-X, Phenylbenzene
Biresal I **bioallethrin + bioresmethrin + piperonyl butoxide** Ameco
Birgin HG **propham** Bayer
Birgin IPC Gorsac G **chlorpropham + propham** Bayer
Birgin Novo G **chlorpropham** Bayer
Birlane IA **chlorfenvinphos** Agrishell, Agroplant, ICI, ICI-Valagro, Shell
bisthiosemi R [39603-48-0] 1,1'-methylenedi(thiosemicarbazide), Kayanex, NK-15561
bitertanol F [55179-31-2] 1-(biphenyl-4-yloxy)-3,3-dimethyl-1-(1*H*-1,2,4-triazol-1-yl)butan-2-ol, Bay KWG 0599, Baycor, Baymat, Biloxazol, Sibutol
Bitterstop F **calcium chloride** Du Pont

Bla-S F **blasticidin-S** Kaken, Kumiai, Nihon Nohyaku
Black Leaf 40 I **nicotine** Black Leaf
Black Leaf loevetanndreper H **2,4-D** Norsk Froe
Black Leaf Lusedreper I **methoxychlor + pyrethrins + rotenone** Norsk Froe
Black-stop F **sulphur** Bourgeois
Blackengranulat spezial H **2,4-D + mecoprop** Pluess-Staufer
Blackengranulat/Antirumex H **dichlobenil** CTA, Leu & Gygax, Sandoz
Blackengranulat/Antirumex granule Shell H **chlorthiamid** Agroplant
Bladafum IA **sulfotep** Agro-Kemi, Agroplant, Bayer, Berner
Bladan IA **parathion** Bayer, Berner
Bladan M IA **parathion-methyl** Bayer
Bladazin H **atrazine + cyanazine** Shell
Bladex H **cyanazine** Radonja, Shell, VCH
Bladex A H **atrazine + cyanazine** Shell
Bladex-combi H **atrazine + cyanazine** Radonja
Bladluisspray N IA **piperonyl butoxide + pyrethrins** Bogena
Bladotyl H **cyanazine + mecoprop** Agrishell, Shell
Blagal H **cyanazine + MCPA** Shell
Blaha-Fliegentod I **methomyl** Blaser
Blaizine H **cyanazine + simazine** Shell
Blanc du rosier Umupro F **bupirimate** Umupro
Blanc P Umupro I **chlorpyrifos** Umupro
Blascide F **tricyclazole** Elanco
blasticidin-S F [2079-00-7] 1-(4-amino-1,2-dihydro-2-oxopyrimidin-1-yl)-4-[(*S*)-3-amino-5-(1 -methylguanidino)valeramido]-1,2,3,4-tetradeoxy-*β*-D-*erythro*-hex-2-enopyranuronic acid, Bla-S, Blasticidin-S-3
Blatat H **cyanazine + MCPA** Shell
4-Blatt IA **piperonyl butoxide + pyrethrins** Graichen
Blattanex I **propoxur** Bayer
Blattanex Aerosol I **dichlorvos + propoxur** Bayer
Blattlaus Spray IA **dimethoate** Barnaengen
Blattlaus- und Spinnmilben-Spray IA **butocarboxim** Dow
Blattlaus-Cit Spray I **piperonyl butoxide + pyrethrins + rotenone** Cit
Blattlaus-Spray W I **butocarboxim** Dow
Blattlaus-Vernichter Nexion I **bromophos** Shell
Blattlausfrei I **permethrin** ASB-Gruenland
Blattol IA **petroleum oils** Schumann
Blaukupfer/Cuivre bleu 50 F **copper oxychloride** Sandoz
Blazer H **acifluorfen-sodium** BASF, Zorka Subotica
Blazon F **chlorothalonil**
Blecar "MN" F **mancozeb** La Littorale
Bledor 3 F **carbendazim + maneb + sulphur** R.S.R.
Blefit H **bromofenoxim + isoproturon** Ciba-Geigy
Blekritt F **carboxin + thiram** Siapa
Blesal MC H **clopyralid + MCPA + mecoprop** Siapa
Blevigor H **bromofenoxim + ioxynil + isoproturon** Ciba-Geigy
Blex IA **pirimiphos-methyl** ICI
blitol Ameisen-Spray I **bendiocarb** Blitol
blitol Ameisenmittel I **trichlorfon** Urania
blitol Graeserfrei fuer Ziergehoelze und Hecken H **EPTC** Urania
blitol Insektenfrei IA **piperonyl butoxide + pyrethrins** Blitol, Spiess, Urania
blitol Insektenfrei Neu I **etrimfos** Urania
blitol Rasenduenger mit Unkrautvernichter H **2,4-D + dicamba** Blitol, Urania
blitol Rasenduenger plus Moosvernichter H **iron sulphate** Blitol, Urania
blitol Rosen-Kombi-Spray FIA **piperonyl butoxide + pyrethrins + sulphur** Blitol
blitol Rosenduenger plus H **monuron** Urania
blitol Schneckenkorn M **metaldehyde** Blitol
blitol Unkrautfrei fuer Rasen Neu H **dicamba + MCPA** Blitol, Spiess, Urania
blitol Unkrautfrei fuer Wege H **atrazine + diuron + simazine** Blitol
Blitox F **copper oxychloride**
Bloc F **fenarimol** Elanco
Bloc "Super Raticide Rastop" R **warfarin** Rastop
Bloc appat EV R **warfarin** Rhodic
Bloc hydrofuge CL R **chlorophacinone** Steininger
Bloc hydrofuge D R **difenacoum** Steininger
Bloc rats souris Mortis R **scilliroside** Sovilo
Bloc'operats R **chlorophacinone** C.N.C.A.T.A.
Blocarat R **warfarin** Sydegal
Blomstra Bladlusspray I **piperonyl butoxide + pyrethrins + rotenone** Wallco

Blosomil O F **2-phenylphenol** Andres
Blossom Set G **(2-naphthyloxy)acetic acid** Phytochemiki
blue copperas LF **copper sulphate**
blue stone LF **copper sulphate**
Blue Viking LF **copper sulphate** Kocide
blue vitriol LF **copper sulphate**
Blue-Shield F **copper hydroxide**
Blumetta Moosvertilger H **iron sulphate** Coop, Parco
Blumetta Rasenduenger mit Moosvernichter H **iron sulphate** Parco
Blumetta Rasenduenger mit Unkrautvernichter H **2,4-D + MCPA** Coop
Blumetta Schneckenkorn M **metaldehyde** Parco
Blumetta-Ameisenmittel I **lindane** Parco
Blusana Pflanzenschutz Spray IA **piperonyl butoxide + pyrethrins** Lenz
BM 2 C FIHN **chloropicrin + methyl bromide** Traital
BM 33 C FIHN **chloropicrin + methyl bromide** Traital
B-Nine G **daminozide** Berner, Uniroyal
BNOA G **(2-naphthyloxy)acetic acid**
BNP 20 H **dinoseb** Hoechst
Bochamp H **linuron + neburon + petroleum oils +trifluralin** R.S.R.
Boden-Schaedlings-frei I **bromophos** Flora-Frey
Bodip IA **lindane** Bogena
Boga-Klip I **permethrin** Bogena
Bogena Insektendood I **bioallethrin + permethrin** Bogena
Bogena Kiemremmer G **chlorpropham + propham** Bogena
Bogena Slakkenkorrels M **metaldehyde** Bogena
Bogena Vliegendood I **methomyl** Bogena
Bol Kiemremmer G **chlorpropham + propham** Bol
Bolda F **carbendazim + maneb + sulphur** Farm Protection
Bolero H **thiobencarb** Chevron
Boliron H **linuron** Bourgeois
Bolls-Eye H **dimethylarsinic acid** Vertac
Bolstar I **sulprofos** Bayer
Bomba Total KB Jardin IAF **dichlone + dicofol + dinocap + lindane +methoxychlor + pyrethrins** Rhone-Poulenc
Bombardier F **chlorothalonil** Universal Crop Protection
Bombe raticide hydrofuge Steininger R **warfarin** Steininger
Bomyl IA **dimethyl 4-(methylthio)phenyl phosphite** Hopkins
Bonalan H **benfluralin** Elanco, Eli Lilly, Radonja, Siapa
Bonebe F **zineb** Bourgeois
Bonitrol FHI **DNOC** Bourgeois
Bonsul FA **sulphur** Ciba-Geigy
Bonzi G **paclobutrazol** ICI
Boots Ant Destroyer IA **lindane** Boots
Boots Caterpillar and Whitefly Killer I **permethrin** Boots
Boots Garden Fungicide F **carbendazim** Boots
Boots Garden Insect Powder FI **carbaryl + rotenone** Boots
Boots Greenfly and Blackfly Killer IA **dimethoate** Boots
Boots Hormone Rooting Powder G **4-indol-3-ylbutyric acid + 1-naphthylaceticacid + thiram** Boots
Boots Lawn Weed and Feed Granules H **2,4-D + dicamba** Boots
Boots Lawn Weed and Feed H **dichlorprop + MCPA** Boots
Boots Lawn Weedkiller H **2,4-D + dichlorprop + mecoprop** Boots
Boots Mouse and Rat Killer H **bromadiolone** Boots
Boots Slow Release Fly Killer I **dichlorvos** Boots
Boots Slug Destroyer Pellets M **metaldehyde** Boots
Boots Total Lawn Treatment F **benazolin + 2,4-D + dicamba + dichlorophen + dichlorprop + mecoprop** Boots
Bora roeggas-kaerte R **sulphur** Kiltin
Boracol 10 Rh L **benzalkonium chloride + borax** Spangenberg
Bordeaux mixture F [8011-63-0] A mixture, with or without stabilizing agents, of calcium hydroxide and copper sulphate, Bordo mixture, pre-reacted neutral Bordeaux, Comac
Bordermaster H **MCPA** Burts & Harvey
Bordil F **copper sulphate + mancozeb** Eli Lilly
Bordocop F **Bordeaux mixture** Ingenieria
Bordo mixture F **Bordeaux mixture**
Bordocritt F **Bordeaux mixture + cymoxanil** Siapa
Bordocure F **Bordeaux mixture** Proval
Bordofix F **Bordeaux mixture + folpet** Burri

Bordoile-alapanyag F **Bordeaux mixture + copper oxychloride + folpet** Orszagos
Bordolex F **Bordeaux mixture** Tecniterra
Bordoman F **Bordeaux mixture + maneb** Tecniterra
Bordovska corba F **copper sulphate** Zorka Sabac, Zupa
Borea H **bromacil**
Borial F **iprodione** Du Pont
Borocil 15 H **bromacil** Anticimex
Borocil 4 H **borax + bromacil** Agriben
Borocil G H **borax + bromacil** Borax
Borodust I **boric acid** Rentokil
Bortene F F **copper sulphate + folpet** Sedagri
Bortene MZ F **copper sulphate + maneb** Sedagri
Bos MH G **maleic hydrazide** Bos
Bosmonam FIHN **metham-sodium** Bos
Boszamet FIHN **dazomet** Bos
Botec F **captan + dicloran** NOR-AM
Botix H **TCA** Van der Boom
Botran F **dicloran** NOR-AM, Schering, Upjohn
Botrilex F **quintozene** ICI
Botrin F **carbendazim** Afrasa
Botrin C F **captan + carbendazim** Afrasa
Botrin M F **carbendazim + maneb** Afrasa
Botrin MBC F **carbendazim** Afrasa
Botrizol F **metiram + vinclozolin** Shell
Botrysan F **anilazine** Bayer
Bottrol DP H **dichlorprop + ioxynil** Rhone-Poulenc
Bottrol PE H **bromoxynil + ioxynil + mecoprop** Hoechst, Rhone-Poulenc
Bouillie Bordelaise F **Bordeaux mixture** Chimac-Agriphar, Cuenoud, Faridis, Rhodiagri-Littorale, R.S.R., Sedagri, Sipcam-Phyteurop, Sovilo, Tarsoulis, Tradi-Agri, UCAR, Umupro
Bouillie Kupfer-Bordo F **Bordeaux mixture** Leu & Gygax
Bouillie Kupferkalkbruehe F **Bordeaux mixture** Burri
Bouillie MOP 20 F **Bordeaux mixture** Marty et Parazols
Bouillie Oreal F **Bordeaux mixture** Gaillard
Bouillie Procida F **Bordeaux mixture** Meoc
Boul'Herb G H **lenacil + neburon** Boul'Herb
Boutanex H **amitrole + diuron** Protex
Boutormone G **4-indol-3-ylbutyric acid** Rhodic
Bovinox I **trichlorfon**
Boxer H **prosulfocarb**
BP 143-25 IA **permethrin** BP
BP Actipron S **petroleum oils** BP
BP Ulvapron S **petroleum oils** BP
BPMC I **fenobucarb**
BPPS A **propargite**
Brabant Amitrol Extra H **amitrole + ammonium thiocyanate** Voorbraak
Brabant Anti-Scabiosum IA **lindane** Voorbraak
Brabant Kiemremmer G **chlorpropham + propham** Voorbraak
Brabant Mixture H **2,4-D + dicamba + MCPA** Voorbraak
Brabant mixture super H **2,4-D + dicamba + MCPA + mecoprop** Voorbraak
Brabant Selective Weedkiller H **dinoseb** Voorbraak
Brabant Slakkendood M **metaldehyde** Voorbraak
Brabant Spuitzwavel F **sulphur** Voorbraak
Brabant Tin Super F **fentin acetate + maneb** Voorbraak
Brabant uitvloeier conc. S **polyglycolic ethers** Voorbraak
Brabant vloeibaar Kiemremmingsmiddel voor aardappelen G **chlorpropham + propham** Voorbraak
Brake H **fluridone** Elanco
Brandol F **1-hydroxy-4-nonyl-2,6-dinitrobenzene** Sariaf
Brasoran H **aziprotryne** Ciba-Geigy
Brassam F **quintozene** Van Wesemael
Brassicol F **quintozene** Hoechst, Pluess-Staufer
Brassix H **trifluralin** Sipcam-Phyteurop
Bravo F **chlorothalonil** BASF, Diamond Shamrock, Grima, Inagra, KenoGard, SDS Biotech
Bravo plus F **carbendazim + chlorothalonil** Siegfried, Sipcam-Phyteurop
Bravocarb F **carbendazim + chlorothalonil** SDS Biotech
Bray's Emulsion FHL **cresylic acid** ICI
Bref C G **chlormequat chloride** Sipcam-Phyteurop
Brek H **chloridazon** Agrimont, Marktredwitz, Montedison, Staehler
Brek H **chloridazon** Montedison
Brennessel Granulat Neu H **MCPA + mecoprop** Spiess, Urania
Brennessel-KO H **dichlobenil** Neudorff

Brestan F **fentin acetate** Hoechst, Roussel Hoechst, Zupa
Brestan F **fentin acetate + maneb** Chromos, Hoechst, Schering
Brestan F **fentin hydroxide** Chromos, Roussel Hoechst
Brestan Combi F **fentin acetate + maneb** Hoechst
Brestan Flow F **fentin hydroxide** Hoechst
Brestan Super F **fentin acetate + maneb** Hoechst
Brestaneb F **fentin acetate + maneb** Hoechst
Brevitox Ino Spray I **dichlorvos** Druchema
Bri-cyde IA **cyanides** Brinkman
Bri-Rem G **chlorpropham** Brinkman
Bri-Spray Super N I **dichlorvos** Tuhamij
Brida H **alachlor + atrazine** Inagra
Brifur IAN **carbofuran** Brichema
Brigade I **bifenthrin** FMC, Rhone-Poulenc
Brill Moosvernichter H **iron sulphate** Brill
Brill Unkrautvernichter H **dicamba + MCPA** Brill
Brillaqua 2 I Z F **imazalil** Brillocera
Brillaqua 2 I Z/O F **imazalil + 2-phenylphenol** Brillocera
Brillaqua 4 I Z F **imazalil** Brillocera
Brillaqua ST F **thiabendazole** Brillocera
Brillaqua ST/2 I Z F **imazalil + thiabendazole** Brillocera
Brillaqua ST/O F **2-phenylphenol + thiabendazole** Brillocera
Brimstone Plus F **di-1-p-menthene + potassium sorbate + sodiummetabisulphite + sodium propionate** Mandops
Brinkman Monam FIHN **metham-sodium** Brinkman
Brior F **carbendazim** Sipcam-Phyteurop
Briotril H **bromoxynil + ioxynil** Agronorden, Leu & Gygax, Pan Britannica
Brioxil H **ioxynil + mecoprop** Aragonesas
Brioxil Doble H **bromoxynil + ioxynil + mecoprop** Aragonesas
Bris FX **thiram** Platecsa
Briten I **trichlorfon** Quimica Estrella
Britex F F **2-phenylphenol** Brogdex
Britex I F **imazalil** Brogdex
Britex T F **thiabendazole** Brogdex
Broadshot H **2,4-D + dicamba + triclopyr** Shell
Brocika-toks H **MCPA + mecoprop + 2,3,6-TBA** Radonja
Brodan I **chlorpyrifos** Planters Products
brodifacoum R [56073-10-0] 3-[3-(4'-bromobiphenyl-4-yl)-1,2,3,4-tetrahydro-1-naphthyl]-4-hydroxy coumarin, Brodifacoum, De-Mice, Havoc, Klerat, Matikus, Mouser, PP 581, Ratak, Talon, Volid, WBA 8119, Weather Blok
Brodifacoum R **brodifacoum** ICI, Sorex
Brodilon mamak R **bromadiolone** Pliva
Brody H **bromacil + diuron** Aragonesas
Brofene I **bromophos**
Brom-o-gas FIHN **chloropicrin + methyl bromide** Great Lakes, Neoquimica, Sobrom, Zootechniki
bromacil H [314-40-9] 5-bromo-3-*sec*-butyl-6-methyluracil, Bromax, Du Pont Herbicide 976, Hyvar X, Nalkil, Rokar X, Rout, Staa-Free, Uragan, Urox B
bromadiolone R [28772-56-7] 3-[3-(4'-bromobiphenyl-4-yl)-3-hydroxy-1-phenylpropyl]-4-hydroxycoumarin, broprodifacoum, Bromatrol, Bromone, Contrac, Contrax, Deadline, LM 637, Maki, Rafix, Ratimus, Rotox, Super-Caid, Super-Rozol, Temus, Topidion
Bromard R **bromadiolone** Rentokil
Bromate H **bromoxynil + MCPA** Rhone-Poulenc
Bromatrol R **bromadiolone** Rentokil
Bromax H **bromacil** Hopkins
Bromazil F **imazalil** Brogdex
bromchlophos IA **naled**
Bromek metylu FIHN **chloropicrin + methyl bromide** Helm
Bromelien-Ethrel G **ethephon** Urania
bromethalin R [63333-35-7] α,α,α-trifluoro-*N*-methyl-4,6-dinitro-*N*-(2,4,6-tribromophenyl)-*o*-toluidine, bromethaline, EL-614, Vengeance
bromethaline R **bromethalin**
Bromethyl 980 FIHN **methyl bromide** SIS
Brometo de metilo FIHN **methyl bromide** Sapec
Bromex IA **bromophos** Anticimex
Bromex R **bromadiolone** Rentokil
Bromex IA **naled** Makhteshim-Agan
Brominal H **bromoxynil** Amchem, Chimac-Agriphar, Pluess-Staufer, Union Carbide
Brominal Flax H **bromoxynil + MCPA** Amchem, Union Carbide
Brominal H H **bromoxynil + mecoprop** Union Carbide
Brominal Mais H **atrazine + bromoxynil** Chimac-Agriphar
Brominal ME 4 H **bromoxynil** Cillus

Brominal MEP H **bromoxynil + mecoprop** Union Carbide
Brominal Plus H **bromoxynil + MCPA** Amchem, Union Carbide
Brominal triple H **bromoxynil + MCPA + mecoprop** Sopra
Brominex H **bromoxynil** Jewin-Joffe
Brominil H **bromoxynil + dicamba + mecoprop** Avenarius
Brominil Lin H **bromoxynil + MCPA** Chimac-Agriphar
Bromo FIHN **chloropicrin + methyl bromide** Maldoy
bromobutide H [74712-19-9] (*RS*)-2-bromo-3,3-dimethyl-*N*-(1-methyl-1-phenylethyl)butyramide, S-4347, S-47, Sumiherb
bromofenoxim H [13181-17-4] 3,5-dibromo-4-hydroxybenzaldehyde 2,4-dinitrophenyloxime, bromophenoxime, C 9122, Dicoprime, Faneron
Bromoflor G **ethephon** Union Carbide
Bromofume NI **ethylene dibromide** Dow
Bromolon H **bromoxynil + clopyralid + dichlorprop** Esbjerg
Bromone R **bromadiolone** Lipha
Bromonil H H **bromoxynil + mecoprop** Serpiol
bromophenoxime H **bromofenoxim**
bromophos I [2104-96-3] *O*-4-bromo-2,5-dichlorophenyl *O,O*-dimethyl phosphorothioate, bromophos-methyl, OMS 658, ENT 27162, Brofene, Nexion, Omexan, S-1942
bromophos-ethyl IA [4824-78-6] *O*-4-bromo-2,5-dichlorophenyl *O,O*-diethyl phosphorothioate, OMS 659, ENT 27258, Filariol, Nexagan, S-2225
bromophos-methyl I **bromophos**
Bromopic FIHN **chloropicrin + methyl bromide** Edefi
bromopropylate A [18181-80-1] isopropyl 4,4'-dibromobenzilate, phenisobromolate, ENT 27552, Acarol, Folbex, GS 19851, Neoron
Bromormone H **bromoxynil + 2,4-D** Schering
Bromosan F **thiophanate + thiram** Cleary
Bromotox I **naled** Aziende Agrarie Trento
Bromotril H **bromoxynil** Makhteshim-Agan
Bromoxan H **bromoxynil** Etisa
bromoxynil H [1689-84-5] 3,5-dibromo-4-hydroxybenzonitrile, ENT 20852, Brominal, Bromotril, Buctril, Certrol B, Litarol, M&B 10064, Merit, Pardner, Sabre, Torch
brompyrazon H [3042-84-0] 5-amino-4-bromo-2-phenylpyridazin-3(2*H*)-one, brompyrazone, BAS 2430H, Basanor
brompyrazone H **brompyrazon**
bromuron H [3408-97-7] *N'*-(4-bromophenyl)-*N,N*-dimethylurea, Faluron, FL 173
Bronate H **bromoxynil + MCPA** Rhone-Poulenc
Bronco H **alachlor + glyphosate** Monsanto
Bronocot B **bronopol** Schering
bronopol B [52-51-7] 2-bromo-2-nitro-1,3-propanediol, Bronocot
Bronox H **linuron + trietazine** Staehler
Bronx F **carbendazim + maneb + sulphur** Sipcam-Phyteurop
Broot IM **trimethacarb** Union Carbide
broprodifacoum R **bromadiolone**
brown copper oxide F **cuprous oxide**
Broxolon H **bromoxynil + clopyralid + mecoprop** Farm Protection
Brozin F **carbendazim** Agriplan
BRP IA **naled**
Bruelex fluessig/liquide HI **DNOC** Siegfried
Bruelex forte H **dinoseb acetate** Siegfried
Brugsen blomsterspray til hus og have I **methoxychlor + piperonyl butoxide + pyrethrins** Esbjerg
Brugsen fluespray I **piperonyl butoxide + resmethrin** Esbjerg
Brugsen gulerodspudder FI **bromophos + captan** Esbjerg
Brugsen insekt og svampepudder FI **bromophos + captan** Esbjerg
Brugsen insektdraeber I **piperonyl butoxide + pyrethrins** Esbjerg
Brugsen insektspray I **piperonyl butoxide + pyrethrins** Esbjerg
Brugsen skimmeldraeber F **maneb** Esbjerg
Brugsen staldfluedraeber I **piperonyl butoxide + resmethrin** Esbjerg
Brugsens combi plaenerens H **2,4-D + dichlorprop** Esbjerg
Brugsens plantedraeber H **simazine** Esbjerg
Brummers vloeibaar Insectendood I **dichlorvos + methoxychlor** Brummer
Brumolin FF R **chlorophacinone + sulphaquinoxaline** Schering
Brumolin Fix Fertig Neu R **bromadiolone + sulphaquinoxaline** Shell

Brumolin-Fix-Fertig R **chlorophacinone + sulphaquinoxaline** Shell
Brumolin-prah R **warfarin** Zorka Sabac
Brumoline R **warfarin** KenoGard
Brumoline R **difenacoum** Haco
Brumoline po R **warfarin** Pinto
Brush ban H **2,4-D + triclopyr** National Chemsearch
Brushfree-Gher H **propyzamide + simazine** Certified Lab.
Brushoff H **metsulfuron-methyl** Du Pont
Brushwood Killer H **2,4,5-T** Marks, Universal Crop Protection
Brution I **methidathion** Du Pont
Bt I **Bacillus thuringiensis**
B.T.F. F **carbendazim + folpet + thiram** Pepro
Btk I **Bacillus thuringiensis**
BTS 14639 A **butacarb**
BTS 27419 AI **amitraz** Schering
BTS 30843 H **epronaz**
BTS 40542 F **prochloraz** Schering
BTS 7901 F **sultropen**
BTS 8684 F **quinazamid**
Buckshot H **dicamba + MCPA** Rhone-Poulenc
Buctril H **bromoxynil** Rhone-Poulenc
Buctril D H **bromoxynil + dicamba + MCPA** Rhone-Poulenc
Buctril M H **bromoxynil + MCPA** Kemira, Ravit
Buctrilin H **bromoxynil + MCPA** Embetec
Bud Nip HG **chlorpropham** PPG
Bueno H **MSMA** Diamond Shamrock, Fermenta, Masso
bufencarb I [8065-36-9] 3-(1-methylbutyl)phenyl methylcarbamate/3-(1-ethylpropyl)phenyl methylcarbamate mixture, bufencarbe, Bux
bufencarbe I **bufencarb**
Bug Gun for Fruit and Vegetables H **pyrethrins** ICI
Bug Gun for Roses and Flowers H **pyrethrins** ICI
BUK 726 H **bentazone + MCPB** BASF
Bullseye CDA H **amitrole + atrazine + diuron** ICI
Bumetran IA **amitraz** Schering
buminafos H [51249-05-9] dibutyl [1-(butylamino)cyclohexyl]phosphonate, Trakephon
Buminal T **hydrolysed proteins** Bayer
Bumuclan F **captafol + folpet** Scam
Bundasol AD 25-25 H **amitrole + diuron** Sandoz
Bundasol S-80 H **simazine** Sandoz
Bundolin Giberelico G **gibberellic acid** Sandoz
bupirimate F [41483-43-6] 5-butyl-2-ethylamino-6-methylpyrimidin-4-yl dimethylsulphamate, Nimrod, PP 588
Buntosan FI **lindane + mancozeb** ICI
buprofezin IA [69327-76-0] 2-*tert*-butylimino-3-isopropyl-5-phenyl-1,3,5-thiadiazinan-4-one , buprofezine, Applaud, NNI-750, PP 618
buprofezine IA **buprofezin**
Buracyl H **lenacil** Organika-Zarow
Burcop F **Burgundy mixture** Agribus, Comac
Burex H **chloridazon** Dimitrova
Burex H **neburon** Protex
Burex D H **chloridazon + dalapon-sodium** Dimitrova
Burex Special H **chloridazon + lenacil** Dimitrova
Burlan H **bromoxynil + dicamba + dichlorprop** Burri
Burtolin G **maleic hydrazide** Burts & Harvey
Burvel P H **dicamba + MCPA + mecoprop** Burri
Busan FNHI **metham-sodium** Buckman
Busan EC-30 F **TCMTB** Zorka Sabac
Bushwacker H **tebuthiuron** Rhone-Poulenc
butacarb A [2655-19-8] 3,5-di-*tert*-butylphenyl methylcarbamate, butacarbe, BTS 14639, RD 14639
butacarbe A **butacarb**
butachlor H [23184-66-9] *N*-butoxymethyl-2-chloro-2',6'-diethylacetanilide, Butanex, Butanox, CP 53619, Lambast, Machete, Pillarsete
Butacide S **piperonyl butoxide** Fairfield
butam H **tebutam**
butamifos H [36335-67-8] *O*-ethyl *O*-6-nitro-*m*-tolyl *sec*-butylphosphoramidothioate, Cremart, S-2864, Tufler
Butamin I **piperonyl butoxide + tetramethrin** Fairfield
Butanex H **butachlor** Makhteshim-Agan
Butanox H **butachlor** Crystal
Butazol H **amitrole + diuron + petroleum oils** Cimelak
Butazol ZX H **amitrole + atrazine** Cimelak

buthidazole H [55511-98-3] 3-(5-*tert*-butyl-1,3,4-thiadiazol-2-yl)-4-hydroxy-1-methyl-2-imidazolidone, Ravage, Vel-5026
buthiobate F [51308-54-4] butyl 4-*tert*-butylbenzyl *N*-(3-pyridyl)dithiocarbonimidate, Denmert, S-1358
Butifos F **tributyl phosphate** URSS
butilate H **butylate**
Butirex H **2,4-DB** L.A.P.A., Protex
Butisan H **metazachlor** Agrolinz, BASF
Butizyl H **MCPB** Chimac-Agriphar
butocarboxim I [34681-10-2] 3-(methylthio)butanone *O*-methylcarbamoyloxime, butocarboxime, Afilene, Co 755, Drawin
butocarboxime I **butocarboxim**
Butoflin I **deltamethrin**
Butormone H **2,4-DB** Universal Crop Protection
Butoss I **deltamethrin** Roussel Uclaf
butoxicarboxim IA **butoxycarboxim**
Butoxone H **2,4-DB** Rhone-Poulenc, Vertac
butoxycarboxim IA [34681-23-7] 3-methylsulphonylbutanone *O*-methylcarbamoyloxime, butoxycarboxime, butoxicarboxim, Co 859, Plant Pin
butoxycarboxime IA **butoxycarboxim**
butralin HG [33629-47-9] *N*-*sec*-butyl-4-*tert*-butyl-2,6-dinitroaniline, butraline, Amchem 70-25, Amchem A-280, Amex, Tamex
butraline HG **butralin**
buturon H [3766-60-7] 3-(4-chlorophenyl)-1-methyl-1-(1-methylprop-2-ynyl)urea, Eptapur, H 95
Butyl-Gelb HI **dinoseb** Staehler
butylate H [2008-41-5] *S*-ethyl di-isobutylthiocarbamate, butilate, diisocarb, Anelda, Genate, R-1910, Sutan
Butyrac H **2,4-DB** EniChem, Union Carbide
Butytox H **MCPB** Agro-Kemi
Buvatox I **fenitrothion + malathion** Budapesti Vegyimuvek
Buvicid F F **folpet** Budapesti Vegyimuvek
Buvicid K F **captan** Budapesti Vegyimuvek
Buvicid porozoszer F **folpet** Budapesti Vegyimuvek
Buvilan H **ethalfluralin** Budapesti Vegyimuvek
Buvilan T H **ethalfluralin + terbutryn** Budapesti Vegyimuvek
Buvinol F **atrazine + fenteracol** Budapesti Vegyimuvek
Buvisild K F **captan** Budapesti Vegyimuvek
Buvisild TR F **oxine-copper + thiophanate-methyl** Budapesti Vegyimuvek
Bux I **bufencarb**
Buxol H **dichlobenil** Cimelak
Byatran FB **organoiodine + thiabendazole** Wheatley
Bygran FG **tecnazene** Wheatley
Bygran F FG **organoiodine + tecnazene** Wheatley
BZ-insektsstoppare I **allethrin** Stroem

3 C-460 G **chlormequat chloride** Benagro
5 C-Benagro G **chlormequat chloride** Phyto
C 1414 IA **monocrotophos** Ciba-Geigy
C 15935 H **parafluron**
C 18898 H **dimethametryn** Ciba-Geigy
C 19490 H **piperophos** Ciba-Geigy
C 1983 H **chloroxuron** Ciba-Geigy
C 2059 H **fluometuron** Ciba-Geigy
C 2242 H **chlorotoluron** Ciba-Geigy
C 2307 I **methocrotophos**
C-3 I **metolcarb** Nihon Nohyaku
C 3126 H **metobromuron** Ciba-Geigy
C-32 G **chlormequat chloride + choline chloride** Pluess-Staufer
C 3470 H **difenoxuron** Ciba-Geigy
C 570 IA **phosphamidon** Ciba-Geigy
C 6313 H **chlorbromuron** Ciba-Geigy
C 6989 H **fluorodifen** Ciba-Geigy
C 7019 H **aziprotryne** Ciba-Geigy
C 709 IA **dicrotophos** Ciba-Geigy
C 8353 I **dioxacarb** Ciba-Geigy
C 8514 AI **chlordimeform** Ciba-Geigy
C 8949 IA **chlorfenvinphos** Ciba-Geigy
C 9122 H **bromofenoxim** Ciba-Geigy
C 9491 IA **iodofenphos** Ciba-Geigy
Cabiet H **chloridazon** Sipcam
Cabor I **carbaryl** Sepran
cacodylic acid H **dimethylarsinic acid**
Caddy GH **chlorsulfuron** Nordisk Alkali
Cadol F **dithianon** Midol
Cafudan F **captan** Diana
Caid R **chlorophacinone** Agriben, Lipha
Calcid I **cyanides** Breymesser, Degesch, DGS
calciferol R [50-14-6] (3*β*,5*Z*,7*E*,22*E*)-9,10-secoergosta-5,7,10(19),22-tetraen-3-ol, vitamin D2, ergocalciferol, Hyperkil, Rodinec, Sorexa

calcium arsenate IH tricalcium arsenate, Pencal, Security, Spra-cal, Turf-Cal
calcium cyanamide HF [156-62-7] calcium carbimide, cyanamide, nitrolime, Alzodef, Cyanamid
calcium polysulphide FIA **lime sulphur**
Calcyan I **cyanides** Breymesser, Cillus, Degesch, DGS
Calda Bordaleza F **copper sulphate** Quimigal, Sapec
Caldan I **cartap hydrochloride** Takeda
Caldo Bordeles F **Bordeaux mixture** Agrocros, Ficoop, Quimica del Valles
Caldon HI **dinoseb** Hoechst
Caldor F **captafol** Calliope
Caliber H **simazine** Ciba-Geigy
Calibre A **hexythiazox** Shell
Calidan F **carbendazim + iprodione** Rhodiagri-Littorale
Calin H **linuron** Marty et Parazols
Calirame F **captan** Calliope
Calirus F **benodanil** BASF
Caliser 20-40 H **amitrole + diuron** Platecsa
Caliser 90 H **amitrole + simazine** Platecsa
Caliser A H **atrazine** Platecsa
Caliser AS H **atrazine + simazine** Platecsa
Caliser S H **simazine** Platecsa
Caliverse G **chlormequat chloride** Calliope
Calixin F **tridemorph** Agrolinz, BASF, Bayer, Chromos, Intrachem, Lucebni zavody
Calixin M F **maneb + tridemorph** BASF
Calixine F **tridemorph** BASF
Calixine M F **maneb + tridemorph** BASF
Calliact H **TCA** Calliope
Callicuivre F **copper oxychloride** Calliope
Callidim IA **dimethoate** Calliope
Callifol A **dicofol** Calliope
Callifon I **trichlorfon** Calliope
Calliherbe H **2,4-D** Calliope
Callimal IA **malathion** Calliope
Calliman F **maneb** Calliope
Callimix F **copper oxychloride + zineb** Calliope
Callinde I **lindane** Calliope
Callindem I **lindane** Calliope, Marty et Parazols
Callineb F **zineb** Calliope
Callio M H **MCPA** Calliope
Callitraz H **atrazine** Calliope
Callizime H **simazine** Calliope
Calmathion IA **malathion** Rhone-Poulenc
Calo-Chlor F **mercurous chloride** Mallinckrodt
Calo-Gran F **mercurous chloride + mercuric chloride** Mallinckrodt
calomel FI **mercurous chloride**
Calron super H **neburon** Calliope
Caltan F **folpet + ofurace** Agroplant, Chevron, Geopharm, Sopra
Caltan Plus F **captafol + folpet + ofurace** Quimigal
Caltir F **thiram** Calliope
Calyram F **captan + metiram** Siegfried
cambendichlor H [56141-00-5] (phenylimino)diethylene bis(3,6-dichloro-*o*-anisate), cambendichlore, Vel 4207
cambendichlore H **cambendichlor**
Camostan FA **captan + quinomethionate** Agro-Kemi
Campagard H **prometryn + propazine** Ligtermoet
Campaprim S H **amitrole + atrazine + simazine** Viopharm
Camparol H **prometryn + simazine** Ligtermoet
Campbell's Destox H **2,4-D** Campbell
Campbell's Dioweed H **2,4-D** Campbell
Campbell's Field Marshall H **dicamba + MCPA + mecoprop** Campbell
Campbell's Grassland Herbicide H **dicamba + MCPA + mecoprop** Campbell
Campbell's M-C-PEX H **dichlorprop + MCPA** Campbell
Campbell's MC Flowable F **carbendazim + maneb** Campbell
Campbell's New Camppex H **2,4-D + dichlorprop + MCPA + mecoprop** Campbell
Campbell's Nico Soap I **nicotine** Campbell
Campbell's Oxystat H **bromoxynil + ioxynil + mecoprop** Campbell
Campbell's Redipon H **dichlorprop** Campbell
Campbell's Redipon Extra H **dichlorprop + MCPA** Campbell
Campbell's Redlegor H **2,4-DB + MCPA** Campbell
Campbell's Solo H **linuron + trifluralin** Campbell
Campbell's Static H **mecoprop** Campbell
Campbell's Sugar Beet Herbicide H **chlorpropham + fenuron + propham** Campbell
Campbell's X-Spor F **maneb** Campbell
camphechlor IRA [8001-35-2] toxaphene,

camphechlore, toxaphene, polychlorcamphene, ENT 9735, Hercules 3956, Melipax, Motox, Phenacide, Phenatox, Strobane-T, Toxakil, Toxaphene
camphechlore IRA **camphechlor**
Campogran F **furmecyclox** BASF
Camporol H **prometryn + simazine** Ligtermoet
Camposan G **ethephon** Bitterfeld
Can-Trol H **MCPB** Rhone-Poulenc
Canarat R **warfarin** Cyanamid
Cancerluq WF **oxine-copper** Luqsa
Cancror F **oxine-copper** Sopepor
Candex H **asulam + atrazine** Rhone-Poulenc
Cannicid H **dalapon-sodium** Sipcam
Canogard IA **dichlorvos** Shell
Canopy H **chlorimuron-ethyl + metribuzin** Du Pont
Caocobre F **cuprous oxide** Sandoz
Cap 50 F **captan** Agrochemiki, Siapa
Caparol H **prometryn** Ciba-Geigy
Capcide F **captan** Aziende Agrarie Trento
Capfos I **fonofos** La Quinoleine
Capidol F **captan** Midol
Capidol-T F **captan + thiram** Midol
Capluq F **captan** Luqsa
Capnebe F **captan + zineb** Agrishell
Capoid FA **dinocap** Gaillard
Caprane F **dinocap** O.P.
Caprecol X **copper naphthenate + odorous substances** Shell
Capriflor H **2,4-D + dicamba** Floralis
Capsidion I **diazinon** Pennwalt France
Capsithrine I **permethrin** Pennwalt France
Capsolane H **dichlormid + EPTC** Agrotec, BASF, La Quinoleine, Ligtermoet, Pepro, Sapec, Siegfried, Stauffer
captab F **captan**
captafol F [2425-06-1] 1,2,3,6-tetrahydro-*N*-(1,1,2,2-tetrachloroethylthio)phthalimide, difolatan, Difolatan, Haipen, Merpafol, Ortho 5865, Pillartan, Sanseal, Sanspor, Santar SM
Captagrex F **captan** Sadisa
Captalon F **captan** Hellenic Chemical
captan F [133-06-2] 1,2,3, 6-tetrahydro-*N*-(trichloromethylthio)phthalimide, captane, captab, ENT26538, Captec, Merpan, Orthocide, Phytocape, Pillarcap, SR 406, Vondcaptan
captane F **captan**
Captano F **captan** Roussel Hoechst
Captanol F **captan** Bourgeois
Captaspor F **captan + zineb** Hermoo
Captazel F **captan** ICI-Zeltia
Captazim F **captan + carbendazim** Hellenic Chemical
Captec F **captan** Griffin
Capteran F **captan** Aragonesas
Captocide F **captan** Visplant
Captol F **captan** Eli Lilly
Captol liquide F **captafol** Interphyto
Captolate F **captan** Prochimagro
Captolate AC FX **anthraquinone + captan** Prochimagro
Captonin FI **captan + diazinon + sulphur** Viopharm
Captoran F **captan** Ellagret
Captosan F **captan** Sanac, Condor
Captosan R F **captan + carbendazim** Condor
Captus F **carbendazim** Shell
Captysem F **captan** Platecsa
Capzin F **captan + carbendazim** Platecsa
Caracolesa M **metaldehyde** Safor
Caracoloro M **metaldehyde** Quimicas Oro
Caracoum R **difenacoum** Valmi
Caragard H **terbumeton + terbuthylazine** Ciba-Geigy, Nitrokemia, OHIS, Viopharm
Caragard invierno H **atrazine + terbumeton + terbuthylazine** Ciba-Geigy
Caragard verano H **terbumeton + terbuthylazine + terbutryn** Ciba-Geigy
Caragarde H **terbumeton + terbuthylazine** Ciba-Geigy
Caraquim M **metaldehyde** Masso
Caratope H **oryzalin + simazine** Ciba-Geigy
Carazol H **amitrole + ammonium thiocyanate + terbuthylazine** Ciba-Geigy
Carba G **carbaryl** Bjoernrud
Carbacryl I **acrylonitrile** Cyanamid
carbam FNHI **metham-sodium**
Carbamag F **maneb** Chimac-Agriphar
Carbamag Plus F **carbendazim + maneb** Chimac-Agriphar
Carbamal I **carbaryl + malathion** Agrotechnica
Carbamate F **ferbam** Pennwalt Holland
Carbamec IG **carbaryl** PBI Gordon
Carbamin I **carbaryl** Margesin
Carbamix F **captan + zineb** Burri
carbamorph F [31848-11-0] morpholinomethyl

dimethyldithiocarbamate, carbamorphe, MC 833
carbamorphe F **carbamorph**
Carbamult I **promecarb** Schering
Carbaphyt F **zineb** Van Wesemael
carbaryl IG [63-25-2] 1-naphthyl methylcarbamate, NAC, sevin, NMC, ENT 23969, Arylam, Carbamec, Cekubaryl, Denapon, Dicarbam, Hexavin, Karbaspray, Murvin, Patrin, Ravyon, Savit, Septene, Sevin, Tercyl, Tricarnam, UC 7744
Carbasol IAN **carbofuran** Aragonesas
Carbate F **carbendazim** Pan Britannica
Carbatene F **metiram** Procida
carbathiin F **carboxin**
Carbathiol F **carbendazim + sulphur** Marty et Parazols
Carbatox I **carbaryl** Azot, SPE
Carbax A **dicofol** Pepro
Carbazinc FX **ziram** Rhodiagri-Littorale, Rhone-Poulenc
Carbeetamide-VLB H **carbetamide** Imex-Hulst
Carben F **carbendazim** SEGE
Carbendagrex F **carbendazim** Sadisa
carbendazim F [10605-21-7] methyl benzimidazol-2-ylcarbamate, carbendazime, carbendazol, MBC, BCM, BAS 346F, Battal, Bavistin, Custos, Delsene, Derosal, Derroprene, Equitdazin, Focal, Hinge, Hoe 17411, Kemdazin, Lignasan, Pillarstin, Stempor, Triticol, Virolex
carbendazime F **carbendazim**
carbendazol F **carbendazim**
Carbenzip F **carbendazim** Interphyto
Carbenzip M F **carbendazim + maneb** Interphyto
Carbetamex H **carbetamide** Embetec, Hortichem, Rhone-Poulenc
carbetamide H [16118-49-3] (*R*)-1-(ethylcarbamoyl)ethyl phenylcarbamate, 11561 RP, Carbetamex, Legurame
Carbetox IA **malathion** Sinteza
Carbevin I **carbaryl** Zootechniki
Carbezal I **carbaryl** Hellenic Chemical
Carbicron IA **dicrotophos** Ciba-Geigy
Carbileen F **zineb** B.B.
Carbina F **zineb** Aziende Agrarie Trento
Carbina TZ F **thiram + zineb** Aziende Agrarie Trento
Carbinex I **carbaryl** Diana
Carbinex 60-40 IA **carbaryl + chlorfenson** Diana
Carbinol I **carbaryl** Geochem
Carbisan I **carbaryl** Formulex
Carbital I **carbaryl** Ital-Agro
Carbo-Craven IHFX **tar oils** Craven
Carbodan IAN **carbofuran** Makhteshim-Agan
Carbodin IA **carbaryl + diazinon** Sofital
Carbofort 3 IA **anthracene oil + DNOC** Siegfried
Carbofort 4 IA **anthracene oil** Siegfried
carbofos IA **malathion**
carbofuran IAN [1563-66-2] 2,3-dihydro-2,2-dimethylbenzofuran-7-yl methylcarbamate, OMS 864, ENT 27164, Bay 70143, Carbodan, Carbosip, Chinufur, Curaterr, FMC 10242, Furadan, Kenofuran, Pillarfuran, Yaltox
Carbol Jaune 3 IA **anthracene oil + DNOC** Leu & Gygax, Pluess-Staufer
Carbomatil I **carbaryl** Platecsa
carbophenothion IA [786-19-6] *S*-(4-chlorophenylthio)methyl *O,O*-diethyl phosphorodithioate, OMS 244, ENT 23708, Dagadip, Garrathion, Nephocarb, R-1303, Trithion
Carbosan I **carbaryl** Sepran
Carbosip IN **carbofuran** Inagra, Chiltern, Sipcam
Carbosol IN **carbofuran** Leu & Gygax
Carbostin IN **carbofuran** Solfotecnica
carbosulfan I [55285-14-8] 2,3-dihydro-2,2-dimethyl-7-benzofuran-7-yl (dibutylaminothio)methylcarbamate, Advantage, FMC 35001, Marshal, Posse
Carbotex HIA **anthracene oil** Protex
Carbotin I **carbaryl** Industrialchimica
Carbotran A **carbophenothion + cyhexatin** Serpiol
Carbovax 2 M F **carboxin + maneb** Agrimont
Carbovis I **carbaryl** Visplant
Carboxide IF **ethylene oxide**
carboxin F [5234-68-4] 5,6-dihydro-2-methyl-1,4-oxathi-ine-3-carboxanilide, carboxine, carbathiin, DCMO, D 735, Kemikar, Kisvax, Vitavax
carboxine F **carboxin**
Carbozyl HIA **anthracene oil** Chimac-Agriphar
Carbrousse H **2,4-D + 2,4,5-T** CARAL
Carbyne H **barban** Efthymiadis, Schering
Cardiamine H **2,4-D** B.B.
Carditol H **2,4-D** B.B.
Caresine H **bentazone + dichlorprop + isoproturon** Agrolinz, BASF, Hoechst

Carezim F **carbendazim** Efthymiadis
Carfene IA **azinphos-methyl** Pepro
Cargol-mat M **metaldehyde** Agriplan
Cargoluq M **metaldehyde** Luqsa
Cariefit F **dodine** Platecsa
Carma I **carbofuran + isofenphos** Bayer
Carman F **carbendazim + maneb** Aragonesas
Carmethin I **dichlorvos** Agrishell
Caroweedex H **petroleum oils** Asepta
Carpathion I **carbaryl + carbophenothion** Hellenic Chemical
Carpathion IA **azinphos-methyl** Tecniterra
Carpene F **dodine** Agrimont, Farmoplant, ICI, Montedison
Carpomon I **parathion-methyl** Agrimont
Carposan I **parathion** Agrimont
Carpul IA **azinphos-methyl + dimethoate** Aragonesas
cartap hydrochloride I [22042-59-7] *S,S'*-2-dimethylaminotrimethylene bis(thiocarbamate) hydrochloride, Caldan, Padan, Patap, Sanvex, Thiobel, TI-1258, Vegetox
Cartex M H **monolinuron + prometryn + propachlor** Sajobabony
Cartouche Hortex R **aluminium powder + calcium phosphate** Sovilo
Cartox IF **ethylene oxide**
Carvil I **fenobucarb** Planters Products
Caryl I **carbaryl** Bredologos
Caryne H **barban** Schering
Carzol AI **formetanate hydrochloride** NOR-AM
Cascade IA **flufenoxuron** Shell
Casoron H **dichlobenil** Argos, Berner, Bjoernrud, Chipman, Du Pont, Duphar, ICI, KVK, La Quinoleine, Quimigal, Schering, Shell, Sodafabrik, Solvay, Synchemicals
Casoron Combi H **dalapon-sodium + dichlobenil** Schering
Castaway Plus IE **lindane + thiophanate-methyl** May & Baker
Castrix R **crimidine** Bayer
CAT H **simazine**
Catt H **benazolin + ioxynil + mecoprop** Du Pont
C-Benagro G **chlormequat chloride** Phyto
C-B-H-O Neu I **lindane** Schacht
CCC G **chlormequat chloride**
C.C.D. F **copper carbonate (basic)** Duclos
C.C.D.T. F **copper carbonate (basic) + sulphur** Duclos
CDAA H **allidochlor**
CDEC H **sulfallate**
Cebeco Dodaal N **1,3-dichloropropene** Cebeco
Cebo Antilimacos M **metaldehyde** Probelte
CeCeCe G **chlormequat chloride** BASF
Cekiuron H **diuron** Cequisa
Ceku Atra H **atrazine** Cequisa
Ceku C.B. F **hexachlorobenzene** Cequisa
Ceku-cobre F **copper oxychloride** Cequisa
Ceku-dine F **dodine** Cequisa
Ceku-Dinoc IA **DNOC** Cequisa
Ceku-Gib G **gibberellic acid** Cequisa
Ceku-Giberelina G **gibberellic acid** Cequisa
Ceku-humectante S **polyglycolic ethers** Cequisa
Ceku-Toxin IA **aluminium phosphide** ERT
Cekubaril I **carbaryl** Cequisa
Cekubaryl IG **carbaryl** Cequisa
Cekucap FA **dinocap** Cequisa
Cekudazim F **carbendazim** Cequisa
Cekudifol A **dicofol** Cequisa
Cekudit A **dicofol + tetradifon** Cequisa
Cekufan IA **endosulfan** Cequisa
Cekufon I **trichlorfon** Cequisa
Cekumal IA **malathion** Cequisa
Cekumeta M **metaldehyde** Cequisa
Cekumethion IA **parathion-methyl** Cequisa
Cekunate H **molinate** Ficoop
Cekuoil 3 P IA **parathion + petroleum oils** Cequisa
Cekuoil V 83 IA **petroleum oils** Cequisa
Cekuper F **copper oxychloride** Cequisa
Cekuquat H **paraquat dichloride** Cequisa
Cekuron H **diuron** Cequisa
Cekusan IA **dichlorvos** Cequisa
Cekusil F **phenylmercury acetate** Cequisa
Cekusil Universal A F **methoxyethylmercury acetate** Cequisa
Cekutan F **captan** Cequisa
Cekuthoate IA **dimethoate** Cequisa
Cekutoato IA **dimethoate** Cequisa
Cekutrothion I **fenitrothion** Cequisa
Cekuzina-T H **atrazine** Cequisa
Cekuzineb F **zineb** Cequisa
Cekuzinfos IA **azinphos-methyl** Cequisa
Cekuzinon I **diazinon** Cequisa
Cela W524 F **triforine** Celamerck
Celamerck Insektenschutz natural I **Bacillus thuringiensis + piperonyl butoxide +pyrethrins** Shell

Celamerck Totalunkrautvernichter Ektorex H **atrazine + diuron + simazine** Shell
Celamerck-Unkrautstab H **2,4-D + mecoprop** Shell
Celathion IA **chlorthiophos** Celamerck, Geopharm, Shell
Celathion Combi IA **chlorthiophos + dimethoate** Shell
Celatox asperges H **diuron** Sovilo
Celatox DP H **dichlorprop** Shell
Celatox gazon H **2,4-D + mecoprop** Sovilo
Celatox Kombi H **2,4-D + MCPA** Shell
Celatox KV H **mecoprop** Epro
Celatox KV-neu H **2,4-D + mecoprop** Epro
Celatox legumes H **linuron** Sovilo
Celatox liseron H **2,4-D** Sovilo
Celatox M H **MCPA** Epro
Celatox rosiers pepinieres H **simazine** Sovilo
Cellcid F **benomyl** Saghegyalja
Cellu-Quin F **oxine-copper**
Celmer F **phenylmercury acetate** Excel
Celmide NI **ethylene dibromide** Excel
Celmone G **1-naphthylacetic acid** Excel
Celphide IR **aluminium phosphide** Excel
Celphine I **aluminium phosphide** Excel, Hellenic Chemical
Celphos IR **aluminium phosphide** Desur, Excel
Celthion IA **malathion** Excel
Celuko H **flurochloridone** Lucebni zavody
Cent-7 H **isoxaben** Elanco, Eli Lilly
Centrol-H H **ioxynil + mecoprop** Aragonesas
Cepanol I **diazinon** Platecsa
Cepedic H **dicamba + mecoprop** Sipcam-Phyteurop
Cepos H **chlorthal-dimethyl** Caffaro
Ceptal H **chlorthal-dimethyl** Sipcam
Ceptral H **2,4-D + picloram + triclopyr** Ciba-Geigy
Ceralsano F **copper oxychloride** Agrocros
Ceranit F **2-methoxyethylmercury chloride** Esbjerg
Ceratex I **malathion** Ellagret
Cerathion I **malathion** Ellagret
Ceratotect F **thiabendazole** MSD AGVET
Cercobin FW **thiophanate-methyl** Agrolinz, BASF, Nippon Soda, Rhone-Poulenc
Cercofen F **fentin acetate** Sivam
Cerealsan F **benomyl + mancozeb** Ravit
Cerebrell G **gibberellic acid** Farmon Agrovia
Cereclair F **carbendazim + chlorothalonil** Du Pont
Cerecons I **lindane** Sivam
Ceredon F **benquinox**
Ceredor F **carboxin + maneb** Ital-Agro
Cereflor F **carbendazim + ditalimfos** Du Pont
Cereline F **benquinox**
Ceresan F **mercury** Bayer, Berner
Ceresol F **phenylmercury acetate** ICI
Ceretal F **captafol + carbendazim** Du Pont
Cerevax F **carboxin + thiabendazole** ICI
Cerevax Extra F **carboxin + imazalil + thiabendazole** ICI
Cerevit I **malathion** Ital-Agro
Ceridal H **chlorotoluron** Grima
Ceridor H **bifenox + mecoprop** Elanco
Ceritox IA **azinphos-ethyl + dimethoate** Prochimagro
Cerone G **ethephon** Berner, Bjoernrud, Chimac-Agriphar, Ciba-Geigy, Du Pont, Embetec, Gullviks, ICI, Imex-Hulst, Luxan, Pepro, Pluess-Staufer, Ravit, Shell, Spiess, Union Carbide, Urania
Certan I **Bacillus thuringiensis** Sandoz
Certox R **strychnine**
Certricide H **amitrole + atrazine** Certified Lab.
Certrol H **ioxynil** Luxan, Spiess, Union Carbide, Urania
Certrol A H **ioxynil + MCPA** Luxan
Certrol B H **bromoxynil** Spiess, Urania
Certrol BL H **bromoxynil + linuron** Avenarius
Certrol Combin S H **ioxynil + MCPA + mecoprop** Union Carbide
Certrol Combin SE H **bromoxynil + MCPA + mecoprop** Luxan, Schering AAgrunol, Union Carbide
Certrol DP H **dichlorprop + ioxynil** Urania
Certrol DS H **2,4-D + ioxynil** C.F.P.I.
Certrol E H **bromoxynil + dichlorprop + ioxynil** Chafer
Certrol G H **bromoxynil + MCPA + mecoprop** Spiess, Urania
Certrol H H **ioxynil + mecoprop** C.F.P.I., Ciba-Geigy, EniChem, Etisa, Spiess, Urania
Certrol NA H **ioxynil** Chimac-Agriphar
Certrol Ox H **bromoxynil + ioxynil** Agronorden

Certrol P H **ioxynil + mecoprop** Chimac-Agriphar, Ciba-Geigy
Certrol PA H **dichlorprop + ioxynil + MCPA** Chafer
Certrol plus H **ioxynil + mecoprop** Ciba-Geigy
Certrol Tetra H **bromoxynil + dichlorprop + ioxynil + MCPA** Gullviks
Certrol-Trippel H **dichlorprop + ioxynil + MCPA** Gullviks
Cervacol X **poly(vinyl acetate)** Avenarius, Ruse
Cesar A **hexythiazox** Procida
Cetrelex H **trifluralin** Sipcam-Phyteurop
Cevex F **carboxin + imazalil** Nordisk Alkali
Cevex Hoest F **carboxin + thiabendazole** Kemi-Intressen
Cevex Vaar F **carboxin + imazalil + thiabendazole** Kemi-Intressen
CF 125 G **chlorflurecol-methyl** Celamerck, Shell, Zupa
CGA 10832 H **profluralin** Ciba-Geigy
CGA 112913 I **chlorfluazuron** Ciba-Geigy
CGA 12223 NI **isazofos** Ciba-Geigy
CGA 13586 G **etacelasil** Ciba-Geigy
CGA 136872 H 2-[3-(4,6-bis(difluoromethoxy)pyrimidin-2-yl)ureidosulfonyl]benzoic acid methyl ester
CGA 15324 IA **profenofos** Ciba-Geigy
CGA 17020 H **dimethachlor** Ciba-Geigy
CGA 18731 H **isoproturon** Ciba-Geigy
CGA 18809 I **azamethiphos** Ciba-Geigy
CGA 20168 IA **methacrifos** Ciba-Geigy
CGA 22598 A **chlormebuform**
CGA 24705 H **metolachlor** Ciba-Geigy
CGA 26351 IA **chlorfenvinphos** Ciba-Geigy
CGA 26423 H **pretilachlor** Ciba-Geigy
CGA 38140 F **furalaxyl** Ciba-Geigy
CGA 41065 G **flumetralin** Ciba-Geigy
CGA 43089 S **cyometrinil**
CGA 48988 F **metalaxyl** Ciba-Geigy
CGA 49104 F **pyroquilon** Ciba-Geigy
CGA 4999 H **mesoprazine**
CGA 61837 H **karbutilate** Ciba-Geigy
CGA 64250 F **propiconazole** Ciba-Geigy
CGA 64251 F **etaconazole** Ciba-Geigy
CGA 71818 F **penconazole** Ciba-Geigy
CGA 72662 I **cyromazine** Ciba-Geigy
CGA 73102 I **furathiocarb** Ciba-Geigy
CGA 92194 S **oxabetrinil** Ciba-Geigy
CGR-811 G **inabenfide** Chugai
Chafer Azotox IA **demeton-S-methyl** Chafer
Challenge H **aclonifen** Pepro
Champion F **copper hydroxide**
Chandor H **linuron + trifluralin** Imex-Hulst
Charabex G **4-CPA** SEGE
Charakini G **4-CPA** Prometheus
Checkmate H **sethoxydim** Embetec
Chefamyl I **piperonyl butoxide + pyrethrins** Chefam
Chefatex I **methoxychlor + piperonyl butoxide + pyrethrins** Chefam
Chelasan F **2-methoxyethylmercury chloride** Pluess-Staufer
Chem ban H **dicamba + mecoprop** Chemsearch
Chem-Fish IA **rotenone** Tifa
Chem-Hoe HG **propham** PPG
Chem-Rice H **propanil** Chimiberg, Diana, Tifa
Chemagro B-1776 G **S,S,S-tributyl phosphorotrithioate**
Chemathoate IA **dimethoate** Cheminova
Chemazin H **atrazine** Bayer
Chemelin H **amitrole + atrazine** Chemsearch, National Chemsearch
Chemian IA **endosulfan** Chemia
Chemifos I **parathion-methyl** Chemia
Chemin I **carbaryl** Chemia
Chemition I **fenitrothion** Chemia
Chemol IA **petroleum oils** Chemia
Chemox HI **dinoseb** Tifa
Chemsect IAHF **DNOC** Tifa
Cheshunt Compound H **ammonium carbonate + copper sulphate** Pan Britannica
Childion A **dicofol + tetradifon** Du Pont
Chim I **phorate** Agrochemiki
Chimac COP H **mecoprop** Agriphyt
Chimac cop special H **2,4-D + mecoprop** Agriphyt
Chimac Dim I **dimethoate** Agriphyt
Chimac Dino H **dinoseb acetate** Agriphyt
Chimac DVP I **dichlorvos** Agriphyt
Chimac Fane H **dinoseb** Agriphyt
Chimac Fol F **folpet** Agriphyt
Chimac L I **lindane** Agriphyt
Chimac Loofdood H **dinoseb** Chimac-Agriphar
Chimac mixte H **2,4-D + MCPA** Agriphyt
Chimac Oil S **petroleum oils** Chimac-Agriphar
Chimac Par H I **parathion + petroleum oils** Agriphyt
Chimac Par M IA **parathion-methyl** Agriphyt

Chimac Tufane H **dinoseb** Agriphyt
Chimac Zin F **zineb** Agriphyt
Chimarix H **glyphosate** Monsanto
Chimasan H **chlorotoluron** Chimac-Agriphar
Chimate FLM **fentin acetate** Chimac-Agriphar
Chimazole TA H **amitrole + ammonium thiocyanate** Chimique
Chimiclor H **alachlor** Chimiberg
Chimigor IA **dimethoate** Chimiberg
Chimiquat H **paraquat dichloride** Chimiberg
Chimition IA **azinphos-methyl** Chimiberg
chinalphos IA **quinalphos**
Chinetrin I **permethrin + piperonyl butoxide + tetramethrin** Chinoin
Chinofur IAN **carbofuran** Key, Procida
Chinoin Fundazol F **benomyl** Chinoin
chinomethionat FAI **quinomethionate**
chinomethionate FAI **quinomethionate**
chinonamid LH **quinonamid**
Chinosol W FW **8-hydroxyquinoline sulphate** Drogenhansa, Hoechst, Hokochemie, Riedel de Haen, Roussel Hoechst, Zupa
chinothionat AF **thioquinox**
Chinufur IAN **carbofuran** Chemolimpex, Chinoin
Chipco 26019 F **iprodione** Rhone-Poulenc
Chipko 4 H **2,4-D + mecoprop** Chipman
Chiptox H **MCPA** Rhone-Poulenc
chlomethoxyfen H [32861-85-1] 5-(2,4-dichlorophenoxy)-2-nitroanisole, chlormethoxynil, Condore, Difenex, Diphenex, Ekkusagoni, Ikkokuso, X-52
Chlor Kil I **chlordane**
chlor-IFC HG **chlorpropham**
chlor-IPC HG **chlorpropham**
Chlor-O-Pic NIFRH **chloropicrin** Great Lakes
chloral hydrate H 2,2,2-trichloro-1,1-ethanediol, Bi 3411
chloralose RX [15879-93-3] (*R*)-1,2-*O*-(2,2,2-trichloroethylidene)-α-D-glucofuranose, alpha-chloralose, glucochloralose, glucochloral, Alfamat, Alphakil, Somio
chloramben H [133-90-4] 3-amino-2,5-dichlorobenzoic acid, chlorambene, ACP M-629, Amiben
chlorambene H **chloramben**
chloramizol F **imazalil**
chloraniformethan F [20856-57-9] *N*-[2,2,2-trichloro-1-(3,4-dichloroanilino)ethyl]formamide, Imugan, Milfaron
chloranil FX [118-75-2] 2,3,5,6-tetrachloro-*p*-benzoquinone, Spergon
chloranocryl H 3',4'-dichloro-2-methylacrylanilide, Dicryl, FMC 4556
chlorazine H [580-48-3] 6-chloro-N^2,N^2,N^4,N^4- tetraethyl-1,3,5-triazine-2,4-diamine, G 30031
chlorbenside A [103-17-3] 4-chlorobenzyl 4-chlorophenyl sulphide, chlorsulphacide, PCBPCPS, Acacide, Chlorocide, Chlorparacide, Mitox
chlorbicyclen I [2550-75-6] 1,2,3,4,7,7-hexachloro-5,6-bis(chloromethyl)-8,9,10-trinorborn-2-ene, chlorbicyclene, ENT 211, ENT 785, Alodan, Hercules 426
chlorbicyclene I **chlorbicyclen**
chlorbromuron H [13360-45-7] 3-(4-bromo-3-chlorophenyl)-1-methoxy-1-methylurea, chlorobromuron, C 6313, Maloran
chlorbufam H [1967-16-4] 1-methylprop-2-ynyl 3-chlorophenylcarbamate, chlorbufame, BIPC, BiPC
chlorbufame H **chlorbufam**
chlordan I **chlordane**
chlordane I [57-74-9; 12789-03-6 for technical grade] 1,2,4,5,6,7,8,8-octachloro-2,3,3a,4,7,7a-hexahydro-4,7-methanoindene, chlordan, OMS 1437, ENT 9932, Belt, Chlor Kil, Chlortox, Corodane, Gold Crest, Intox, Kypchlor, Octa-Klor, Octachlor, Sydane, Synklor, Termi-Ded, Topiclor, Velsicol 1068
chlordecone IF [143-50-0] decachloropentacyclo[5.2.1.0 ^{2,6}.0 ^{3,9}.0 ^{5,8}]decan-4-one, Kepone
chlordimeform AI [6164-98-3; 19750-95-9 (hydrochloride)] N^2-(4-chloro-*o*-tolyl-N^1,N^1- dimethylformamidine, chlordimeforme, chlorophenamidine, chlorodimeform, OMS1209, Bermat, C 8514, Fundal, Galecron, Ovatoxin
chlordimeforme AI **chlordimeform**
chlorethephon G **ethephon**
chlorfenac H [85-34-7] (2,3,6-trichlorophenyl)acetic acid, fenac, Fenac, Fenatrol
chlorfenazole F [3574-96-7] 2-(2-chlorophenyl)benzimidazole, CUR 616

chlorfenethol A [80-06-8] 1,1-bis(4-chlorophenyl)ethanol, DCPC, BCPE, DMC, Dimite, Qikron
chlorfenidim H **monuron**
chlorfenprop-methyl H [14437-17-3] methyl 2-chloro-3-(4-chlorophenyl)propionate, Bidisin
chlorfenson A [80-33-1] 4-chlorophenyl 4-chlorobenzenesulphonate, chlorofenizon, CPCBS, ovex, ENT 16358, K 6451, Ovex, Ovotran, Sappiran, Trichlorfenson
chlorfensulphide A [2274-74-0] [(4-chlorophenyl)thio](2,4,5-trichlorophenyl)diazene, CPAS
chlorfenvinphos IA [470-90-6 ((*Z*)-isomer: 18708-87-7; (*E*)-isomer: 18708-86-6)] 2-chloro-1-(2,4-dichlorophenyl)vinyl diethyl phosphate, CVP, OMS 1328, ENT 24969, Apachlor, Birlane, C 8949, CGA 26351, Haptarax, Haptasol, Sapecron, SD 7859, Steladone, Supona, Vinylphate
chlorfluazuron I [71422-67-8] 1-[3,5-dichloro-4-(3-chloro-5-trifluoromethyl-2-pyridyloxy)phenyl]-3- (2,6-difluorobenzoyl)urea, Aim, Atabron, CGA 112913, IKI-7899, Jupiter
chlorflurazole H [3615-21-2] 4,5-dichloro-2-trifluoromethylbenzimidazole, NC 3363
chlorflurecol-methyl GH [2536-31-4] methyl 2-chloro-9-hydroxyfluorene-9-carboxylate, chlorflurenol-methyl, chloroflurenol-methyl, CF 125, Curbiset, IT 3456, Maintain, Morphactin, Multiprop
chlorfluren G [24539-66-0] 2-chlorofluorene-9-carboxylic acid, chlorflurene, IT-5732
chlorflurene G **chlorfluren**
chlorflurenol-methyl GH **chlorflurecol-methyl**
chlorfonium chlorure G **chlorphonium chloride**
chlorhydrate du propamocarbe F **propamocarb hydrochloride**
chloridazon H [1698-60-8] 5-amino-4-chloro-2-phenylpyridazin-3(2*H*)-one, chloridazone, pyrazon, PAC, PCA, Brek, Curbetan, H 119, Pyramin
chloridazone H **chloridazon**
chlorimuron-ethyl H ethyl 2-[[[[(4-chloro-6-methoxypyrimidin-2-yl) amino]carbonyl]amino]sulfonyl] benzoate, Classic, DPX-F6065
chlorinat H **barban**
Chlorizyl CP H **chlorpropham** Agriphyt
chlormebuform A [37407-77-5] *N*[1]-butyl-*N*[2]-(4-chloro-*o*-tolyl)-*N*[1]-methylformamidine, chlormebuforme, CGA 22598, Ektomin
chlormebuforme A **chlormebuform**
chlormephos I [24934-91-6] *S*-chloromethyl *O,O*-diethyl phosphorodithioate, Dotan, MC 2188
chlormequat chloride G [999-81-5] 2-chloroethyltrimethylammonium chloride, CCC, chlorocholine chloride, AC 38555, Barleyquat, Bettaquat, CeCeCe, Cycocel, Cycogan, Farmacel, Hyquat, Increcel, Titan
chlormethoxynil H **chlomethoxyfen**
Chlormite A **chloropropylate** Ciba-Geigy
chlornitrofen H [1836-77-7] 2,4,6-trichloro-1-(4-nitrophenoxy)benzene, CNP, MO, MO 338
Chloro FIHN **chloropicrin** Maldoy
chloro-IPC HG **chlorpropham**
Chloroben A **chlorobenzilate** Hellenic Chemical
chlorobenzilate A [510-15-6] ethyl 4,4'-dichlorobenzilate, ENT 18596, Acaraben, Akar, Benzilan, Folbex, G 23992, Kopmite
Chloroble fort superfix D FIX **endosulfan + lindane + oxine-copper** Pepro
Chloroble M total superfix FIX **anthraquinone + lindane + maneb** Pepro
Chloroble 'M' superfix F **maneb** Pepro
chlorobromuron H **chlorbromuron**
chlorocholine chloride G **chlormequat chloride**
Chlorocide A **chlorbenside**
chlorodimeform AI **chlordimeform**
Chlorofen F **iron compounds (unspecified)** Lucebni zavody
chlorofenizon A **chlorfenson**
chloroflurenol-methyl GH **chlorflurecol-methyl**
chlorofos I **trichlorfon**
chloroneb F [2675-77-6] 1,4-dichloro-2,5-dimethoxybenzene, chloronebe, Demosan, Soil Fungicide 1823, Teremec, Terraneb, Tersan SP
chloronebe F **chloroneb**
chlorophacinone R [3691-35-8] 2-[2-(4-chlorophenyl)-2-phenylacetyl]indan-1,3-dione, Caid, Drat, Droot, KarRATe, Lepit, Liphadione, LM 91, Microzul, Quick, Ramucide, Ratomet, Raviac, Redentin, Rozol, Topitox
Chloropham G **chlorpropham + propham** SEGE

chlorophenamidine AI **chlordimeform**
chloropicrin NIFRH [76-06-2] trichloronitromethane, chloropicrine, Acquinite, Chlor-O-Pic, Dojyopicrin, Dorochlor, Pic-Clor, Picrin-80, Tri-Chlor
chloropicrine NIFRH **chloropicrin**
chloropropylate A [5836-10-2] isopropyl 4,4'-dichlorobenzilate, ENT 26999, Chlormite, G 24163, Rospin
Chlororat R **chlorophacinone** L'Hygiene
chlorothalonil F [1897-45-6] tetrachloroisophthalonitrile, TPN, chlorthalonil, Bombardier, Bravo, Clortocaffaro, Daconil 2787, Exotherm-Termil, Faber, Notar, Repulse
chlorotoluron H [15545-48-9] 3-(3-chloro-*p*-tolyl)-1,1-dimethylurea, chlortoluron, C 2242, Clortokem, Deltarol, Dicuran, Higaluron, Highuron, Tolurex
chloroxifenidim H **chloroxuron**
chloroxuron H [1982-47-4] 3-[4-(4-chlorophenoxy)phenyl]-1,1-dimethylurea, chloroxifenidim, chlorphencarb, C 1983, Gesamoos, Tenoran
Chlorozon H **paraquat dichloride** Spyridakis
Chlorparacide A **chlorbenside**
Chlorpham H **chlorpropham** Sipcam-Phyteurop
chlorphencarb H **chloroxuron**
chlorphonium chloride G [115-78-6] tributyl(2,4-dichlorobenzyl)phosphonium chloride, chlorfonium chlorure, Phosfleur, Phosfon
chlorphoxim I [14816-20-7] 2-(2-chlorophenyl)-2-(diethoxyphosphinothioyloxyimino) acetonitrile, chlorphoxime, OMS 1197, Bay SRA 7747, Baythion C
chlorphoxime I **chlorphoxim**
chlorprocarb H [23121-99-5] methyl 3-[1-(chloromethyl) propylcarbamoyloxy] carbanilate, chlorprocarbe, BAS 379H
chlorprocarbe H **chlorprocarb**
chlorpropham HG [101-21-3] isopropyl 3-chlorophenylcarbamate, chlorprophame, IPC, chlor-IFC, CIPC, chloro-IPC, chlor-IPC, Bud Nip, Elbanil, Furloe, Prevenol, Sprout Nip, Spud-Nic, Taterpex, Triherbide CIPC
chlorprophame HG **chlorpropham**
chlorpyrifos I [2921-88-2] *O,O*-diethyl *O*-3,5,6-trichloro-2-pyridyl phosphorothioate, chlorpyriphos, chlorpyriphos-ethyl, chlorpyrifos-ethyl, OMS 971, ENT 27311, Brodan, Crossfire, Detmol, Dursban, Eradex, Lorsban, Loxiran, Pyrinex, Spannit, Stipend, Zidil
chlorpyrifos-ethyl I **chlorpyrifos**
chlorpyrifos-methyl IA [5598-13-0] *O,O*-dimethyl *O*-3,5,6-trichloro-2-pyridyl phosphorothioate, chlorpyriphos-methyl, OMS 1155, ENT 27520, Dowco 214, Graincote, Reldan, Tumar, Zertell
chlorpyriphos I **chlorpyrifos**
chlorpyriphos-ethyl I **chlorpyrifos**
chlorpyriphos-methyl IA **chlorpyrifos-methyl**
chlorquinox F [3495-42-9] 5,6,7,8-tetrachloroquinoxaline, Lucel
chlorsulfuron H [64902-72-3] 2-chloro-*N*-[[(4-methoxy-6-methyl-1,3,5-triazin-2-yl)amino]carbonyl]benzenesulfonamide, DPX 4189, Glean, Telar
chlorsulphacide A **chlorbenside**
chlorthal-dimethyl H [1861-32-1] dimethyl tetrachloroterephthalate, DCPA, TCTP, chlorthal-methyl, DAC 893, Dacthal, Dacthalor
chlorthal-methyl H **chlorthal-dimethyl**
chlorthalonil F **chlorothalonil**
chlorthiamid H [1918-13-4] 2,6-dichlorothiobenzamide, chlortiamide, DCBN, Prefix, WL 5792
chlorthion I [500-28-7] *O,O*-dimethyl *O*-(3-chloro-4-nitrophenyl) phosphorothioate, Chlorthion
chlorthiophos IA [60238-56-4] *O,O*-diethyl *O*-[dichloro(methylthio)phenyl] phosphorothioate, Celathion
chlortiamide H **chlorthiamid**
Chlortiepin IA **endosulfan**
Chlortoluree H **chlorotoluron** Interphyto
chlortoluron H **chlorotoluron**
Chlortox I **chlordane** All-India Medical
chlorure mercureux FI **mercurous chloride**
chlozolinate F [72391-46-9] ethyl (±)-3-(3,5-dichlorophenyl)-5-methyl-2,4-dioxo-oxazolidine-5-carboxylate, dichlozolinate, M 8164, Manderol, Serinal
Chock Flugspray N I **piperonyl butoxide + pyrethrins + resmethrin** Ewos
cholecalciferol R [67-97-0] (3*β*,5*Z*,7*E*)-9,10-secocholesta-5,7,10(19)-trien-3-ol, activated 7-dehydrocholesterol, Quintox, Rampage
Chopper H **imazapyr** Cyanamid
Christmann-Cuma Fertigkoeder R **coumatetralyl** Christmann
Chromodin F **dodine** Chromos

Chromolur H **trifluralin** Chromos
Chromoneb F **propineb** Chromos
Chromorel-D I **chlorpyrifos + cypermethrin** Chromos
Chrysal-AV B G **benazolin + bromoxynil + mecoprop** Bendien
Chryson I **resmethrin** Sumitomo
Chrysopon G **4-indol-3-ylbutyric acid** Puteaux
Chrysotop G **4-indol-3-ylbutyric acid** Puteaux
Chryzoplus G **4-indol-3-ylbutyric acid** ACF, Grower, Huebecker, Puteaux, Turba Projar
Chryzopon G **4-indol-3-ylbutyric acid** ACF, Turba Projar
Chryzosan G **4-indol-3-ylbutyric acid** ACF, Puteaux, Turba Projar
Chryzotek G **4-indol-3-ylbutyric acid** ACF, Grower, Puteaux, Turba Projar
Chryzotop G **4-indol-3-ylbutyric acid** ACF
Chwastox D H **dicamba + MCPA** Organika-Sarzyna
Chwastox DF H **dicamba + flurecol-butyl + MCPA** Organika-Sarzyna
Chwastox extra H **MCPA** Organika-Sarzyna
Chwastox F H **flurecol-butyl + MCPA** Organika-Sarzyna
Chwastox M H **MCPA + mecoprop** Organika-Sarzyna
C.I. 77402 F **cuprous oxide**
ciafos I **cyanophos**
Ciatral H **alachlor + atrazine + cyanazine** Radonja
Ciba 2446 IA **amidithion**
Cibac F **copper oxychloride + zineb** OHIS
Cibral H **chlorotoluron + isoxaben** Ciba-Geigy
Cicatral W **pine oil** Rhodic
Cicatrisant W **pine oil** Umupro
Cidax H **amitrole + ammonium thiocyanate** Sedagri
Cidial IA **phenthoate** Agrimont, Chromos, Condor, Diana, Farmoplant, Montedison
Cidial Olio I **petroleum oils + phenthoate** Agrimont
Cidiol IA **petroleum oils + phenthoate** ERT
Cidokor H **glyphosate** Radonja
Cidorel F **nuarimol** Shell
Cikloat H **cycloate** Zorka Subotica
Cilezol H **amitrole + bromacil** LHN
Ciluan F **cymoxanil + mancozeb** Epro, Shell
Cimeka 33 H **dicamba + MCPA** Cimelak
Cimeka GR H **atrazine + 2,4-D + diuron** Cimelak
Cimeka TX H **amitrole + 2,4-D + diuron** Cimelak
Cimeka TX Super H **amitrole + bromacil + diuron** Cimelak
Cimofol F **cymoxanil + folpet** OHIS
Cimozin F **cymoxanil + zineb** OHIS
Cinabran double FX **anthraquinone + methoxyethylmercury silicate** Agriben
Cinabran triple FIX **anthraquinone + lindane + methoxyethylmercury silicate** Chassart
Cincturo G **4-CPA** Efthymiadis
Cindy IA **dimethoate** Aerosol Service, Annen-Chemie
Cineb F **zineb**
cinerin IA **pyrethrins**
Cinkfosfid R **zinc phosphide**
Cinkgalic F **zinc sulphate** PVV
Ciodrin I **crotoxyphos** Shell
CIPC HG **chlorpropham**
Cipertran I **cypermethrin + piperonyl butoxide +tetramethrin** Sepran
Cipotril H **ioxynil** Ravit
Ciptal F **captan** Elanco
Ciram F **ziram** Zorka Sabac
Cirex OP H **amitrole + diuron** Wagner
Cirrus Flow FA **sulphur** Platecsa
Cirtoxin H **clopyralid** Schering
Cislin I **deltamethrin** Wellcome
Cit-80-G-Staub I **lindane** Cit
Citan IA **dimethoate** Inagra
Cito Maeuseweizen R **zinc phosphide** Propfe
Cito Wuehlmaustod R **zinc phosphide** Propfe
Citowett S **polyglycolic ethers** BASF, Delis, Shell
Citramin F **2-aminobutane** Serpis
Citrashine SF **polyethylene wax + fungicide** Decco-Roda, Hellafarm, Pennwalt
Citrashine IMZ SF **imazalil + polyethylene wax** Pennwalt
Citrashine N-PE SF **2-phenylphenol + polyethylene wax** Pennwalt
Citrashine NT SF **thiabendazole + polyethylene wax** Pennwalt
Citrashine NT IMZ SF **imazalil + thiabendazole + polyethylene wax** Pennwalt

Citrazon A **benzoximate** Nippon Soda, Sandoz, SEGE
Citrex I Z F **imazalil** Brillocera
Citrex I Z F **imazalil** Brillocera
Citrex I Z/O F **imazalil + 2-phenylphenol** Brillocera
Citrex ST F **thiabendazole** Brillocera
Citrex ST IZ F **imazalil + thiabendazole** Brillocera
Citrex ST/O F **2-phenylphenol + thiabendazole** Brillocera
Citripor I **parathion + petroleum oils** Sopepor
Citrofix G **2,4-D** Inagra
Citrolane I **mephosfolan** Lapapharm
Citropeste I **petroleum oils** Ciba-Geigy
Citrosol A IMAD F **imazalil** Serpis
Citrosol A Tecto F **thiabendazole** Serpis
Citrosol A Tecto O F **2-phenylphenol + thiabendazole** Serpis
Citrosol A Tecto/Imad F **imazalil + thiabendazole** Serpis
Citrosol E S **polyethylene wax** Citrosol
Citrosol R Imad F **imazalil + polyethylene wax** Serpis
Citrosol Tecto/Imad F **imazalil + polyethylene wax + thiabendazole** Serpis
Citrus Fix H **2,4-D** Amvac
CL 11,344 H **pyridate** Agrolinz, Agrichem
CL 12,503 IA **pyridafenthion** Cyanamid
CL 252,214 H **imazaquin** Cyanamid
CL 252,925 H **imazapyr** Cyanamid
CL 263,499 H **imazethapyr** Cyanamid
CL 7521 F **dodine** Cyanamid
CL 85 H **cycloate + lenacil** Agriben, Condor, Pepro
Clairsol H **amitrole + diuron + simazine** BP, R.S.R.
Clanex H **propyzamide** Shell
Clap H **atrazine + pyridate** Agrolinz
Clar-Frut G **1-naphthylacetic acid** Luqsa
Clarosan H **terbutryn** Ciba-Geigy
Classic H **chlorimuron-ethyl** Du Pont
Clean-Boll H **dimethylarsinic acid** Vineland
Clean-Up H **tar oils** ICI
Cleanazole TA H **amitrole + ammonium thiocyanate** Commercia
Cleansweep H **diquat dibromide + paraquat dichloride** Shell
Cleanweed H **amitrole + atrazine** Commercia
Clear-Sol FIHN **metham-sodium** Diana
Clearcide H **fluothiuron**
Clearway H **amitrole + ammonium thiocyanate + simazine** Rhone-Poulenc
Cleaval H **cyanazine + mecoprop** Shell
Clemencuaje G **gibberellic acid** Argos
Clenecorn H **mecoprop** FCC
Clerdone H **clopyralid + dimefuron** Rhodiagri-Littorale
climbazol F [38083-17-9] 1-(4-chlorophenoxy)-1-(imidazol-1-yl)-3,3-dimethylbutanone, Baysan
Clinex I **diazinon + dimethoate + methoxychlor** Protex
Clingspray G **1-naphthylacetic acid** Protex
cliodinate H [69148-12-5] 2-chloro-3,5-di-iodo-4-pyridyl acetate, ASD 2288
Clipper G **paclobutrazol** ICI
Clix H **chlorthiamid** Chemia
Clobber H **cypromid**
Cloca IA **endosulfan** Chimac-Agriphar
cloethocarb IN [51487-69-5] 2-(2-chloro-1-methoxyethoxy)phenyl methylcarbamate, BAS 263I, Lance
clofentezine A [74115-24-5] 3,6-bis(2-chlorophenyl)-1,2,4,5-tetrazine, Acaristop, Apollo, Apolo, NC 21314, Panatac
clofop-isobutyl H [51337-71-4] isobutyl 2-[4-(4-chlorophenoxy)phenoxy]propionate, Alopex, Hoe 22870
clomazone H **dimethazone**
Clomitane F **captan** Chimiberg
clonitralid M **niclosamide**
clopropoxydim H **cloproxydim**
cloproxydim H [95480-33-4] (±)-2-[1-(3-chloroallyloxy)iminobutyl]-5-(2-ethylthiopropyl)-3-hydroxycyclohex-2-enone, cloproxydime, clopropoxydim, RE-36290, Selectone
cloproxydime H **cloproxydim**
clopyralid H [1702-17-6] 3,6-dichloropyridine-2-carboxylic acid, 3,6-dichloropicolinic acid, 3,6-DCP, Cirtoxin, Cyronal, Dowco 290, Format, Lontrel, Matrigon, Shield
Clorane H **chlorpropham** Top
Clorat R **chlorophacinone** Colkim
Clordyn A **tetradifon** Visplant
Cloresene I **lindane** Caffaro
Cloretil H **atrazine** Ital-Agro
Clorimid F **captan** Du Pont
Clorizol H **chloridazon** Sipcam-Phyteurop
Clorocarb F **captan** Inagra
Clorofix F **copper oxychloride** Gaillard
Clorofos I **trichlorfon** Sojuzchimexport

Clortedin I **trichlorfon** Inagra
Clortocaffaro F **chlorothalonil** Caffaro
Clortokem H **chlorotoluron** Kemichrom
Clortosip F **chlorothalonil** Caffaro, Sipcam
Clortosip R F **chlorothalonil + copper oxychloride** Caffaro, Sipcam
Clorturex H **chlorotoluron** Aragonesas
Clorturex ter H **chlorotoluron + terbutryn** Aragonesas
Clorval H **chlorotoluron** Serpiol
Clout H **alloxydim-sodium** Hortichem, Rhone-Poulenc
Clovacorn Extra H **2,4-DB + linuron + MCPA** FCC
Clovotox H **mecoprop** Rhone-Poulenc
Club IM **methiocarb** ICI
Club Root Control F **mercurous chloride** ICI
Club Root Dip F **thiophanate-methyl** Murphy
CM 400 H **chlorotoluron + mecoprop** Interphyto
CM Rosenspray Saprol Plus FIA **diazinon + tetradifon + triforine** Shell
CM-4D Dicopur H **2,4-D + mecoprop** Leu & Gygax
CMA H **pentanochlor**
CME 127 H **aclonifen** Celamerck
CME 134 I **teflubenzuron** Celamerck
CME 7002 I **trichlophenidin**
CMMP H **pentanochlor**
CMPP H **mecoprop**
CMU H **monuron**
CN-11-2936 H **prodiamine**
CNA F **dicloran**
CNC FX **copper naphthenate**
CNP H **chlornitrofen**
Co 755 I **butocarboxim** Wacker
Co 859 IA **butoxycarboxim** Wacker
Co-Rax R **warfarin** Prentiss
Cobex H **dinitramine** Agrochemiki, Borax, Condor, Ravit, Rhone-Poulenc, Wacker-Dow
Cobexo H **dinitramine**
Cobox F **copper oxychloride** BASF, Intrachem
Cobra Azufre F **copper oxychloride + sulphur** Ficoop
Cobre Nordox F **cuprous oxide** Masso
Cobre Sandoz F **cuprous oxide** Sandoz
Cobre Sandoz MZ F **cuprous oxide + manganese sulphate + zinc sulphate** Sandoz
Cobre-Zineb F **copper oxychloride + zineb** Shell, Ficoop
Cobreline F **copper oxychloride** Masso
Cobreline bordeles F **Bordeaux mixture** Masso
Cobreluq F **copper oxychloride** Luqsa
Cobrever F **Bordeaux mixture + maneb** Ciba-Geigy
Coccidol E IA **petroleum oils** Caffaro, Ergex
Cockroach Control Agent I **boric acid** Borax
Codal H **metolachlor + prometryn** Ciba-Geigy, Zorka Subotica
Code D2 I **trichlorfon** Denka
Codicap F **captafol + captan** Protex
Codicap F **captafol + captan + folpet** Schering
Codlemone T **pheromones** Sipcam
Cofol F **Bordeaux mixture** Argos
Cohrs Pyrethrum Spritzmittel IA **piperonyl butoxide + pyrethrins** Cohrs
Cohrs Raupenvernichter Dipel I **Bacillus thuringiensis** Cohrs
Coldan H **metribuzin** RACROC
Collego H **Colletotrichum gloeosporioides f. sp. aeschynomene** NOR-AM
Colloidox F **copper oxychloride** Mechema
Colloisol AF **sulphur** Eli Lilly
Colpor IA **dimethoate** Platecsa
Coltal Konz F **thiram + ziram** Chimiberg
Colvite M **metaldehyde** Sopepor
Colzanet H **trifluralin** L.A.P.A.
Comac F **Bordeaux mixture** La Cornubia, McKechnie
Comac Cutonic F **copper hydroxide** Comac
Comadienochloor A **dienochlor** Maasmond
Comando H **flamprop-M-isopropyl** Shell
Combat H **xylachlor**
Combat I **hydramethylnon** Cyanamid
Combathrin I **alphacypermethrin** Tuhamij
Combi fluide H **2,4-D + mecoprop** CTA
Combiform-Antimoos H **iron sulphate** Coswig
Combiram F **copper compounds (unspecified) + zineb** Sofital
Cometox I **DDT + lindane** Dudesti
Comfuval F **thiabendazole** Compo
Comite A **propargite** Uniroyal
Command H **dimethazone** FMC

Commando H **flamprop-M-isopropyl** Shell
Commerce H **fenoxan + trifluralin** FMC
Como S **alkyl phenol ethoxylate** Ellagret
Comodor H **tebutam** ICI, La Quinoleine
Comodor T H **tebutam + trifluralin** Maag
Compact H **desmedipham + phenmedipham** Schering
Compass F **iprodione + thiophanate-methyl** Rhone-Poulenc
Compete H **fluoroglycofen-ethyl** Supacryl
Compitox H **mecoprop** Rhone-Poulenc
Complesal gazon H **dicamba + mecoprop** Procida
Complexane F **Bordeaux mixture + mancozeb** Chimiberg
Complexugec H **2,4-D + MCPA** Sipcam-Phyteurop
Compo Ameisen-Mittel I **bromophos** Compo
Compo Ameisenvernichter I **lindane** BASF
Compo Antilimaces M **metaldehyde** BASF
Compo Antimousse Gazon H **iron sulphate** BASF
Compo Antimousse Gazon Super H **chloroxuron** BASF
Compo Antisouris R **difenacoum** BASF
Compo Compron I **ethiofencarb** Compo
Compo Debroussaillant H **triclopyr** BASF
Compo Embark G **mefluidide** BASF
Compo Erdbeerschutz F **thiophanate-methyl** BASF
Compo Gartenunkrautvernichter H **dichlobenil** Compo
Compo Gazon Regulateur Embark Special G **mefluidide** BASF
Compo Herbicide Gazon H **2,4-D + dicamba + MCPA + mecoprop** BASF
Compo Herbicide Jardin H **simazine** BASF
Compo Insecticide Jardin I **piperonyl butoxide + pyrethrins + rotenone** BASF
Compo Insectspray I **allethrin + piperonyl butoxide** BASF
Compo Insekten-Spray Neu IA **piperonyl butoxide + pyrethrins** Compo
Compo Insektenmittel I **permethrin** Compo
Compo Insektenvernichter IA **dimethoate** BASF
Compo Mehltaumittel F 238 Meltatox F **dodemorph acetate** Compo
Compo Mehltauspray F **dodemorph acetate** Compo, BASF
Compo Moosvernichter Neu H **iron sulphate** BASF, Compo
Compo Pflanzenschutzspray IA **piperonyl butoxide + pyrethrins** BASF, Compo
Compo Pilzfrei F **metiram** BASF
Compo Rasen-Floranid mit Moosvernichter H **iron sulphate** Compo
Compo Rasen-Unkrautvernichter Combi-Fluid H **dicamba + MCPA** Compo
Compo Rasenduenger mit Moosvernichter H **iron sulphate** BASF
Compo Rasenduenger mit Unkrautvernichter Neu H **2,4-D + dicamba** BASF
Compo Rasenunkrautvernichter H **fenoprop + MCPA** BASF
Compo Rosen-Spray FIA **piperonyl butoxide + pyrethrins + sulphur** Compo
Compo Rosenduenger mit Unkrautvernichter H **simazine** BASF, Compo
Compo Rosenschutz F **dodemorph acetate + dodine** BASF
Compo Rosenspray FA **dodemorph acetate + fenitrothion + tetradifon** BASF
Compo Rosenspritzmittel F **metiram + sulphur** Compo
Compo Schneckenkorn M **metaldehyde** BASF, Compo
Compo spezial Unkrautvernichter Filatex H **glyphosate** Compo
Compo Spray Roses FIA **dicofol + malathion + thiophanate-methyl** BASF
Compo Super Herbicide Total H **amitrole + chlorbromuron + simazine** BASF
Compo Super Insecticide Jardin I **pirimiphos-methyl** BASF
Compo Talpigram R **monochlorobenzene** Compo
Compo Tannen-Schutz IA **endosulfan** Compo
Compo Total-Unkraut-Spray H **amitrole + atrazine + 2,4-D** Compo
Compo Total-Unkrautmittel H **amitrole + diuron + simazine** Compo
Compo Total-Unkrautvernichter H **bromacil + diuron** Compo
Compound 118 I **aldrin** Shell
Compound 1081 R **fluoroacetamide**
Compound 497 I **dieldrin**
Compound 711 I **isodrin**
Concep S **cyometrinil**
Concep II S **oxabetrinil** Ciba-Geigy
Concord I **alphacypermethrin** Shell

Condor F **triflumizole** Kwizda
Condore H **chlomethoxyfen** Ishihara Sangyo
Conquest H **atrazine + cyanazine** Du Pont
Conservasept G **chlorpropham + propham** Asepta, Phytopharmaceutiki
Conservat C G **chlorpropham** Bourgeois
Conservat-leger G **chlorpropham + propham** Bourgeois
Conservat-mix G **chlorpropham + propham** Bourgeois
Conserve H **isouron** Elanco
Conservo G **chlorpropham + propham** Eurofyto
Conservor G **ethoxyquin** Chemia
Contrac R **bromadiolone** Lipha
Contraven IN **terbufos** Cyanamid
Contrax R **bromadiolone** Frowein
Contrax-cuma R **warfarin** Frowein
Contrax-fit R **warfarin** Frowein
Contrax-fit bloc R **warfarin** Frowein
Contrax-fluessig R **warfarin** Frowein
Contrax-P R **pindone** Motomco
Contrax-top Koeder H R **bromadiolone** Frowein
Contrax-top Konzentrat R **bromadiolone** Frowein
Contreverse G **chlormequat chloride** Tradi-agri
Controlumaca M **metaldehyde** Caffaro
Coop Faltan F **folpet** Ficoop
Coopazinfos IA **azinphos-methyl** Ficoop
Cooper Garden Spray F **resmethrin** Temana
Cooper Graincote IA **chlorpyrifos-methyl** Wellcome
Cooper Multispray I **fenitrothion + lindane + pyrethrins** Wellcome
Cooper puutarhasumute IA **piperonyl butoxide + pyrethrins + rotenone** Berner
Cooper Pybuthrin IA **piperonyl butoxide + pyrethrins** Wellcome
Cooper Regular Flykiller IA **piperonyl butoxide + pyrethrins** Wellcome
Cooper Residual Insecticide I **fenitrothion** Wellcome
Cooper Smite IA **chlorpyrifos-methyl** Wellcome
Cooper Super Strength Flykiller IA **piperonyl butoxide + pyrethrins** Wellcome
Cooper's nr. 65 I **piperonyl butoxide + pyrethrins** Abrahamson
Cooper-pyretriiniruiskute IA **piperonyl butoxide + pyrethrins** Berner
Cooper-sisahyonteisruiskute IA **piperonyl butoxide + pyrethrins** Berner
Coopermatic Flykiller Aerosol IA **piperonyl butoxide + pyrethrins** Wellcome
Coopermatic insektdraeber aerosol 88 I **piperonyl butoxide + pyrethrins** Mortalin
Coopertox DC I **dichlorvos** Cooper
Coopex I **permethrin** Wellcome
Coopfuradan IAN **carbofuran** Ficoop
Copac F **cuprammonium** BASF
Copac-E F **copper sulphate** BASF
C. Operats R **chlorophacinone** C.N.C.A.T.A.
Coperval F **copper oxychloride** Zootechniki
Copezin F **copper oxychloride + zineb** Inagra
copper 8-hydroxyquinolate F **oxine-copper**
copper 8-quinolinolate F **oxine-copper**
Copper Count-N F **cuprammonium** Anthokipouriki
Copper Green FW **copper naphthenate** Van Wesemael
copper hydroxide F [20427-59-2] Blue-Shield, Champion, Criscobre, Cudrox, Cuidrox, Cupravit Blue, Cuproxide, Kocide, Parasol
copper naphthenate FX [1338-02-9] copper naphthenate, naphthenate de cuivre, CNC, Copper Uversol, Cuprinol, Troysan, Wiltz-65
Copper Nordox F **cuprous oxide** Nordox
copper oxinate F **oxine-copper**
copper oxychloride F [1332-40-7] dicopper chloride trihydroxide, oxychlorure de cuivre, Blitox, Cekuper, Cobox, Coprantol, Coptox, Cupravit, Cuprocaffaro, Cuprokylt, Cuprosan, Cuprovinol, Cuprox, Fytolan, Kauritil, Recop, Viricuivre, Vitigran
copper sulphate LF [7758-98-7] copper sulphate, blue vitriol, copper vitriol, blue stone, blue copperas, cupric sulphate, Blue Viking, Phyton-27, Sulfacop, Triangle, Vencedor
Copper Uversol FX **copper naphthenate**
copper vitriol LF **copper sulphate**
Copper-ox F **copper oxychloride** Gefex
Copral F **copper sulphate + cymoxanil** Du Pont
Coprantol F **copper oxychloride** Ciba-Geigy
Coprarex F **copper oxychloride** Du Pont
Coptox F **copper oxychloride**

Coratop F **pyroquilon** Ciba-Geigy
Corbel F **fenpropimorph** BASF, Galenika, La Quinoleine, Maag, Schering, Spiess, Urania
Corbel Duo F **carbendazim + fenpropimorph** BASF
Corbel epi F **captafol + fenpropimorph** La Quinoleine
Corbel Prim F **carbendazim + fenpropimorph** Maag
Corbel star F **chlorothalonil + fenpropimorph** BASF
Corbel Top F **captafol + fenpropimorph** Maag
Corbel triple F **carbendazim + chlorothalonil + fenpropimorph** BASF
Corben H **MCPA** ERT
Corbit X **anthraquinone** Bayer
Corborid X **anthraquinone** Liro
Cormaison FI FX **anthraquinone + captan** La Quinoleine
Cormaison simple F.I. F **captan** La Quinoleine
Cormaison ST F **thiram** La Quinoleine
Cormaison STX F **carboxin + thiram** La Quinoleine
Cormaison T FX **anthraquinone + thiram** La Quinoleine
Cormaison TX FX **anthraquinone + carboxin + thiram** La Quinoleine
Cormaison X FX **anthraquinone + captan + carboxin** La Quinoleine
Cornal H **bromoxynil + dicamba + mecoprop** Shell
Cornar H **MCPA** Afrasa
Cornox CWK H **benazolin** Schering
Cornox M H **MCPA** Asepta
Cornox RK HG **dichlorprop**
Cornufera mit Moosvernichter H **iron sulphate** Guenther
Cornufera Rasenduenger mit Unkrautvernichter H **chlorflurecol-methyl + MCPA** Link & Reimann
Cornufera Rasenduenger mit Moosvernichter H **iron sulphate** Cornufera
Cornufera UV Rasenduenger mit Unkrautvernichter H **chlorflurecol-methyl + MCPA** Guenther
Corodane I **chlordane** PPG
Corothion IA **parathion**
Corozate FX **ziram** PPG
Corry's Hormone Rooting Powder F **captan + 1-naphthylacetic acid** Synchemicals
Corry's Medo FW **cresylic acid** Synchemicals
Corsair I **permethrin** Rhone-Poulenc, Serpiol
Cortilan I **chlorpyrifos** Maag
Cortilan-Neu I **lindane** Staehler
Cortix Dust I **carbaryl + malathion** Geochem
Corvet CM F **carbendazim + fenpropimorph + mancozeb** La Quinoleine
Corzine F **carbendazim** Tradi-agri
Cosan FA **sulphur** Hoechst, Permutadora, Siapa, Zorka, Zorka Sabac
Cosban I **XMC** Hodogaya
Cosmic F **carbendazim + maneb + tridemorph** BASF
Coteran H **fluometuron** Ciba-Geigy
Cotip I **chlordimeform + profenofos** Ciba-Geigy
Cotnex I **azinphos-ethyl** Permutadora
Cotnion IA **azinphos-methyl** Agronova
Cotnion M I **azinphos-methyl** Sapec
Cotnion-Ethyl IA **azinphos-ethyl** Makhteshim-Agan
Cotnion-Ethyl-Methyl I **azinphos-ethyl + azinphos-methyl** Alpha
Cotnion-Methyl IA **azinphos-methyl** Makhteshim-Agan
Cotodon H **dipropetryn + metolachlor** Ciba-Geigy
Cotofor H **dipropetryn** Ciba-Geigy, OHIS
Cotogard H **fluometuron + metolachlor** Ciba-Geigy
Cotolita Tio IA **endosulfan** KenoGard
Cotoran H **fluometuron** Ciba-Geigy, Viopharm
Cotton-Dust FI **carbaryl + sulphur** Diana, Ellagret, Geochem, Hellenic Chemical
Cotton-Pro H **prometryn** Griffin
Cottonex H **fluometuron** Edifi, Makhteshim-Agan
Cougar H **diflufenican + isoproturon** Rhone-Poulenc
coumachlor R [81-82-3] 3-[1-(4-chlorophenyl)-3-oxobutyl]-4-hydroxycoumarin, coumachlore, G 23133, Ratilan, Tomorin
coumachlore R **coumachlor**
coumafene R **warfarin**
coumafuryl R [117-52-2] 3-[1-(2-furanyl)-3-oxobutyl]-4-hydroxy-2*H*-1-benzopyran-2-one, furmarin, tomarin, Fumarin, Fumasol, Krumkil, Lurat, Rat-A-Way, Ratafin

coumatetralyl R [5836-29-3] 4-hydroxy-3-(1,2,3,4-tetrahydro-1-naphthyl)coumarin, Racumin
Coumatox R **warfarin** C.F.P.I.
Counter I **terbufos** BASF, Cyanamid, Lapapharm, Opopharma, Procida, Siegfried, Wacker, ZorkaSubotica
Coupler H **clopyralid + cyanazine** Shell
Courte paille G **chlormequat chloride** Tradi-agri
Courte paille C5 G **chlormequat chloride + choline chloride** Tradi-agri
Cov-R-Tox R **warfarin** Hopkins
Covid F **copper oxychloride + zineb** Aporta
Covinex F **copper oxychloride + zineb** Agriplan
Covinex Forte F **copper oxychloride + copper sulphate + maneb + zineb** Agriplan
Coxidante FA **potassium permanganate + sulphur** Abelte
3-CP G **3-CPA**
3-CPA G [5825-87-6] 2-(3-chlorophenoxy)propionic acid, 3-CP, Amchem 64-50, Fruitone CPA
4-CPA G [122-88-3] *p*-chlorophenoxyacetic acid, PCPA, Sure-Set, Tomato Fix, Tomato Hold, Tomatotone
CP 15336 H **di-allate** Monsanto
CP 23426 H **tri-allate** Monsanto
CP 31393 H **propachlor** Monsanto
CP 40 H **chlorpropham** EniChem
CP 41845 G **glyphosine**
CP 46358 H **terbuchlor**
CP 50144 H **alachlor** Monsanto
CP 52223 H **delachlor**
CP 53619 H **butachlor** Monsanto
CPAS A **chlorfensulphide**
CPBS A **fenson**
CPCBS A **chlorfenson**
CP-Etokap I **codlelure** Chemika
CR-1693 FA **dinocap** Rohm & Haas
Crab-E-Rad H **DSMA**
Crab-Kleen H **DSMA**
Crackdown I **deltamethrin** Wellcome
Crag Fungicide 341 F **glyodin**
Crag Fungicide 974 NFHI **dazomet** Union Carbide
Crag Herbicide I H **2,4-DES-sodium**
Cremart H **butamifos** Sumitomo
Cresofin IAHF **DNOC** Craven
Cresol IAF **DNOC** Terranalisi
Cresopur H **benazolin** Agrolinz
Cresopur H **benazolin** Schering
Cresylex Creme H **DNOC** Protex
crimidine R [535-89-7] 2-chloro-4-dimethylamino-6-methylpyrimidine, Castrix
Crioram F **copper sulphate + mancozeb** L.A.P.A.
Crioram F Combi F **Bordeaux mixture + cymoxanil + folpet** Siapa
Criptoser F **copper oxychloride + zineb** Serpiol
Criscobre F **copper hydroxide**
Crisodrin IA **monocrotophos** Crystal
Crisquat H **paraquat dichloride** Crystal
Crisuron H **diuron** Crystal
Crittam F **ziram** Agrochemiki, Siapa
Crittan F **dodine** Montedison
Critteb F **maneb** Siapa
Crittomet FIHN **dazomet** Siapa
Crittox F **zineb** Siapa
Crittox MZ F **mancozeb** Siapa
Croak H **fluometuron + MSMA** Drexel
Croneton I **ethiofencarb** Agro-Kemi, Agroplant, Bayer
Crop Saver I **malathion + permethrin** Pan Britannica
Cropotex A **flubenzimine** Agroplant, Bayer
Cropspray 11E S **petroleum oils** SOPRO, Tripart
Croptex Amber H **propachlor** Hortichem
Croptex Chrome H **chlorpropham + fenuron** Hortichem
Croptex Fungex F **cuprammonium** Hortichem
Croptex Onyx H **bromacil** Hortichem
Croptex Pewter H **cetrimide + chlorpropham** Hortichem
Croptex Steel H **sodium monochloroacetate** Hortichem
Crosacina ST H **atrazine + simazine** Agrocros
Crosacina T H **atrazine** Agrocros
Crosazina S H **simazine** Agrocros
Crosfluid F **sulphur** R.S.R.
Crosmaneb F **maneb** Agrocros
Crossbow H **triclopyr** Dow
Crossfire I **chlorpyrifos** Rhone-Poulenc
Croszineb F **zineb** Agrocros
Croszintox IA **azinphos-methyl** Agrocros
Crosziram F **ziram** Agrocros
Crotofen FA **dinocap** Inagra
Crotofit IA **monocrotophos** Visplant
Crotopec FA **dinocap** Sapec
Crotos IA **monocrotophos** Siapa
Crotothan F **dinocap** Rhone-Poulenc

Crotothane FA **dinocap** Agriben, Nordisk Alkali, Rhone-Poulenc
Crotovid IA **monocrotophos** Sadisa
crotoxyfos I **crotoxyphos**
crotoxyphos I [7700-17-6] dimethyl (*E*)-1-methyl-2-(1-phenylethoxycarbonyl)vinyl phosphate, crotoxyfos, OMS 239, ENT 24717, Ciodrin, Cypona, Decrotox, SD 4294
Crusader S H **bromoxynil + clopyralid + fluroxypyr + ioxynil** Dow
Cruscagro M **metaldehyde** Ital-Agro
Crynebe F **maneb** Sedagri
Cryptolina F **8-hydroxyquinoline sulphate** Fivat
Cryptonol F **8-hydroxyquinoline sulphate** Argos, Hellafarm, La Qunioleine, Sanac
Cryptonol special E F **quintozene** La Quinoleine
Cryptosan F **maneb + sulphur + ziram** Van Wesemael
Crysthyon IA **azinphos-methyl**
CS 8890 F **dichlozoline**
36 C Tipo G **(2-naphthyloxy)acetic acid** Gobbi
Cu-Pri-Mix F **copper oxychloride** Midol
cube IA **rotenone**
Cubelte F **copper oxychloride + sulphur** Abelte
Cuberol IA **rotenone** Umupro
Cudrox F **copper hydroxide**
Cufolan F **copper oxychloride + cymoxanil + sulphur** Dow
Cufram Z FA **cufraneb** Universal Crop Protection
cufraneb FA [11096-18-7] cufraneb, Cufram Z
Cuidrox F **copper hydroxide**
Cuivrochim F **copper oxychloride** Prochimagro
Cuivrolite F **copper oxychloride** Sedagri
Cuivroneb F **copper oxychloride + zineb** Bourgeois
Cuivrozan FB **copper oxychloride** Bourgeois
Cultar G **paclobutrazol** ICI
Cuman FX **ziram** Ciba-Geigy
Cumarax R **warfarin** Spiess, Urania
Cumflinex I **dichlorvos** Bitterfeld
Cumulus S F **sulphur** Delis
Cunilate 2472 F **oxine-copper** Ventron
Cunitex X **thiram** Agrotec, Hortichem, Kwizda, May & Baker, Rhodiagri-Littorale, Rhone-Poulenc
Cupagrex F **copper oxychloride** Sadisa
Cuper-zine F **copper oxychloride + zineb** Quimagro
Cuperit Oxychlorure de Cuivre F **copper oxychloride** Protex
Cupertane F **copper oxychloride + zineb** Permutadora
Cupertex F **copper oxychloride + zineb** Permutadora
Cupertine F **Bordeaux mixture + maneb** Quimicas del Valles
Cupertine Folpet F **Bordeaux mixture + folpet** Quimicas del Valles
Cupertine M F **copper sulphate + maneb** Zootechniki
Cupertine Super F **Bordeaux mixture + cymoxanil** Quimicas del Valles
Cuperzate F **copper oxychloride + zineb** Reis
Cuprablau-Z F **calcium copper oxychloride + zinc sulphate** Cinkarna
Cuprachlor F **copper oxychloride** Ellagret
Cupraflow F **calcium copper oxychloride + zinc sulphate** Cinkarna
Cuprafol F **copper oxychloride + folpet** Chromos
Cuprafor F **copper oxychloride** Safor
Cupragrol FA **copper compounds (unspecified) + dicofol + sulphur + zineb** CTA
Cupramina F **Bordeaux mixture** EniChem
Cupramix F **calcium copper oxychloride + metiram** Cinkarna
Cupranorg F **copper oxychloride** Anorgachim
Cuprargos F **copper oxychloride** Argos
Cuprasol F **copper oxychloride** Spiess, Urania
Cuprasol Z F **copper oxychloride + zineb** Margesin
Cupravit F **copper oxychloride** Bayer
Cupravit Z F **copper oxychloride + zineb** Bayer
Cuprazin F **copper oxychloride + zineb** Geochem, Solfotecnica
Cuprazufre F **copper oxychloride + sulphur** Pallares
Cuprazyl F **copper oxychloride** Agrofarm
Cupreclor F **copper oxychloride** Agrocros
Cuprenox F **copper oxychloride** Chimiberg
Cuprex F **copper oxychloride** B.B.
cupric sulphate LF **copper sulphate**
Cuprin F **copper oxychloride** Agrochemiki, Mormino

Cuprinebe F **copper oxychloride + zineb** Agrishell
Cuprinol FX **copper naphthenate**
Cuprital F **copper sulphate + maneb** Vital
Cuprix F **Bordeaux mixture** Vital
Cuprizol F **copper oxychloride + sulphur** EniChem, Mormino
Cupro Antracol F **copper oxychloride + propineb** Bayer
Cupro Dithane F **copper sulphate + mancozeb** Serpiol
Cupro dithane fuerte F **copper sulphate + mancozeb** Serpiol
Cupro Euparene F **copper oxychloride + dichlofluanid** Bayer
Cupro Zinebe F **copper oxychloride + zineb** ICI-Valagro
Cuproben F **copper oxychloride** ERT
Cuprocaffaro F **copper oxychloride** Caffaro
Cuprocal F **Bordeaux mixture** Shell
Cuprocar F **copper oxychloride + mancozeb** Visplant
Cuprocarbina F **copper oxychloride + zineb** Aziende Agrarie Trento
Cuprocid F **Bordeaux mixture + maneb** SPE
Cuprocin F **copper oxychloride + zineb** ERT
Cuprocure F **copper oxychloride** Proval
Cuprodithane F **copper oxychloride + mancozeb** Rohm & Haas
Cuprofal F **copper oxychloride + folpet** Du Pont
Cuprofix F **copper oxychloride** Maag
Cuprofix CZ F **copper sulphate + cymoxanil + zineb** R.S.R.
Cuprofix F F **copper sulphate + folpet** R.S.R.
Cuprofix F active F **copper sulphate + cymoxanil + folpet + zineb** R.S.R.
Cuprofix M F **copper sulphate + maneb** R.S.R., Tarsoulis
Cuprofix Z F **copper sulphate + zineb** R.S.R., Tarsoulis
Cuprofog F **copper bicarbonate + sulphur** Procida
Cuprofol P FA **copper compounds (unspecified) + cyhexatin + fenarimol + mancozeb** Maag
Cuprofolpet F **copper oxychloride + folpet** Sivam
Cuproford F **cuprammonium** Ziegler
Cuprofrut F **copper oxychloride** Visplant
Cuprogarda F **copper oxychloride** Aziende Agrarie Trento
Cuprokane F **copper oxychloride + zineb** Sandoz
Cuprokylt F **copper oxychloride** Universal Crop Protection
Cuprol F **copper oxychloride** Hellafarm
Cuprolate corbeaux liquide FX **anthraquinone + oxine-copper** Prochimagro
Cuprolate FSB F **oxine-copper** Gen. Rep.
Cuprolate plus F **oxine-copper** Prochimagro
Cuprolate plus corbeaux FX **anthraquinone + oxine-copper** Prochimagro
Cuprolate plus MG 3 FIX **endosulfan + lindane + oxine-copper** Prochimagro
Cuprolate plus T1 F **oxine-copper** Prochimagro
Cuprolate plus T2 FX **anthraquinone + oxine-copper** Prochimagro
Cuprolate plus T3 FIX **anthraquinone + lindane + oxine-copper** Prochimagro
Cuprolate Plus T4 FIX **anthraquinone + endosulfan + lindane + oxine-copper** Prochimagro
Cuprolate plus triple FIX **anthraquinone + lindane + oxine-copper** Prochimagro
Cuprolex F **copper oxychloride + copper sulphate + folpet** Rhone-Poulenc
Cupromaag F **copper oxychloride** Maag
Cuproman 17 + 17 F **copper sulphate + mancozeb** Sivam
Cupromat F **copper oxychloride + sulphur + zineb** Ellagret
Cupromix F **copper oxychloride + sulphur** Mormino
Cuproneb F **copper sulphate + maneb** Hellenic Chemical
Cupronebe F **copper oxychloride + propineb** Gaillard
Cuprophyt F **copper oxychloride** Chimac-Agriphar
Cuprorganico F **copper oxychloride + zineb** Mormino
Cuprosagrex Triple F **copper oxychloride + maneb + zineb** Sadisa
Cuprosan F **copper oxychloride** Rhone-Poulenc
Cuprosan F **copper oxychloride + zineb** Maag
Cuprosan 3 F F **copper oxychloride + fenarimol + folpet** Maag
Cuprosan 311 super D F **copper**

oxychloride + maneb + zineb Condor, Pepro
Cuprosan Extra F **copper oxychloride + cymoxanil + zineb** Rhone-Poulenc
Cuprosan F1 F **copper oxychloride + copper sulphate + folpet** Pepro
Cuprosan fluid F **copper oxychloride + folpet** Maag
Cuprosan Super A F **copper oxychloride + zineb** Rhone-Poulenc
Cuprosan U F **copper oxychloride + folpet** Du Pont
Cuprosan Ultra F **copper oxychloride + folpet** Maag, Viopharm
Cuprosana H F **copper oxychloride** UCP
Cuprosariaf F **copper oxychloride** EniChem
Cuproscam F **copper oxychloride + zineb** Scam
Cuproscam Bleu F **copper oxychloride + zineb** Scam, Sivam
Cuprosoufre M FA **copper compounds (unspecified) + cyhexatin + sulphur + zineb** Gaillard
Cuprospor Z F **copper oxychloride + zineb** Sivam
Cuprossil F **copper oxychloride** Scam
Cuprossil 8 F **copper oxychloride + sulphur + zineb** Scam
Cuprossina F **copper oxychloride** Agrimont
Cuprosul Extra F **copper oxychloride + sulphur** ERT
Cuprosulf F **copper oxychloride + sulphur** Geochem
Cuprothiol F **copper sulphate + sulphur** R.S.R., UCAR
cuprous oxide F [1317-39-1] copper(I) oxide, oxyde cuivreux, brown copper oxide, red copper oxide, C.I. 77402, Kupferoxid, Caocobre, Cobre Sandoz, Copper Nordox, Copper-Sandoz, Fungi-Rhap, Kupfer Sandoz, Perenox, Yellow Cuprocide
Cuprovinol F **copper oxychloride**
Cuprox F **copper oxychloride** ICI-Zeltia
Cuproxat F **copper oxysulphate** Agrolinz, Leu & Gygax
Cuproxi F **copper oxychloride** Aragonesas
Cuproxide F **copper hydroxide**
Cuproxin F **copper oxychloride** Montedison
Cuproxyde Macclesfield F **copper hydroxide** La Cornubia
Cuprozan F **copper oxychloride + zineb** Viopharm
Cuprozin F **copper oxychloride + zineb** Caffaro, Safor
Cuprozin D F **copper oxychloride + sulphur + zineb** Bredologos
Cuprozineb F **copper oxychloride + zineb** Marba
Cupryl F **copper oxychloride** Visplant
Cupzin F **copper oxychloride + zineb** Visplant
CUR 616 F **chlorfenazole**
Curacron IA **profenofos** Ciba-Geigy
Curamil F **pyrazophos** Hoechst
Curamil F **mancozeb** Permutadora
Curamix F **copper oxychloride + zineb** Proval
Curanil F **chlorothalonil + cymoxanil** Kwizda
Curater I **carbofuran** Bayer
Curaterr IN **carbofuran** Agro-Kemi, Agroplant, Bayer
Curaterr Forte IN **carbofuran + fenamiphos** Bayer
Curattin R **warfarin** Hentschke & Sawatzki
Curb X **aluminium ammonium sulphate** Dreyfus
Curbetan H **chloridazon** Dow, Du Pont, Huels-Kemi, PLK, Wacker-Dow
Curbiset G **chlorflurecol-methyl** Celamerck, Sovilo
Curenox F **copper oxychloride** Quimicas del Valles
Curit Blau F **copper oxychloride + zineb** Schering
Curit K F **copper oxychloride + zineb** Schering
Curit N F **chlorothalonil + cymoxanil** Schering
Curit R F **copper oxychloride + cymoxanil** Schering
Curit Zeb F **cymoxanil + mancozeb** Schering
Curital F **dicloran** Schering
Curitan F **dodine** Rhone-Poulenc
Curlone I **chlordecone** Laguarigue
Curoxan triple FIX **anthraquinone + lindane + oxine-copper** Bourgeois
Curpo-Phynebe F **copper oxychloride + zineb** Bayer
Curthiram F **thiram** Proval
Curzate F **cymoxanil** Du Pont
Curzate C F **Bordeaux mixture + cymoxanil** Du Pont
Curzate Combi F **copper oxychloride + cymoxanil + folpet** Du Pont

Curzate Cu F **copper oxychloride + cymoxanil** Organika-Azot
Curzate F F **cymoxanil + folpet** Du Pont
Curzate K F **copper oxychloride + cymoxanil** Spolana
Curzate M F **cymoxanil + mancozeb** Du Pont
Curzate MF F **cymoxanil + fentin acetate + mancozeb** Du Pont
Curzate R F **copper oxychloride + cymoxanil** Du Pont
Curzate SM F **cymoxanil + mancozeb** Spolana
Curzate Super CZ F **copper oxychloride + cymoxanil + zineb** PVV
Curzate Super Z F **cymoxanil + zineb** PVV
Custos F **carbendazim** Du Pont, Rustica
Cuteryl H **iron sulphate** Bayer
Cutlass G **dikegulac-sodium** ICI
Cutless G **flurprimidol** Eli Lilly
Cutralin H **amitrole + 2,4-D + TCA** Bayer
CVMP IA **tetrachlorvinphos**
CVP IA **chlorfenvinphos**
CX 99 I **bromophos** Shell
Cyaforgel I **hydramethylnon** Cyanamid
Cyalane IA **phosfolan** Cyanamid
cyanamide HG [420-04-2] calcium cyanamide, carbamonitrile, amidocyanogen, hydrogen cyanamide, cyanogenamide, Alzodef
Cyanater IN **terbufos** Siapa
cyanatryn H [21689-84-9] 2-(4-ethylamino-6-methylthio-1,3,5-triazin-2-ylamino)-2-methylpropionitrile, cyanatryne, WL 63611
cyanatryne H **cyanatryn**
cyanazine H [21725-46-2] 2-(4-chloro-6-ethylamino-1,3,5-triazin-2-ylamino)-2-methylpropionitrile, Bladex, Fortrol, SD 15418, WL 19805
cyanofenphos I [13067-93-1] *O*-4-cyanophenyl *O*-ethyl phenylphosphonothioate, Cyp, S-4087, Surecide
cyanophos I [2636-26-2] *O*-4-cyanophenyl *O,O*-dimethyl phosphorothioate, CYAP, ciafos, OMS 226, OMS 869, Cyanox, Cynock, S-4084
Cyanosil R **hydrogen cyanide** DGS
Cyanotril I **dimethoate + flucythrinate** Cyanamid
Cyanox I **cyanophos** Sumitomo
cyanthoate IA [3734-95-0] *S*-[*N*-(1-cyano-1-methylethyl)carbamoylmethyl] *O,O*-diethyl phosphorothioate, M 1568, Tartan
CYAP I **cyanophos**
Cybet H **cycloate** L.A.P.A.
Cybolt I **flucythrinate** Cyanamid, Sapec, Zorka Sabac
cyclafuramid F [34849-42-8] *N*-cyclohexyl-2,5-dimethyl-3-furamide, cyclafuramide, BAS 327F
cyclafuramide F **cyclafuramid**
cycloate H [1134-23-2] *S*-ethyl *N*-cyclohexyl-*N*-ethyl(thiocarbamate), hexylthiocarbam, R-2063, Ro-Neet, Sabet
Cyclobet H **cycloate** Hellenic Chemical
Cyclodan IA **endosulfan** Hoechst
cycloheximide FG [66-81-9] 4-[(2*R*)-2-[(1*S*,3*S*,5*S*)-(3,5-dimethyl-2-oxocyclo hexyl)]-2-hydroxyethyl]piperidine-2,6-dione, naramycin, Acti-dione, Actispray, Hizarocin
Cyclon IR **hydrogen cyanide**
cycloprothrin I [63935-38-6] (*RS*)-α-cyano-3-phenoxybenzyl (*RS*)-2,2-dichloro-1-(4-ethoxyphenyl)cyclopropanecarboxylate, phencyclate, Cyclosal, NK 8116
Cyclosal I **cycloprothrin** Nippon Kayaku
Cyclosan F **mercurous chloride** Hortichem, May & Baker
cycloxydim H [101205-02-1] (±)-2-[1-(ethoxyimino)butyl]-3-hydroxy-5-thian-3-ylcyclohex-2-enone, cycloxydime, BAS 517H, Focus, Laser, Stratos
cycloxydime H **cycloxydim**
cycluron H [2163-69-1] 3-cyclooctyl-1,1-dimethylurea, OMU
Cycocel G **chlormequat chloride** BASF, Chimac-Agriphar, Cyanamid, Drugtrade, Du Pont, ICI, Lapapharm, Procida, Ravit, Rhodiagri-Littorale, Sandoz, Sipcam, Spiess, Urania
Cycocel Extra G **chlormequat chloride + choline chloride** BASF, Burri, Ciba-Geigy, Collett, CTA, Cyanamid, Delis, Hokochemie, Leu & Gygax, Ligtermoet, Liro, Lonza, Luxan, Maag, Nordisk Alkali, Opopharma, Pluess-Staufer, Schering-AAgrunol, Shell, Siapa, Siegfried
Cycofrem G **chlormequat chloride** Eurofyto
Cycogan G **chlormequat chloride** COPCI, Edefi, Makhteshim-Agan
Cycoquat G **chlormequat chloride** Siapa

Cycosin FW **thiophanate-methyl** Cyanamid
Cydexine H **simazine** Sedagri
Cydexone H **2,4-D** Sedagri
Cydexone double H **2,4-D + MCPA** Sedagri
Cydexone special H **dicamba + MCPA** Sedagri
Cydexone super H **dicamba + mecoprop** Sedagri
Cyduron H **linuron** Sedagri
cyfluthrin I [68359-37-5] (*RS*)-*α*-cyano-4-fluoro-3-phenoxybenzyl (1*RS*,3*RS*:1*RS*,3*SR*)-3-(2,2-dichlorovinyl)-2,2-dimethylcyclopropanecarboxylate, cyfluthrine, Bay FCR 1272, Baythroid, Responsar, Solfac, Tempo
cyfluthrine I **cyfluthrin**
Cygon IA **dimethoate** Cyanamid
cyhalothrin I [68085-85-8] (*RS*)-*α*-cyano-3-phenoxybenzyl (*Z*)-(1*RS*,3*RS*)-(2-chloro-3,3,3-trifluoropropenyl)-2,2-dimethylcyclopropanecarboxylate, cyhalothrine, Grenade, PP 563
cyhalothrine I **cyhalothrin**
cyhexatin A [13121-70-5] tricyclohexyltin hydroxide, tricyclohexyltin hydroxide, ENT 27395-X, Acarstin, Dowco 213, Mitacid, Plictran
Cylan IA **phosfolan** Cyanamid
Cymag I **sodium cyanide** ICI, Prometheus
Cymbigon IA **cypermethrin** Kwizda
Cymbush I **cypermethrin** ICI, ICI-Valagro, ICI-Zeltia, Maag, Sopra, Zorka Subotica
Cymbusz I **cypermethrin** ICI
cymoxanil F [57966-95-7] 1-(2-cyano-2-methoxyiminoacetyl)-3-ethylurea, Curzate, DPX 3217
Cymperator I **cypermethrin** ICI
Cynem NI **thionazin** Cyanamid
Cynkomiedzian F **copper oxychloride + zineb** Organika-Azot
Cynkotox F **zineb** Organika-Azot
Cynock I **cyanophos** Sumitomo
Cyolan IA **phosfolan** Cyanamid
Cyolane IA **phosfolan** Cyanamid
cyometrinil S [63278-33-1] (*Z*)-cyanomethoxyimino(phenyl)acetonitrile, CGA 43089, Concep
Cyp I **cyanofenphos**
cypendazole F [28559-00-4] 1-(5-cyanopentylcarbamoyl)benzimidazol-2-ylcarbamate, DAM 18654, Folicidin
Cypercopal I **cypermethrin** Gilmore
Cyperkil I **cypermethrin** Agrolinz
Cyperkill I **cypermethrin** Mitchell Cotts, Tuhamij
cypermethrin I [52315-07-8] (*RS*)-*α*-cyano-3-phenoxybenzyl (1*RS*)-*cis,trans*-3-(2,2-dichlorovinyl)-2,2-dimethylcyclopropanecarboxylate, cypermethrine, OMS 2002, Ambush C, Ammo, Arrivo, Barricade, Cymbush, Cymperator, Cypercopal, Cyperkill, Cyrux, Demon, Fenom, Imperator, Kafil Super, LE 79600, NRDC 149, Nurelle, Polytrin, PP 383, Ripcord, Sherpa, Siperin, Toppel, Ustaad, WL 43467
cypermethrine I **cypermethrin**
cyperquat chloride H [39794-99-5] 1-methyl-4-phenylpyridinium chloride, S 21634
Cypex I **cypermethrin** Burri
cyphenothrin I [39515-40-7] (*RS*)-*α*-cyano-3-phenoxybenzyl (1*R*)-*cis,trans*-chrysanthemate, Gokilaht
Cypon-Cumarin R **warfarin** Van Loosen
Cypon-Fertigkoeder R **warfarin** Van Loosen
Cypona I **crotoxyphos**
cyprazine H [22936-86-3] 6-chloro-N^2-cyclopropyl-N^4-isopropyl-1,3,5-triazine-2,4-diamine, Outfox, S 6115
cyprazole H [42089-03-2] *N*-[5-(2-chloro-1,1-dimethylethyl)-1,3,4-thiadiazol-2-yl]cyclopropanecarboxamide, S 19073
Cyprex F **dodine** Cyanamid
cyprofuram F [69581-33-5] (±)-*α*-[*N*-(3-chlorophenyl)cyclopropanecarboxamido]-*γ*-butyrolactone, SN 78314, Vinicur
cypromid H [2759-71-9] 3',4'-dichlorocyclopropanecarboxanilide, cypromide, Clobber, S 6000
cypromide H **cypromid**
cyromazine I [66215-27-8] *N*-cyclopropyl-1,3,5-triazine-2,4,6-triamine, Armor, CGA 72662, Larvadex, Neporex, Trigard, Vetrazine
Cyronal H **clopyralid** Schering
Cyrux I **cypermethrin** United Phosphorus
Cytel I **fenitrothion** Cyanamid
Cythion IA **malathion** Cyanamid
Cythrin I **flucythrinate** Cyanamid
Cytro-Lane IA **mephosfolan** Cyanamid
Cytrol H **amitrole** Cyanamid
Cytrolane IA **mephosfolan** Cyanamid

1,3 D Soil Fumigant N **1,3-dichloropropene** Sipcam

2,4-D H [94-75-7] (2,4-dichlorophenoxy)acetic acid, 2,4-PA, Agroxone, Aqua-Kleen, Citrus Fix, Dacamine, Ded-Weed, Desormone, Dikamin, Dikonirt, Embamine, Emulsamine, Fernimine, Fernoxone, Hedonal, Hi-Dep, Netagrone, Planotox, Spritz-Hormin, U 46 D, Weed-B-Gon, Weed-Rhap, Weedar, Weedone, Weedtrine-II
2,4-D Combi H **2,4-D + MCPA** Berghoff
D 85 LAPA H **dalapon-sodium** L.A.P.A.
D-264 IA **diazinon** Drexel
D 735 F **carboxin** Uniroyal
D'Operats R **difenacoum** Agrinet, C.N.C.A.T.A.
D-D N [8003-19-8] 1,2-dichloropropane with 1,3-dichloropropene, DD, ENT 8420, D-D
d-phenothrin I **phenothrin**
D-propionat H **dichlorprop** Nordisk Alkali
D-Z-N IA **diazinon** Ciba-Geigy
D. Souryl R **difenacoum** C.N.C.A.T.A.
DAC 893 H **chlorthal-dimethyl** SDS Biotech
Dacamox I **thiofanox** Agriben, Agrotec, BASF, Diamond Shamrock, Pepro, Rhone-Poulenc, Spiess
Dacol IA **dimethoate** Caffaro
Daconate H **MSMA** Diamond Shamrock, Fermenta
Daconil F **chlorothalonil** BASF, Chimac-Agriphar, Ciba-Geigy, Diamond Shamrock, Elanco, Ligtermoet, Masso, Schering AAgrunol, SDS Biotech, Siegfried, Sipcam
Daconil M F **chlorothalonil + maneb** Schering AAgrunol
Daconil MS F **chlorothalonil + maneb + sulphur** Schering AAgrunol
Daconil-Combi F **chlorothalonil + cymoxanil** Siegfried
Daconyl F **chlorothalonil** La Quinoleine
Dacostar F **chlorothalonil + copper oxychloride** Ravit
Dacthal H **chlorthal-dimethyl** Diamond Shamrock, Ravit, Schering AAgrunol, SDS Biotech, Siegfried, Sipcam-Phyteurop
Dacthalor H **chlorthal-dimethyl** SDS Biotech
Dacus T **hydrolysed proteins**
Dacuser IA **dimethoate** Serpiol
Dacutrin IA **dimethoate** Enotria
Dafene IA **dimethoate** Condor
Dafenil I **dimethoate** Rhone-Poulenc
Dagadip IA **carbophenothion**
Dagger H **imazamethabenz-methyl** Cyanamid
Daioh H **pyrazosulfuron-ethyl** Nissan
Dakogal F **chlorothalonil** Galenika
Daktal H **chlorthal-dimethyl** Galenika
Dakuron H **pentanochlor**
Dal-E-Rad H **MSMA** Vineland
Dalacide H **dalapon-sodium** Chimiberg
Dalaphyt H **dalapon-sodium** Sipcam-Phyteurop
dalapon-sodium H [127-20-8] sodium 2,2-dichloropropionate, DPA, proprop, Basfapon, Basinex P, Dowpon, Gramevin, Radapon, SYS 67 Omnidel
Dalascam H **dalapon-sodium** Scam
Dalaved H **dalapon-sodium** Agronova
Dalbit H **isocarbamid + lenacil** Grima
DAM 18654 F **cypendazole**
Damex H **2,4-D + MCPA** Protex
Damfin IA **methacrifos** Ciba-Geigy
daminozide G [1596-84-5] *N*-dimethylaminosuccinamic acid, SADH, Alar, B-995, B-Nine, Dazide, Kylar
Dana Gamax FI **lindane + thiram** Nordisk Alkali
Danatex F **thiram** Nordisk Alkali
Danbrex H **atrazine** Visplant
Dancozan F **mancozeb** Ellagret
Danex I **trichlorfon** Alpha, Edefi, Makhteshim-Agan
Danicut AI **amitraz** Nissan
Danitol AI **fenpropathrin** Agrishell, Nitrokemia, Sumitomo, Shell, Siapa, Zupa
Dantox I **dimethoate** Nordisk Alkali
Dantril H **bromoxynil + dichlorprop + ioxynil + MCPA** Agronorden
DAPA F **fenaminosulf**
Dapacryl AF **binapacryl** Hoechst
Daphene IA **dimethoate** Rhodic
Dardo H **glyphosate + simazine** Sipcam
Darlem IAN **phorate** Du Pont
Dart I **teflubenzuron** Celamerck, Rhodiagri-Littorale
Dartone I **phosalone + teflubenzuron** Rhodiagri-Littorale
Daryl F **carbendazim + folpet + sulphur** Pepro
Daryline F **carbendazim** Sedagri
Daryline M F **carbendazim + maneb** Sedagri

Das H **amitrole + diuron + simazine + sodiumthiocyanate** Hellafarm
Dasanit NI **fensulfothion** Bayer
Daskor I **chlorpyrifos-methyl + cypermethrin** Dow
Datapron H **amitrole + diuron** BP
Dathion I **fenitrothion + lindane** Afrasa
Davlan Super F **mancozeb + thiram** Agrotechnica
Davline F **quintozene** Ellagret
Davlinex F **quintozene** Agrofarm
Davlitine Nea F **copper sulphate + mancozeb** Hellenic Chemical
Davlitoktono PCNB F **quintozene** Gefex
Davlitox F **quintozene** Geochem
Davlitoxan B F **quintozene** Geopharmaceutiki
Davlotan F **maneb** Mylonas
Davlotox F **maneb** Nitrofarm
Davloxan Super F **TCMTB** Diana
Daxtron H **pyriclor**
Dazide G **daminozide** Fine Agrochemicals, NCK, Research & Consulting
Dazo-Fum-Perlat FIHN **dazomet** Sivam
Dazoberg FIHN **dazomet** Chimiberg
Dazom FIHN **dazomet** Chemia
dazomet FIHN [533-74-4] 3,5-dimethyl-1,3,5-thiadiazinane-2-thione, tiazon, DMTT, Basamid, Crag Fungicide 974, Fongosan, Mylone, N-521, Salvo
Dazomil FIHN **dazomet** Eli Lilly
Dazosar FIHN **dazomet** EniChem
Dazoscam FIHN **dazomet** Scam
Dazzel IA **diazinon**
2,4-DB H [94-82-6] 4-(2,4-dichlorophenoxy)butyric acid, Butirex, Butormone, Butoxone, Butyrac, Embutone, Embutox, M&B 2878
371 DBA H **amitrole + atrazine + 2,4-D + simazine** Urania
DBCP N **dibromochloropropane**
371 DBH H **amitrole + atrazine + bromacil + simazine** Urania
DBN H **dichlobenil**
DCBN H **chlorthiamid**
DCIP N [108-60-1] 2,2'-oxybis[1-chloropropane], Nemamort
DCMO F **carboxin**
DCMOD F **oxycarboxin**
DCMU H **diuron**
DCNA F **dicloran**
3,6-DCP H **clopyralid**
DCP 50 N **D-D** Solvay-Werke
DCPA H **chlorthal-dimethyl**
DCPA H **propanil**
DCPC A **chlorfenethol**
DCS 2 AB F **2-aminobutane** Brogdex
DCS-A F **2-phenylphenol** Brogdex
DCU H [116-52-9] *N,N'*-bis(2,2,2-trichloro-1-hydroxyethyl)urea, dichloralurea
DDD I **TDE**
DDOD F **dichlozoline**
DDT I [50-29-3] 1,1,1-trichloro-2,2-bis(4-chlorophenyl)ethane, *p,p'*-DDT, zeidane, Digmar, Hildit
DDVP IA **dichlorvos**
De-Cut G **maleic hydrazide** Fair Products
De-Fend IA **dimethoate**
De-Fol-Ate H **sodium chlorate**
De-Green G **S,S,S-tributyl phosphorotrithioate** Mobay
De-Kan F **dinocap** Midol
De-Mice R **brodifacoum** Rigo
Deadline M **metaldehyde** Epro
Deadline R **bromadiolone** Rentokil
Debantic IA **tetrachlorvinphos** SDS Biotech, PLK
Debresol I **piperonyl butoxide + pyrethrins** Debrella
Debrousol H **2,4,5-T** ICI
Debroussaillant 4323 H **picloram + 2,4,5-T** C.F.P.I.
Debroussaillant 4323 DP H **dichlorprop + picloram** C.F.P.I.
Debroussaillant 72 H **triclopyr** Prochimagro
Debroussaillant concentre H **2,4,5-T** Pepro
Debroussaillant G 2 H **2,4-D + 2,4,5-T** Ciba-Geigy
Debroussaillant granule 3309 H **picloram** C.F.P.I.
Debroussex H **ammonium sulphamate** LHN
Debrouxal H **2,4-D + triclopyr** La Quinoleine
Decabane H **dichlobenil** Shell
decafentin F [15652-38-7] decyltriphenylphosphonium bromochlorotriphenylstannate(IV), A 36, Stanoram
decan-1-ol G [112-30-1] decan-1-ol, Antak, Fair-Tac, Paranol
Decarol F **carbendazim** Kwizda
Decaropron H **petroleum oils** R.S.R.
Decco F **thiabendazole** Pennwalt
Decco Antimousse S **dimethylpolysiloxane** Pennwalt

Decco TO F F **2-phenylphenol** Pennwalt
Deccoklor F **sodium hypochlorite** Pennwalt
Deccoprozil F **imazalil + iprodione** Pennwalt
Deccoquin G **ethoxyquin** Pennwalt
Deccosil F **imazalil** Pennwalt France
Deccosol F **2-phenylphenol** Pennwalt
Deccotane F **2-aminobutane** Pennwalt
Deccozil F **imazalil** Pennwalt
Decemtion IA **phosmet** Dimitrova, Duslo
Decilaz D H **2,4-D** Chimac-Agriphar
Decilaz DP H **dichlorprop** Chimac-Agriphar, Schwebda
Decimate H **chlorthal-dimethyl + propachlor** SDS Biotech
Decimax H **diuron** Agriphyt, Chimac-Agriphar
Decis I **deltamethrin** Bayer, Hoechst, OHIS, Pluess-Staufer, Procida, Quimigal, RousselHoechst, Roussel Uclaf
Decis B I **deltamethrin + heptenophos** Procida
Decis PS FI **deltamethrin + sulphur** Procida
Decisquick I **deltamethrin + heptenophos** Hoechst, Rouseel Hoechst
Decrotox I **crotoxyphos**
Dedevap I **dichlorvos** Bayer, Berner
Dedisol C N **1,3-dichloropropene** La Littorale
DEET X **diethyltoluamide**
DEF G **S,S,S-tributyl phosphorotrithioate** Bayer, Mobay
Def-6 G **ethion** Bayer
Defanex PDT H **dinoseb** Bourgeois
Defanol forte H **dinoseb** CTA
defenuron H [1007-36-9] *N*-methyl-*N'*-phenylurea
Defoal H **magnesium chlorate** Aragonesas
Defol H **sodium chlorate** Drexel, Sadisa
Deftor H **metoxuron** Sandoz
Degesch Magtoxin IAR **magnesium phosphide** Degesch
Deherban A H **2,4-D** Chromos
Deherban combi-MD H **2,4-D + mecoprop** Chromos
Deherban fluid H **mecoprop** Chromos
Deherban forte H **2,4-D + MCPA** Chromos
Deherban M H **MCPA** Chromos
Deherban special H **2,4-D + MCPA + mecoprop** Chromos
deiquat dibromide H **diquat dibromide**
delachlor H [24353-58-0] 2-chloro-*N*-(isobutoxymethyl)acet-2',6'-xylidide, delachlore, CP 52223
delachlore H **delachlor**
Delan F **dithianon** BASF, Celamerck, Duphar, Epro, Geopharm, Kemira, Margesin, Permutadora, Pinus, Prochimagro, Sandoz, Seigfried, Shell
Delan K F **copper oxychloride + dithianon** Margesin
Delan MZ F **dithianon + mancozeb** Margesin
Delan-Col F **dithianon** ICI
Delancol F **dithianon** Agrolinz
Delex I **carbofuran** Sopra
Delicia-Beutel I **aluminium phosphide** Delicia
Delicia-csigaolo szer M **metaldehyde** Delicia
Delicia-Delitex I **lindane** Delicia
Delicia-Fribal IR **camphechlor** Delicia
Delicia-Gastoxin I **aluminium phosphide** Delicia
Delicia-Giftgetreide R **zinc phosphide** Delicia
Delicia-Kornkaeferbegasungspraeparat I **aluminium phosphide** Delicia
Delicia-Milon I **malathion** Delicia
Delicia-Pulvis I **aluminium phosphide** Delicia
Delicia-Py I **piperonyl butoxide + pyrethrins** Delicia
Delicia-Ratron R **warfarin** Delicia
Delicia-Rattekal R **zinc phosphide** Delicia
Delicia-Schnecken-Ex M **metaldehyde** Delicia
Delicia-Sperlingsweizen R **strychnine** Delicia
Delicia-Spezial I **lindane + piperonyl butoxide + pyrethrins** Delicia
Delicia-tasak I **aluminium phosphide** Delicia
Delicia-Texyl-Druckzerstaeuber I **camphechlor + lindane** Delicia
Delicia-Texyl-Spray I **lindane + permethrin** Delicia
Delicia-tipp fix I **lindane + piperonyl butoxide + pyrethrins** Delicia
Delicia-tipp fix B I **bioallethrin + lindane + piperonyl butoxide** Delicia
Delicia-Wuehlmauspraeparat R **zinc phosphide** Delicia
Delindol I **lindane** Pepro
Delnav IA **dioxathion** NOR-AM

Deloxil H **bromoxynil + ioxynil** Hoechst, Pluess-Staufer
Delsan F F **carbendazim + folpet** Du Pont
Delsekte I **deltamethrin** Wellcome
Delsene F **carbendazim** Du Pont
Delsene M F **carbendazim + maneb** Du Pont
Delta R **chlorophacinone** Fleurix
deltamethrin I [52918-63-5] [1*R*-[1α(*S**),3α]]-cyano(3-phenoxyphenyl)methyl 3-(2,2-dibromoethenyl)-2,2-dimethylcyclopropanecarboxylate, deltamethrine, OMS1998, Butoss, Cislin, Crackdown, Decis, Delsekte, K-Othrin, NRDC 161, RU 22974
deltamethrine I **deltamethrin**
Deltanet I **furathiocarb** Ciba-Geigy
Deltarol H **chlorotoluron** Ciba-Geigy
Deltastop CP I **codlelure** JZD Slusovice
Deltic IA **dioxathion** NOR-AM
Deltox-Lack I **lindane + permethrin** Delicia
DELU-Schneckenkorn M **metaldehyde** Geistler
DELU-Wuehlmausgas R **phosphine** Geistler
DELU-Wuehlmauskoeder R **zinc phosphide** Geistler
Delvolan F **pimaricin** Gist-Brocades
Demecor I **dimethoate + endosulfan** Protex
demeton IA [8065-48-3; 298-03-3 (demeton-O); 126-75-0 (demeton-S)] *O,O*-diethyl *O*-[2-(ethylthio)ethyl] phosphorothioate, mixture with *O,O*-diethyl *S*-[2-(ethylthio)ethyl] phosphorothioate, demeton-O + demeton-S, mercaptofostion + mercaptostiol, ENT 17295, Bay 10756, E-1059, Systemox, Systox
demeton-O + demeton-S IA **demeton**
demeton-S-methyl IA [919-86-8] *S*-2-ethylthioethyl *O,O*-dimethyl phosphorothioate, methyl demeton, methylmercaptofostiol, DSM, Bay 18436, Bay 25/154, Demetox, Duratox, Metasystox i, Mifatox
demeton-S-methyl sulphone IA [17040-19-6] *S*-2-ethylsulphonylethyl *O,O*-dimethyl phosphorothioate, demeton-S-methylsulphone, demeton-S-methylsulfone, Bay 20315, E 158, M3/158, Metaisosystoxsulfon, Phosulphon
demeton-S-methyl sulphoxide IA **oxydemeton-methyl**
demeton-S-methylsulfone IA **demeton-S-methyl sulphone**
demeton-S-methylsulphone IA **demeton-S-methyl sulphone**
Demetox I **demeton-S-methyl** Hellenic Chemical, Phytorgan
Demetra R **zinc phosphide** Korleti
Demon I **cypermethrin** ICI
Demos IA **dimethoate** Montedison
Demosan F **chloroneb** Du Pont
Demuscan I **dichlorvos + dimethoate** Zucchet
Denapon IG **carbaryl**
Denarin F **triforine** Celamerck, Schering
Dendrosan plus W **fenpropimorph + propiconazole** International Tree Service
Dendroxyl I **petroleum oils** Hellenic Chemical
Dendrozal Z-70 H **amitrole + diuron** Katzilakis
Denet H **sodium chlorate** De Weerdt
Denka algendoder L **quaternary ammonium** Denka
Denka Anti Bladluis Nieuw IA **piperonyl butoxide + pyrethrins** Denka
Denka Anti Klaver H **mecoprop** Denka
Denka Anti Meeldauw Spray F **pyrazophos** Denka
Denka Anti Onkruid Mix H **2,4-D + dicamba + MCPA** Denka
Denka Antimos H **iron sulphate** Denka
Denka Antispruit G **chlorpropham + propham** Denka
Denka mierengietmiddel I **deltamethrin** Denka
Denka Moestuin I **diazinon** Denka
Denka Mollendood R **strychnine** Denka
Denka Onkruidkorrels H **simazine** Denka
Denka Onkruidverdelger P H **diquat dibromide + paraquat dichloride** Denka
Denka Slakkenkorrels M **metaldehyde** Denka
Denka spray away I **deltamethrin** Denka
Denka Veewasmiddel IA **lindane** Denka
Denkamethrin I **permethrin** Denka
Denkaphon I **trichlorfon** Denka
Denkarin Grains R **zinc phosphide** Denka
Denkarin Paste R **thallium sulphate** Denka
Denkatex N F **copper carbonate (basic) + sulphur** Denka
Denkavepon IA **dichlorvos** Denka
Denmert F **buthiobate** Sumitomo

DEP I **trichlorfon** Jin Hung
DEP Vapor I **dichlorvos** C.G.I.
Depethan IA **piperonyl butoxide + pyrethrins + S421** Staehler
Deration R **bromadiolone** Colkim
Derby F **mancozeb + ofurace** ICI-Zeltia
Deril P I **permethrin** Maag
Deroman F **carbendazim + maneb + zineb** Hoechst
Derosal F **carbendazim** Hoechst
Derosal M F **carbendazim + maneb** Hoechst
Derosalin F **carbendazim** Roussel Hoechst
derris IA **rotenone**
Derroprene F **carbendazim** ICI, Sopra
Derrosan I **lindane** Aziende Agrarie Trento
Derrothan-Neu IA **piperonyl butoxide + pyrethrins** Staehler
Dervan H **sodium chlorate** Caffaro
2,4-DES-sodium H [136-78-7] sodium 2-(2,4-dichlorophenoxy)ethyl sulphate, sesone, disul-sodium, Crag Herbicide I, SES
Des-I-Cate HLG **endothal** Dow, Pennwalt
Desgorgogil I **carbon disulphide + carbon tetrachloride** Agrocros
Desherbant H **linuron** Agriphyt, Agrishell, SCAC-Fisons, Umupro
Desherbant gazon H **2,4-D + mecoprop** Umupro
Desherbant ideal S H **amitrole + bromacil + diuron** SCAC-Fisons
Desherbant liquide 50 50 H **amitrole + bromacil** LHN
Desherbant liseron H **2,4-D** Truffaut
Desherbant oignons H **chlorpropham** Umupro
Desherbant Total EDFP H **sodium chlorate** Debauche
Desherbant total granule S H **amitrole + atrazine + simazine** SCAC-Fisons
Desherbant total granule H **amitrole + diuron + sodium thiocyanate** Umupro
Desherbant Total RD H **sodium chlorate** Eurofyto
Desherbant total S H **amitrole + simazine** Umupro
Desherbant W1 H **petroleum oils** Agrishell
Desinfectante CL F **sodium hypochlorite** Brillocera
Desinfectante de Semillas IF **lindane + maneb** Ficoop
desmedipham H [13684-56-5] ethyl 3-phenylcarbamoyloxyphenylcarbamate, desmediphame, Betanal AM, Betanex, SN 38107
desmediphame H **desmedipham**
Desmel F **propiconazole** Ciba-Geigy
desmetryn H [1014-69-3] 2-isopropylamino-4-methylamino-6-methylthio-1,3,5-triazine, desmetryne, G 34360, Samuron, Semeron, Topusyn
desmetryne H **desmetryn**
Desormona H **2,4-D** Condor
Desormone H **2,4-D** Pepro, Ravit, Rhone-Poulenc
Desormone prairies H **2,4-D + dichlorprop + picloram** Pepro
Desormone total concentre H **2,4-D + dichlorprop** Pepro
Despirol I **kelevan** Sandoz
Dessin AF **dinobuton** Union Carbide
Desteral H **2,4-D** Mitsotakis
Destox H **2,4-D** Campbell
Destralin I **chlorpyrifos** Riwa
Destructor "88" H **sodium chlorate** Kolvoet-Lecomte
Destun H **perfluidone** Efthymiadis, Young Il
Detamide X **diethyltoluamide**
Dethion I **parathion** Denka
Dethmor R **warfarin** Ellagret
Dethnel R **warfarin**
Detia Alfos R **aluminium phosphide** Nicholas-Mepros
Detia Ameisenkoederdose 'TC' I **trichlorfon** Detia Freyberg
Detia Ameisenpuder I **diazinon** Detia Freyberg
Detia Ameisenpuder I **lindane** Delitia, Detia Freyberg
Detia Ameisenpuder Neu I **diazinon** Delitia
Detia Beutelrolle I **aluminium phosphide** Delitia, Vorratsschutz
Detia BIO Universal-Staub IA **piperonyl butoxide + pyrethrins** Delitia, Detia Freyberg
Detia Dimecron IA **phosphamidon** Delitia, Detia Freyberg
Detia Ex-B I **aluminium phosphide** Verfaillie-Elsig
Detia fosfin-pelete IA **aluminium phosphide** Radonja
Detia Frass-Ratron R **warfarin** Delitia, Detia Freyberg
Detia Gas-Ex-B I **aluminium phosphide**

Delitia, Detia Freyberg, Grima, ICI-Zeltia, Viopharm, Vorratsschutz
Detia Gas-Ex-M I **methyl bromide** Vorratsschutz
Detia Gas-Ex-P I **aluminium phosphide** Delitia, Viopharm, Vorratsschutz
Detia Gas-Ex-T IR **aluminium phosphide** Ciba-Geigy, Delitia, Protekta, Viopharm, Vorratsschutz
Detia Giftkoerner R **zinc phosphide** Delitia, Detia Freyberg
Detia Giftox R **zinc phosphide** Delitia, Detia Freyberg
Detia Insekten-Strip I **dichlorvos** Delitia, Vorratsschutz
Detia Kartoffel Keimfrei G **propham** Delitia
Detia Kornmotten-Gas-Ex I **aluminium phosphide** Delitia, Vorratsschutz
Detia Maeusegiftkoerner R **zinc phosphide** Detia Freyberg
Detia Maeusekoeder R **warfarin** Delitia, Detia Freyberg
Detia Pellets I **aluminium phosphide** Grima, ICI-Zeltia, Toufruits
Detia Pflanzen-Schutzoel IA **petroleum oils** Delitia, Detia Freyberg
Detia Pflanzol-Spray IA **dichlorvos + piperonyl butoxide + pyrethrins** Delitia, Detia Freyberg
Detia Phosphine pellets IA **aluminium phosphide** Protekta
Detia Pilzol SZ F **sulphur + zineb** Delitia, Detia Freyberg
Detia Pyrethrum-Emulsion I **piperonyl butoxide + pyrethrins + rotenone** Detia Freyberg
Detia Rasenrein H **dicamba + MCPA** Delitia, Detia Freyberg
Detia Ratron R **warfarin** Delitia, Detia Freyberg
Detia Rattekal Giftpaste R **zinc phosphide** Delitia, Detia Freyberg
Detia Ratten-Frassratron R **warfarin** Detia Freyberg
Detia Raumnebel P I **piperonyl butoxide + pyrethrins** Delitia, Vorratsschutz
Detia Raumnebel V I **dichlorvos** Delitia, Vorratsschutz
Detia Rosen- und Zierpflanzenspray FI **butocarboxim + fenarimol** Delitia
Detia Schneckenband M **metaldehyde** Delitia, Detia Freyberg
Detia Schneckenkorn M **metaldehyde** Delitia
Detia Staeubol-Kombi Puder I **bromophos** Delitia, Detia Freyberg
Detia Total H **bromacil + diuron** Delitia, Detia Freyberg
Detia Universal IA **piperonyl butoxide + pyrethrins** Delitia, Detia Freyberg
Detia Werrenpraeparat I **lindane** Delitia, Detia Freyberg
Detia Wuehlmaus-Gas R **aluminium phosphide** Detia Freyberg
Detia Wuehlmauskiller R **aluminium phosphide** Delitia, Detia Freyberg
Detia Wuehlmauskoeder R **zinc phosphide** Delitia, Detia Freyberg
Detia Zierpflanzenspray FI **butocarboxim + fenarimol** Detia Freyberg
Detiaphos IR **magnesium phosphide** Ciba-Geigy, Delitia, Protekta, Vorratsschutz
Detmol I **dichlorvos + piperonyl butoxide + pyrethrins** Frowein, Leu & Gygax, Riwa
Detmol M I **methoxychlor + piperonyl butoxide + pyrethrins** Riwa
Detmol MD I **fenpropimorph + propiconazole** Riwa
Detmol-concentraat DU I **fenpropimorph + propiconazole** Riwa
Detmol-concentraat MA I **malathion** Riwa
Detmol-concentraat PER I **permethrin + pyrethrins** Riwa
Detmol-Konzentrat PY I **pyrethrins** Frowein
Detmol-long I **permethrin + piperonyl butoxide + pyrethrins** Riwa
Detmol-mal IA **malathion** Frowein
Detmol-Rauch I **lindane** Frowein, Muehlberger
Detmolin F I **dichlorvos + piperonyl butoxide + pyrethrins** Riwa
Detmolin M I **dichlorvos + malathion** Frowein
Detmolin P I **piperonyl butoxide + pyrethrins** Frowein, Riwa
Detruirats Agricola R **warfarin** Mourao
Devatern G **chlorpropham + propham** Shell
Devep locaux I **dichlorvos** C.G.I.
Devep mala IA **dichlorvos + malathion** C.G.I.
Devevap dampemiddel I **dichlorvos** Bayer
Devigon IA **dimethoate**
Devrinol H **napropamide** Agriben, Avenarius, Cillus, Efthymiadis, OHIS, Pepro, Serpiol, Siegfried, Spiess, Stauffer, Urania

Devrinol Goal H **napropamide + oxyfluorfen** Serpiol
Devrinol plus H **metazachlor + napropamide** Siegfried
Devrinol super H **napropamide + trifluralin** Serpiol
Dextrone X H **paraquat dichloride** Chipman
Dexuron H **diuron + paraquat dichloride** Chipman
DeZaeta F **mancozeb** Rohm & Haas
DFF H **diflufenican**
Di focap F **captafol + captan + folpet** Chevron
di-allate H [2303-16-4] *S*-2,3-dichloroallyl di-isopropyl(thiocarbamate), diallate, Avadex, CP 15336
Di-Farmon H **dicamba + mecoprop** Farm Protection
Di-on H **diuron** Makhteshim-Agan
Di-Syston IA **disulfoton** Bayer
Di-Thios I **parathion** Chimac-Agriphar
Di-Trapex FIHN **D-D + methyl isothiocyanate** Gullviks, Huhtamaki, Kwizda, Schering
Di-Trapex CP F **chloropicrin + D-D + methyl isothiocyanate** Schering
Dia-min H **amitrole + diuron** Safor
Diacide IA **diazinon** Gaillard
Diacit IA **diazinon** Sivam
Diacon I **methoprene** Sandoz
Diadon H **atrazine + diuron + picloram** Chipman
Diagrenon I **diazinon** Sadisa
Diagrex H **diuron** Sadisa
Dial IA **dimethoate** Aziende Agrarie Trento
dialifor IA **dialifos**
dialifos IA [10311-84-9] *S*-2-chloro-1-phthalimidoethyl *O,O*-diethyl phosphorodithioate, dialiphos, dialifor, ENT 27320, Hercules 14503, Torak
dialiphos IA **dialifos**
Diamal I **malathion** Diana
Diametan F **cymoxanil + propineb** Bayer
Diametan B F **cymoxanil + propineb + triadimefon** Bayer
diamidafos N [1754-58-1] phenyl *N,N'*-dimethylphosphorodiamidate, Dowco 169, Nellite
Diaminoagrex H **amitrole + diuron** Sadisa
Dianal H **phenmedipham** Hermoo
dianat H **dicamba**
Dianazil I **diazinon** Diana
Dianex I **methoprene** Sandoz
Dianex H **diuron** Diana
Dianoil I **parathion + petroleum oils** Diana
Dianoram H **molinate** Diana
Dianosin H **ametryn + atrazine** Diana
Diantin F **oxycarboxin** Du Pont
Diapadrin IA **dicrotophos** KenoGard
Diaprop F **propineb** Montedison
Diaprop cobra F **calcium copper oxychloride + propineb** Montedison
Diaract I **teflubenzuron** Celamerck
Diaryl H **amitrole + ammonium thiocyanate + diuron** Hellenic Chemical
Diasaat FI **captan + diazinon + thiram** Leu & Gygax
Diaser IA **diazinon** Sepran
Diastinon I **diazinon** Solfotecnica
Diater H **diuron** Condor, Rhone-Poulenc
Diater I **diazinon** Visplant
Diathio I **fenitrothion** Diana
Diaz I **diazinon** Chemia
Diaziben IA **diazinon** ERT
Diazicoop I **diazinon** Ficoop
Diazide G **daminozide** Brinkman
Diazikey IA **diazinon** Key
Diazilan I **diazinon** Aragonesas
Diazin IA **diazinon** Chemie, Eurofyto, Gefex
Diazinate HI **diazinon + molinate** Siapa
Diazinfor IA **diazinon** Safor
diazinon IA [333-41-5] *O,O*-diethyl *O*-2-isopropyl-6-methylpyrimidin-4-yl phosphorothioate, OMS 469, ENT 19507, Basudin, D-264, D-Z-N, Dazzel, Diazitol, Diazol, Disonex, Drawizon, G 24480, Gardentox, Kayazinon, Kayazol, Knox-out, Neocidol, Nipsan, Nucidol, Sarolex, Spectracide
Diazipol IA **diazinon** Afrasa
Diazithion I **diazinon** Sipcam
Diazithrine IA **diazinon** Pennwalt France
Diazitol IA **diazinon** Ciba-Geigy
Diazol IA **diazinon** Alpha, Budapesti Vegyimuvek, Edefi, Makhteshim-Agan, Terranalisi
Diazole TA H **amitrole + ammonium thiocyanate** J.S.B.
Diazolin I **diazinon** Eurofyto, Hellenic Chemical
Diazon IA **diazinon** Aveve
Diban H **dicamba + dichlorprop + MCPA** Maataloispalvelu
Diblecar "MN" FI **lindane + mancozeb** La Littorale
Dibrom IA **naled** Chevron, Geopharm

Dibrome NI **ethylene dibromide** United Phosphorus
dibromochloropropane N [96-12-8] 1,2-dibromo-3-chloropropane, DBCP, Fumazone, Nemagon, Nematocide
dibromure d'ethylene NI **ethylene dibromide**
Dibudrin I **dinoseb** Kwizda
Dibutagrex AF **dinobuton** Sadisa
Dibutagrex T AF **dinobuton + tetradifon** Sadisa
Dibutox H **dinoseb** Fagaras
S,S-di-sec-butyl O-ethyl phosphorodithioate IN [95465-99-9] Apache, FMC 67825, Rugby, Taredan
Dicalon H **dicamba + dichlorprop + MCPA** Esbjerg
dicamba H [1918-00-9] 3,6-dichloro-*o*-anisic acid, MDBA, dianat, Banvel, Fallowmaster, Mediben, Velsicol 58-CS-11
Dicamba M H **dicamba + MCPA** Shell
Dicamba P H **dicamba + mecoprop** Shell
Dicamex H **dicamba + mecoprop** Protex
Dicap F **dinocap** Scam
dicapthon I [2463-84-5] *O*-2-chloro-4-nitrophenyl *O,O*-dimethyl phosphorothioate, OMS 214, ENT 17035, Dicapton, Experimental Insecticide 4124
Dicapton I **dicapthon**
Dicarbam I **carbaryl** BASF, Intrachem, Sapec
Dicarfol F **captafol + folpet** Visplant
Dicaron A **dicofol** Bimex
Dicarzol AI **formetanate hydrochloride** Schering
Dicazin H **atrazine + dicamba** Siegfried
dichlobenil H [1194-65-6] 2,6-dichlorobenzonitrile, DBN, Barrier, Casoron, Decabane, Du Cason, Dyclomec, H 133, Norosac, Prefix D
dichlofenthion IN [97-17-6] *O*-(2,4-dichlorophenyl) *O,O*-diethyl phosphorothioate, ECP, dichlorofenthion, dichlofention, Diclophenthion, Tri VC 13, VC 13 Nemacide
dichlofention IN **dichlofenthion**
dichlofluanid F [1085-98-9] *N*-dichlorofluoromethylthio-*N',N'*-dimethyl-*N*-phenylsulphamide, dichlofluanide, dichlorfluanid, Bay 47531, Elvaron, Euparen, Euparene, Ku 13-032-c
dichlofluanide F **dichlofluanid**
dichlone FL [117-80-6] 2,3-dichloro-1,4-naphthoquinone, ENT 3776, Phygon, Quintar, USR 604
Dichlor mala IA **dichlorvos + malathion** C.G.I.
dichloralurea H **DCU**
dichloran F **dicloran**
dichlorane F **dicloran**
dichlorfenidim H **diuron**
dichlorfluanid F **dichlofluanid**
dichlorflurecol-methyl G [21634-96-8] 2,7-dichloro-9-hydroxyfluorene-9-carboxylic acid, dichlorflurenol-methyl, dichloroflurenol-methyl, IT-5733
dichlorflurenol-methyl G **dichlorflurecol-methyl**
dichlorfos IA **dichlorvos**
dichlormate H [1966-58-1] 3,4-dichlorobenzyl methylcarbamate, Rowmate, UC 22463
dichlormid S [37764-25-3] *N,N*-diallyl-2,2-dichloroacetamide, R-25788
dichlorofenthion IN **dichlofenthion**
dichloroflurenol-methyl G **dichlorflurecol-methyl**
dichlorophen LFB [97-23-4] 4,4'-dichloro-2,2'-methylenediphenol, dichlorophene, antiphen, Bio Moss Killer, G4, Mosstox, Mostox, Panacide, Super-Mosstox
dichlorophene LFB **dichlorophen**
3,6-dichloropicolinic acid H **clopyralid**
1,3-dichloropropene N [542-75-6] 1,3-dichloropropene, dichloro-1,3-propene, Dorlone II, Telone II
dichlorprop HG [120-36-5] 2-(2,4-dichlorophenoxy)propionic acid, 2,4-DP, Cornox RK, Dikofag DP, Dyprop, Hedonal, Polymone, RD 406, U 46 DP Fluid
dichlorvos IA [62-73-7] 2,2-dichlorovinyl dimethyl phosphate, DDVP, dichlorfos, OMS 14, ENT 20738, Apavap, Atgard, Bay 19149, Benfos, Canogard, Cekusan, Dedevap, Denkavepon, Divipan, Equigard, Mafu, Marvex, Nerkol, Nogos, Nuvan, Oko, Phosvit, Task, Vapona, Vaponite
Dichlotox I **dichlorvos** Bourgeois
dichlozolinate F **chlozolinate**
dichlozoline F [24201-58-9] 3-(3,5-dichlorophenyl)-5,5-dimethyl-1,3-oxazolidine-2,4-dione, DDOD, CS 8890, Ortho 8890, Sclex
Dicid FIHN **metham-sodium** Bitterfeld
Dicil H **bromacil + diuron** Resoco
diclobutrazol F [75736-33-3] (2*RS*,3*RS*)-1-(2,4-dichlorophenyl)-4,4-dimethyl-2-(1*H*-1,2,4-triazol-1-yl)pentan-3-ol, PP 296, Vigil

diclofop-methyl H [51338-27-3] methyl (*RS*)-2-[4-(2,4-dichlorophenoxy)phenoxy]propionate, Hoe 23408, Hoe-Grass, Hoegrass, Hoelon, Illoxan, Iloxan
Diclophenthion IN **dichlofenthion** Sintesul
dicloran F [99-30-9] 2,6-dichloro-4-nitroaniline, dichloran, dichlorane, CNA, DCNA, ditranil, Allisan, Botran, Kiwi Lustr, Marisan, RD 6584, Resisan, Scleran, Sclerosan
Diclorcal I **dichlorvos** Calliope
Diclordon H **2,4-D** Borzesti
Diclorfan I **trichlorfon** Aragonesas
Diclotir F **dicloran + thiram** Sivam
Dico-Banvel-M H **dicamba + MCPA** Agro-Kemi
Dicofen I **fenitrothion** Pan Britannica
Dicofluid-DP H **dichlorprop** Zorka Subotica
Dicofluid MP-combi H **2,4-D + mecoprop** Ruse
dicofol A [115-32-2] 2,2,2-trichloro-1,1-bis(4-chlorophenyl)ethanol, kelthane, ENT 23648, Acarin, Cekudifol, FW-293, Hilfol, Kelthane, Mitigan
Dicofol Doble A **dicofol + tetradifon** Shell
Diconal H **phenisopham**
Diconirt H **2,4-D** Phytorgan
Diconox F **copper oxychloride + mancozeb** Masso
Diconox Plus F **chlorothalonil + maneb** Masso
Dicontal I **fenitrothion + trichlorfon** Agroplant, Bayer
Dicophyt A **dicofol** Sipcam-Phyteurop
Dicoprime H **bromofenoxim** Ciba-Geigy
Dicopur H **2,4-D** Agrolinz
Dicopur DP H **dichlorprop** Agrolinz
Dicopur M H **MCPA** Agrolinz, S/48
Dicopur U 46 KV H **mecoprop** Agrolinz
Dicopur U 46 KV neu H **2,4-D + mecoprop** Agrolinz
Dicotex H **2,4-D + dicamba + MCPA + mecoprop** Protex
Dicotex royal H **dicamba + ioxynil + MCPA + mecoprop + 2,4,5-T** Graines Loras
Dicothion IA **dicofol + parathion-methyl** Interphyto
Dicotox Extra H **2,4-D** May & Baker
Dicotox-D H **2,4-D** Agro-Kemi
Dicotox-M H **MCPA** Agro-Kemi
Dicoveex A **dicofol** Agriplan
Dicron IA **dicrotophos** Hui Kwang
dicrotophos IA [141-66-2] (*E*)-2-dimethylcarbamoyl-1-methylvinyl dimethyl phosphate, OMS 253, ENT 24482, Bidrin, C 709, Carbicron, Diapadrin, Dicron, Ektafos, SD 3562
Dicryl H **chloranocryl**
Dictril H **ioxynil + mecoprop** Rhodic
Dicuran H **chlorotoluron** Budapesti Vegyimuvek, Chromos, Ciba-Geigy, Ligtermoet, Liro, Scam, VCH
Dicuran Combi H **chlorotoluron + dipropetryn** Viopharm
Dicuran extra H **chlorotoluron + terbutryn** Ciba-Geigy
Dicuran Prop H **chlorotoluron + dichlorprop** Budapesti Vegyimuvek
Dicuran Special H **chlorotoluron + methoprotryne + simazine** Ciba-Geigy
Dicurane Duo H **bifenox + chlorotoluron** Ciba-Geigy
Dicusat R **warfarin** Chimiberg, Ferrosan
Dicusat M R **chlorophacinone** Chimiberg
dicyclopentadiene X [77-73-6] 3a,4,7,7a-tetrahydro-4,7-methano-1*H*-indene
Didifum FIHN **D-D** Chemia
Didivane I **dichlorvos** Chimiberg
Diefesan FB **captan + carboxin + 8-hydroxyquinoline sulphate** Sepran
Diefital FB **captan + carboxin + 8-hydroxyquinoline sulphate** Ital-Agro
Dieldrex I **dieldrin** Shell
dieldrin I [60-57-1] 1,2,3,4,10,10-hexachloro-1*R*,4*S*,4a*S*,5*R*,6*R*,7*S*,8*S*,8a*R*-octahydro-6,7-epoxy-1,4:5,8-dimethanonaphthalene, HEOD, dieldrine, OMS 18, ENT 16225, Compound 497, Dieldrex, Dieldrite, Octalox, Panoram D-31
dieldrine I **dieldrin**
Dieldrite I **dieldrin** Shell
dienochlor A [2227-17-0] perchloro-1,1'-bicyclopenta-2,4-dienyl, dienochlore, ENT 25718, Pentac
dienochlore A **dienochlor**
Dienox A **dienochlor** Protex
diethamquat H 1,1'-bis(diethylcarbamoylmethyl)-4,4'-bipyridinium, PP 831
diethatyl-ethyl H [38727-55-8] *N*-(chloroacetyl)-*N*-(2,6-diethylphenyl)glycine ethyl ester, Antor, Hercules 22234

diethion AI **ethion**
diethofencarb F [87130-20-9] isopropyl 3,4-diethoxycarbanilate, dietofencarb, S-1605, S-32165
diethyltoluamide X [134-62-3] *N,N*-diethyl-*m*-toluamide, DEET, ENT 20218, Detamide, Metadelphene, Off
dietofencarb F **diethofencarb**
Dietol IA **dimethoate** Sivam
Difar I **dichlorvos** Zorka Sabac
Difen H **diphenamid** Sipcam
difenacoum R [56073-07-5] 3-(3-biphenyl-4-yl-1,2,3,4-tetrahydro-1-naphthyl)-4-hydroxycoumarin, Neosorexa, Ratak, Ratrick, Vorex
difenamide H **diphenamid**
Difenex H **chlomethoxyfen** Ishihara Sangyo
Difenkryl olie emulsion AF **petroleum oils** Midol
difenoxuron H [14214-32-5] 3-[4-(4-methoxyphenoxy)phenyl]-1,1-dimethylurea, C 3470, Lironion
difenzoquat methyl sulphate H [43222-48-6] 1,2-dimethyl-3,5-diphenylpyrazolium methyl sulphate, AC 84777, Avenge, Finaven, Yeh-Yan-Ku
Differat R **difenacoum** Duratox
Diffetox R **difenacoum** Duratox
diflubenzuron I [35367-38-5] 1-(4-chlorophenyl)-3-(2,6-difluorobenzoyl)urea, difluron, OMS 1804, ENT 29054, Astonex, Dimilin, DU 112307, Larvakil, PDD 60-40-I, PH 60-40, TH6040
diflufenican H [83164-33-4] 2',4'-difluoro-2-(*α,α,α*-trifluoro-*m*-tolyloxy)nicotinanilide, diflufenicanil, DFF, M&B 38544
diflufenicanil H **diflufenican**
Diflur H **linuron + trifluralin** Visplant
difluron I **diflubenzuron**
Difol A **dicofol** Hellenic Chemical
Difolatan F **captafol** Chemia, Chevron, Geopharm, Maag, Terranalisi
Difolpet F **captafol + folpet** Terranalisi
Difoltox F **captafol** Caffaro
Difon A **dicofol + tetradifon** Peppas
Difonat I **fonofos** OHIS
Difontan F **carbendazim + pyrazophos** Hoechst
Diforam F **captafol + copper oxychloride** EniChem
Difosan F **captafol + maneb** Agriben
Difosan FLO F **captafol** Pepro
Digatox H **bromacil + diuron** Schering
Digermin H **trifluralin** Agrimont, La Littorale, Masso, Montedison, Staehler
Digitex Agrumi H **carbaryl + propanil** Enotria
Digmar I **DDT** All-India Medical
Digor I **dimethoate** Hoechst
Digrain I **dichlorvos** Lodi
Digrain 4 IA **dichlorvos + malathion** Lodi
Digrain 7 I **dichlorvos** Lodi
Digrain 360 I **bromophos** Lodi
Digrain concentre IA **dichlorvos + malathion** Lodi
diisocarb H **butylate**
Dikamba-Trippel H **dicamba + MCPA + mecoprop** Lantmaennen
Dikamin D H **2,4-D** Nitrokemia
Dikar F **dinocap + mancozeb** Rohm & Haas
dikegulac-sodium G [52508-35-7] sodium 2,3:4,6-di-*O*-isopropylidene-*α*-L-*xylo*-2-hexulofuranosonate, Atrimmec, Atrinal, Cutlass, Ro 07-6145
Diklo-Hormo H **dichlorprop + MCPA** Maatalouspalvelu
Dikocid H **2,4-D** Radonja
Dikofag H **2,4-D** Hoechst
Dikofag DP HG **dichlorprop** Hoechst
Dikofag Kombi H **2,4-D + MCPA** Hoechst
Dikofag KV H **mecoprop** Hoechst
Dikofag KV universal H **2,4-D + mecoprop** Hoechst
Dikofag MP Kombi fluessig H **2,4-D + mecoprop** Hoechst
Dikofag P H **mecoprop** Hoechst
Dikogren H **MCPA + terbutryn** Dimitrova
Dikogren Special H **MCPA + mecoprop + terbutryn** Dimitrova
Dikolen I H **MCPA** Dimitrova
Dikonirt H **2,4-D** Chemolimpex, Nitrokemia
Dikopan H **dalapon-sodium** Pinus
Dikotex H **MCPA** Dimitrova
Dilethol IA **dimethoate** Truffaut
Dillex H **dinoseb acetate** Kwizda
Dilpex F **carbendazim** Sipcam-Phyteurop
Diluq IA **dimethoate** Luqsa
Dim IA **dimethoate** Chemia
Dim-pynamin I **piperonyl butoxide + tetramethrin** Maansson
Dimafir I **phosphamidon** EniChem
Dimafon I **trichlorfon** Agrocros
Dimaneb F **maneb** Foret
Dimanin L **quaternary ammonium** Bayer

Dimate IA **dimethoate** Bourgeois

Dimatrol H **amitrole + diuron** Formulex

Dimatrol Special H **amitrole + 2,4-D + diuron** Formulex

Dimax IA **dichlorvos + malathion** Hecquet

Dimeaat IA **thiometon** Duphar

Dimecron IA **phosphamidon** Ciba-Geigy, Du Pont, Ligtermoet, Liro, Nitrokemia, OHIS, Spolana

Dimefos I **dimethoate** Hellenic Chemical

dimefox IA [115-26-4] tetramethylphosphorodiamidic fluoride, Hanane, Pestox XIV, Terra-Systam, Wacker S 14/10

dimefuron H [34205-21-5] *N'*-[3-chloro-4-[5-(1,1-dimethylethyl)-2-oxo-1,3,4-oxadiazol-3(2*H*)-ylphenyl]-*N,N*-dimethylurea, 23465 RP

Dimelfan I **dimethoate + endosulfan** Scam

dimelon-methyl H **methyldymron**

Dimepax H **dimethametryn** Ciba-Geigy

dimephenthoate IA **phenthoate**

dimepiperate H [61432-55-1] *S*-(1-methyl-1-phenylethyl) 1-piperidinecarbothioate, MUW 1193, MY 93, Yukamate

Dimepor I **dimethoate** Sopepor

Dimet I **parathion-methyl** Eli Lilly

Dimetate IA **dimethoate**

dimethachlor H [50563-36-5] 2-chloro-*N*-(2-methoxyethyl)acet-2',6'-xylidide, dimethachlore, CGA 17020, Teridox

dimethachlore H **dimethachlor**

dimethametryn H [22936-75-0] N^2-(1,2-dimethylpropyl)-N^4-ethyl-6-methylthio - 1,3,5-triazine-2,4-diamine, dimethametryne, C 18898, Dimepax

dimethametryne H **dimethametryn**

dimethazone H [81777-89-1] 2-(2-chlorobenzyl)-4,4-dimethyl-1,2-oxazolidin-3-one, fenoxan, clomazone, Command, FMC 57020

dimethipin HG [55290-64-7] 2,3-dihydro-5,6-dimethyl-1,4-dithi-ine 1,1,4,4-tetraoxide, oxydimethin, Harvade, N 252

dimethirimol F [5221-53-4] 5-butyl-2-dimethylamino-6-methylpyrimidin-4-ol, dimethyrimol, Milcurb, PP 675

dimethoate IA [60-51-5] *O,O*-dimethyl *S*-methylcarbamoylmethyl phosphorodithioate, fosfamid, OMS 94, OMS 111, ENT 24650, Bi 58, Cekuthoate, Chemathoate, Cygon, Daphene, De-Fend, Demos L, Devigon, Dimetate, Dimethoate, Dimethogen, EI 12880, L 395, Perfekthion, Rebelate, Rogor, Roxion, Trimetion

dimethoate-met IA **omethoate**

Dimethogen IA **dimethoate**

Dimethol I **dimethoate** Ellagret

Dimethon IA **dimethoate** Key

Dimethoxal I **dimethoate** Peppas

Dimethugec IA **dimethoate** Sipcam-Phyteurop

dimethyl phthalate X [131-11-3] dimethyl 1,2-benzenedicarboxylate, DMP, ENT 262

dimethylarsinic acid H [75-60-5] dimethylarsinic acid, cacodylic acid, Bolls-Eye, Clean-Boll, Dutch-Treat, Ezy Pickin, Kack, Phytar, Rad-E-Cate, Salvo

dimethylvinfos I **dimethylvinphos**

dimethylvinphos I [2274-67-1] 2-chloro-1-(2,4-dichlorophenyl)ethenyl dimethyl phosphate, dimethylvinfos, Rangado, SD 8280

dimethyrimol F **dimethirimol**

dimetilan I [644-64-4] 1-dimethylcarbamoyl-5-methylpyrazol-3-yl dimethylcarbamate, Snip

Dimetiox I **parathion-methyl** Caffaro

Dimetol I **parathion-methyl** Terranalisi

Dimetrone IA **dimethoate** Agronova

Dimevur IA **dimethoate** Sinteza

Dimezyl I **dimethoate** Agriphyt

dimidazon H [3295-78-1] 4,5-dimethoxy-2-phenylpyridazin-3(2*H*)-one, dimidazone, BAS 255H

dimidazone H **dimidazon**

Dimilin I **diflubenzuron** Agroplant, Agros, Du Pont, Duphar, Ewos, Galenika, ICI, KVK, LaQuinoleine, Maag, Quimigal, Schering, Sodafabrik, Zootechniki

Dimite A **chlorfenethol** Sherwin-Williams

Dimitrox I **dimethoate** Eurofyto

Dimop A **dicofol** Marty et Parazols

Dimox I **dioxacarb** Chimitox

dimuron H **dymron**

Dinethon H **etinofen**

dinex AIM [131-89-5] 2-cyclohexyl-4,6-dinitrophenol, DNCP, DNOCHP, pedinex, DN-111

diniconazole F [76714-88-0] (*E*)-(*RS*)-1-(2,4-dichlorophenyl)-4,4-dimethyl-2-(1*H*)-1,2,4-triazol-1-yl)pent-1-en-3-ol, Ortho Spotless, S-3308L, Spotless, Sumi-8, Sumi-eight, XE-779L

Dinitiol I **malathion** Caffaro

dinitramine H [29091-05-2] N^1,N^1-diethyl-2,6-dinitro-4-trifluoromethyl-*m*-phenylenediamine, Cobex, Cobexo, USB 3584

Dinitrex H **bromoxynil** Burri
Dinitrex Combi H **bromoxynil + ioxynil** Burri
Dinitro HI **dinoseb** Vertac
Dinitro HI **DNOC** CTA
Dinitrodendroxal I **DNOC + petroleum oils** Hellenic Chemical
Dinitrol I **DNOC** Staehler
dinobuton AF [973-21-7] 2-*sec*-butyl-4,6-dinitrophenyl isopropyl carbonate, OMS 1056, ENT 27244, Acrex, Dessin, Dinofen, Drawinol, MC 1053, Talan
dinocap FA [39300-45-3] 2(or 4)-(1-methylheptyl)-4,6(or 2,6)-dinitrophenyl crotonate, DPC, DNOCP, ENT 24727, Caprane, Cekucap, CR-1693, Crotothane, Ezenosan, Fenocap, Karathane
Dinocruz FA **dinocap** KenoGard
dinocton AF [32534-96-6; 32535-08-3] methyl 2,6-dinitro-4-octylphenyl carbonates/methyl 2,4-dinitro-6-octylphenyl carbonates, MC 1945, MC 1947
Dinofane H **dinoseb** Pepro
Dinofen AF **dinobuton** Margesin
Dinogil FA **dinocap** Rhone-Poulenc
Dinograne H **chlomethoxyfen + neburon** Sopra
Dinokar F **dinocap** Visplant
Dinokey F **dinocap** Key
Dinopec I **dinoseb** Sapec
Dinopron A **dinobuton** BP
Dinoquat H **paraquat dichloride** Diana
dinoseb HI [88-85-7] 2-*sec*-butyl-4,6-dinitrophenol, dinosebe, DNBP, ENT 1122, Basanite, Butyl-Gelb, Caldon, Chemox, Dinitro, DN 289, Dynamyte, Gebutox, Haulmone, Hivertox, Phytoxone, Premerge, Subitex
dinoseb acetate H [2813-95-8] 2-*sec*-butyl-4,6-dinitrophenyl acetate, dinosebe acetate, DNBPA, Aretit, Hoe 02904, Ivosit, Phenotan
dinoseb methacrylate AF **binapacryl**
dinosebe HI **dinoseb**
Dinosip F **dinocap** Sipcam
Dinosivam F **dinocap** Sivam
dinoterb H [1420-07-1] 2-*tert*-butyl-4,6-dinitrophenol, dinoterbe, DNTBP, Herbogil
dinoterb acetate HA [3204-27-1] 2-*tert*-butyl-4,6-dinitrophenyl acetate, dinoterbe-acetate, P 1108
dinoterbe H **dinoterb**
Dinovap I **dichlorvos** Diana
Dinoveex FA **dinocap** Agriplan
Dinox super H **2,4-D + mecoprop** Chimac-Agriphar
Dinugec H **dinoseb** Sipcam-Phyteurop
Dinurex H **diuron** Sipcam-Phyteurop
Dinzinon I **diazinon** Zorka Subotica
Dioksakarb I **dioxacarb** Zorka Sabac
Diostop IA **dimethoate** Schering
Dioweed H **2,4-D** Campbell
dioxabenzofos I [3811-49-2] 2-methoxy-4*H*-1,3,2λ[5]-benzodioxaphosphinine 2-sulphide, salithion, fenphosphorin, Salithion
dioxacarb I [6988-21-2] 2-(1,3-dioxolan-2-yl)phenyl methylcarbamate, dioxacarbe, OMS 1102, ENT 27389, C 8353, Elocron, Famid
dioxacarbe I **dioxacarb**
dioxane phosphate IA **dioxathion**
dioxathion IA [78-34-2] *S,S'*-(1,4-dioxane-2,3-diyl) *O,O,O',O'*-tetraethyl di(phosphorodithioate), dioxane phosphate, ENT 22879, Delnav, Deltic, HerculesAC258
Dipagrex I **trichlorfon** Sadisa
Diparat H **paraquat dichloride** Chemia
Dipel I **Bacillus thuringiensis** Abbott, English Woodlands, Ledona, Perifleur, Permutadora, Schering, Schering AAgrunol, Siegfried, Sipcam, Staehler
Dipet F **folpet** Caffaro
diphacin R **diphacinone**
diphacinone R [82-66-6] 2-(diphenylacetyl)indan-1,3-dione, diphacins, diphenadione, diphacin, Parakakes, PCQ, Ramik, Rodent Cake
diphacins R **diphacinone**
diphenadione R **diphacinone**
diphenamid H [957-51-7] *N,N*-dimethyl-2,2-diphenylacetamide, difenamide, Dymid, Enide, Fenam, L-34314, Rideon
Diphenex H **chlomethoxyfen** Ishihara Sangyo
diphenyl sulphone A [127-63-9] , diphenylsulphone, DPS
diphenylsulphide A **tetrasul**
Dipher F **zineb**
Dipiril H **paraquat dichloride** Afrasa
Dipro H **dichlorprop + MCPA** Kemira
Dipron H **diuron + propham** Caffaro
Dipronat H **dichlorprop** S/48
Diprop H **dichlorprop**

dipropetryn H [4147-51-7] 6-ethylthio-*N*2,*N*4-di-isopropyl-1,3,5-triazine -2,4-diyldiamine, dipropetryne, Cotofor, GS 16068, Sancap
dipropetryne H **dipropetryn**
Dipsol I **trichlorfon** Luqsa
Dipter plus I **naled + trichlorfon** Chromos
Dipterex I **trichlorfon** Agroplant, Bayer, Nihon Tokushu Noyaku Seizo, Pinus
Dipterex MR I **oxydemeton-methyl + trichlorfon** Bayer
Diptersivam I **trichlorfon** Sivam
Diptyl H **dicamba + MCPA + mecoprop** Agriphyt
diquat dibromide H [85-00-7; 6385-62-2 (monohydrate)] 1,1'-ethylene-2,2'-dipyridylium dibromide, deiquat dibromide, FB 2, Midstream, Reglex, Reglone, Reglox, Weedtrine-D
Dirado G **1-naphthylacetic acid** Tecniterra
Dirafix G **1-naphthylacetic acid** Margesin
Dirax R **antu**
Dirazol H **amitrole + diuron** Resoco
Dirazol X H **amitrole + bromacil** Resoco
Direfon G **ethephon** EniChem
Direx F **anilazine** Bayer
Direxin G **1-naphthylacetamide + (2-naphthyloxy)acetic acid** Farmer
Direz F **anilazine** Bayer
Dirigol N G **1-naphthylacetamide** Margesin, Siegfried
Dirimal H **oryzalin** Elanco
Disan H **bensulide** Stauffer
Disap H **EPTC** Agrimont
Disecat H **diquat dibromide** Visplant
Disefos I **dichlorvos** Sepran
Diserbagro MCP/30 H **MCPA** Ital-Agro
Diserbante H **mecoprop** Caffaro
Diserbante Omnia S H **sodium chlorate** Ital-Agro
2,4-Diserbin H **2,4-D** Caffaro
Diserbo Canali H **dalapon-sodium** Siapa
Diserbo Mais H **atrazine** Caffaro
Diserbo-STI H **dalapon-sodium** Solfotecnica
Diserbone E H **2,4-D** Chemia
Diserbone KN H **2,4-D + MCPA** Chemia
Disermix Cereali H **2,4-D + MCPA** Caffaro
Disitrin H **diuron + prometryn + simazine** Agrochemiki
Disonex IA **diazinon** Protex
Disseccante H **paraquat dichloride** Caffaro
Dissil X H **bromacil + diuron** Bayer
Distan D H **2,4-D + picloram** Shell
Disteron H **2,4-D + 2,4,5-T** Mitsotakis
disulfoton IA [298-04-4] *O,O*-diethyl *S*-2-ethylthioethyl phosphorodithioate, ethylthiodemeton, M-74, thiodemeton, dithiodemeton, ENT 23347, Bay 19639, Di-Syston, Disyston, Dithiosystox, Frumin AL, S 276, Solvigran, Solvirex
Disyston IA **disulfoton** Bayer
Disyston-S IA **oxydisulfoton**
Ditacap F **captan + ditalimfos** Dow
ditalimfos F [5131-24-8] *O,O*-diethyl phthalimidophosphonothioate, ditalimphos, Dowco 199, Farmil, Frutogard, Laptran, Leucon, Millie, Plondrel
ditalimphos F **ditalimfos**
Ditargos F **zineb** Argos
Ditene S **polyglycolic ethers** Key
Dithane 945 F **mancozeb** Chafer, Pan Britannica, Rohm & Haas
Dithane A-40 FL **nabam** Rohm & Haas
Dithane C 90 F **mancopper** Rohm & Haas
Dithane C 90 semences FI **lindane + mancopper** Rohm & Haas
Dithane Cupromix F **copper sulphate + mancozeb** Rohm & Haas
Dithane D-8 F **mancozeb** Osszefogas
Dithane D-8, K-50 F **mancozeb + sulphur** Osszefogas
Dithane DG F **mancozeb** La Quinoleine, Rohm & Haas
Dithane FL F **mancozeb** Rohm & Haas
Dithane flo F **mancozeb** Agrocros, Serpiol
Dithane LF F **mancozeb** KVK, Rhodiagri-Littorale, Rohm & Haas
Dithane LFT F **mancozeb** Rohm & Haas
Dithane M-22 F **maneb** Protex, Rohm & Haas
Dithane M-45 F **mancozeb** Argos, Broeste, Du Pont, Farmos, ICI-Valagro, KVK, La Quinoleine, Permutadora, Pinus, Plantevern-Kjemi, Protex, Quimagro, Ravit, Rhodiagri-Littorale, Rohm & Haas
Dithane M-70 F **mancozeb** Galenika
Dithane plavi F **mancozeb** Galenika
Dithane S-60 F **mancozeb** Galenika, ICI, Rohm & Haas
Dithane Ultra F **mancozeb** BASF, Ciba-Geigy, Dow, Hoechst, Maag, Spiess, Urania
Dithane Z-78 F **zineb** Permutadora, Protex, Rohm & Haas

dithianon F [3347-22-6] 5,10-dihydro-5,10-dioxonaphtho[2,3-*b*]-1,4-dithiin-2,3-dicarboni trile, Delan, Delan-Col, IT-931, MV 119A
dithio IA **sulfotep**
dithiodemeton IA **disulfoton**
Dithiomal IA **malathion** Agrocros
dithiomethon IA **thiometon**
dithione IA **sulfotep**
Dithiosystox IA **disulfoton** Mobay
Dithiozin F **zineb** Phytorgan
Ditil I **parathion-methyl** Sofital
Ditiozin F **zineb** Visplant
Ditiozol F **sulphur + zineb** Visplant
Ditiver C F **copper oxychloride** KenoGard
Ditiver Captan F **captan** KenoGard
Ditiver Doble F **copper oxychloride + zineb** KenoGard
Ditiver M-45 F **mancozeb** KenoGard
Ditiver MX F **dodine** KenoGard
Ditiver T FX **thiram** KenoGard
Ditiver Z F **zineb** KenoGard
Ditiver ZR F **ziram** KenoGard
Ditoneb F **maneb** Visplant
ditranil F **dicloran**
Ditreen N **D-D** Shell
Ditrifon I **trichlorfon** Budapesti Vegyimuvek, Chemolimpex
Ditrin I **carbaryl** Candilidis, Eli Lilly
Ditrix A **dicofol** Sipcam-Phyteurop
Ditrol H **amitrole + diuron** Hermoo
Ditrozan H **chlorotoluron** Eurofyto
Diucisar H **chlorpropham + diuron** EniChem
Diuranex H **diuron** Edefi
Diurex H **diuron** Makhteshim-Agan
Diurokey H **diuron** Key
diuron H [330-54-1] 3-(3,4-dichlorophenyl)-1,1-dimethylurea, DCMU, dichlorfenidim, DMU, Cekiuron, Crisuron, Decimax, Di-on, Diater, Diurex, Dynex, Karmex, Marmer, Ronex, Unidron, Vonduron
Diuron Super H **amitrole + diuron** Eurofyto
Diurosaf H **diuron** Safor
Diutrin H **diuron** Enotria
Divapan Potasio FIHN **metham-potassium** Foret
Divipan IA **dichlorvos** Makhteshim-Agan
Divopan* H **MCPB** Maag
Divutox I **dichlorvos** Terranalisi
Dizan H **chlorotoluron** Hoechst
Dizine H **diuron + simazine** Protex
Dizineb F **zineb** Foret
Diziram F **ziram** Foret
Dlapon H **dalapon-sodium** Chemia
DLG combi plaenerens H **2,4-D + dichlorprop** Esbjerg
DLG D-acetate 50 H **2,4-D** Esbjerg
DLG D-propcombi H **2,4-D + dichlorprop** Esbjerg
DLG D-propionat H **dichlorprop** Esbjerg
DLG D-prop-mix H **dichlorprop + MCPA** Esbjerg
DLG fluespray I **piperonyl butoxide + resmethrin** Esbjerg
DLG Fungazil F **imazalil** Esbjerg
DLG Kobber-oxychlorid F **copper oxychloride** Esbjerg
DLG Lindanbejdse I **lindane** Esbjerg
DLG M-acetat H **MCPA** Esbjerg
DLG M-propacid H **2,4-D + mecoprop** Esbjerg
DLG M-propionat H **mecoprop** Esbjerg
DLG plantedraeber H **simazine** Esbjerg
DLG Pyramin H **chloridazon** Esbjerg
DLG Roxion I **dimethoate** Esbjerg
DLG Staldfluedraeber I **piperonyl butoxide + resmethrin** Esbjerg
DM 68 H **dinoterb + mecoprop** Agriben, Pepro, Van Wesemael
DMC A **chlorfenethol**
DMDT I **methoxychlor**
DMP X **dimethyl phthalate**
DMPA H [299-85-4] *O*-2,4-dichlorophenyl *O*-methyl isopropylphosphoramidothioate, Dowco 118, K 22023, Zytron
DMSP NI **fensulfothion**
DMTP IA **methidathion**
DMTT NFHI **dazomet**
DMU H **diuron**
DN 65 F **dodine** EniChem
DN 111 AIM **dinex**
DN 289 HI **dinoseb** Dow
DNBP HI **dinoseb**
DNBPA H **dinoseb acetate**
DNC IAHF **DNOC**
DNCP AIM **dinex**
DNOC IAHF [534-52-1] 4,6-dinitro-*o*-cresol, DNC, Antinonnin, Chemsect, Cresofin, Elgetol, Extar-A, Extar-Lin, Nitrador, Sandoline, Selinon, Sinox, Trifina, Trifocide
DNOCHP AIM **dinex**
DNOCP FA **dinocap**
DNTBP H **dinoterb**
DO 14 A **propargite** Uniroyal

Docklene H **dicamba + MCPA + mecoprop** Schering
Docnol H **DNOC** Agronova
Doctor F **chlorothalonil + nuarimol** Eli Lilly
dodemorph acetate F [31717-87-0] 4-cyclododecyl-2,6-dimethylmorpholinium acetate, acetate de dodemorphe, BAS 238F, BASF-Mehltaumittel, Meltatox, Milban
Dodene F **dodine** Sipcam, Truchem
Dodex F **dodine** Agriplan, Aziende Agrarie Trento
Dodiagrex F **dodine** Sadisa
Dodiben F **dodine** ERT
Dodil F **dodine** Caffaro
Dodilan F **dodine** Hellenic Chemical
Dodim F **dodine** Quimigal
Dodina F **dodine** CAG, Chemia
dodine F [2439-10-3] 1-dodecylguanidinium acetate, doguadine, tsitrex, AC 5223, Carpene, CL 7521, Curitan, Cyprex, Efuzin, Melprex, Syllit, Venturol, Vondodine
Dodinol F **dodine** Sapec
Dodinox F **dodine** Organika-Azot
Dodival F **dodine** ICI-Valagro
Dodoscam F **dodine** Scam
Doff Ant Killer I **lindane** Doff-Portland
Doff Hormone Rooting Powder F **captan + 1-naphthylacetic acid** Doff-Portland
doguadine F **dodine**
Dojyopicrin NIFRH **chloropicrin**
Dokirin F **oxine-copper** Nihon Nohyaku
Doli Schneckenkorn M **metaldehyde** Melchior
Doluq F **dodine** Luqsa
Dom Saatschutzmittel F **thiram** Dom-Samen-Fehlemann
Dom Schneckenkorn M **metaldehyde** Dom-Samen-Fehlemann
Domatol H **amitrole + simazine** Urania
Domatol Spezial H **amitrole + MCPA + simazine** Urania
Dominex I **alphacypermethrin** FMC
Donatuskoerner R **zinc phosphide** Drogenhansa
Donatus-Schneckweg M **metaldehyde** Drogenhansa
Donatus Wuehlmausfertigkoeder R **warfarin** Drogenhansa
Dopax H **ametryn + metolachlor** Ciba-Geigy
Dorado F **pyrifenox** La Quinoleine, Maag
Dorex H **diuron** Bredologos
Dorin F **triadimenol + tridemorph** Bayer
Dorine Bio I **bioallethrin + resmethrin** Hoechst, Procida
Dorital IA **carbaryl + endosulfan** Ital-Agro
Dorlone II N **1,3-dichloropropene** Pepro
Dormone H **2,4-D** Burts & Harvey
Dorochlor NIFRH **chloropicrin** Mitsui Toatsu
Doruplant H **ametryn** Bitterfeld
Dorvert A **cyhexatin + tetradifon** Condor, Du Pont, Sandoz
Dorvos I **dichlorvos** Chemia
Dory-Mildiou pulverisation concentre FI **carbaryl + maneb** Sovilo
Dosaflo H **metoxuron** Farm Protection, ICI, Sandoz
Dosamix H **metoxuron + simazine** Urania
Dosanex H **metoxuron** Bjoernrud, Farmos, Hanson & Moehring, Organika-Azot, Sandoz, Schering, Shell, Spiess, Urania, VCH, Zupa
Dotan I **chlormephos** Condor, Pepro, Pliva, Ravit, Rhone-Poulenc
Dotix I **fenitrothion** Sedagri
Double A **dicofol + tetradifon** Agro-Kanesho
DoubleDown I **disulfoton + fonofos** ICI
Doublet H **bromoxynil + ioxynil + isoproturon** Agronorden
Dow Shield H **clopyralid** Dow
Dowco 118 H **DMPA**
Dowco 169 N **diamidafos**
Dowco 199 F **ditalimfos** Dow
Dowco 213 A **cyhexatin** Dow
Dowco 214 IA **chlorpyrifos-methyl** Dow
Dowco 217 I **fospirate**
Dowco 233 H **triclopyr** Dow
Dowco 269 F **pyroxychlor**
Dowco 290 H **clopyralid** Dow
Dowco 356 H **tridiphane** Dow
Dowco 417 I **fenpirithrin**
Dowco 433 H **fluroxypyr** Dow
Dowco 453 EE H **haloxyfop-ethoxyethyl** Dow
Dowfume NI **ethylene dibromide** Dow
Dowfume MC-2 FIHN **chloropicrin + methyl bromide** Dow
Dowpon H **dalapon-sodium** Aragonesas, Argos, Bjoernrud, Dow, Du Pont, Epro, Prochimagro, Procida, Schering, Shell
Dozeb F **dodine + mancozeb** Chemia
Dozer H **fenuron-TCA** Hopkins
DP H **dichlorprop** Dow, Europhyto, Schering
2,4-DP H **dichlorprop**

DPA H **dalapon-sodium**
DPC FA **dinocap**
DPS A **diphenyl sulphone**
DPX 1108 H **fosamine-ammonium** Du Pont
DPX 1410 IAN **oxamyl** Du Pont
DPX 3217 F **cymoxanil** Du Pont
DPX 3674 H **hexazinone** Du Pont
DPX 4189 H **chlorsulfuron** Du Pont
DPX 5648 H **sulfometuron-methyl** Du Pont
DPX 6376 H **metsulfuron-methyl** Du Pont
DPX F5384 H **bensulfuron-methyl** Du Pont
DPX F6065 H **chlorimuron-ethyl** Du Pont
DPX H6573 F **flusilazol** Du Pont
DPX M6316 H **thiameturon-methyl** Du Pont
DPX T6376 H **metsulfuron-methyl** Du Pont
DPX Y6202 H **quizalofop-ethyl** Du Pont
Draca A **cyhexatin + tetradifon** Sipcam-Phyteurop
Dragon I **permethrin** ICI
Drat R **chlorophacinone** Embetec, May & Baker
Dratex I **lindane** Bitterfeld
Drawifol F **metomeclan** Pluess-Staufer
Drawigran plus F **fenfuram + imazalil + thiabendazole** Dow
Drawigran Spezial F **maneb + quintozene + thiabendazole** Dow
Drawin I **butocarboxim** Argos, Efthymiadis, Wacker
Drawinol AF **dinobuton** Wacker-Dow
Drawipas W **captafol + thiabendazole** Wacker
Drawirit H **dalapon-sodium + MCPA + monolinuron + simazine** Dow
Drawisal F **carbendazim** Dow
Drawisan F **fenarimol** Elanco
Drawizon IA **diazinon** Wacker-Dow
Draza IM **methiocarb** Bayer
drazoxolon F [5707-69-7] 4-(2-chlorophenylhydrazono)-3-methyl-5-isoxazolone, Ganocide, Mil-Col, PP 781, SAIsan
DRB 91 H **amitrole + 2,4-D + MCPA + TCA** CARAL
DRB 147 H **atrazine + 2,4-D + diuron** CARAL
DRB 153 H **amitrole + atrazine** CARAL
Drepamon H **tiocarbazil** Agrimont, Montedison, Quimigal, Scam, Sipcam
Drexar H **MSMA** Drexel
Drifene AP IA **endosulfan + parathion** Pepro
Driftless S **polyacrylamide** Quintet
Drinox I **heptachlor**
Droot R **chlorophacinone** May & Baker
Drop-Leaf H **sodium chlorate**
Dropp H **thidiazuron** Schering
Drozin S **dazomet** Chimac-Agriphar
DRP 642532 FX **thiram**
Drupina F **ziram** Aziende Agrarie Trento, Visplant
DRW 1139 H **metamitron** Bayer
DS 15647 IA **thiofanox** Diamond Shamrock
DSM IA **demeton-S-methyl**
DSMA H [144-21-8] disodium methylarsonate, Ansar-8100, Arrhenal, Arsinyl, Crab-E-Rad, Crab-Kleen, Methar, Namate, Weed-E-Rad 360
DU 112307 I **diflubenzuron** Duphar
Du-Cam Combi H **dicamba + MCPA + mecoprop** Duphar
Du Cason H **dichlobenil** Siapa
Du-Cryl IG **carbaryl** Duphar
Du-Dim I **diflubenzuron** Siapa
Du-Dusit H **bromacil + dichlobenil** Siapa
Du-Mazin F **maneb + zineb** Duphar
Du Pont 634 H **lenacil** Du Pont
Du Pont 1179 IA **methomyl** Du Pont
Du Pont 1318 H **siduron** Du Pont
Du Pont 1991 F **benomyl** Du Pont
Du Pont Desherbant H **linuron** Du Pont
Du Pont Flax Herbicide H **lenacil + linuron** Du Pont
Du Pont Herbicide 82 H **isocil**
Du Pont Herbicide 326 H **linuron** Du Pont
Du Pont Herbicide 732 H **terbacil** Du Pont
Du Pont Herbicide 976 H **bromacil** Du Pont
Du Pont Manzate F **maneb** Du Pont
Du Pont Vydate N **oxamyl** Du Pont
Du-Ter F **fentin hydroxide** Duphar, ICI, KVK, La Quinoleine, Schering
Du Ter-M F **fentin hydroxide + maneb** Agroplant, Sodafabrik
Du-Ter M F **fentin acetate + maneb** Duphar
Du-Ter/Mancozeb F **fentin hydroxide + mancozeb** Duphar

Du-Ter/maneb F **fentin hydroxide + maneb** Duphar
Duacil H **lenacil + metolachlor** Budapesti Vegyimuvek
Dual H **metolachlor** Budapesti Vegyimuvek, Ciba-Geigy, Dimitrova, Dudesti, Ligtermoet, Liro, Ruse, Sinteza, Viopharm
Dual Mix H **methoprotryne + metolachlor** VCH
Dualenox H **atrazine + simazine** Schering
Dualin H **linuron + metolachlor** Ciba-Geigy
Dubarol S **alkyl phenol ethoxylate** Petrochema
Dubon F **captan** Safor
Ducason H **dichlobenil** La Quinoleine
Dudulex H **dalapon-sodium + dichlobenil** Siapa
Due Zeta S F **sulphur + zineb** Siapa
Duelor H **metolachlor** Ciba-Geigy
Dumex I **chlorpyrifos** Anticimex
Duo Top F **triflumizole** Siegfried
Duocide R **pindone**
Duogran H **bromoxynil + pyridate** Agrolinz
Duopan H **diuron + oryzalin** Maag
Duos IA **dimethoate** Fivat
Duotop F **triflumizole** Siegfried
Duplitox 5 + 3 I **DDT + lindane** Dudesti
Duplosan H **mecoprop** BASF
Duplosan DP HH **dichlorprop-P** BASF
Duplosan KV H **mecoprop** BASF
Duplosan KV-Combi H **2,4-D + mecoprop-P** BASF
Duplosan KW H **mecoprop-P** BASF
Duplosan New System CMPP H **mecoprop-P** BASF
Duplosan Super H **dichlorprop-P + MCPA + mecoprop-P** BASF
Duraphos IA **mevinphos** Amvac
Duratox IA **demeton-S-methyl** Shell
Duratox Super I **dichlorvos** Shell
Dursban I **chlorpyrifos** Agrocros, BASF, Chimac-Agriphar, Chromos, Condor, Dow, Farmos, FarmProtection, ICI-Zeltia, La Quinoleine, Luxan, Maag, Prochimagro, Quimigal, Sandoz, Schering, Schering AAgrunol, Shell
Durvos I **dichlorvos** Dreyfus
Dusturan-Kornkaeferpuder I **piperonyl butoxide + pyrethrins** Urania
Dutch-Treat H **dimethylarsinic acid** Vineland
Duter F **fentin hydroxide** Du Pont, Shell
Duter extra F **fentin hydroxide** Shell
Duter M F **fentin hydroxide + maneb** Shell
Dutom H **pentanochlor** Duphar
Duvit WK Fluessigbeize F **thiabendazole** Kwizda
Dwell F **etridiazole** Uniroyal
Dyanap H **dinoseb + naptalam** Uniroyal
Dybar H **bromacil + diuron + hexazinone** Du Pont
Dybar H **fenuron**
Dybar flo H **bromacil + diuron + hexazinone** Du Pont
Dycarb I **bendiocarb** Schering
Dyclomec H **dichlobenil** PBI Gordon
Dyfonat I **fonofos** OHIS
Dyfonate I **fonofos** Geopharm, Kwizda, Ligtermoet, Liro, Maag, Procida, Sapec, Serpiol, Shell, Stauffer
Dylox I **trichlorfon** Bayer
Dymefos I **acephate** Visplant
Dymid H **diphenamid** Elanco, Eli Lilly
Dymox I **azamethiphos** Ciba-Geigy
dymron H [42609-52-9] *N*-(4-methylphenyl)-*N'*-(1-methyl-1-phenylethyl)urea, dymrone, dimuron, K 223, Showrone
dymrone H **dymron**
Dyna-Form FB **formaldehyde**
Dynafos-Nova I **dichlorvos** Formulex
Dynamec IA **abamectin** MSD AGVET
Dynamyte HI **dinoseb** Drexel
Dynex H **diuron** Vertac
Dynone F **prothiocarb**
Dyprop HG **dichlorprop** May & Baker
Dyrene F **anilazine** Bayer
Dytrol IA **DNOC + petroleum oils** Shell

E-48 I **isoxathion** Sankyo
E 158 IA **demeton-S-methyl sulphone** Bayer
E 601 I **parathion-methyl** Bayer
E 605 I **parathion** Bayer
E 605 Combi IA **oxydemeton-methyl + parathion** Bayer
E-1059 IA **demeton** Bayer
E 3314 I **heptachlor**
Earina I **cypermethrin** ERT
Early Impact F **carbendazim + flutriafol** ICI
Earthcide F **quintozene** Nissan
Easout FW **thiophanate-methyl** Ciba-Geigy

Easytec G **tecnazene** MSD AGVET
Ecadion IA **parathion** Sedagri
Ecadion methyl IA **parathion-methyl** Sedagri
Ecatin IA **thiometon** Sandoz
Ecatox I **parathion** Sandoz
Ecatox Metil I **parathion-methyl** Sandoz
echlomezol F **etridiazole**
Eclipse H **glyphosate** Sovilo
Ecombi IA **oxydemeton-methyl + parathion** Bayer
ECP IN **dichlofenthion**
Ectiban I **permethrin** ICI
Ectoral I **fenchlorphos**
Ectrin IA **fenvalerate** SDS Biotech
Edabrom IN **ethylene dibromide** Geopharm
EDB NI **ethylene dibromide**
EDDP F **edifenphos**
Edge H **ethalfluralin** Elanco
edifenphos F [17109-49-8] *O*-ethyl *S,S*-diphenyl phosphorodithioate, EDDP, Bay 78418, Hinosan
Edil buitenboel L **quaternary ammonium** Intradal
Edrizar IA **amitraz** Siapa
Edron G **vegetable extracts** Aifar
Efdaton I **dimethoate** Efthymiadis
Efdiazon I **diazinon** Efthymiadis
Effican F **copper oxychloride + folpet** Schering
Effican Mix F **copper oxychloride + cymoxanil + folpet** Schering
Effican S F **copper oxychloride + cymoxanil + folpet** Schering
Effican Ultra F **copper oxychloride + folpet** Schering
Effix H **flamprop-M-isopropyl** Shell
Eflurin H **trifluralin** Efthymiadis
Eforol I **parathion** Kwizda
Eftetrex A **chlorobenzilate** Efthymiadis
Eftol IA **parathion** Spiess, Urania
Eftol-Oel I **parathion + petroleum oils** Spiess, Urania
Efuzin F **dodine** Chemolimpex, Szovetkezet
Egesa Ameisenmittel I **bromophos** Egesa
Egesa Ameisentod I **lindane** Egesa
Egesa Pflanzenspray IA **dichlorvos + dinocap + lindane** Egesa
Egesa Pflanzenspray-Extra IA **lindane + petroleum oils** Egesa
Egesa Rasenduenger mit Moosvernichter neu H **iron sulphate** Egesa
Egesa Rasenduenger mit Unkrautvernichter Neu H **2,4-D + dicamba** Egesa
Egesa Schneckenkorn M **metaldehyde** Egesa
Egesa Total-Unkrautvernichter H **amitrole + diuron** Egesa
eglinazine-ethyl H [6616-80-4] *N*-(4-chloro-6-ethylamino-1,3,5-triazin-2-yl)glycine ethyl ester, MG-06
Egodan Parathion I **parathion** Nordisk Alkali
EGYT 2250 G **heptopargil** EGYT
Ehrenpreis-Vernichter Anicon H **dichlorprop + ioxynil** Shell
EI 4049 IA **malathion** Cyanamid
EI 12880 IA **dimethoate** Cyanamid
EI 18682 AI **prothoate**
EI 47031 IA **phosfolan** Cyanamid
EI 47470 IA **mephosfolan** Cyanamid
Eimue D H **bromacil + diuron** Eimermacher
Eimue-Grasstop G **maleic hydrazide** Eimermacher
Eimue-zin H **simazine** Eimermacher
EK 480 H **clopyralid + dicamba + MCPA** Esbjerg
EK insektspray I **piperonyl butoxide + pyrethrins** Esbjerg
Ekadonal I **lindane + thiometon** Sandoz
Ekadrine IA **endosulfan + parathion** Sandoz
Ekalux IA **quinalphos** Chinoin, Chromos, Geopharm, Sandoz
Ekamet I **etrimfos** Du Pont, Farmos, Sandoz, Schering, Spiess, Urania
Ekatin I **thiometon** Chromos, Farm Protection, Geopharm, Sandoz, Schering
Ekatin M I **morphothion**
Ekatine IA **thiometon** Sandoz
Ekatox IA **parathion** Sandoz
Ekatox Acaricida A **parathion + tetradifon** Sandoz
Ekavinol IA **parathion + petroleum oils** Sandoz
Ekkusagoni H **chlomethoxyfen** Nihon Nohyaku
Eksmin I **permethrin** Sumitomo
Ektafos IA **dicrotophos** Ciba-Geigy
Ektomin A **chlormebuform**
EL-103 H **tebuthiuron** Elanco
EL-107 H **isoxaben** Elanco
EL-110 H **benfluralin** Elanco
EL-119 H **oryzalin** Elanco
EL-161 H **ethalfluralin** Elanco
EL-171 H **fluridone** Elanco

EL-177 H 1-*tert*-butyl-5-cyano-*N*-methylpyrazole-4-carboxamide
EL-179 H **isopropalin** Elanco
EL-222 F **fenarimol** Elanco
EL-228 F **nuarimol** Elanco
EL-273 F **triarimol**
EL-291 F **tricyclazole** Elanco
EL-500 G **flurprimidol** Eli Lilly
EL-531 G **ancymidol** Elanco
EL-614 R **bromethalin** Eli Lilly
Elam H **propanil** Ellagret
Elam Special H **fenoprop + propanil** Ellagret
ELANCO Beize F **imazalil + nuarimol** Elanco
Elancolan H **trifluralin** Elanco
Elancolan K H **napropamide + trifluralin** Elanco
Elbacim H **lenacil + proximpham** Fahlberg-List
Elbanil HG **chlorpropham** Fahlberg-List
Elbanox H **chlorpropham + propham + proximpham** Fahlberg-List
Elbarex H **DCU + lenacil** Fahlberg-List
Elbatan H **lenacil** Fahlberg-List
Elburon H **atrazine + fenuron** Fahlberg-List
Eldol H **amitrole + simazine** Rhone-Poulenc
Elefant Unkrautvertilger H **sodium chlorate** Epple
Elefant-Sommeroel IA **petroleum oils** Epple
Eleo-Cidial I **petroleum oils + phenthoate** Diana
Eleophenothion I **carbophenothion + petroleum oils** Hellenic Chemical
Elgetiver active FI **anthracene oil + DNOC** Truffaut
Elgetol IAHF **DNOC** FMC
Elital F **fenarimol** Elanco
Elite H **aclonifen + oxadiazon** Rhodiagri
Elmagran H **ioxynil + isoproturon + mecoprop** Agriben, Hoechst
Elocril IA **iodofenphos** Ciba-Geigy
Elocron I **dioxacarb** Chromos, Ciba-Geigy, Spolana
Elosal FA **sulphur** Argos, Hoechst, Pluess-Staufer, Roussel Hoechst
Elsan IA **phenthoate** Nissan
Eltarin I **carbaryl** Ellagret
Eltox I **parathion-methyl** Ellagret
Elvaron F **dichlofluanid** Bayer
E-M I **azinphos-ethyl + azinphos-methyl** Agrotechnica
EM 379 FX **guazatine acetates**
Embark G **mefluidide** 3M, Berner, C.F.P.I., Chimac-Agriphar, Chimiberg, Cillus, Kwizda, Ligtermoet, Sepran
Embutone AS H **2,4-DB** Rhone-Poulenc, Rhodiagri-Littorale
Embutox H **2,4-DB** Kwizda
Emelcar I **carbaryl** Roussel Hoechst
Emeldac IA **dimethoate** Roussel Hoechst
Emeldor H **linuron + trifluralin** Roussel Hoechst
Emeltenmiddel I **temephos** Bayer
Emelvit Z F **copper oxychloride + zineb** Roussel Hoechst
Emelzir F **ziram** Roussel Hoechst
Emerald GS **di-1-p-menthene** Intracrop
E-Methylbromid FIHN **methyl bromide** Potsdam
Emetres I **carbaryl + dimethoate + lindane** Luqsa
Emmatos IA **malathion**
Emmedi H **2,4-D + MCPA** Agrimont
Emol F **thiram** Polifarb
Empal H **MCPA** Universal Crop Protection
empenthrin I [54406-48-3] (*RS*)-1-ethynyl-2-methyl-2-pentenyl (1*R*)-*cis,trans*-chrysanthemate, Nexalotte, S-2852 Forte, Vaporthrin
Emtebe H **mecoprop** Chimac-Agriphar, Schwebda
Emulsamine E-3 H **2,4-D** Union Carbide
Emulsol S **alkyl phenol ethoxylate** Sipcam-Phyteurop
Endamon IA **endosulfan** Du Pont
Endaven H **benzoylprop-ethyl** Shell
Endocel IA **endosulfan** Excel
Endocide I **endothion**
Endodan F **ethylene thiuram monosulphide** Diana
Endofex I **endosulfan** Permutadora
Endogerme CP G **chlorpropham** Agriphyt
Endogor I **dimethoate + endosulfan** Tecniterra
Endomed IA **endosulfan** Agrimont
Endomosyl T **hydrolysed proteins** Hoechst
Endomozal I **malathion** Hellenic Chemical
Endosan FA **binapacryl** Hoechst, Roussel Hoechst
Endosivam I **endosulfan** Sivam
endosulfan IA [115-29-7]; α-endosulfan [959-98-8], β-endosulfan [33213-65-9]

C,C'-(1,4,5,6,7,7-hexachloro-8,9,10-trinorborn-5-en-2,3-ylene)(dimethyl sulphite), thiodan, benzoepin, OMS 570, ENT 23979, Beosit, Chlortiepin, Cyclodan, Endocel, FMC 5462, Hildan, Hoe 2671, Insectophene, Malix, Thifor, Thimul, Thiodan, Thionex, Thiosulfan
Endosulfanol IA **endosulfan + petroleum oils** Siegfried
Endoter IA **endosulfan** Terranalisi
endothal HLG [145-73-3] 7-oxabicyclo[2.2.1]heptane-2,3-dicarboxylic acid, endothall, Accelerate, Aquathol, Des-I-Cate, Hydrothol, Pennout
endothall HLG **endothal**
endothion I [2778-04-3] *S*-5-methoxy-4-oxo-4*H*-pyran-2-ylmethyl *O,O*-dimethyl phosphorothioate, ENT 24653, 7175 RP, AC 18737, Endocide, FMC 5767
Endotrin IA **endosulfan** Enotria
Endovis IA **endosulfan** Visplant
Endox H **dicamba + mecoprop** FCC
Endrex IR **endrin** Shell
endrin IR [72-20-8] 1,2,3,4,10,10-hexachloro-1*R*,4*S*,4a*S*,5*S*,6,7*R* ,8*R*,8a*R*-octahydro-6,7-epoxy-1,4:5,8-dimethanonaphthalene, endrine, nendrin, Endrex, Hexadrin
EndSpray HF **fentin hydroxide + metoxuron** Pan Britannica
Endurance H **prodiamine** Sandoz
Engrais desherbant gazon S H **2,4-D + dicamba** Umupro
Engrais desherbant rosiers H **simazine** SCAC-Fisons, Tradi-Agri
Enhance S **alkyl phenol ethoxylate** Midkem
Enide H **diphenamid** Eli Lilly, Grima, NOR-AM, Sapec, Schering, Shell, Upjohn
enilconazole F **imazalil**
Enkil F **maneb** Baslini
Eno Acar A **dicofol + tetradifon** Enotria
Eno K S **hydroxyethyl cellulose + lauryl alcoholethoxylate** Enotria
Enocaptan F **captan** Enotria
Enocid E I **methidathion** Enotria
Enocomplex F **calcium copper oxychloride + sulphur + zineb** Enotria
Enodene F **dodine** Enotria
Enodom F **dodine** Candilidis
Enoflur H **trifluralin** Enotria
Enoil IA **petroleum oils** Enotria
Enolofos I **chlorfenvinphos** Organika-Azot
Enonoder S **hydroxyethyl cellulose + lauryl alcoholethoxylate** Scam
Enothiram F **thiram** Enotria
Enotiol Ramato F **calcium copper oxychloride + sulphur** Enotria
Enotox I **dichlorvos** Enotria
Enovit M FW **thiophanate-methyl** Inagra, Siegfried, Sipcam
Enoxal H **pendimethalin** Visplant
Enozin F **zineb** Enotria, Visplant
Enstar I **kinoprene** KenoGard, Zoecon
ENT 133 IA **rotenone**
ENT 211 I **chlorbicyclen**
ENT 262 X **dimethyl phthalate**
ENT 785 I **chlorbicyclen**
ENT 987 FX **thiram**
ENT 1122 HI **dinoseb**
ENT 1716 I **methoxychlor**
ENT 3776 FL **dichlone**
ENT 4225 I **TDE**
ENT 7796 I **lindane**
ENT 8420 N **D-D**
ENT 9735 IRA **camphechlor**
ENT 9932 I **chlordane**
ENT 14250 S **piperonyl butoxide**
ENT 14689 F **ferbam**
ENT 14874 F **zineb**
ENT 14875 F **maneb**
ENT 15108 IA **parathion**
ENT 15152 I **heptachlor**
ENT 15349 NI **ethylene dibromide**
ENT 15949 I **aldrin**
ENT 16225 I **dieldrin**
ENT 16273 IA **sulfotep**
ENT 16358 A **chlorfenson**
ENT 16894 I **O,O,O',O'-tetrapropyl dithiopyrophosphate**
ENT 17034 IA **malathion**
ENT 17035 I **dicapthon**
ENT 17292 IA **parathion-methyl**
ENT 17295 IA **demeton**
ENT 17298 IA **EPN**
ENT 17510 I **allethrin**
ENT 18596 A **chlorobenzilate**
ENT 18771 AI **TEPP**
ENT 19507 IA **diazinon**
ENT 19763 I **trichlorfon**
ENT 20218 X **diethyltoluamide**
ENT 20738 IA **dichlorvos**
ENT 20852 H **bromoxynil**
ENT 20871 S **sesamex**
ENT 22014 IA **azinphos-ethyl**
ENT 22374 IA **mevinphos**
ENT 22879 IA **dioxathion**
ENT 23233 IA **azinphos-methyl**

ENT 23347 IA **disulfoton**
ENT 23648 A **dicofol**
ENT 23708 IA **carbophenothion**
ENT 23737 A **tetradifon**
ENT 23969 IG **carbaryl**
ENT 23979 IA **endosulfan**
ENT 24042 IAN **phorate**
ENT 24105 AI **ethion**
ENT 24482 IA **dicrotophos**
ENT 24650 IA **dimethoate**
ENT 24652 AI **prothoate**
ENT 24653 I **endothion**
ENT 24717 I **crotoxyphos**
ENT 24727 FA **dinocap**
ENT 24945 NI **fensulfothion**
ENT 24964 IA **oxydemeton-methyl**
ENT 24969 IA **chlorfenvinphos**
ENT 24980-X IA **amiton**
ENT 24988 IA **naled**
ENT 25208 FLM **fentin acetate**
ENT 25445 H **amitrole**
ENT 25506 IA **ethoate-methyl**
ENT 25515 IA **phosphamidon**
ENT 25540 I **fenthion**
ENT 25545 I **isobenzan**
ENT 25606 FAI **quinomethionate**
ENT 25647 IA **oxydeprofos**
ENT 25670 I **isoprocarb**
ENT 25671 I **propoxur**
ENT 25705 IA **phosmet**
ENT 25712 I **trichloronat**
ENT 25715 I **fenitrothion**
ENT 25718 A **dienochlor**
ENT 25726 MIAX **methiocarb**
ENT 25784 I **aminocarb**
ENT 25793 AF **binapacryl**
ENT 25830 IA **phosfolan**
ENT 25841 IA **tetrachlorvinphos**
ENT 25991 IA **mephosfolan**
ENT 26058 F **anilazine**
ENT 26538 F **captan**
ENT 26613 IA **vamidothion**
ENT 26999 A **chloropropylate**
ENT 27093 IAN **aldicarb**
ENT 27115 A **tetrasul**
ENT 27129 IA **monocrotophos**
ENT 27160 IA **amidithion**
ENT 27162 I **bromophos**
ENT 27163 IA **phosalone**
ENT 27164 IAN **carbofuran**
ENT 27165 I **temephos**
ENT 27193 IA **methidathion**
ENT 27226 A **propargite**
ENT 27244 AF **dinobuton**
ENT 27257 IA **formothion**
ENT 27258 IA **bromophos-ethyl**
ENT 27300 I **promecarb**
ENT 27311 I **chlorpyrifos**
ENT 27320 IA **dialifos**
ENT 27386 IA **phenthoate**
ENT 27389 I **dioxacarb**
ENT 27394 IA **quinalphos**
ENT 27395-X A **cyhexatin**
ENT 27396 IA **methamidophos**
ENT 27520 IA **chlorpyrifos-methyl**
ENT 27521 I **fospirate**
ENT 27552 A **bromopropylate**
ENT 27566 AI **formetanate hydrochloride**
ENT 27738 A **fenbutatin oxide**
ENT 27766 I **pirimicarb**
ENT 27822 I **acephate**
ENT 27920 IN **terbufos**
ENT 27967 AI **amitraz**
ENT 27972 I **phenothrin**
ENT 28009 F **fentin hydroxide**
ENT 29054 I **diflubenzuron**
ENT 29117 I **kadethrin**
ENT 29261 NIA **aldoxycarb**
ENT 51762 R **norbormide**
Entex I **fenthion** Bayer
Entobakterin I **Bacillus thuringiensis** MIU
Entomofin AI **endosulfan** Agrocros
Entomozyl T **hydrolysed proteins** Hoechst
Envert H **2,4-D + dichlorprop** Union Carbide
Envoy H **cyanazine + MCPA** Shell
Enxoframil FA **sulphur** Permutadora
Enxofre FA **sulphur** Quimigal
Epame Gamma I **lindane** Platecsa
Epargol G **sulphur + oligomers** Ravit, Rhone-Poulenc
Ephaneb F **maneb** Efthymiadis
Ephmazin H **simazine** Efthymiadis
Epho Emulsion I **parathion** Siegfried
Epic F **furmecyclox** Gustafson
Epidor P F **carbendazim + mancozeb** La Quinoleine
Epigon I **permethrin** Kwizda
Epinat H **benzthiazuron + lenacil** RACROC
EPN IA [2104-64-5] *O*-ethyl *O*-4-nitrophenyl phenylphosphonothioate, OMS 219, ENT 17298, EPN
epofenonane I [57342-02-6] 6,7-epoxy-3-ethyl-7-methylnonyl 4-ethylphenyl ether, Ro 10-3108
epoxyethane IF **ethylene oxide**
Epro-Schneckenkorn M **metaldehyde** Epro

epronaz H [59026-08-3] *N*-ethyl-*N*-propyl-3-propylsulphonyl-1*H*-1,2,4-triazole-1-carboxamide, BTS 30843
Eprozin H **atrazine** Epro
EPS-Ziram F **ziram** ERT
Eptam H **EPTC** Collett, Geopharm, KenoGard, Leu & Gygax, Ligtermoet, Liro, Pluess-Staufer, Rhone-Poulenc, Serpiol, Spiess, Stauffer, Urania
Eptane IA **endosulfan** Scam
Eptapur H **buturon** BASF
EPTC H [759-94-4] *S*-ethyl dipropylthiocarbamate, R-1608, Alirox, Eptam, Genep, Niptan, Witox
Equigard IA **dichlorvos** Shell
Equitdazin F **carbendazim** Equitable Trading
Era mosedreper H **iron sulphate** Skramstad
Eradex I **chlorpyrifos** Planters Products
Eradex AF **thioquinox**
Eradic R **bromadiolone** C.N.C.A.T.A.
Eradic blocs R **scilliroside** C.N.C.A.T.A.
Eradic-Taupe R **chloralose** C.N.C.A.T.A.
Eradicane H **dichlormid + EPTC** Agritec, Ciba-Geigy, Du Pont, Geopharm, Ligtermoet, Liro, Lucebnizavody, OHIS, Sapec, Serpiol, Stauffer, Wacker
Eradicane 50-11 F H **atrazine + dichlormid + EPTC** Serpiol
Eradicane A10/2.5 G H **atrazine + dichlormid + EPTC** Serpiol
Eradicane Extra H **dichlormid + O,O-diethyl O-phenylphosphorothioate + EPTC** Stauffer
Eraditon AF **thioquinox**
Erazidon AF **thioquinox**
Erbacid Total H **dalapon-sodium + MCPA + simazine** EniChem
Erban H **propanil** Sintesul
Erbanil H **propanil** Visplant
Erbastop H **diuron** Rhodic
Erbazone H **bentazone** Siapa
Erbifos H **metolachlor** Ravit
Erbital H **2,4-D + dalapon-sodium + simazine** Visplant
Erbitox Bietole H **chloridazon** Siapa
Erbitox E H **MCPA** Siapa
Erbitox Giavone H **molinate** Siapa
Erbitox Grano H **dicamba + MCPA** Siapa
Erbitox LV 4 H **2,4-D** Siapa
Erbitox Riso H **propanil** Siapa
Erbitox S H **2,4-D** Siapa
Erbitox T H **TCA** Siapa
Erbivin AP H **atrazine + paraquat dichloride** Diana
Erbivin SP H **paraquat dichloride + simazine** Diana
erbon H [136-25-4] 2-(2,4,5-trichlorophenoxy)ethyl 2,2-dichloropropionate, Baron, Novege
Erbotan H **thiazafluron** Ciba-Geigy, Plantevern-Kjemi, Spiess, Urania
Erbotrin H **dalapon-sodium** Enotria
Erdbeer-Spritzmittel Rovral F **iprodione** Shell
Erditotal H **amitrole + atrazine** Research Development
Erfel H **atrazine** Enotria
ergocalciferol R **calciferol**
Ergoplant G **N-acetylthiazolidin-4-carboxylic acid + folic acid** Montedison
Ergostim G **N-acetylthiazolidin-4-carboxylic acid + folic acid** Agrimont, Montedison, Sopra
Ergoton P I **piperonyl butoxide + pyrethrins** Zucchet
Ergovit G **N-acetylthiazolidin-4-carboxylic acid + folic acid** Pepro
Erisan F **captan + mancozeb** Pepro
Erisan super F **captan + fenarimol + mancozeb** Pepro
Eritox IA **monocrotophos** Terranalisi
Erkovan I **dichlorvos** Viopharm
Ermatol H **amitrole + simazine** ERT
Ermazina H **simazine** ERT
Ermetox I **methoxychlor** ERT
Erp-Actril H **ioxynil + MCPA + mecoprop** Maag
Erpanol H **2,4-D + mecoprop** Maag
Erranca H **glyphosate** Herbex
Ertacilo H **lenacil + TCA** ERT
Ertalin I **lindane** ERT
Ertan H **propanil** ERT
Ertane A **dicofol** ERT
Ertane Compuesto A **dicofol + tetradifon** ERT
Ertazina H **atrazine** ERT
Ertazinfos IA **azinphos-ethyl** ERT
Ertefon I **trichlorfon** ERT
Ertevin I **carbaryl** ERT
Ertidan IA **endosulfan** ERT
Ertimix F **cymoxanil + folpet + mancozeb** ERT
Ertimix cuprico F **copper sulphate + cymoxanil + mancozeb** ERT
Erturon H **chlorotoluron** ERT
Erturon Extra H **chlorotoluron + terbutryn** ERT

Erunit H **acetochlor + atrazine** Chemolimpex
Eruzin I **lindane** Marktredwitz
Ervax H **amitrole + simazine** Quimigal
Ervisan H **amitrole + simazine** Sandoz
ESA-Unkrautsalz H **sodium chlorate** Kober
Esbiol I **allethrin** Roussel Uclaf, Wellcome
Esca Proteica Siapa T **hydrolysed proteins** Siapa
Esca Regina I **lindane** Ital-Agro
Escacide M **metaldehyde** Chemia
Escanex M **metaldehyde** Tecniterra
Escargol M **metaldehyde** ERT
Escort H **metsulfuron-methyl** Du Pont
Escort H **clopyralid + ioxynil** Dow
Escuran H **chlorotoluron + trifluralin** Ciba-Geigy, Liro
esfenvalerate I [66230-04-4] (*S*)-α-cyano-3-phenoxybenzyl (*S*)-2-(4-chlorophenyl)-3-methylbutyrate, Asana, Halmark, S-1844, S-5620Aα, Sumi-alfa, Sumi-alpha
Esgram H **paraquat dichloride**
ESP IA **oxydeprofos**
Espot H **alachlor + atrazine** ICI
Espumer O F **2-phenylphenol** Brillocera
Essanvit H **mecoprop** Agriben
Essevi 50 I **carbaryl** Schering
Estermilo F **captan + carbendazim** Probelte
Estox IA **oxydeprofos**
Estratto di tabacco I **nicotine** Brissago
etacelasil G [37894-46-5] 2-chloroethyltris (2-methoxyethoxy)silane, Alsol, CGA 13586
etaconazole F [60207-93-4] (±)-1-[2-(2,4-dichlorophenyl)-4-ethyl-1,3-dioxolan-2-ylmethyl]-1*H*-1,2,4-triazole, Benit, CGA 64251, Sonax, Vangard
Etaldina S **polyglycolic ethers** Condor
Etaldyn S **polyglycolic ethers** Ravit
Etaldyne S **polyglycolic ethers** Agriben, Pepro, Rhone-Poulenc
Etalene I **fenitrothion** Chimiberg
Etan I **lindane** Chimiberg
Etanyde A **cyhexatin** Interphyto
Etazin F **zineb** Terranalisi
Etazin 50 H **secbumeton** Budapesti Vegyimuvek, Ciba-Geigy, Viopharm
Etazin 80 F **zineb** Terranalisi
Etazin 3585 H **secbumeton + simazine** Ciba-Geigy
Etazin R F **copper oxychloride + zineb** Terranalisi
Etazina liquida H **secbumeton + simazine** Ciba-Geigy
Etazine H **secbumeton** Ciba-Geigy
Etazine autosuspensible H **secbumeton + simazine** Ciba-Geigy
ETCMTD F **etridiazole**
ethalfluralin H [55283-68-6] *N*-ethyl-α,α,α-trifluoro-*N*-(methylallyl)-2,6-dinitro-*p*-toluidine, ethalfluraline, Edge, EL- 161, Sonalan, Sonalen
ethalfluraline H **ethalfluralin**
Ethane F **dinocap** Tecniterra
Ethanox AI **ethion** Rhone-Poulenc
ethazol F **etridiazole**
ethazole F **etridiazole**
ethephon G [16672-87-0] 2-chloroethylphosphonic acid, chlorethephon, Bromoflor, Camposan, Cerone, Etheverse, Ethrel, Flordimex, Florel, Prep, Tomathrel
Etheverse G **ethephon** C.F.P.I., Ciba-Geigy
Ethiargos IA **ethion** Argos
ethidimuron H [30043-49-3] 1-(5-ethylsulphonyl-1,3,4-thiadiazol-2-yl)-1,3-dimethylurea, MET 1486, Ustilan
Ethil Cotnion IA **azinphos-ethyl** Edefi
Ethiluq IA **ethion** Luqsa
ethiofencarb I [29973-13-5] 2-ethylthiomethylphenyl methylcarbamate, ethiophencarbe, Croneton, HOX 1901
Ethiol AI **ethion** Rhone-Poulenc
ethiolate H [8071-40-7] *S*-ethyl diethylthiocarbamate, Prefox, S-15076
ethion AI [563-12-2] *O,O,O',O'*-tetraethyl *S,S'*-methylene bis(phosphorodithioate), diethion, ENT 24105, Ethanox, Ethiol, FMC 1240, Hylemox, Rhodiacide, Rhodocide, RP-Thion, Vegfru Fosmite
Ethion Dormant Volck IA **ethion + petroleum oils** Agrocros
Ethion Oil Foret IA **ethion + petroleum oils** Foret
Ethion superior Volck IA **ethion + petroleum oils** Agrocros
Ethionyl I **parathion** Aveve
ethiophencarbe I **ethiofencarb**
ethirimol F [23947-60-6] 5-butyl-2-ethylamino-6-methylpyrimidin-4-ol, ethyrimol, Milcurb Super, Milgo, Milstem, PP 149
ethoate-methyl IA [116-01-8] *S*-ethylcarbamoylmethyl *O,O*-dimethyl phosphorodithioate, ethoate-methyle, OMS 252, ENT 25506, B/77, Fitios

ethofenprox I [80844-07-1] 2-(4-ethoxyphenyl)-2-methylpropyl 3-phenoxybenzyl ether, ethophenprox, MTI 500, Trebon
ethofumesate H [26225-79-6] (±)-2-ethoxy-2,3-dihydro-3,3-dimethylbenzofuran-5-yl methanesulphonate, NC 8438, Nortron, Prograss, Tramat
Ethokem S **tallow amine ethoxylate** Midkem
ethophenprox I **ethofenprox**
ethoprop NI **ethoprophos**
ethoprophos NI [13194-48-4] *O*-ethyl *S,S*-dipropyl phosphorodithioate, ethoprop, Mocap, VC9-104
Ethoraz I **ethion** Agrochemiki
ethoxyquin G [91-53-2] 6-ethoxy-1,2-dihydro-2,2,4-trimethylquinoline, ethoxyquine, Deccoquin, Nix-Scald, Santoquin, Stop-Scald, Stopscald, Xedaquine
ethoxyquine G **ethoxyquin**
Ethrel G **ethephon** Amchem, Bjoernrud, C.F.P.I., Chimac-Agriphar, Cillus, EniChem, Etisa, Gullviks, KVK, ICI, Luxan, Pluess-Staufer, Ravit, Rhone-Poulenc, Spiess, Tuhamij, Union Carbide, Urania
ethyl 1-naphthyl acetate G [2122-70-5] ethyl 1-naphthyl acetate, Tipoff, Tre-Hold
Ethyl Guthion IA **azinphos-ethyl** Bayer
ethyl parathion IA **parathion**
ethyl pyrophosphate AI **TEPP**
ethylan I **perthane**
ethylene dibromide NI [106-93-4] 1,2-dibromoethane, EDB, dibromure d'ethylene, ENT 15349, Bromofume, Celmide, Dibrome, Dowfume, Edabrom
ethylene oxide IF [75-21-8] oxirane, epoxyethane, ETO, Carboxide, Cartox, Etox, T-Gas
ethylthiodemeton IA **disulfoton**
ethyrimol F **ethirimol**
Etil Mangan F **maneb** Eli Lilly
Etilon IA **parathion**
etinofen H [2544-94-7] α-ethoxy-4,6-dinitro-*o*-cresol, etinofene, Dinethon
etinofene H **etinofen**
Etiol IA **malathion** Galenika
Etisso Balkonpflanzenspray Combi FI **butocarboxim + fenarimol** Shell
Etisso Combi-Duengerstaebchen I **dimethoate** Shell
Etisso-insecticide I **bioallethrin + piperonyl butoxide + pyrethrins** Shell
Etisso-Insektenvernichter IA **dimethoate** Shell
Etisso Pflanzenschutz IA **piperonyl butoxide + pyrethrins** Shell
Etisso Rasenduenger mit Unkrautvernichter H **2,4-D + dicamba** Shell
Etisso Rasenunkraut-Vernichter H **dicamba + MCPA** Shell
Etisso Schneckenkorn M **metaldehyde** Shell
Etisso Unkrautfrei Giess H **bromacil + diuron** Shell
Etizol TL H **amitrole + ammonium thiocyanate** Etisa, Ficoop
ETMT F **etridiazole**
ETO IF **ethylene oxide**
Etoc I **prallethrin** Sumitomo
Etolin IN **ethoprophos + lindane** Rhone-Poulenc
Etox I **ethylene oxide** Geopharm, Heydt Bona
Etoxiat FBI **ethylene oxide** Balux
Etoxipron G **ethoxyquin** BP
Etravon S **alkyl phenol ethoxylate** Ciba-Geigy
etridiazole F [2593-15-9] 5-ethoxy-3-trichloromethyl-1,2,4-thiadiazole, echlomezol, ethazol, ethazole, ETCMTD, ETMT, AAterra, Dwell, Koban, OM 2424, Pansoil, Terrazole, Truban
etrimfos IA [38260-54-7] *O*-6-ethoxy-2-ethylpyrimidin-4-yl *O,O*-dimethyl phosphorothioate, OMS 1806, Ekamet, SAN 197I, Satisfar
Etrofolan I **isoprocarb** Bayer
Etrolene I **fenchlorphos**
Etzel HG **DNOC** Du Pont, Wacker
Eucid IA **diazinon** Sofital
Eucritt F **mancozeb + metalaxyl** Siapa
Eucritt F F **folpet + metalaxyl** Siapa
Eudim IA **dimethoate** Sofital
Eulim M **metaldehyde** Sofital
Eupareen F **dichlofluanid** Bayer
Eupareen M FA **tolylfluanid** Bayer
Eupareen poeder AF **tolylfluanid** Bayer
Eupareen-stuif F **dichlofluanid** Bayer
Euparen F **dichlofluanid** Bayer
Euparen-Kupfer F **copper oxychloride + dichlofluanid** Agroplant
Euparen M F **tolylfluanid** Agro-Kemi, Bayer
Euparen-Ruiskutejauhe F **dichlofluanid** Berner

Euparene F **dichlofluanid** Agroplant, Bayer
Euparene-cuivre F **copper oxychloride + dichlofluanid** Agroplant
Euphytane I **petroleum oils** Sandoz
Euramine H **2,4-D** Eurofyto
Euro 11-E S **petroleum oils** Eurofyto
Euroate I **dimethoate** Phyto
Eurobron H **neburon** Phyto
Eurocarb I **carbaryl** Sofital
Euroil IA **petroleum oils** Sofital
Eurokar D A **dicofol** Phyto
Eurokill super H **dinoseb** Phyto
Euromite A **propargite** Sofital
Europron H **dalapon-sodium** Phyto
Euroram F **copper oxychloride** Sofital
Eurosan M F **maneb** Phyto
Eurosan Z F **zineb** Phyto
Eurostop H **mecoprop** Phyto
Eurotin F **fentin acetate** Eurofyto
Eurotox H **TCA** Phyto
Eurotra H **atrazine** Phyto
Euroxone H **MCPA** Eurofyto
Ever 3 R IA **parathion + petroleum oils** Platecsa
Ever 35 R I **parathion-methyl** Platecsa
Ever 72 IA **petroleum oils** Platecsa
Ever F I **fenitrothion** Platecsa
Evercyn IR **hydrogen cyanide** PCK
Evershield CM FI **captan + malathion** Gunson
Evik H **ametryn** Ciba-Geigy
Evisect I **thiocyclam hydrogen oxalate** Geopharm, Sandoz, SDS Biotech, Spiess, Urania
Evisekt I **thiocyclam hydrogen oxalate** Sandoz
Evital H **norflurazon** Sandoz
Evitoxal M **metaldehyde** Burri
EWAK-Rattenbekaempfungsmittel R **coumatetralyl** Klinkenberg
Exagama I **lindane** Rhone-Poulenc
Exagamma I **lindane** Condor
Exan R **calcium phosphide** Hentschke & Sawatzki
Exation IA **malathion** Condor
Excellent Gazonmest H **2,4-D + dicamba** Moreels-Guano
Excello F **sulphur** SEGE
Exel H **bifenox + mecoprop** Pepro
Exel 3 H **bifenox + clopyralid + mecoprop** Pepro
Exell S **tallow amine ethoxylate** Agstock, Siegfried
Exit H **amitrole + atrazine + diuron** May & Baker
Exotherm Termil F **chlorothalonil** Diamond Shamrock, Kwizda, Rigo
EXP 3864 H **quizalofop-ethyl**
Expand H **sethoxydim** Ewos
Experimental Fungicide 328 F **milneb**
Experimental Insecticide 4124 I **dicapthon**
Exporsan H **bensulide**
Express Flora Insecten Plantenspray IA **piperonyl butoxide + pyrethrins** Eurofill
Exsol I **lindane** Melchior
Extar A HIAF **DNOC** Sandoz
6208 extra F **copper sulphate + maneb** Rhodiagri-Littorale
42 Extra Jyvasyotti R **warfarin** Farmos
42 Extra Nestesyotti R **warfarin** Farmos
Extar Forte H **dinoterb** Sandoz
Extar lin H **DNOC** Sandoz
Extra select H **dicamba + MCPA** Sipcam-Phyteurop
Extra Strength Ant Killer FI **lindane + pyrethrins** Secto
Extra-Cobre F **copper oxychloride** Permutadora
Extragri H **amitrole + diuron** Sipcam-Phyteurop
Extramex H **atrazine + simazine** Sipcam-Phyteurop
Extraminol H **amitrole + atrazine** Sipcam-Phyteurop
Extramitrol H **amitrole + ammonium thiocyanate** Sipcam-Phyteurop
Extranurex H **diuron** Sipcam-Phyteurop
Extrasim H **simazine** Sipcam-Phyteurop
Extratex H **atrazine + simazine** Sipcam-Phyteurop
Extravon S **alkyl phenol ethoxylate** Ciba-Geigy, Viopharm
Extravril H **amitrole + simazine** Sipcam-Phyteurop
Extrazine H **atrazine + cyanazine** Du Pont
Exuberone G **4-indol-3-ylbutyric acid** Condor, Rhodiagri-Littorale
Ezenosan FA **dinocap** Agrotec
Ezine F **zineb** Mormino
Ezitan H **secbumeton** Ciba-Geigy
E-Z-off D G **S,S,S-tributyl phosphorotrithioate** Mobay
Ezy Pickin H **dimethylarsinic acid** Drexel
Ezzin H **atrazine + metolachlor** Ciba-Geigy
66 F G **1-naphthylacetic acid** Gobbi

3336-F F **thiophanate** Cleary
F238 F **dodemorph acetate** BASF
F-319 F **hymexazol** Sankyo
F-461 F **oxycarboxin** Uniroyal
F.20 IA **prothoate** Scam
F.C.H. 60 I X **petroleum oils** Perfor
F.P. 40 IA **propargite + prothoate** Scam
Faber F **chlorothalonil** Tripart
Fac AI **prothoate** Montedison
Facar IA **prothoate** Visplant
Faciron R **chlorophacinone** Pliva
Facmor IA **chlorbenside + prothoate** Aziende Agrarie Trento
Rademorf F **trimorphamide** Chemika, Duslo, Vuagt
Fagal H **ioxynil + isoproturon + mecoprop** Ciba-Geigy
Fair-2 G **maleic hydrazide** Fair Products
Fair 85 G **fatty alcohols** Fair Products
Fair-Plus G **maleic hydrazide** Fair Products
Fair-Tac G **decan-1-ol** Fair Products
Fajt I **pirimicarb** ICI
FALI-Curan H **chlorotoluron** FALI
Falibetan H **lenacil + propham + proximpham** Fahlberg-List
Falicarben F **carbendazim** Fahlberg-List
Falidazon H **chloridazon + lenacil + proximpham** Fahlberg-List
Faliherban H **atrazine + fenuron + thiazafluron** Fahlberg-List
Falimilbon FA **dinobuton** Fahlberg-List
Falimorph F **aldimorph** Fahlberg-List
Falisan-CX-Universal-Trockenbeize F **carboxin + phenylmercury acetate** Fahlberg-List
Falisan-Saatgut-Nassbeize F **phenylmercury acetate** Fahlberg-List
Falisan-Universal-Feuchtbeize F **di(methylmercury) p-toluenesulphonamide** Fahlberg-List
Falisan-Universal-Fluessigbeize F **phenylmercury acetate** Fahlberg-List
Falisan-Universal-Trockenbeize F **phenylmercury acetate** Fahlberg-List
Falisilvan H **fenuron** Fahlberg-List
Falisolan F **bronopol + carbendazim** Fahlberg-List
Falitomal G **azoluron + gibberellic acid** Fahlberg-List
Falitox-CMPP H **mecoprop** Ciba-Geigy
Falitox-D H **2,4-D** Ciba-Geigy
Falitox-DP H **dichlorprop** Ciba-Geigy
Falitox-Kombi H **2,4-D + MCPA** Ciba-Geigy
Falitox-MP-Kombi H **2,4-D + mecoprop** Ciba-Geigy
Fall H **sodium chlorate**
Fallowmaster H **dicamba** Sandoz
Falnet Maissaatgutpuder IX **diazinon** Spiess, Urania
Faltex F **folpet** Terranalisi
Faltocure F **Bordeaux mixture + folpet** Proval
Faluron H **bromuron** Fahlberg-List
Famid I **dioxacarb** Ciba-Geigy
Faneron H **bromofenoxim** Chromos, Ciba-Geigy, Ligtermoet, Liro, Plantevern-Kjemi, Spolana
Faneron combi H **bromofenoxim + terbuthylazine** Chromos, Ciba-Geigy, Scam
Faneron D H **bromofenoxim + 2,4-D** Nitrokemia
Faneron extra H **bromofenoxim + terbuthylazine** Ciba-Geigy
Faneron Multi H **bromofenoxim + terbuthylazine** Ciba-Geigy, Nitrokemia
Faneron P H **bromofenoxim + mecoprop** Liro
Faneron P H **bromofenoxim + mecoprop + terbuthylazine** Ciba-Geigy
Faneron plus H **bromofenoxim + mecoprop + terbuthylazine** Ciba-Geigy
Faneron spezial H **bromofenoxim + terbuthylazine** Ciba-Geigy
Fanicide H **dinoseb** La Littorale
Fanoil IA **parathion + petroleum oils** Inagra
Fanoprim H **atrazine + bromofenoxim** Ciba-Geigy, Ligtermoet, Liro
Fanulen I **parathion-methyl** Inagra
Fapeltar F **maneb + thiophanate-methyl** LHN
Farabin H **amitrole + atrazine + 2,4-D** Ligtermoet
Faran IA **prothoate + tetradifon** Baslini
Farex double concentre H **amitrole + simazine** Farcor
Far-Go H **tri-allate** Monsanto
Farmacel G **chlormequat chloride** Farm Protection
Farmaneb F **maneb** Farm Protection
Farmatin F **fentin hydroxide** Farm Protection
Farmil F **ditalimfos** Dow
Farmiprop H **MCPA + mecoprop** Farmipalvelu
Farmix Vliegerin I **methomyl** Farmix

Farmon Blue S **alkyl phenol ethoxylate** Farm Protection
Farmon Condox H **dicamba + mecoprop** Farm Protection
Farmon MCPB Plus H **MCPA + MCPB** Farm Protection
Farmon PDQ H **diquat dibromide + paraquat dichloride** Farm Protection
Farorid I **methoprene** Mortalin
Fasnet H **phenmedipham** Sipcam-Phyteurop
Fast-Fruit G **2,4-D** ERT
Fastac I **alphacypermethrin** Agrishell, Ciba-Geigy, Margesin, Pliva, Shell
Fatex H **chlorfenprop-methyl** Duslo
Fatox royal H **siduron** LHN
Fazor G **maleic hydrazide** Chiltern, Uniroyal
FB 2 H **diquat dibromide** ICI
FB 7 FI **ethylmercury chloride + lindane** Borzesti
FBC-32197 H **quizalofop-ethyl** Schering
FD-olja FA **petroleum oils + thiram** Hoechst
Febor R **chlorophacinone** Umupro
Fecundal F **imazalil** Janssen
Feeli Rosen-Spray FIA **piperonyl butoxide + pyrethrins + sulphur** Flora-Frey
Fegesol X **fatty acids** Staehler
Fegol X **animal meal + bitumen** Forst-Chemie, Skovkontor
Fekama AT I **butonate** Fettchemie
Fekama-Haftmittel S **latex** Fettchemie
Fekama-Spezial neu I **dichlorvos + lindane** Leuna-Werke
Fekama SVM X **animal oil** Fettchemie
Fekama-Tribuphon I **butonate** Fettchemie
Fekama-WM 308 X **tar oils** Fettchemie
fenac H **chlorfenac**
Fenam H **diphenamid** Siapa
Fenamin H **atrazine**
fenaminosulf F [140-56-7] sodium 4-dimethylaminobenzenediazosulphonate, phenaminosulf, DAPA, Bay 22555, Bay 5072, Lesan
fenamiphos N [22224-92-6] ethyl 4-methylthio-*m*-tolyl isopropylphosphoramidate, phenamiphos, Bay 68138, Nemacur
fenapanil F [61019-78-1] (±)-2-(imidazol-1-ylmethyl)-2-phenylhexanenitrile, RH-2161, Sisthane
fenarimol F [60168-88-9] (±)-2,4'-dichloro-α-(pyrimidin-5-yl)benzhydryl alcohol, Bloc, EL-222, Rimidin, Rubigan
Fenasip F **fenarimol** Sipcam
Fenasip Combi F **fenarimol + sulphur** Sipcam
Fenatrol H **chlorfenac** Union Carbide
fenazaflor A [14255-88-0] phenyl 5,6-dichloro-2-trifluoromethylbenzimidazole-1-carboxylate, fenoflurazole, Lovozal
fenazox AI [495-48-7] diphenyldiazene 1-oxide, Fentoxan
fenbutatin oxide A [13356-08-6] bis[tris(2-methyl-2-phenylpropyl)tin] oxide, fenbutatin oxyde, hexakis, ENT 27738, Neostanox, Osadan, SD 14114, Torque, Vendex
fenbutatin oxyde A **fenbutatin oxide**
fenchlorphos I [299-84-3] *O,O*-dimethyl *O*-(2,4,5-trichlorophenyl) phosphorothioate, ronnel, Ectoral, Etrolene, Korlan, Nankor, Ronnel, Trolene, Viozene
Fender H **phenmedipham** SDS Biotech
Fendona I **alphacypermethrin** Shell, Sorex
fenethacarb I [30087-47-9] 3,5-diethylphenyl methylcarbamate, phenetacarbe, BAS 235I
fenfuram F [24691-80-3] 2-methyl-3-furanilide, fenfurame, Pano-ram, Panoram, WL 22361
fenfurame F **fenfuram**
Fengib G **gibberellic acid + MCPA-thioethyl** Inagra
Fenibel I **fenitrothion** Sipcam-Phyteurop
Fenican H **diuron + terbuthylazine** Ciba-Geigy
Fenifid I **fenitrothion** Sofital
Fenilan A **dicofol + tetradifon** Caffaro
Fenilene F **fentin acetate** Scam
Fenion I **fenitrothion** Inagra
Feniter I **fenitrothion** Terranalisi
Fenitex I **fenitrothion** Anticimex
Fenithion K I **fenitrothion** Key
Fenition I **fenitrothion** Bang, Ficoop, Lantmaennen
Fenito IA **fenitrothion** Farmipalvelu
Fenitox I **fenitrothion** All-India Medical, Visplant
Fenitrogard I **fenitrothion** Sumitomo
Fenitron IA **fenitrothion** Farmos, KVK
Fenitrosivam I **fenitrothion** Sivam
fenitrothion I [122-14-5] *O,O*-dimethyl *O*-4-nitro-*m*-tolyl phosphorothioate, MEP,

OMS 43, ENT 25715, AC 47300, Accothion, Agrothion, Bay 41831, Cekutrothion, Cytel, Dicofen, Fenitox, Fenitrogard, Folithion, Novathion, S-1102A, S-5660, Sumipower, Sumithion

Fenitrotion I **fenitrothion** Galenika, Zorka Sabac, Zorka Subotica

fenizon A **fenson**

Fenkill IA **fenvalerate** United Phosphorus

Fenmedifam H **phenmedipham** Imex-Hulst, Zorka Sabac

Fenobit H **MCPA-thioethyl** Hokko

fenobucarb I [3766-81-2] 2-*sec*-butylphenyl methylcarbamate, BPMC, Bassa, Bay 41367c, Baycarb, Carvil, Hopcin, Osbac

Fenocap F **dinocap** Chemia

Fenodit H **2,4-D** Sipcam

fenoflurazole A **fenazaflor**

Fenogran H **2,4-D + MCPA** Scam

Fenom I **cypermethrin** Ciba-Geigy

fenophosphon I **trichloronat**

fenoprop HG [93-72-1] (±)-2-(2,4,5-trichlorophenoxy)propionic acid, silvex, 2,4,5-TP, AquaVex, Fruitone T, Kuron, Kurosal

Fenoris H **dichlorprop + MCPA** EniChem

Fenormone H **2,4,5-T** Pepro

fenothiocarb A [62850-32-2] *S*-4-phenoxybutyl dimethylthiocarbamate, B1-5452, KCO-3001, Panocon

Fenotral H **ioxynil + mecoprop** Sipcam

Fenotri H **2,4,5-T** Chimac-Agriphar

Fenox S H **dicamba + dichlorprop + MCPA** Schering

fenoxan H **dimethazone**

fenoxaprop-ethyl H [66441-23-4] ethyl (±)-2-[4-(6-chloro-2-benzoxazol-2-yloxy)phenoxy]propionate, Acclaim, Furore, Hoe 33171, Whip

Fenoxilene H **MCPA** Sipcam

fenoxycarb I [79127-80-3] ethyl 2-(4-phenoxyphenoxy)ethylcarbamate, ACR-2807 B, Insegar, Logic, Pictyl, Torus, Varikill

fenphosphorin I **dioxabenzofos**

fenpirithrin I [68523-18-2] (*RS*)-cyano(6-phenoxy-2-pyridyl)methyl (1*RS*,3*RS*:1*RS*,3*SR*)-3-(2,2-dichlorovinyl)-2,2-dimethylcyclopropanecarboxylate, fenpirithrine, Dowco 417, Vivithrin

fenpirithrine I **fenpirithrin**

fenpropathrin AI [64257-84-7 (racemate); 39515-41-8 (unstated stereochemistry)] (*RS*)-α-cyano-3-phenoxybenzyl 2,2,3,3-tetramethylcyclopropanecarboxylate, fenpropathrine, Danitol, Herald, Kilumal, Meothrin, Ortho Danitol, Rody, S-3206

fenpropathrine AI **fenpropathrin**

fenpropidin F [67306-00-7] (*RS*)-1-[3-(4-*tert*-butylphenyl)-2-methylpropyl]piperidine, fenpropidine, Patrol, Ro 12-3049

fenpropidine F **fenpropidin**

fenpropimorph F [67564-91-4] (±)-*cis*-4-[3-(4-*tert*-butylphenyl)-2-methylpropyl]-2,6 -dimethylmorpholine, fenpropimorphe, BAS421F, Corbel, Mistral

fenpropimorphe F **fenpropimorph**

fenson A [80-38-6] 4-chlorophenyl benzenesulphonate, fenizon, CPBS, PCPBS, Murvesco

fensulfothion NI [115-90-2] *O,O*-diethyl *O*-4-methylsulphinylphenyl phosphorothioate, DMSP, OMS 37, ENT 24945, Bay 25141, Dasanit, S-767, Terracur P

Fentan I **fenitrothion** Bimex

fenthiaprop-ethyl H [66441-11-0] ethyl (±)-2-[4-(6-chlorobenzothiazol-2-yloxy)phenoxy]propionate, Hoe 35609, Joker, Taifun

fenthion I [55-38-9] *O,O*-dimethyl *O*-4-methylthio-*m*-tolyl phosphorothioate, MPP, mercaptophos, OMS 2, ENT 25540, Bay 29493, Baycid, Baytex, Entex, Lebaycid, Queletox, S-1752, Tiguvon

Fentifen I **fenitrothion + fenthion** Bayer

Fentin A Super F **fentin acetate + maneb** Van Wesemael

fentin acetate FLM [900-95-8] triphenyltin acetate, fentine acetate, TPTA, asetat fenolovo, OMS 1020, ENT 25208, Batasan, Brestan, Chimate, Fenilene, Fentol, Hoe 2824, HokkoSuzu, Liro-Tin, Pandar, Phenostan-A, Stanex, Suzu, Tinestan, Ucetin, VP 1940

fentin hydroxide F [76-87-9] triphenyltin hydroxide, fentine hydroxyde, TPTH, TPTOH, gidrookie fenolovo, OMS 1017, ENT 28009, Brestan Flow, Du-Ter, Duter, Farmatin, Flo Tin, Flotin, Haitin, Phenostat-H, Super-Tin, Suzu-H, Triple Tin, Tubotin

Fentin Supra F **fentin acetate + maneb** CTA

fentine acetate F **fentin acetate**

fentine hydroxyde F **fentin hydroxide**

Fentol F **fentin acetate** Bayer

Fentospor F **fentin acetate** Solfotecnica
Fentoxan AI **fenazox** Fahlberg-List
Fentrex A **difenacoum** Rentokil
fentrifanil A [62441-54-7] *N*-(6-chloro-α,α,α-trifluoro-*m*-tolyl)-α,α,α-trifluoro-4,6-dinitro-*o*-toluidine, hexafluoramin, PP 199
Fentrol R **difenacoum** Rentokil
Fentron I **fenitrothion** Efthymiadis
fenulon H **fenuron**
fenuron H [101-42-8; 4482-55-7 (trichloroacetate)] 1,1-dimethyl-3-phenylurea, PDU, fenulon, Dozer, Dybar, Falisilvan, GC-2603, Urab
fenvalerate IA [51630-58-1] (*RS*)-α-cyano-3-phenoxybenzyl (*RS*)-2-(4-chlorophenyl)-3-methylbutyrate, OMS 2000, Belmark, Ectrin, Fenkill, Pydrin, S-5602, Sanmarton, Sumibac, Sumicidin, Sumifleece, Sumifly, Sumitick, WL 43775
Fenzalate corbeaux FX **anthraquinone + fenfuram + imazalil** Prochimagro
Fenzalate triple FIX **anthraquinone + fenfuram + imazalil + lindane** Prochimagro
Fenzol F **fenarimol + sulphur** Siapa
ferbam F [14484-64-1] iron tris(dimethyldithiocarbamate), ferbame, ENT 14689, AAfertis, Carbamate, Ferbam, Ferberk, Hexaferb, Knockmate, Trifungol
Ferbamate F **ferbam** Bourgeois
ferbame F **ferbam**
Ferberk F **ferbam**
Fermax H **linuron + trifluralin** Sipcam-Phyteurop
Fernesta H **2,4-D** ICI-Zeltia
Fernex I **pirimiphos-ethyl** ICI
Fernide F **thiram** ICI, ICI-Valagro, ICI-Zeltia
Fernimine H **2,4-D** ICI-Zeltia
Ferniram F **ziram** ICI-Zeltia
Fernos I **pirimicarb** ICI
Ferox R **warfarin** C.G.I.
Ferox bloc raticide R **coumatetralyl** C.G.I.
Ferox hydrofuge R **difenacoum** C.G.I.
Ferox rat super R **difenacoum** C.G.I.
Ferox souris super R **difenacoum** C.G.I.
Ferrax FX **anthraquinone + ethirimol + flutriafol + oxine-copper** Sopra
Ferrax F **ethirimol + flutriafol + thiabendazole** ICI
Fertigkoeder R **warfarin** Schaumann
Fertosan-Schneckenkiller M **aluminium sulphate** Austrosaat
Fervin H **alloxydim-sodium** Schering, Schering AAgrunol
Fervinal H **sethoxydim** Schering, Schering AAgrunol
Fettel H **dicamba + mecoprop + triclopyr** Farm Protection
Fibacol M **metaldehyde** Ficoop
Fibaril I **carbaryl** Ficoop
Fibarroz H **propanil** Ficoop
Fibenzol F **benomyl** Agrocros
Fiberelico G **gibberellic acid** Ficoop
Ficam I **bendiocarb** Schering, NOR-AM
Ficlorfon I **trichlorfon** Ficoop
Ficobre F **copper oxychloride** Ficoop
Ficotradifol A **dicofol + tetradifon** Ficoop
Ficozeb F **mancozeb** Ficoop
Ficsan H **hexazinone** Gullviks
Fifluralin H **trifluralin** Ficoop
Filariol IA **bromophos-ethyl**
Filex F **propamocarb hydrochloride** Schering, Schering AAgrunol
Filitox IA **methamidophos** Bitterfeld
Fimatrazina H **atrazine + simazine** Ficoop
Fimazina H **simazine** Ficoop
Fimojal S **polyglycolic ethers** Ficoop
Fina ugrasolje H **petroleum oils** Fina
Fina vestan ema S **petroleum oils** Fina
Final X **fenvalerate** Nordisk Alkali
Final H **glufosinate-ammonium** Hoechst
Finale H **glufosinate-ammonium** Argos, Bayer, Hoechst
Finale onkruidruimer H **glufosinate-ammonium** Luxan
Finaven H **difenzoquat methyl sulphate** Cyanamid
Finesse H **chlorsulfuron + metsulfuron-methyl** Du Pont
Finetyl D I **chlorpyrifos + dimethoate** Prochimagro
Finimouse R **zinc phosphide** Bogena
Finocap FA **dinocap** Ficoop
Finotox I **diazinon + malathion** Mylonas
Fiodina F **dodine** Ficoop
Firestop F **flumequine** 3M
Firodal H **bromofenoxim + dicamba** Ciba-Geigy
Firodal C F **carbendazim + propiconazole** Siapa
Fish Tox IA **rotenone**
Fisons Ameisen-Spray I **bendiocarb** Fisons
Fisons Ameisen-Staub I **bendiocarb** Fisons
Fisons Ant Killer R **phoxim** Fisons

Fisons araignees rouges A **dicofol** SCAC-Fisons
Fisons blanc du rosier Oidium F **bupirimate** SCAC-Fisons
Fisons bouillie bordelaise F **Bordeaux mixture** SCAC-Fisons
Fisons Brennessel-Vernichter H **dicamba + MCPA** Fisons
Fisons courtilieres I **lindane** SCAC-Fisons
Fisons desherbant bulbes H **chlorpropham** SCAC-Fisons
Fisons desherbant gazon liquide H **2,4-D + mecoprop** SCAC-Fisons
Fisons desherbant total S liquide H **amitrole + atrazine** SCAC-Fisons
Fisons engrais antimousse gazon A H **iron sulphate** SCAC-Fisons
Fisons engrais desherbant gazon H **MCPA + mecoprop** SCAC-Fisons
Fisons engrais desherbant rosiers et arbustes H **propyzamide + simazine** SCAC-Fisons
Fisons Evergreen H **2,4-D + dicamba** Fisons
Fisons Filex F **propamocarb hydrochloride** Fisons
Fisons Granular Herbazin Total H **atrazine** Fisons
Fisons Greenfly and Blackfly Killer I **dimethoate + lindane + malathion** Fisons
Fisons Herbazin H **simazine** Fisons
Fisons Herbazin Plus H **amitrole + simazine** Fisons
Fisons Herbazin Special H **amitrole + atrazine + 2,4-D** Fisons
Fisons Insect Spray for Houseplants IA **permethrin** Fisons
Fisons Lawn Sand H **iron sulphate** Fisons
Fisons Lawn Spot Weeder H **2,4-D + dicamba** Fisons
Fisons Limaclor Schneckenkorn M **metaldehyde** Fisons
Fisons mauvaises herbes H **amitrole + ammonium thiocyanate** SCAC-Fisons
Fisons Mosskil H **iron sulphate** Fisons
Fisons New Improved Problem Weeds Killer H **amitrole + MCPA** Fisons
Fisons Path Weeds Killer H **amitrole + MCPA + simazine** Fisons
Fisons Peltar F **maneb + thiophanate-methyl** SCAC-Fisons
Fisons Pucerons S I **dimethoate** Scac-Fisons
Fisons Rasenduenger mit Moosvernichter H **iron sulphate** Fisons
Fisons Rasenduenger mit Unkrautvernichter H **2,4-D + dicamba** Fisons
Fisons Rasenunkraut-Spray H **2,4-D + dicamba** Fisons
Fisons Rasenunkraut-Vernichter H **dicamba + MCPA** Fisons
Fisons silmine R **scilliroside** SCAC-Fisons
Fisons Slug and Snail Killer M **metaldehyde** Fisons
Fisons Soil Pest Killer R **phoxim** Fisons
Fisons stop mousses H **dichlorophen** Scac-Fisons
Fisons Total Unkraut Spray H **amitrole + MCPA** Fisons
Fisons Total-Unkrautvernichter H **amitrole + MCPA + simazine** Fisons
Fisons Turf Weeds Killer H **dicamba + MCPA** Fisons
Fisons Whitefly and Caterpillar Killer IA **permethrin** Fisons
Fisulfan IA **endosulfan** Ficoop
Fitanebe F **maneb + zineb** Sapec
Fitanol I **petroleum oils** Sapec
Fitexion I **malathion** Sapec
Fitios IA **ethoate-methyl**
Fitobar FI **barium polysulphide** Eli Lilly
Fitocid H **sodium chlorate** Caffaro
Fitocor G **1-naphthylacetic acid** Visplant
Fitodith F **zineb** Chimiberg
Fitofen I **fenitrothion** Ital-Agro
Fitofil S **polyglycolic ethers** Agrimont
Fitogamma Inodoro I **lindane** Agrimont
Fitogold G **borax** Du Pont
Fitolin H **linuron** Aziende Agrarie Trento
Fitomyl F **benomyl** Chimiberg
Fitonex Combi H **2,4-D + MCPA** Terranalisi
Fitonex D H **2,4-D** Terranalisi
Fitonex MC H **MCPA** Terranalisi
Fitonil F **zineb** Ciba-Geigy
Fitoran-Gruen F **copper oxychloride** Ciba-Geigy
Fitorex F **mancozeb** Ciba-Geigy
Fitormin G **2,4-D** Afrasa
Fitosan I **malathion** Scam
Fitosan-Extra F **copper oxychloride + zineb** Sapec
Fitothion I **fenitrothion** ERT
Fitracina liquida H **atrazine** Ficoop
Fivalin F **zineb** Fivat
Fix Schneckenkorn M **metaldehyde** Alschu
Fixofruit G **dichlorprop** C.F.P.I., Etisa
FK-Kombi Flytende H **dichlorprop + MCPA** Felleskjoepet

FK Kvikksoelvbeis F **methoxyethylmercury acetate** Felleskjoepet
FK Kvikksoelvbeis VT F **methoxyethylmercury acetate** Felleskjoepet
FL-80 + Karnak F **captan + zineb** Lainco
FL-80 Fuerte Mn F **zineb** Lainco
FL 173 H **bromuron** Fahlberg-List
flamprop-M-isopropyl H [57973-67-8] isopropyl *N*-benzoyl-*N*-(3-chloro-4-fluorophenyl)-D-alaninate, L-flamprop-isopropyl, Barnon, Commando, Suffix BW, WL 43425
flamprop-methyl H [52756-25-9] methyl *N*-benzoyl-*N*-(3-chloro-4-fluorophenyl)-DL-alaninate, Lancer, Mataven, WL 29761
Fleur-Ameisen-Giess I **bromophos** Schering
Fleur-bladluis I **pirimicarb** Schering AAgrunol
Fleur granulaat I **chlorfenvinphos** Schering AAgrunol
Fleur-Insekten I **bromophos** Schering
Fleur-Moos-frei H **iron sulphate** Schering
Fleur Rozen spuitbus IAF **dicofol + malathion + thiophanate-methyl** Schering AAgrunol
Fleur rupsin I **permethrin** Schering AAgrunol
Fleur-Unkraut-Giess H **bromacil + diuron** Schering
Fleur-Unkraut-Streu H **dichlobenil** Schering
Flex H **fomesafen** ICI
Flexidor H **isoxaben** Elanco
Flibol I **trichlorfon** Fettchemie, Leuna-Werke
Flibol PE I **butonate + dichlorvos** Leuna-Werke
Flibol-Extra I **naled** Fettchemie
Fliefos I **chlorfenvinphos** Leu & Gygax
Fliegen/Ungeziefer-Spray Jacutin I **bromophos + pyrethrins** Shell
Flo Tin F **fentin hydroxide** Agtrol
flocoumafen R [90035-08-8] 4-hydroxy-3-[1,2,3,4-tetrahydro-3-[4-(4-trifluoromethylbenzyloxy)phenyl]-1-naphthyl]coumarin, flocoumafene, Storm, Stratagem, WL 108366
flocoumafene R **flocoumafen**
Flora-hyonteistikku IA **butoxycarboxim** Farmos
Florabella antibladluis I **pirimicarb** Klasmann
Florabella Rasenduenger mit Unkrautvernichter H **2,4-D + dicamba** Beckmann Duenger
Floradix N G **benomyl + captan + 1-naphthylacetic acid** Glowacki
Floramon G **1-naphthylacetic acid** Danydea, Novotrade
Floranid H **2,4-D + dicamba** BASF
Florasan F **imazalil** Siapa
Flordimex G **ethephon** Bitterfeld
Florel G **ethephon** Union Carbide
Floristella Kenpor FA **sulphur** Kenorlo uzem
Flornet H **paraquat dichloride** Sofital
Florovin plant-pin stapici I **butoxycarboxim** Pliva
Flortis Diserbante Totale H **atrazine + simazine + TCA** Orvital
Flotin F **fentin hydroxide**
Flotox FA **sulphur** O.P., Ortho, Propfe
Flozin F **zineb** SEGE
fluazifop-butyl H [69806-50-4] butyl (*RS*)-2-[4-(5-trifluoromethyl-2-pyridyloxy)phenoxy]propionate, Fusilade, Hache Uno Super, IH 773B, Onecide, PP 009, TF1169
fluazifop-P-butyl H [79241-46-6] butyl (*R*)-2-[4-(5-trifluoromethyl-2-pyridyloxy)phenoxy]propionate, Fusilade 5, PP 005
Flubalex H **benfluralin** Chemolimpex
Fluben I **diflubenzuron** Solfotecnica
flubenzimine AF [37893-02-0] (2*Z*,4*E*,5*Z*)-N^2,3-diphenyl-N^4,N^5-bis(trifluoromethyl)-1,3-thiazolidine-2,4,5-triylidenetriamine, Bay SLJ 0312, Cropotex
fluchloralin H [33245-39-5] *N*-(2-chloroethyl)-2,6-dinitro-*N*-propyl-4-(trifluoromethyl)benzenamine, fluchloraline, BAS 3920, BAS 3921, BAS 3922, BAS 392H, Basalin
fluchloraline H **fluchloralin**
flucythrinate I [70124-77-5] (*RS*)-*α*-cyano-3-phenoxybenzyl (*S*)-2-(4-difluoromethoxyphenyl)-3-methylbutyrate, OMS 2007, AC 222,705, Cybolt, Cythrin, Pay-Off
Flue-kvit I **piperonyl butoxide + pyrethrins** Aeropak
Fluegol weiss X **fatty acids** Fluegel
fluenethyl I **fluenetil**
fluenetil I [4301-50-2] 2-fluoroethyl biphenyl-4-ylacetate, fluenethyl, Lambrol, M 2060

flufenoxuron IA 1-[4-(2-chloro-α,α,α-trifluoro-*p*-tolyloxy)-2-fluoro phenyl]-3-(2,6-difluorobenzoyl)urea, Cascade, WL 115110
Fluidosoufre F **sulphur** R.S.R., Tarsoulis, UCAR
Fluizol AF **sulphur** Caffaro
flumetralin G [62924-70-3] *N*-(2-chloro-6-fluorobenzyl)-*N*-ethyl-α,α,α-tri fluoro-2,6-dinitro-*p*-toluidine, flumetraline, CGA 41065, Prime, Prime+
flumetraline G **flumetralin**
flumezin H [25475-73-4] 2-methyl-4-(α,α,α-trifluoro-*m*-tolyl)-1,2,4-oxadiazine-3,5-dione, flumezine, BAS 348H
flumezine H **flumezin**
fluometuron H [2164-17-2] 1,1-dimethyl-3-(α,α,α-trifluoro-*m*-tolyl)urea, C 2059, Cotoran, Cottonex, Higalcoton
Fluorakil R **fluoroacetamide** Rentokil
fluorbenside A [405-30-1] 4-chlorobenzyl 4-chlorophenyl sulphide, Fluorparacide, Fluorsulphacide, HRS 924, RD 2454
fluoroacetamide R [640-19-7] 2-fluoroacetamide, Baran, Compound 1081, Fluorakil, Fussol, Navron, Rodex, Yanock
fluorochloridone H **flurochloridone**
fluorodifen H [15457-05-3] 4-nitrophenyl α,α,α-trifluoro-2-nitro-*p*-tolyl ether, fluorodifene, C 6989, Preforan
fluorodifene H **fluorodifen**
fluoroglicofene-ethyl H **fluoroglycofen-ethyl**
fluoroglycofen-ethyl H [77501-90-7] ethyl *O*-[5-(2-chloro-α,α,α-trifluoro-*p*-tolyloxy)-2-nitrobenzoyl]glycolate, fluoroglicofene-ethyl, Compete, RH-0265
fluoromide F [41205-21-4] 2,3-dichloro-*N*-4-fluorophenylmaleimide, MK-23, Spartcide
fluoromidine H [13577-71-4] 6-chloro-2-trifluoromethyl-3*H*-imidazo[4,5-*b*] pyridine, fluromidine, NC 4780
Fluorparacide A **fluorbenside**
Fluorsulphacide A **fluorbenside**
fluothiuron H [33439-45-1] 3-[3-chloro-4-(chlorodifluoromethylthio)phenyl]-1,1-dimethylurea, Bay KUE 2079A, Clearcide
fluotrimazole F [31251-03-3] 1-(3-trifluoromethyltrityl)-1*H*-1,2,4-triazole, Bay BUE 0620, Persulon
Flural H **trifluralin** Key
Fluralex H **trifluralin** Protex
Fluralor H **trifluralin** Tradi-agri
Fluran H **trifluralin** Afrasa
flurecol H **flurecol-butyl**
flurecol-butyl H [2314-09-2] butyl 9-hydroxyfluorene-9-carboxylate, flurenol-butyl, flurenol, flurecol, Aniten, Anitop, IT 3233
Flurene H **trifluralin** Chimiberg, Sepran
flurenol H **flurecol-butyl**
flurenol-butyl H **flurecol-butyl**
fluridone H [59756-60-4] 1-methyl-3-phenyl-5-(α,α,α-trifluoro-*m*-tolyl)-4-pyr idone, Brake, EL-171, Pride, Sonar
Flurin H **trifluralin** SPE
flurochloridone H [61213-25-0] 3-chloro-4-chloromethyl-1-(3-trifluoromethylphenyl)-2-pyrrolidinone, fluorochloridone, R-40244, Racer
fluromidine H **fluoromidine**
fluroxypyr H [69377-81-7; 81406-37-3 (1-methylheptyl ester)] 4-amino-3,5-dichloro-6-fluoro-2-pyridyloxyacetic acid, Dowco 433, Starane
flurprimidol G [56425-91-3] (*RS*)-2-methyl-1-pyrimidin-5-yl-1-(4-trifluoromethoxy)phenylpropan-2-ol, Cutless, EL-500
flusilazol F [85509-19-9] 1-[[bis(4-fluorophenyl)methylsilyl]methyl]-1*H*-1,2,4-triazole, flusilazole, DPX H6573, Nustar, Olymp, Punch
flusilazole F **flusilazol**
flutolanil F [66332-96-5] α,α, α-trifluoro-3'-isopropoxy-*o*-toluanilide, Moncut, NNF-136
Flutox I **piperonyl butoxide + pyrethrins** Superfos
flutriafol F [76674-21-0] (*RS*)-2,4'-difluoro-α-(1*H*-1,2,4-triazol-1-ylmethyl)benzhydryl alcohol, Impact, PP 450
Flutrix H **trifluralin** Sinteza
fluvalinate IA [102851-06-9] cyano(3-phenoxyphenyl)methyl *N*-[2-chloro-4-(trifluoromethyl)phenyl]-D-valinate, Klartan, Mavrik, Spur
Fluweed H **trifluralin** Visplant
Flux F **sulphur** Marty et Parazols
Fluxol I **lindane** Sedagri
Fly-Toxol I **diazinon** Arkovet, Ciba-Geigy
Flymaster I **piperonyl butoxide + pyrethrins** Sanerings
Flytrol IA **diazinon** Rentokil
FMC 1240 AI **ethion** FMC
FMC 2995 H **swep**
FMC 4512 H **pentanochlor** FMC
FMC 4556 H **chloranocryl**

FMC 5462 IA **endosulfan** FMC
FMC 5767 I **endothion**
FMC 9102 F **metiram** FMC
FMC 9260 I **tetramethrin**
FMC 10242 IAN **carbofuran** FMC
FMC 11092 H **karbutilate** FMC
FMC 17370 I **resmethrin** FMC
FMC 33297 I **permethrin** FMC
FMC 35001 I **carbosulfan** FMC
FMC 54800 IA **bifenthrin** FMC
FMC 57020 H **dimethazone** FMC
FMC 67825 NI **S,S-di-sec-butyl O-ethyl phosphorodithioate** FMC
Foamer F **2-phenylphenol** Fomesa
Foamex F **2-phenylphenol** Brogdex
Focal F **carbendazim** Bayer, Schering
Focus H **cycloxydim** BASF
FOE 1976 H **mefenacet** Bayer
FOG 2 I **dichlorvos + piperonyl butoxide + pyrethrins** Frowein
FOG 3 I **dichlorvos + malathion** Frowein
Fogard H **atrazine** Siapa
Fogard S H **atrazine + simazine** Siapa
Fogox 3 HFI I **piperonyl butoxide + pyrethrins** Hecquet
Fogstral G **chlorpropham** Dreyfus
Folafan F **folpet** Afrasa
Folane I **fonofos + lindane** Rhodiagri-Littorale
Folbex A **chlorobenzilate** Ciba-Geigy
Folbex VA A **bromopropylate** Ciba-Geigy, Liro
Folcap F **dinocap + folpet** Sovilo
Folcarb Combi F **cymoxanil + folpet** Siapa
Foldic A **dicofol** Peppas
Foldicryl F **copper carbonate (basic) + copper sulphate +folpet** Prochimagro
Folex G **merphos** Rhone-Poulenc, Sandoz
Folgan F **carbendazim + folpet** Du Pont
Folicidin F **cypendazole**
Folicote S **petroleum oils** Agrolinz, Efthymiadis, Resoco
Folicur F **terbuconazole**
Folidol IA **parathion** Bayer
Folidol E 605 I **parathion + parathion-methyl** Bayer
Folidol M I **parathion-methyl** Bayer
Folidol oil I **parathion + petroleum oils** Bayer
Folimat IA **omethoate** Bayer, Pinus
Folimat T IA **omethoate + tetradifon** Bayer
Folimat TK AI **dicofol + omethoate + tetradifon** Bayer
Folimate IA **omethoate** Bayer
Foliol IA **petroleum oils** Shell
Folithion I **fenitrothion** Bayer, Pinus
Folition IA **fenitrothion** Berner
Folosan F **quintozene** Uniroyal
Folpan F **folpet** Anorgachim, Chromos, Diana, Edefi, Makhteshim-Agan
Folpax F **folpet** Agriben
Folpec F **folpet** Sapec
folpel F **folpet**
folpet F [133-07-3] *N*-(trichloromethylthio)phthalimide, folpel, Folpan, Fungitrol, Phaltan, Thiophal, Vinicoll
Folpet Ramato Blu F **copper oxychloride + folpet** Bayer
Folpet-Bordo F **copper oxychloride + folpet** Leu & Gygax
Folpetan F **folpet** Eli Lilly
Folpete F **folpet** Ciba-Geigy
Folplan F **folpet** Agriplan
Folpomix F **copper oxychloride + folpet + sulphur** Leu & Gygax
Folprame F **copper oxychloride + folpet** Caffaro
Folsystem F **folpet** Platecsa
Foltamin F **folpet** Du Pont
Foltan F **folpet** Scam
Foltane F **folpet** Sipcam-Phyteurop
Foltapet F **captafol + folpet** ICI
Foltapet Ramato F **captafol + copper oxychloride + folpet** ICI
Foltazip F **folpet** Interphyto
Foltene F **folpet** Inagra
Folticryl F **Bordeaux mixture + copper carbonate (basic) + folpet** Gen. Rep.
Folticuivre F **copper oxychloride + folpet + maneb** Gen. Rep., Prochimagro
Foltimil F **copper sulphate + folpet** Sipcam-Phyteurop
fomesafen H [72178-02-0] 5-(2-chloro-α,α,α-trifluoro-*p*-tolyloxy)-*N*-methyl sulphonyl-2-nitrobenzamide, fomesafene, Flex, PP 021, Reflex
fomesafene H **fomesafen**
Foncar F **mancozeb** Aragonesas
Fonex I **trichlorfon** Burri
Fonganil F **furalaxyl** Ciba-Geigy, Shell
Fongarid F **furalaxyl** Ciba-Geigy, Ligtermoet
Fongaride F **furalaxyl** Ciba-Geigy
Fonginebe F **propineb** Gaillard
Fongoren F **pyroquilon** Ciba-Geigy
Fongorene F **pyroquilon** Ciba-Geigy

Fongosan NFHI **dazomet** Kwizda, Pluess-Staufer, Prochimagro
fonofos I [944-22-9 (unstated stereochemistry); 66767-39-3 (racemate)] *O*-ethyl *S*-phenyl (*RS*)-ethylphosphonodithioate, OMS 410, Dyfonate, N-2790
Fopet F **folpet** Chemia
For-ester H **2,4-D** Synchemicals
Foral I **lindane + phorate** Zorka Sabac
Forasip IAN **phorate** Sipcam
Forate IAN **phorate** Caffaro, Chemia, Visplant
Foray I **Bacillus thuringiensis** Microbial Resources
Forbel F **fenpropimorph** KenoGard
Forca I **tefluthrin** ICI
Force I **tefluthrin** ICI, Sopra
Fore F **mancozeb** Rohm & Haas
Foresite H **oxadiazon** May & Baker
Foretox 1 I **DDT + lindane** Borzesti
Foretox 3 II **DDT + lindane + terpene polychlorinates** Borzesti
formaldehyde FB [50-00-0] formaldehyde, aldehyde formique, formalin, Dyna-Form
formalin FB **formaldehyde**
Formalina FB **formaldehyde** Tarnow-Zaklady, Victoria
Format H **clopyralid** Dow
Formec F **mancozeb** PBI Gordon
formetanate hydrochloride AI [23422-53-9] 3-dimethylaminomethyleneiminophenyl methylcarbamate hydrochloride, ENT 27566, Carzol, Dicarzol, SN 36056
Formiclor I **bioallethrin + methoxychlor + piperonyl butoxide** Zucchet
Formiclor 40 I **diazinon + piperonyl butoxide + pyrethrins** Zucchet
Formiclor 80 I **bioallethrin + diazinon + methoxychlor + piperonyl butoxide** Zucchet
Formilex I **sodium dimethylarsinate** Sovilo
formothion IA [2540-82-1] *S*-(*N*-formyl-*N*-methylcarbamoylmethyl) *O,O*-dimethyl phosphorodithioate, OMS 698, ENT 27257, Aflix, Anthio, J-38, SAN 6913I
Forst Nexen I **lindane** Epro
Forstgranulat H **hexazinone** Avenarius
Forstmausstop R **chlorophacinone** Schering
Forte P 65 H **chloridazon** Oxon
Forti gazon H **dicamba + mecoprop** LHN
Fortox I **pyrethrins + S421** Lambert
Fortrol H **cyanazine** Shell
Forza I **tefluthrin** ICI
Fos-Fall A G **S,S,S-tributyl phosphorotrithioate** Mobay
fosamine-ammonium H [25954-13-6] ammonium ethyl carbamoylphosphonate, DPX 1108, Krenite
Fosatox I **phosalone** Visplant
Fosazin IA **azinphos-methyl** Agriplan
Foschlor I **trichlorfon** Organika-Azot
Fosdan I **phosmet** Quiminor
Fosdon I **parathion-methyl** Agrotechnica
Fosdrin IA **mevinphos** Zupa, Kemira
fosetyl Al F **fosetyl-aluminium**
fosetyl-aluminium F [39148-24-8] aluminium tris(ethyl phosphonate), phosethyl Al, fosetyl Al, aluminium phosethyl, Aliette, LS 74783
Fosfamid IA **dimethoate** Zorka, Zorka Sabac
Fosfaran IA **malathion** Aragonesas
Fosferno IA **parathion** ICI
Fosferno M I **parathion-methyl** ICI
Fosfet I **parathion** Sivam
Fosfomal I **malathion** Terranalisi
Fosforol I **parathion** Margesin
Fosfosol I **parathion** Eli Lilly
Fosfotion IA **malathion** Dimitrova
Fosfura de Zinc RI **zinc phosphide** Delicia, Galenika
Fosfuro di Zinco R **zinc phosphide** Ital-Agro
Fosmagrex I **phosmet** Sadisa
Fosmal I **malathion** Sivam
Fosmetile I **parathion-methyl** Aziende Agrarie Trento
Fospar I **parathion-methyl** Agriplan
fospirate I [5598-52-7] dimethyl 3,5,6-trichloro-2-pyridyl phosphate, OMS 1168, ENT 27521, Dowco 217
Foster I **phosalone** Visplant
fosthietan IN [21548-32-3] diethyl 1,3-dithietan-2-ylidenephosphoramidate, AC 64475, Acconem, Geofos, Nem-a-tak
Fostion AI **prothoate** Montedison
Fostox I **parathion** Agrochemiki, Siapa
Fostox Metil I **parathion-methyl** Agrochemiki, Siapa
Fosulan IA **endosulfan** Agriplan
Fosvan IA **azinphos-methyl** Argos
Foszfotion IA **malathion** Dimitrova
Foxal H **bifenox + mecoprop** May & Baker
Foxpro H **bifenox + ioxynil + mecoprop** Pepro
Foxpro H **bifenox + dichlorprop + isoproturon** Agrotec

Foxstar H **bifenox + isoproturon + mecoprop** May & Baker
Foxtar H **bifenox + isoproturon + mecoprop** Rhone-Poulenc
Foxtar H **bifenox + dichlorprop + isoproturon** Agriben, Agrotec
Foxto H **bifenox + isoproturon + neburon** Pepro
Foxtril H **bifenox + dichlorprop + ioxynil** Agriben
F-Permetriini IA **permethrin** Farmos
Foxtril H **bifenox + ioxynil + mecoprop** Agrotec, Rhone-Poulenc
Framed H **simazine** Agrimont
Frankol-Combi Neu H **dicamba + MCPA** Franken-Chemie
Frankol-forte H **bromacil + diuron** Franken-Chemie
Frankol-i-Granulat H **borax + bromacil** Franken-Chemie
Frankol-i-Granulat Neu H **amitrole + diuron + ethidimuron** Franken-Chemie
Frankol-Paranol IA **petroleum oils** Franken-Chemie
Frankol-prompt H **diuron + paraquat dichloride** Franken-Chemie
Frankol-spezial H **diuron + methabenzthiazuron** Franken-Chemie
Frankol-vollaktiv H **sodium chlorate** Franken-Chemie
Freinherbe G **maleic hydrazide** JSB
Frescon M **trifenmorph**
Freshgard F **imazalil** FMC
Frigate S **tallow amine ethoxylate** SDS Biotech, Sipcam, Sipcam-Phyteurop
Frioul F **diclobutrazol + sulphur** Sopra
Frubel I **lindane** Sepran
Fructil B **flumequine** Chimac-Agriphar
Fructyben G **gibberellic acid** ERT
Frufix G **1-naphthylacetic acid** Inagra, Maag
Frugold F **mancozeb + sulphur** Visplant
Frugon G **1-naphthylacetic acid** Terranalisi
Fruit V IA **petroleum oils** Agriplan
Fruitdo F **oxine-copper** Sankyo
Fruitel G **ethephon** Serpiol
Fruitgard 2 AB F **2-aminobutane** Fomesa
Fruitgard P F **2-phenylphenol** Fomesa
Fruitone G **1-naphthylacetamide + 1-naphthylacetic acid + (2-naphthyloxy)acetic acid** EniChem
Fruitone G **1-naphthylacetamide + 1-naphthylacetic acid** Meoc
Fruitone Anticascola G **dichlorprop** EniChem
Fruitone DP G **dichlorprop** C.F.P.I.
Fruitone-NA G **1-naphthylacetic acid** Luxan
Fruitone T HG **fenoprop** Union Carbide
Fruitseal 2 I Extra F **imazalil** Fomesa
Fruitseal 2 P Extra F **2-phenylphenol** Fomesa
Fruitseal 2 I/28 Extra F **imazalil + 2-phenylphenol** Fomesa
Fruitseal 3 I Extra F **imazalil** Fomesa
Fruitseal ST Extra F **thiabendazole** Fomesa
Fruitseal T Extra F **thiabendazole** Fomesa
Fruitseal T/2 P Extra F **2-phenylphenol + thiabendazole** Fomesa
Frumex 24 H **2,4-D** Scam
Frumidor F **maneb + thiophanate-methyl** Sipcam
Frumidor M F **mancozeb + thiophanate-methyl** Inagra
Frumin AL IA **disulfoton** Sandoz
Frunax R **warfarin** Frunol Chemie
Frut Hormon G **1-naphthylacetic acid** Agrocros
Frutapon IA **petroleum oils** Agrolinz
Frutassa F **thiram** Sopepor
Frutogard F **ditalimfos** Spiess
Fruttal I **carbaryl** EniChem
Fruttene F **ziram** Sipcam
Fruttor G **gibberellic acid + 1-naphthylacetamide + (2-naphthyloxy)acetic acid** EniChem
Fruvit F **oxadixyl + propineb** Bayer
FT-2 F **calcium copper oxychloride** L.A.P.A., Siapa
FT-2F F **copper sulphate + folpet** L.A.P.A., Protex
Ftalinol F **captafol** EniChem
Ftalofos IA **phosmet** Sojuzchimexport
fthalide F [27355-22-2] 4,5,6,7-tetrachlorophthalide, phthalide, TCP, KF-32, Rabcide
fuberidazole F [3878-19-1] 2-(2'-furyl)benzimidazole, furidazol, Bay 33172, Voronit
Fubol F **mancozeb + metalaxyl** Ciba-Geigy
Fubotran F **dicloran** Grima
Fuclasin F **ziram** Hoechst, Permutadora, Schering
Fudiolan FI **isoprothiolane** Nihon Nohyaku
Fudrat R **warfarin** Siapa

Fuego H **glyphosate + simazine** Ravit
Fuji-one FI **isoprothiolane** Nihon Nohyaku
Fulit F **dodine + ziram** Kwizda
Fulkil IA **parathion-methyl** May & Baker
Full H **bromoxynil + pyridate** Ravit
Fulmit IA **petroleum oils** KenoGard
Fulmit Especial IA **DNOC + petroleum oils** KenoGard
Fulset G **(2-naphthyloxy)acetic acid** Asepta
Fulton H **napropamide + nitralin** Pepro
Fulvax F **cymoxanil + mancozeb** La Quinoleine
Fulvax C F **copper sulphate + cymoxanil + mancozeb** La Quinoleine
Fulvin Attivato TMTD F **sulphur + thiram** Sofital
Fulvin Attivato Ziram F **sulphur + ziram** Sofital
Fumarin R **coumafuryl** Union Carbide
Fumasol R **coumafuryl**
Fumathane FIHN **metham-sodium** Argos, Rohm & Haas
Fumazone N **dibromochloropropane**
Fumi-Cel IR **magnesium phosphide** Degesch, Goncalves
Fumi-Strip IR **magnesium phosphide** Degesch
Fumical FIHN **metham-sodium** Calliope
Fumicel I **magnesium phosphide** Degesch
Fumicid H **bromacil + diuron** Schacht
Fumigam FIHN **metham-sodium** Bourgeois, Lambert
Fumigan I **dichlorvos** Zupa
Fumigrain I **dichlorvos + malathion** L.C.B.
Fumisect 10 I **dichlorvos** L.C.B.
Fumisect 99 I **dichlorvos + malathion** L.C.B.
Fumisect 664 TR I **dichlorvos + malathion + pyrethrins** Tripette & Renaud
Fumispore F **4-hydroxyphenylsalicylamide** Brogdex
Fumistrip I **magnesium phosphide** Degesch
Fumite General Purpose Greenhouse Insecticide Smokes I **pirimiphos-methyl** ICI
Fumite Tecnalin Smoke Generators FI **lindane + tecnazene** ICI
Fumite Whitefly Greenhouse Insecticide Smokes I **permethrin** ICI
Fumitoxin I **aluminium phosphide** Desinsectisation, Mayr, Pestcon
Fumo R **potassium nitrate + sulphur** Adroka
Funaben 3 F **carbendazim** Organika-Fregata, Organika-Sarzyna
Funaben 4 F **captafol + carbendazim** Organika-Fregata
Funaben 50 F **carbendazim** Organika-Sarzyna
Funaben T F **carbendazim + thiram** Organika-Sarzyna
Funbas F **fenpropimorph** BASF
Fundal AI **chlordimeform** Schering, NOR-AM
Fundazol F **benomyl** Chemolimpex, ICI-Zeltia, Zeltia
Funedin FIHN **dazomet** Enotria
Fungaflor F **imazalil** Aragonesas, Ciba-Geigy, Duphar, Hortichem, Janssen, NordiskAlkali, Schering, Shell
Fungaflor TZ F **imazalil + thiabendazole** Schering
Fungapor F **maneb + thiram + zineb** Aporta
Fungatop F **imazalil + thiophanate-methyl** Duphar
Fungazil F **imazalil** Cillus, Duphar, Janssen
Fungazil-TBZ F **imazalil + thiabendazole** Cillus
Funghitan F **copper oxychloride + zineb** Bimex
Fungi MZ F **mancozeb** Sivam
Fungi TH F **thiram** Sivam
Fungicap F **captan** Van Wesemael
Fungi-cid Orto F **2-phenylphenol** Serpis
Fungicide Cupra F **copper oxychloride + zineb** Tarsoulis
Fungiclor F **dicloran** Visplant
Fungicombi F **thiram + ziram** Sivam
Fungilon F **dodine** Bayer
Fungiman F **maneb** Gullviks, Sivam
Funginex F **triforine** Celamerck, Chimac-Agriphar, FMC, Maag, Pennwalt, Rhodic, Sandoz, Sovilo, Synchemicals
Funginex Plus FIA **malathion + tetradifon + triforine** Chimac-Agriphar
Fungi-Rhap F **cuprous oxide** CP Chemicals
Fungisan F **ziram** Tecniterra
Fungisem FI **lindane + maneb** KenoGard
Fungitan F **captan** Sivam
Fungitec F **thiabendazole** Avitec
Fungitex F **thiram** Protex
Fungitox F **maneb + zineb** Gullviks
Fungitrol F **folpet** Chevron

Fungizid/fongicide-Gesal F **captafol + mancozeb + sulphur** Reckitt & Colman
Fungizid/fongicide-Pirox F **copper oxychloride + sulphur** Maag
Fungizid/Fongicide-Stop F **carbendazim + sulphur** Wyss Samen & Pflanzen
Fungizir F **ziram** Sivam
Fungochrom F **benomyl** Chromos
Fungoro F **captan** Quimicas Oro
Fungosan FIHN **dazomet** Visplant
Fungostop F **ziram** Visplant
Funguran F **copper oxychloride** Urania
Fungus Fighter F **thiophanate-methyl** May & Baker
Furacide IN **carbofuran** Visplant
Furacon I **benfuracarb** Siapa
Furadan IN **carbofuran** Chromos, Cillus, Condor, FMC, Inca, Kwizda, Rhone-Poulenc
Furado F **mancozeb + pyrifenox** La Quinoleine
Furagrex IAN **carbofuran** Sadisa
furalaxyl F [57646-30-7] methyl *N*-(2-furoyl)-*N*-(2,6-xylyl)-DL-alaninate, CGA 38140, Fonganil, Fongarid
Furalin I **carbofuran + lindane** Chromos
furathiocarb I [65907-30-4] butyl 2,3-dihydro-2,2-dimethylbenzofuran-7-yl *N,N*-dimethyl-*N,N*-thiodicarbamate, CGA 73102, Deltanet, Promet
furavax F **methfuroxam**
furcarbanil F [28562-70-1] 2,5-dimethyl-*N*-phenyl-3-furancarboxamide, BAS 319F
furidazol F **fuberidazole**
Furloe HG **chlorpropham** PPG
furmarin R **coumafuryl**
furmecyclox F [60568-05-0] methyl *N*-cyclohexyl-2,5-dimethylfuran-3-carbohydroxamate, BAS 389F, Campogran, Epic, Xyligen B
furophanate F [53878-17-4] methyl 4-(2-furfurylideneaminophenyl)-3-thioallophanate, RH-3928
Furore H **fenoxaprop-ethyl** Hoechst, Pluess-Staufer
Fusar F **8-hydroxyquinoline sulphate** Chemia
Fusarex FG **tecnazene** ICI
Fusariol-Neu-Universal F **carbendazim + imazalil** Marktredwitz
Fusatox Royal F **anilazine + benomyl + chlorothalonil** Graines Loras
Fusee AR **sulphur** Chimac-Agriphar, Protex
Fusee Sap No 3 IR **barium nitrate + sulphur** Franco-Belge
Fusee Top No 3 IR **barium nitrate + sulphur** Pyragric
Fusilade H **fluazifop-butyl** Agrolinz, AVG, Berner, ICI, ICI-Valagro, ICI-Zeltia, Ishihara Sangyo, Pinus, Sopra
Fusilade 5 H **fluazifop-P-butyl** ICI
Fusilade Extra H **fluazifop-P-butyl** Agrolinz, Ciba-Geigy
Fusilade W H **fluazifop-butyl** Bayer, ICI
Fusiman F **maneb** Kwizda
Fussol R **fluoroacetamide** Sankyo
Futschikato-Radikal-Unkrautvertilger H **sodium chlorate** Propfe
FW-293 A **dicofol**
FW-734 H **propanil** Rohm & Haas
FW-925 H **nitrofen** Rohm & Haas
Fydulan H **dalapon-sodium + dichlobenil** Chipman, Duphar, La Quinoleine, Shell
Fydumas H **dichlobenil + simazine** Du Pont, Galenika, Schering, Shell
Fydusit H **bromacil + dichlobenil** Shell
Fydutrix H **dalapon-sodium + dichlobenil + simazine** Duphar
Fyfanon IA **malathion** Cheminova
Fytolan F **copper oxychloride**
Fytospore F **cymoxanil + mancozeb** Farm Protection
Fytostrep B **streptomycin** Gist-Brocades
Fyzol S **petroleum oils** Schering Ready

G4 LFB **dichlorophen**
G 13006 I **athidathion**
G 20072 F **quinacetol sulphate**
G 23133 R **coumachlor** Ciba-Geigy
G 23992 A **chlorobenzilate** Ciba-Geigy
G 24163 A **chloropropylate** Ciba-Geigy
G 24480 IA **diazinon** Ciba-Geigy
G 27692 H **simazine** Ciba-Geigy
G 27901 H **trietazine** Ciba-Geigy
G 30028 H **propazine** Ciba-Geigy
G 30031 H **chlorazine**
G 30031 H **ipazine**
G 30044 H **simeton**
G 31435 H **prometon** Ciba-Geigy
G 32911 H **simetryn** Ciba-Geigy
G 34161 H **prometryn** Ciba-Geigy
G 34162 H **ametryn** Ciba-Geigy
G 34360 H **desmetryn** Ciba-Geigy
G 34690 H **methometon**
G 34698 H **mesoprazine**

G 36393 H **methoprotryne**
GA3 G **gibberellic acid**
Gabi Antimoos H **iron sulphate** GABI-Biochemie
Gabi Antimoos U Kombi H **2,4-D + iron sulphate + MCPA** GABI-Biochemie
Gabi Pflanzenspray IA **dimethoate** GABI-Biochemie
Gabi Rasenduenger mit UV H **2,4-D + MCPA** GABI-Biochemie
Gabi Rasenunkrautvernichter H **dicamba + MCPA** GABI-Biochemie
Gabi Schneckenkorn M **metaldehyde** GABI-Biochemie
Gabi Unkrautvernichter H **bromacil + diuron** GABI-Biochemie
Gabonil H **dicamba + MCPA** Budapesti Vegyimuvek
Gadisan H **linuron + trifluralin** Grima
Gafex F **copper oxychloride** Bayer
Gailletox H **mecoprop** Bourgeois
Galar H **bromacil + diuron** Siapa
Galation I **fenitrothion** Galenika
Galaxy F **cymoxanil + mancozeb** Maag
Galben F **benalaxyl** Agrimont, Margesin, Montedison
Galben F F **benalaxyl + folpet** Agrimont, La Littorale, Margesin, Montedison, Pluess-Staufer, Serpiol, Shell, Sipcam
Galben M F **benalaxyl + mancozeb** Agrimont, Dow, ICI-Valagro, La Littorale, Margesin, Montedison, Pluess-Staufer, Serpiol, Shell, Sipcam
Galben R F **benalaxyl + copper oxychloride** Montedison, Sipcam
Galecron AI **chlordimeform** Ciba-Geigy
Galepron-DT H **2,4-D + 2,4,5-T** Galenika
Galex H **metobromuron + metolachlor** Chromos, Ciba-Geigy
Galinet H **bentazone + dichlorprop** Leu & Gygax
Galiox H **bifenox + mecoprop** Eli Lilly
Galipan H **benazolin** Maag
Galition I **fenitrothion + malathion** Galenika
Galition plus I **fenitrothion + lindane** Galenika
Galium A H **mecoprop** Sipcam-Phyteurop
Galium extra H **MCPA + mecoprop** Sipcam-Phyteurop
Gallant H **haloxyfop-ethoxyethyl** Bayer, Chimac-Agriphar, Dow, Zorka Sabac
Gallogama I **lindane** Rhone-Poulenc
Gallusane H **amitrole + atrazine** Progallus
Galluzole TA H **amitrole + ammonium thiocyanate** Progallus
Galmin IA **petroleum oils** Galenika
Galokson H **paraquat dichloride** Galenika
Galoprop H **mecoprop** Galenika
Galpar I **parathion + petroleum oils** Galenika
Galtak H **benazolin** Ciba-Geigy
Galtox I **diazinon** Ziegler
Gamacid I **lindane** Pliva
Gamakarbatox I **carbaryl + lindane** Organika-Azot
Gamalo I **lindane** Maag
Gamametox I **lindane + methoxychlor** Organika-Azot
Gamaphex I **lindane**
Gamasat FI **captan + lindane + thiram** Maag
Gamasat L I **lindane** Maag
Gamaterr I **lindane** Propfe, Staehler
Gamatin I **lindane + thiram** Kemira
Gamazid I **lindane** Leu & Gygax
Gameron FIX **anthraquinone + lindane + methoxyethylmercury silicate** Liro
Gamex I **lindane** Burri
Gamexane I **lindane** ICI
Gamma I **lindane** Bayer, Epro, Nordisk Alkali, Shell, Tecniterra
gamma benzene hexachloride I **lindane**
gamma BHC I **lindane**
Gammacide I **lindane** Interphyto
Gamma-Col IA **lindane** ICI, ICI-Zeltia
Gammacol I **lindane** Sopra
Gammactif I **lindane** EMTEA, Procida
Gammagro I **lindane** Ital-Agro
gamma-HCH I **lindane**
gamma-HKhTsH I **lindane**
Gammakarbatox I **carbaryl + lindane** Organika-Azot
Gammalex FI **captan + lindane** ICI
Gammalex Liquid FI **carbendazim + lindane + thiram** ICI
Gammalin I **lindane** ICI
Gammalo-K-forte I **lindane** Agro
Gammamul I **lindane** Fattinger
Gammapuder I **lindane** Agro
Gammarol-Supra I **lindane** Kwizda
Gammasan 5 I **diazinon** Eli Lilly
Gammasan 30 I **lindane** ICI
Gammaterr I **lindane** Agro
Gammatox I **copper sulphate + lindane** Cooper
Gammex I **lindane**
Gammexane I **lindane** ICI
Gammexide IA **lindane** AUV

Gammexine I **lindane** Sopra
Gamoan I **lindane** Aragonesas
Gamoline I **lindane** ICI
Gandural F **nuarimol** Quimigal
Ganerdon I **endosulfan + parathion** Inagra
Ganocide F **drazoxolon** ICI
Ganon H **benzthiazuron** Bayer
Garbol I **petroleum oils** Hoechst
Gardcide IA **tetrachlorvinphos** Shell
Garden Insect Powder I **fenitrothion** Doff-Portland
Garden Insect Spray I **lindane + pyrethrins** Secto
Gardena perfect H **chlorflurecol-methyl + MCPA + mecoprop** Du Pont
Gardentox IA **diazinon**
Gardenurs H **oxyfluorfen + propyzamide** Procida, Rohm & Haas
Gardol Ameisentod I **lindane** Floralis
Gardol Pflanzenspray IA **dichlorvos + dinocap + lindane** Bauhaus
Gardol Schneckentod M **metaldehyde** Floralis
Gardol Spezial Rasenduenger mit UKV H **2,4-D + dicamba** Bauhaus
Gardol Spezial Rasenduenger mit Moosvernichter H **iron sulphate** Bauhaus
Gardomil H **metolachlor + terbuthylazine** Ciba-Geigy
Gardona I **tetrachlorvinphos** Agroplant, EniChem, Shell, Zupa
Gardopax H **ametryn + terbuthylazine** Ciba-Geigy
Gardoprim H **terbuthylazine** Ciba-Geigy, Plantevern-Kjemi, Shell
Gardoprim F H **bromofenoxim + terbuthylazine** Ciba-Geigy
Gardoprim M H **chlorbromuron + terbuthylazine** Ciba-Geigy
Gardoprim plus H **metolachlor + terbuthylazine** Ciba-Geigy
Garlon H **triclopyr** Bayer, Burts & Harvey, Chimac-Agriphar, Chipman, Ciba-Geigy, Dow, ICI, La Quinoleine, Maag, Prochimagro, Schering, Spiess, Urania
Garlon D H **2,4-D + triclopyr** Bayer, Rhodic, Sovilo, Umupro
Garlon L H **clopyralid + triclopyr** Prochimagro
Garlon micron H **2,4-D + triclopyr** Ciba-Geigy
Garlozor H **triclopyr** Zorka Subotica
Garrathion IA **carbophenothion** Stauffer
Garten-Cit I **piperonyl butoxide + pyrethrins** Cit
Garten-Cit Staub I **piperonyl butoxide + pyrethrins + rotenone** Cit
Garten-Pflanzen-Spray N IA **omethoate** Flora-Frey
Garten-Substral Rasenduenger mit Unkrautvernichter H **2,4-D + dicamba** Lonza
Gartenbau-Cycocel G **chlormequat chloride** BASF, Compo, Spiess, Urania
Gartenkrone Rasenduenger mit Moosvernichter H **iron sulphate** Buchmann, Coop
Gartenkrone Rasenduenger mit Unkrautvernichter H **2,4-D + MCPA** Buchmann, Coop
Gartenland Rasenduenger mit Moosvernichter H **iron sulphate**
Gartenland Rasenduenger mit Unkrautvernichter H **2,4-D + dicamba**
Gartenperle Rosen-Spray FIA **piperonyl butoxide + pyrethrins + sulphur** Flora-Frey
Gartenperle Unkraut-frei Giess- und Spritzmittel H **bromacil + diuron** Flora-Frey
Gartenpracht Rasenduenger mit Moosvernichter H **iron sulphate** Gartenpracht
Gartenpracht Rasenduenger mit Unkrautvernichter H **2,4-D + MCPA** Gartenpracht
Gartenspray Parexan IA **piperonyl butoxide + pyrethrins** Shell
Gartrel H **MCPA + propanil + triclopyr** Ravit
Garvox I **bendiocarb** Efthymiadis, Grima, Pluess-Staufer, Schering, Schering AAgrunol
Garvoxin I **bendiocarb** Schering, Staehler
Garwin H **glyphosate** Monsanto
Gastoxin I **aluminium phosphide** Soprochim
Gastrotox M **metaldehyde** Sipcam
Gatnon H **benzthiazuron** Bayer
GAU 1356 F **rabenzazole**
Gauntlet F **nuarimol** Shell
Gazon-kuur H **iron sulphate** Mommersteeg
Gazon Net H **2,4-D + dicamba + MCPA + mecoprop** Bayer
Gazon Plus H **benazolin + dicamba + MCPA** Bayer
Gazonan H **dicamba + MCPA** Aseptan

Gazonfloranid met onkruidverdelger H **2,4-D + dicamba** BASF
GB Insecticide IA **malathion + piperonyl butoxide + pyrethrins** GB-Inno-BM
GC-2603 H **fenuron**
Gebutox HI **dinoseb** Chromos, Hoechst
Gehoelze-Unkraut-frei H **dichlobenil** Flora-Frey
Geigy total ukrudtsmiddel H **terbuthylazine** Ciba-Geigy
Geigy ukrudtsmiddel H **simazine** Ciba-Geigy, KVK
Gela-Rasenlangzeitduenger mit Moosvernichter H **iron sulphate** GELA-Duenger
Gela-Rasenlangzeitduenger mit Unkrautvernichter H **dicamba + MCPA** GELA-Duenger
Gelb Tox IAF **DNOC** Sivam
Gelbkarbol IA **anthracene oil + DNOC** Leu & Gygax, Pluess-Staufer
Gelon Immergruen Rasenduenger mit Moosvernichter H **iron sulphate** Dormann & Preuss
Gemafos I **parathion** Scam
Gemini H **chlorimuran-ethyl + linuron** Du Pont
Gemuese-Spritzmittel Polyram-Combi F **metiram** Shell
Genamin S **tallow amine ethoxylate** Monsanto
Genate H **butylate** PPG
Gencor I **hydroprene** Sandoz
Genep H **EPTC** PPG
General Weedkiller H **dinoseb** Protex
Genol S **petroleum oils** Ciba-Geigy
Genoxone ZX H **2,4-D + triclopyr** La Quinoleine
Geo Ameisenfrei I **lindane** Globol-Werk
Geo Bio Gartenspray FIA **lecithin + piperonyl butoxide + pyrethrins** Globol-Werk
Geo Gartenspray FIA **piperonyl butoxide + pyrethrins + sulphur** Globol-Werk
Geo Insektenpuder I **lindane** Globol-Werk
Geo Pflanzenspray I **dichlorvos** Globol-Werk
Geo Ratten- und Maeusefrei R **warfarin** Globol-Werk
Geo Schneckenfrei Neu M **metaldehyde** Globol-Werk
Geo Unkraut-Frei H **bromacil + diuron** Globol-Werk
Geobilan I **lindane** OHIS
Geocid I **carbofuran** Chromos
Geodan I **chlormephos** Margesin
Geodinfos I **chlorpyrifos** Siapa
Geofos I **parathion** Siapa
Geofos IN **fosthietan**
Geofos D I **diazinon** Siapa
Geohalkos F **copper oxychloride** Bredologos
Geolin I **lindane** Zorka, Zorka-Sabac
Geomalatox I **malathion** Bredologos
Geomet IAN **phorate** Cyanamid
Geometil I **parathion-methyl** Sofital
Geonter H **terbacil** Chemolimpex, Kobanyai
Geoptan F **captan** Bredologos
Geor FX **anthraquinone + ethirimol + flutriafol + oxine-copper** La Quinoleine
Geort FIHN **metham-sodium** Siapa
Geosan I **lindane** Aziende Agrarie Trento
Geosep I **malathion** Sepran
Geotan I **malathion** Bimex
Geotion I **malathion** Bimex
Geozineb F **zineb** Bredologos
Geramid-Neu G **1-naphthylacetamide** Gobbi
Geriko F **diniconazole** Pepro
Geriko Double F **diniconazole + iprodione** Agriben
Geriko super FX **anthraquinone + diniconazole + iprodione** Pepro
Gerilyn G **gibberellic acid** Farmer
Germate FI **carboxin + lindane + maneb**
Germex G **propham** Reis
Germex G **chlorpropham + propham** Maag
Germidorm C G **chlorpropham + propham** BASF
Germidorm C Extra GI **chlorpropham + piperonyl butoxide + propham +pyrethrins** BASF
Germidorm Extra GI **piperonyl butoxide + propham + pyrethrins** BASF
Germilate G **chlorpropham + propham** Agrocros
Germinate CM FIX **anthraquinone + ethion + oxine-copper** Pepro
Germinate CSP FX **anthraquinone + captan + carbendazim** Sedagri
Germinate double FX **anthraquinone + oxine-copper** Pepro
Germinate MG FIX **endosulfan + lindane + oxine-copper** Pepro
Germinate special mais FX **anthraquinone + captan** Pepro
Germinate T3 FIX **anthraquinone + lindane + oxine-copper** Pepro

Germinate T4 FIX **anthraquinone + endosulfan + lindane + oxine-copper** Pepro
Germinate TD F **thiram** Pepro
Germinate TD AC FX **anthraquinone + thiram** Pepro
Germinate triple liquide FIX **anthraquinone + lindane + oxine-copper** Pepro
Germino TS F **carbendazim + iprodione** Pepro
Germinol F **captan + carbendazim** Pepro
Germipro V5 X FX **anthraquinone + carboxin + iprodione + maneb** Pepro
Germisan F **carboxin + maneb** Sofital
Germisan GF F **carboxin + imazalil** Ciba-Geigy
Germisan spezial F **methfuroxam + thiabendazole** Ciba-Geigy
Germitan G **chlorpropham** Bimex
Germon G **1-naphthylacetic acid** Gobbi
Germostop G **chlorpropham + propham** Liro
Germotect FG **chlorpropham + propham + thiabendazole** Ligtermoet, Liro
Gerox B **streptomycin**
Gesabal H **ipazine**
Gesacral H **prometryn** Agrotechnica
Gesafloc H **trietazine** Ciba-Geigy
Gesafor H **diuron** Ciba-Geigy
Gesafram H **prometon** Ciba-Geigy
Gesagard H **prometryn** Ciba-Geigy, Dimitrova, Organika-Azot, Pinus, Plantevern-Kjemi, Viopharm
Gesagarde H **prometryn** Ciba-Geigy
Gesagram H **atrazine + metolachlor** Ciba-Geigy
Gesal M **metaldehyde** Ciba-Geigy
Gesal Ameisenmittel I **cypermethrin** Ciba-Geigy
Gesal Antimehltau F **bentaluron** Ciba-Geigy
Gesal antimousse H **chloroxuron + iron sulphate** Ciba-Geigy
Gesal Gazonmest en Onkruidverdelger H **2,4-D + dicamba + mecoprop** Ligtermoet
Gesal Insektizid IA **diazinon** Ciba-Geigy
Gesal Mosfjerner H **chloroxuron + iron sulphate** Ciba-Geigy
Gesal myremiddel I **diazinon** Ciba-Geigy
Gesal Rasenduenger mit Unkrautvernichter H **2,4-D + dicamba + mecoprop** Ciba-Geigy
Gesal Rasenpflege H **2,4-D + dicamba + mecoprop** Ciba-Geigy
Gesal Rosenspray FI **chloropropylate + dichlone + dimethoate + dinocap + piperonylbutoxide + pyrethrins** Ciba-Geigy
Gesal Rosenspritzmittel FIA **dinocap + dodine + monocrotophos** Ciba-Geigy
Gesal rosesproejtemiddel FI **dinocap + dodine + monocrotophos** Ciba-Geigy
Gesal Schneckenkoerner M **metaldehyde** Airwick
Gesal sneglekorn M **metaldehyde** Ciba-Geigy
Gesal ukrudtsskum H **2,4-D + MCPA** Ciba-Geigy
Gesal Unkrautvertilger H **2,4-D + secbumeton + simazine** Ciba-Geigy
Gesal Zimmerpflanzenspray IA **diazinon** Ciba-Geigy
Gesamil H **propazine** Ciba-Geigy, Liro
Gesamoos H **chloroxuron** Ciba-Geigy
Gesamoos Plus H **chloroxuron + iron sulphate** Airwick
Gesapax H **ametryn** Ciba-Geigy
Gesaprim H **atrazine** Ciba-Geigy, Ewos, Ligtermoet, Liro, Organika-Azot, Viopharm
Gesaprim Combi H **atrazine + terbutryn** Ciba-Geigy
Gesaprim H H **atrazine + 2,4-D** Kwizda
Gesaprim M H **atrazine + simazine** Ciba-Geigy
Gesaprim S H **atrazine + simazine** Ciba-Geigy
Gesaprime H **atrazine** Ciba-Geigy
Gesaran H **methoprotryne**
Gesaran 2079 H **methoprotryne + simazine** Ciba-Geigy
Gesastop H **simazine** Ciba-Geigy
Gesatamin H **atraton**
Gesaten H **ametryn + prometryn** Viopharm
Gesatene H **ametryn + prometryn** Ciba-Geigy
Gesatop H **simazine** Ciba-Geigy, Ewos, Ligtermoet, Liro, Organika-Sarzyna, Plantevern-Kjemi, Schering AAgrunol, Viopharm, Zerpa
Gesatope H **simazine** Ciba-Geigy
Gesik IA **monocrotophos** EniChem
Gesin H **2,4-D** Ciba-Geigy
Gesopral H **amitrole** Ciba-Geigy
Getreidehalmfestiger G **chlormequat chloride** Stinnes
Getreideherbizid H **dichlorprop + flurecol-butyl + ioxynil + MCPA** Shell

Gher H **propyzamide + simazine** Du Pont, LHN, Procida, Rohm & Haas
Gi-Tre G **gibberellic acid** Ital-Agro
Giallo Spray IA **DNOC + petroleum oils** EniChem
Giallolio IA **DNOC + petroleum oils** Siapa
Giav H **molinate** Chemia
Giavotox H **molinate** Visplant
Gibaifar G **gibberellic acid** Aifar
gibberellic acid G [77-06-5] (3*S*,3a*S*,4*S*,4a*S*,7*S*,9a*R*,9b*R*,12*S*)-7,12-dihydroxy-3-methyl-6-methylene-2-oxoperhydro-4a,7-methano-9b,3-propeno[1,2-*b*]furan-4-carboxylic acid, acide gibberellique, gibberellinA3, GA3, Activol, Berelex, Brellin, Ceku-Gib, Gibefol, Gibrel, Grocel, Pro-Gibb, Regulex
gibberellin A3 G **gibberellic acid**
Gibbons Pelleted Rat-Bait R **diphacinone** Tegok
Gibefol G **gibberellic acid** Eli Lilly, Inagra
Giber Fruit G **gibberellic acid** Afrasa
Giberefor G **gibberellic acid** Safor
Giberine G **1-naphthylacetamide + 1-naphthylacetic acid + thiourea** Agriphyt
Giberkey G **gibberellic acid** Key
Giberlan G **gibberellic acid** Scam
Giberluq G **gibberellic acid** Luqsa
Giberol G **gibberellic acid** Agrocros
Gibrel G **gibberellic acid** MSD AGVET
Gibrelex G **gibberellic acid** Biolchim, SEGE
Gibrescol G **gibberellic acid** Efthymiadis, Polfa-Kutno
Gibsan G **gibberellic acid** Grima
gidrookie fenolovo F **fentin hydroxide**
Giftweizen R **zinc phosphide** Neudorff, Staehler, Wuelfel
Giustiziere IA **pirimiphos-methyl** ICI
Gladiator A **clofentezine + cyhexatin** Du Pont
Gladiator H **chloridazon** Tripart
Glean H **chlorsulfuron** Du Pont, Farmos, Siapa
Glean C H **chlorsulfuron + methabenzthiazuron** Du Pont
Glean T H **chlorsulfuron + methabenzthiazuron** Du Pont
Glean TP H **bromoxynil + chlorsulfuron + ioxynil** Du Pont
Glialka H **glyphosate** AVG
Glifocoop H **glyphosate** Ficoop
Glifonox H **glyphosate** Crystal
Glifosate-STI H **glyphosate** Solfotecnica
Glitz-Cumarin Fertigkoeder R **warfarin** Schaeffner
Glodi rosa- og skrautplontuudi I **carbaryl + folpet + piperonyl butoxide + pyrethrins + rotenone** Kirk
glucochloral RX **chloralose**
glucochloralose RX **chloralose**
glufosinate-ammonium H [77182-82-2] ammonium 4-[hydroxy(methyl)phosphinoyl]-DL-homoalaninate, Basta, Final, Finale, Hoe 39866, Total
Glycel H **glyphosate** Excel
glyodin F [556-22-9] 2-heptadecyl-2-imidazolinium acetate, Crag Fungicide 341, Glyoxalidin, Glyoxide
Glyoxalidin F **glyodin**
Glyoxide F **glyodin**
glyphosate H [1071-83-6] *N*-(phosphonomethyl)glycine, Azural, Glifonox, Glycel, MON-0573, Muster, Polado, Rodeo, Roundup, Sonic, Spasor, Sting, Tumbleweed
glyphosine G [2439-99-8] *N,N*-bis(phosphonomethyl)glycine, CP 41845, Polaris
Gnatisid I **methoprene** Ewos
Go-Go-San H **pendimethalin** Cyanamid
Goal H **oxyfluorfen** Du Pont, Rohm & Haas, Serpiol, Siapa
Goalapon H **dalapon-sodium + oxyfluorfen** Rohm & Haas
Goemar B G **boric acid + gibberellin** Liro
Goemar BM G **auxins + gibberellin** Liro
Goemar MG G **gibberellin + magnesium oxide** Liro
Goeta Raattgift R **warfarin** Goeta Lantmaen
Gokilaht I **cyphenothrin** Sumitomo
Golclair F **sodium pentaborate + sulphur** EniChem
Golclair special F **oligomers + sulphur** Du Pont
Gold Crest I **chlordane** Sandoz
Goldcorn H **mecoprop** FCC
Golden Dew FA **sulphur** Wilbur-Ellis
Golden M I **dichlorvos + fenchlorphos** Hellafarm
Golden Malrin I **methomyl**
Golden Muscamone vliegendoder I **methomyl + pheromones** Ceva
Golden NT I **methomyl + pheromones** Sipcam
Golden Scab 'Rosso' F **copper compounds (unspecified) + folpet + maneb** Tecniterra

Golden Scab 'Verde' F **dodine + maneb + thiram** Tecniterra
Goldenon F **captan** Margesin
Goldibor F **borax + mancozeb + sulphur** Duphar
Goldion F **mancozeb + sulphur** Duphar
Golf-Rasenspray H **dicamba + MCPA** Agrolinz
Golf-Totalspray H **atrazine + 2,4-D** Agrolinz
Goliath H **phenmedipham** ABM Chemicals
Goltix H **metamitron** Agro-Kemi, Agroplant, Bayer, Berner, Pinus
Good-rite n.i.x. H **proxan-sodium**
Gopha-Rid R **zinc phosphide** Bell Labs
Gophacide R **phosacetim**
Gopher-Gitter R **strychnine**
Gorfos IA **dimethoate** Agriplan
Gorgosem I **malathion** KenoGard
Gori algefjerner L **benzalkonium chloride + tributyltin naphthenate** Gorivaerk
Gori desinfektion L **benzalkonium chloride + tributyltin naphthenate** Gorivaerk
Gori permetrol IA **permethrin** Gorivaerk
Gorsit H **sodium chlorate** Bayer
Gorsol S **polyglycolic ethers** Bayer, Tensia
Goudron Vegetal FW **natural resins** Franco-Belge
GP Weedkiller H **diuron + tar oils** Killgerm
GPC-5544 H **tebutam**
GR 222 H **propyzamide + simazine** Urania
Gradix Clor H **chlorotoluron** Inagra
Graesmattegoedsel Kombi H **2,4-D + dicamba** Svaloef
Grafam H **bromofenoxim + isoproturon + mecoprop** Liro
Grain quick R **chlorophacinone** Chromos
Grain Store Smoke I **lindane** Dow
Graincote IA **chlorpyrifos-methyl** Wellcome
Grakill H **MCPA-thioethyl + simetryn**
Gralam H **asulam** Agroplant
Gralat H **paraquat dichloride** Visplant
Graluq I **lindane** Luqsa
Gramazin H **paraquat dichloride + simazine** Scam
Gramevin H **dalapon-sodium** Shell
Graminacid H **dalapon-sodium** Caffaro
Graminex H **TCA** Condor, Van Wesemael
Graminex B H **diuron + paraquat dichloride** ICI
Graminon H **isoproturon** Ciba-Geigy, Ligtermoet, Liro
Graminon Extra H **isoproturon + methoprotryne** Ciba-Geigy
Graminon Plus H **bentazone + dichlorprop + isoproturon** BASF, Ciba-Geigy, Ligtermoet
Gramipon H **dalapon-sodium** Van Wesemael
Gramisan H **terbutryn** Zorka, Zorka Sabac
Gramit H **alloxydim-sodium** BASF
Gramix H **paraquat dichloride** Sopra
Gramix-Super H **dichlorprop + MCPA + mecoprop** Hermoo
Gramixel H **paraquat dichloride** Siapa, Sopra
Gramocil H **diuron + paraquat dichloride** ICI
Gramonol H **monolinuron + paraquat dichloride** Hoechst, ICI
Gramox H **diquat dibromide** ICI
Gramoxone H **paraquat dichloride** AVG, Dillen, ICI, ICI-Valagro, ICI-Zeltia, Organika-Sarzyna, Pinus, Schering, Sinteza, Sopra, Spolana
Gramoxone Plus H **diquat dibromide + paraquat dichloride** Sopra
Gramuron H **diuron + paraquat dichloride** ICI
Granador I **permethrin** Rhodiagri-Littorale
Granamide H **chlorthiamid** C.F.P.I.
Grananet-T H **amitrole + atrazine + diuron + picloram** Ciba-Geigy
Grananit-F F **fuberidazole + imazalil** Esbjerg
Granaplouse H **2,4-D + dicamba** Sedagri
Granater I **lindane** Sedagri
Grandor H **2,4-D** Eli Lilly
Granebe triple FIX **anthraquinone + lindane + maneb** Agrishell
Granet I **piperonyl butoxide + pyrethrins** Masso
Granex F **mancozeb** Protex
Granforza F **carboxin + maneb** Du Pont
Grangrano H **neburon + pendimethalin** ICI
Granisan F **copper oxychloride** Aporta
Granix H **MCPA** Van der Boom
Granock H **naproanilide + thiobencarb**
Granol FI **lindane + maneb** Chipman
Granoplus F **captan + mancozeb** Chemia

Granox P-F-M F **captan + maneb** Chipman
Granox Plus F **maneb + thiabendazole** Chipman
Granozan D F **carbendazim + maneb** Du Pont
Granstar H **metsulfuron-methyl** Du Pont
Granular Naptol H **chloramben** Synchemicals
Granule 2000 R **difenacoum** Salomez
Granurex H **neburon** Agriben, Pepro, Rhone-Poulenc
Granusol total H **amitrole + diuron + sodium thiocyanate** C.F.P.I.
Granutox IAN **phorate** Cyanamid
Grapamone T **pheromones** Sipcam
Grapol F **diclobutrazol + sulphur** Sopra
Grasex H **chloral hydrate** Bitterfeld
Grasidim H **sethoxydim** Sipcam
Grasip H **alloxydim-sodium** Siegfried, Sipcam
Grasipan H **alloxydim-sodium** Pinus, Sipcam
Graslan H **tebuthiuron** Elanco
Grasmat H **alloxydim-sodium** Sipcam
Grasmoord H **sodium chlorate** Scaldiswerken
Grasp H **tralkoxydim** ICI, Sopra
Graspaz H **alloxydim-sodium** Sipcam
Grassat H **glyphosate** Monsanto
Grassedge H **2,4-D + thiobencarb**
Grasskill H **paraquat dichloride** Geochem
Grazon H **picloram** Dow
Grazon 90 H **clopyralid + triclopyr** Dow
Green limit G **mefluidide** C.F.P.I.
Green Up Lawn Weed and Feed H **2,4-D + dicamba** Synchemicals
Green Up Mossfree H **iron sulphate** Synchemicals
Green Up Weedfree Lawn Weedkiller H **2,4-D + dicamba** Synchemicals
Green Up Weedfree Spot Weedkiller H **2,4-D + dicamba** Synchemicals
Greenhouse Smoke Crawling Pest Killer I **lindane** May & Baker
Greenhouse Smoke Disease Killer F **tecnazene** May & Baker
Greenhouse Smoke Whitefly Killer I **permethrin** May & Baker
Greenkeeper H **2,4-D + dicamba** Asef Fison, Moreels-Guano
Greenmaster and Mosskiller H **iron sulphate** Fisons
Greensand H **iron sulphate** Chimac-Agriphar
Greffix W **petroleum oils + petroleum wax** R.S.R.
Gregarix G **chlormequat chloride** La Quinoleine
Grelite Viritox G **propham** Figueiredo
Grelutin H **naptalam** Fahlberg-List
Grenade I **cyhalothrin** ICI, Coopers
Grex F **carbendazim + maneb** Condor
Grex TX F **carbendazim + maneb** Pepro
Griffex H **atrazine** Griffin
Grillkiller I **fenitrothion** Bimex
Grillosep I **malathion** Sepran
Grima GP G **gibberellic acid** Grima
Grimetilo FIHN **methyl bromide** Grima
Grittox F **zineb** Agrochemiki
Gro-Stop G **chlorpropham + propham** PLK
Grocel G **gibberellic acid** ICI
Groen-ex L **quaternary ammonium** Schering AAgrunol
Gropper H **metsulfuron-methyl** Du Pont
Groundhog H **amitrole + diquat dibromide + paraquat dichloride + simazine** ICI
Grovex S 295 I **piperonyl butoxide + pyrethrins** Groves
Grovex S 300 I **fenitrothion** Groves
Grovex ULV 400 H **pyrethrins** Groves
Grovex ULV 500 I **phenothrin + tetramethrin** Groves
Gruen 35 H **2,4-D + dicamba** Terrasan
Gruen/Vert M **metaldehyde** Pluess-Staufer
Gruenkupfer 'Linz' F **copper oxychloride** Agrolinz
Gruenkupfer Marktredwitz F **copper oxychloride** CFM, Marktredwitz
Gruenkupfer Schirm F **copper oxychloride** Sideco-Chemie
GS 12968 I **lythidathion**
GS 13005 IA **methidathion** Ciba-Geigy
GS 13010 A **prothidathion**
GS 13529 H **terbuthylazine** Ciba-Geigy
GS 14254 H **secbumeton** Ciba-Geigy
GS 14259 H **terbumeton** Ciba-Geigy
GS 14260 H **terbutryn** Ciba-Geigy
GS 16068 H **dipropetryn** Ciba-Geigy
GS 19851 A **bromopropylate** Ciba-Geigy
GS 29696 H **thiazafluron** Ciba-Geigy
GTA FX **guazatine acetates**
Guanidol F **dodine** Agrochemiki, Siapa
Guardaton F **chlorothalonil + nuarimol** Grima
Guarditox R **difenacoum** Esoform
guazatine acetates FX [39202-40-9] (for

triacetate of 1,1'-iminodi(octamethylene)diguanidine) acetates of a mixture of the reaction products from polyamines [comprising mainly octamethylenediamine, iminodi(octamethylene)diamine andoctamethylenebis(imino-octamethylene)diamine] and carbamonitrile, GTA, EM 379, Kenopel, MC 25, Panoctine, Panolil, Radam
Guenther's Moosvernichter H **iron sulphate** Cornufera, Guenther
Guidex H **chloridazon** L.A.P.A., Protex
Gumisan F **copper oxychloride + maneb + zineb** ERT
Gupol IA **azinphos-methyl** Aziende Agrarie Trento
Gusadeen I **azinphos-methyl + propoxur** Bayer
Gusafan IA **azinphos-methyl** Aragonesas
Gusagrex IA **azinphos-methyl** Sadisa
Gusagrex 2 IA **azinphos-ethyl** Sadisa
Gusagrex 20 IA **azinphos-methyl** Sadisa
Gusagrex 20 E IA **azinphos-ethyl** Sadisa
Gusagrex 3 M IA **azinphos-methyl** Sadisa
Gusapor I **azinphos-ethyl** Sopepor
Gusathion I **azinphos-methyl** Bayer, Budapesti Vegyimuvek, Pinus
Gusathion A IA **azinphos-ethyl** Bayer
Gusathion K forte I **azinphos-ethyl** Bayer, Schering
Gusathion M I **azinphos-methyl** Agro-Kemi, Bayer
Gusathion MS IA **azinphos-methyl + demeton-S-methyl sulphone** Bayer
Gusathion PB IA **azinphos-methyl** Bayer
Gusathion perfekt I **azinphos-methyl** Bayer
Gusation ruiskutejauhe IA **azinphos-methyl** Berner
Gusatox MS I **azinphos-methyl + demeton-S-methyl sulphone** Agroplant
Guset I **azinphos-ethyl** Filocrop
Gusethyl I **azinphos-ethyl** Hellenic Chemical
Gusmaton IA **azinphos-methyl** Marba
Gusmethyl I **azinphos-methyl** Hellenic Chemical
Gusto H **phenmedipham** Farm Protection
Gutam IA **azinphos-methyl** Aziende Agrarie Trento
Gutex IA **azinphos-ethyl** Aziende Agrarie Trento, Visplant
Guthiben IA **azinphos-methyl** ERT
Guthiben R IA **azinphos-methyl + dimethoate** ERT
Guthion IA **azinphos-methyl** Bayer
Guthyl Extra I **azinphos-ethyl + azinphos-methyl** Hellenic Chemical
Guver I **trichlorfon** Afrasa
Gy-bon H **simetryn** Ciba-Geigy, Hokko, Nihon Nohyaku, Nippon Kayaku, Sank
Gyumolcsfaolaj I **petroleum oils** KKV

H-69 I **XMC** Hodogaya
H 95 H **buturon** BASF
H 119 H **chloridazon** BASF
H 133 H **dichlobenil** Duphar
H 321 MIAX **methiocarb** Bayer
H 1244 G **pydanon**
H 9789 H **norflurazon** Sandoz
H 52143 H **norflurazon** Sandoz
H Ovicida A **fenson** Tecniterra
H-Stop G **dikegulac-sodium** ICI
Hache Uno Super H **fluazifop-butyl** Ishihara Sangyo
Haft Kupper-Z F **copper oxychloride + zineb** Agrotex
Haft Vitigran F **copper oxychloride** Hoechst
Haftkupfer Linz F **copper oxychloride** Agrolinz
HAG 107 I **tralomethrin** Roussel Uclaf
haiari IA **rotenone**
Haipen F **captafol** Chevron
Haistick S **alkyl phenol ethoxylate** Gefex
Haiten S S **alkyl phenol ethoxylate** Anthokipouriki
Haitin F **fentin hydroxide** Nihon Nohyaku
Hakusap I **fenvalerate + malathion**
halacrinate F [34462-96-9] 7-bromo-5-chloroquinolin-8-yl acrylate
Hallizan M **metaldehyde** Tamogan
Halloween G **chlormequat chloride + di-1-p-menthene** Mandops
Halmark I **esfenvalerate** Shell
Halmfest G **chlormequat chloride** Bayer
Halmverstaerker CCC G **chlormequat chloride** BASF, Du Pont, Wacker
haloxydine H [2693-61-0] 3,5-dichloro-2,6-difluoropyridin-4-ol, PP 493
haloxyfop-ethoxyethyl H [87237-48-7] 2-ethoxyethyl (*RS*)-2-[4-(3-chloro-5-trifluoromethyl-2-pyridyloxy)phenoxy]propionate, Dowco 453 EE, Gallant, Zellek

Haltox FHIN **methyl bromide** Degesch, DGS
Hanane IA **dimefox**
Hantrex X **anthraquinone** Ital-Agro
Haptarax IA **chlorfenvinphos**
Haptasol IA **chlorfenvinphos**
Hare-Rid R **strychnine**
Harilak H **amitrole + atrazine + simazine** Ellagret
Harmony H **thiameturon-methyl** Du Pont
Harmony M H **metsulfuron-methyl + thiameturon-methyl** Du Pont
Harness H **acetochlor** Monsanto
Harvade HG **dimethipin** Uniroyal
Harvest-Aid H **sodium chlorate** Wilbur-Ellis
Ha Te X **odorous substances** Epro, Shell
Hataclean F **trichlamide** Nippon Kayaku
Haulmone HI **dinoseb** Universal Crop Protection
Havoc R **brodifacoum** ICI
Hazodrin IA **monocrotophos** Hui Kwang
HCB F **hexachlorobenzene**
Heclotox 3 I **lindane** Borzesti
Hedapur DP H **dichlorprop** Kwizda
Hedapur KV H **mecoprop** Schering
Hedapur KV kombi H **2,4-D + mecoprop** Schering
Hedapur M H **MCPA** Schering
Hedarex H **2,4-D** Kwizda
Hedit Neu H **amitrole + diuron** Hoechst
Hedolit H **DNOC** Bitterfeld
Hedonal H **2,4-D** Bayer
Hedonal BI H **2,4-D + MCPA** Bayer
Hedonal DP H **dichlorprop** Bayer
Hedonal Gazon H **2,4-D + dicamba + MCPA + mecoprop** Bayer
Hedonal KV universal H **2,4-D + mecoprop** Bayer
Hedonal M H **MCPA** Bayer
Hedonal MCPP H **mecoprop** Bayer
Hedonal MP-D H **2,4-D + mecoprop** Bayer
Hedonal S H **MCPA** Bayer
Hedonal vloeibaar H **mecoprop** Bayer
Hedonal-4 MCPA H **MCPA** Bayer
Hedonal-neste H **MCPA** Berner
Hedro M **metaldehyde** Hehl Emil
Helarion M **metaldehyde** Agrolinz, Fisons
Hele Stone G **chlormequat chloride + di-1-p-menthene** Mandops
Helicide Granule Umupro M **metaldehyde** Chimac-Agriphar
Helimat M **metaldehyde** KenoGard
Helimax M **metaldehyde** de Sangosse
Heliosol S **terpene alcohols** DRT
heliotropin acetal S **piprotal**
Helitox M **metaldehyde** Quimagro
Hellacap NI **ethoprophos** Hellafarm
Hellapam FIHN **metham-sodium** Hellafarm
Hellapol I **parathion + petroleum oils** Hellafarm
Hellatan Extra F **copper oxychloride + zineb** Hellafarm
Hellatox I **parathion** Hellafarm, Filocrop
Hello F **carbendazim + maneb** Hellafarm
Helothion I **sulprofos** Bayer
Helugec M **metaldehyde** Sipcam-Phyteurop
Hema Gazonmestkorrel met onkruidverdelger H **2,4-D + dicamba** Hema
Hema mosbestrijder H **iron sulphate** Hema
HEOD I **dieldrin**
heptachlor I [76-44-8] 1,4,5,6,7,8,8-heptachloro-3a,4,7,7a-tetrahydro-4,7-methanoindene, heptachlore, OMS 193, ENT 15152, Drinox, E 3314, Heptagran, Heptamul, Heptox
heptachlore I **heptachlor**
Heptagran I **heptachlor**
Heptamul I **heptachlor**
heptenophos I [23560-59-0] 7-chlorobicyclo[3.2.0]hepta-2,6-dien-6-yl dimethyl phosphate, OMS 1845, Hoe 02982, Hostaquick
heptopargil G [73886-28-9] (*E*)-(1*RS*,4*RS*)-bornan-2-one *O*-prop-2-ynyloxime, EGYT 2250, Limbolid
Heptox I **heptachlor**
Herald AI **fenpropathrin**
Herald H **chloridazon + chlorpropham + fenuron + propham** Rhone-Poulenc
Herba-Banvel H **dicamba + MCPA** Bang, KVK
Herba total H **amitrole + atrazine + simazine** C.G.I.
Herba-Vetyl IA **piperonyl butoxide + pyrethrins** Vetyl-Chemie
Herbadex D H **2,4-D** S/48
Herbadox H **pendimethalin** Cyanamid
Herbadro H **sodium chlorate** Lambert
Herbalon H **clopyralid + MCPA + mecoprop** KVK
Herbalt H **neburon + nitrofen** Pepro
Herbamill H **molinate** Agriplan
Herbamix-DM H **2,4-D + MCPA** KVK

Herbamix-DPD H **2,4-D + dichlorprop** KVK
Herbamix-DPM H **dichlorprop + MCPA** KVK
Herbamix-MPD H **2,4-D + mecoprop** KVK
Herbamout H **trifluralin** Argos
Herban H **noruron**
Herban H **amitrole + diuron** SEGE
Herban combi-MD H **2,4-D + mecoprop** Chromos
Herbaphen H **phenmedipham** KVK
Herbaprop H **mecoprop** KVK
Herbaquat H **amitrole + ammonium thiocyanate** Rhodic
Herbaron B H **bromoxynil + dicamba + mecoprop** La Littorale
Herbasol H **dinoseb** KVK
Herbastop H **2,4-D + MCPA** Agriplan
Herbatop H **glyphosate** Zorka Sabac
Herbatox H **bentazone + dichlorprop + isoproturon** BASF, Ciba-Geigy
Herbatox APT H **atrazine + picloram + prometryn** Kwizda
Herbatox combi 3 H **2,4-D + dichlorprop + MCPA** KVK
Herbatox-D H **2,4-D** Bang, KVK
Herbatox-DP H **dichlorprop** KVK
Herbatox-M H **MCPA** Bang, KVK
Herbatox-MP H **mecoprop** Kirk, KVK
Herbatox-S H **bentazone + dichlorprop + isoproturon** Ciba-Geigy
Herbatoxol-S H **2,4-D + simazine** Organika-Azot
Herbatranex H **atrazine** Land-Forst
Herbavex Plus H **dicamba + MCPA + mecoprop** Kemira
Herbax H **propanil** Agro-Quimicas de Guatemala
Herbazid S H **sodium chlorate** Frowein
Herbazid UG H **bromacil + diuron** Riwa
Herbazol N H **paraquat dichloride** Burri
Herbec H **tebuthiuron** Elanco
Herbex DPM H **dichlorprop + MCPA** KVK
Herbexan-D H **2,4-D** KVK
Herbexan-DP H **dichlorprop** KVK
Herbexan-M H **MCPA** KVK
Herbexan-MP H **mecoprop** KVK
Herbexit H **2,4-D** CTA
Herbexol H **petroleum oils** CTA
Herbiace H **bialaphos** Meiji Seika
Herbiagrex H **2,4-D** Sadisa
Herbicer H **2,4-D** Afrasa
Herbicid Leuna H **allyl alcohol** Leuna-Werke
Herbicida AS H **amitrole + diuron** Inagra
Herbicida BE H **2,4-D** Inagra
Herbicida BG H **2,4-D** Probelte
Herbicida C H **2,4-D + MCPA** Probelte
Herbicida D H **2,4-D** KenoGard
Herbicida M H **MCPA** KenoGard
Herbicida Total H **diuron** Marba
Herbicide Total H **sodium chlorate** Donck, Fontaine Beauvois, Vervier
Herbiclor H **chlorotoluron** Sadisa
Herbiclor extra H **chlorotoluron + terbutryn** Sadisa
Herbicoop H **2,4-D** Ficoop
Herbicruz H **2,4-D** KenoGard
Herbicruz arbustos A-35 H **2,4,5-T** KenoGard
Herbicruz atilon H **2,4-D** KenoGard
Herbicruz Atrazina H **atrazine** KenoGard
Herbicruz doble sal H **2,4-D + MCPA** KenoGard
Herbicruz Du H **diuron** KenoGard
Herbicruz duat H **amitrole + diuron** KenoGard
Herbicruz grama H **dalapon-sodium** KenoGard
Herbicruz Jardin H **glyphosate** KenoGard
Herbicruz LI H **linuron** KenoGard
Herbicruz magapol H **2,4-D + dicamba + MCPA** KenoGard
Herbicruz MH G **maleic hydrazide** KenoGard
Herbicruz Simatra H **atrazine + simazine** KenoGard
Herbicruz Simazin H **simazine** KenoGard
Herbicruz Stam H **propanil** KenoGard
Herbicruz trifluralin H **trifluralin** KenoGard
Herbicurane H **chlorotoluron + terbutryn** Agriplan
Herbidens H **MCPA** Key
Herbifruit H **simazine** Luqsa
Herbifruit doble H **amitrole + simazine** Luqsa
Herbiland AT H **atrazine** Afrasa
Herbiland doble H **atrazine + simazine** Afrasa
Herbiland SM H **simazine** Afrasa
Herbilane H **amitrole + diuron** Afrasa, Sopepor
Herbiluq H **atrazine** Luqsa
Herbiluq-Amina H **2,4-D** Luqsa
Herbiluq doble H **2,4-D** Luqsa
Herbimoor H **sodium chlorate** ICI

Herbin H **dichlorprop + MCPA** Farmipalvelu, Kemira
Herbin Plus H **dicamba + dichlorprop + MCPA** Farmipalvelu
Herbinet H **sodium chlorate** Hoechst
Herbinex H **diuron** Ravit
Herbinexa H **MCPA** Permutadora
Herbinexa-MS H **amitrole + simazine** Permutadora
Herbinil H **atrazine** Rhone-Poulenc
Herbion H **2,4-D** Argos
Herbion MCPA H **MCPA** Argos
Herbioro H **diuron** Quimicas Oro
Herbioro doble H **amitrole + diuron** Quimicas Oro
Herbipec H **chlorotoluron** Sapec
Herbipron H **terbumeton + terbuthylazine** Probelte
Herbipron invierno H **atrazine + terbumeton + terbuthylazine** Probelte
Herbipron verano H **terbumeton + terbuthylazine + terbutryn** Probelte
Herbirail TX H **amitrole + atrazine + diuron + dimefuron** Rhodic
Herbirroz H **propanil** Quiminor
Herbisan AD H **amitrole + diuron** Ellagret
Herbit H **sodium chlorate** Protex
Herbit H **MCPA-thioethyl** Hokko
Herbitex H **sodium chlorate** B.B.
Herbitref H **trifluralin** Zorka Subotica
Herbitrol H **amitrole + simazine** Hoechst
Herbitrol 90 H **amitrole** Reis
Herbivit H **MCPA** Agriben
Herbivit CMPP H **mecoprop** Agrotec
Herbivit DP H **dichlorprop** Agrotec
Herbivit M H **MCPA** Agrotec
Herbivit MPD H **2,4-D + mecoprop** Agrotec
Herbivit Supra H **2,4-D + MCPA** Agriben
Herbivorax H **glyphosate** Umupro
Herbizid D H **2,4-D** Du Pont, Marks
Herbizid DP H **dichlorprop** Du Pont, Elsner
Herbizid-Combi H **2,4-D + MCPA** Elsner
Herbizid 'ES' H **petroleum oils** PCZ
Herbizid Granulat 8102 H **dichlobenil** Bail
Herbizid H H **MCPB** De Weerdt
Herbizid Kombi DM H **2,4-D + MCPA** Du Pont
Herbizid M H **MCPA** Du Pont
Herbizid Marks DP H **dichlorprop** Marks
Herbizid Marks Kombi DM H **2,4-D + MCPA** Marks
Herbizid Marks M H **MCPA** Marks
Herbizid Marks MP H **mecoprop** Marks
Herbizid Marks MPD H **2,4-D + mecoprop** Marks
Herbizid MP H **mecoprop** Du Pont, Elsner
Herbizid MPD H **2,4-D + mecoprop** Du Pont
Herbizid TM H **MCPA + 2,4,5-T** De Weerdt
Herbizid-Zusatzoel S **petroleum oils** Leu & Gygax, Pluess-Staufer
Herbizides Spritzpulver FL 106 H **diuron** Fahlberg-List
Herbizole H **amitrole** Fair Products
Herbocid H **2,4-D** Pinus
Herbofital H **MCPA** Sapec
Herbogex A H **atrazine** Quimigal
Herbogex S H **simazine** Rhone-Poulenc
Herbogil H **dinoterb** Agriben, Pepro, Rhone-Poulenc
Herbogil DP-D H **2,4-D + dichlorprop** Wacker
Herbolex H **glyphosate** Aragonesas
Herbolex pesado H **2,4-D** Aragonesas
Herbon Gold H **chlorpropham + fenuron + propham** Agribus
Herbonex H **amitrole + simazine** Rhodic
Herbopin H **atrazine** Pinus
Herboprop H **MCPA + mecoprop** KVK
Herborol H **diuron** Inagra
Herbosate H **glyphosate** Sovilo
Herbotal H **amitrole + diuron** Sandoz
Herbotal Diklo H **dichlorprop + MCPA** BASF
Herbotal-neste H **MCPA** BASF
Herbotal Plus H **MCPA + mecoprop** Farmos
Herbotifal H **2,4-D + MCPA** Sapec
Herboxan H **2,4-D + mecoprop** Rhodic
Herboxan sport H **2,4-D + dicamba** Rhodic
Herboxone H **paraquat dichloride** Crystal
Herbozina H **simazine** Quimigal
Herbrak H **metamitron** Bayer
Herburon H **diuron** Hoechst
Herbutrina H **terbutryn** Sadisa
Hercip H **chlorpropham** Agriplan
Hercules 426 I **chlorbicyclen**
Hercules 3956 IRA **camphechlor**
Hercules 7531 H **noruron**
Hercules 9573 H **terbucarb**
Hercules 14503 IA **dialifos**
Hercules 22234 H **diethatyl-ethyl**
Hercules AC258 IA **dioxathion**
Hercynia Gelb H **DNOC** Chimiberg

Hercynol IA **DNOC + petroleum oils** Chimiberg
Herflane H **trifluralin** Agriplan
Hergaben H **2,4-D** Inagra
Hergaflan H **trifluralin** Inagra
Hergaprim H **atrazine** Inagra
Hergaprim S H **atrazine + simazine** Inagra
Hergaroz H **amitrole + atrazine + dalapon-sodium** Inagra
Hergazina H **simazine** Inagra
Heritage H **trifluralin** Elanco
Heritrol H **amitrole + simazine** Agriplan
Heritrol Forte H **amitrole + MCPA + simazine** Agriplan
Herking H **amitrole + MCPA + simazine** Agriplan
Herli-Unkrautvertilger H **sodium chlorate** Feldmann
Herlin H **linuron** Agriplan
Hermais H **alachlor + atrazine** Agriplan
Hermenon H **MCPA** Inagra
Hermilhor H **alachlor + atrazine** Hoechst
Hermosan F **thiram** Hermoo
Hermotrip H **methabenzthiazuron** Hermoo
Hermovit F **sulphur** Hermoo
Herrifex H **mecoprop** Bayer
Herrisol H **dicamba + MCPA + mecoprop** Bayer
Hersan H **2,4-D** Key
Hertac H **amitrole + diuron + simazine** Phytosan
Hertazin Flow H **atrazine** Agriplan
Hertin atrami H **amitrole + atrazine** Phytosan
Hertin Selectif gazon H **2,4-D** Phytosan
Hertin Sitri H **atrazine + simazine** Phytosan
Hertin Spring H **amitrole + simazine** Phytosan
Hertog 2 H **atrazine + 2,4-D + diuron** Phytosan
Hertox H **amitrole + simazine** Phytosan
Hertrasine H **atrazine + simazine** Phytosan
Hertrial H **trifluralin** ERT
Hertyl H **amitrole + 2,4-D** Phytosan
Herzina H **simazine** Agriplan
Herzol Forte H **amitrole + MCPA** Agriplan
heteroauxin G **indol-3-ylacetic acid**
hexachlorobenzene F [118-74-1] hexachlorobenzene, HCB, perchlorobenzene, Anticarie, Ceku C.B., No Bunt
hexaconazole F [79983-71-4] (*RS*)-2-(2,4-dichlorophenyl)-1-(1*H*-1,2,4-triazol-1-yl)hexan-2-ol, Anvil, PP 523
Hexadrin IR **endrin**
Hexaferb F **ferbam**
hexafluoramin A **fentrifanil**
hexaflurate H [17029-22-0] potassium hexafluoroarsenate, Nopalmate
Hexagon I **lindane** FCC
hexakis A **fenbutatin oxide**
Hexathane F **zineb**
Hexathir FX **thiram**
Hexavin IG **carbaryl**
hexazinone H [51235-04-2] 3-cyclohexyl-6-dimethylamino-1-methyl-1,3,5-triazine-2,4(1*H*,3*H*)-dione, DPX 3674, Pronone, Velpar
Hexazir FX **ziram**
Hexevax F **oxycarboxin** Cillus
Hexilur H **lenacil** Sojuzchimexport
Hexol S **petroleum oils** Bayer
Hexyl I **lindane + rotenone + thiram** Pan Britannica
hexylthiocarbam H **cycloate**
hexythiazox A [78587-05-0] *trans*-5-(4-chlorophenyl)-*N*-cyclohexyl-4-methyl-2-oxothiazolidine-3-carboxamide, Acarflor, Acorit, Calibre, Cesar, Matacar, NA-73, Nissorun, Savey, Stopper, Trevi, Zeldox
HHDN I **aldrin**
Hickstor FG **tecnazene** Hickson & Welch
Hidrofertil G **chlormequat chloride** Platecsa
Higalcoton H **fluometuron** Hightex
Higalfon I **trichlorfon** Hightex
Higalmetox I **methoxychlor** Hightex
Higalnate H **molinate** Hightex
Higaluron H **chlorotoluron** Hightex
Hildan IA **endosulfan** Hindustan Insecticides
Hildit I **DDT** Hindustan Insecticides
Hilfol A **dicofol** Hindustan Insecticides
Hilthion IA **malathion** Hindustan Insecticides
Hinge F **carbendazim**
Hinochloa H **mefenacet** Bayer
Hinopoly H F **edifenphos + polyoxins**
Hinosan F **edifenphos** Bayer, Nihon Tokushu Noyaku Seizo
Hispor F **carbendazim + propiconazole** Ciba-Geigy
Hit I S **alkyl phenol ethoxylate** Kosmos
Hivertox HI **dinoseb** Spiess
Hizarocin FG **cycloheximide**

HL-Spritz- und Giessmittel I **lindane** Bitterfeld
Hobane H **bromoxynil + ioxynil** Farm Protection
Hodina F **dodine** Hoechst
Hoe 02671 IA **endosulfan** Hoechst
Hoe 02747 H **monolinuron** Hoechst
Hoe 02784 AF **binapacryl** Hoechst
Hoe 02810 H **linuron** Hoechst
Hoe 02824 FLM **fentin acetate** Hoechst
Hoe 02873 F **pyrazophos** Hoechst
Hoe 02904 H **dinoseb acetate** Hoechst
Hoe 02960 IAN **triazophos** Hoechst
Hoe 02982 I **heptenophos** Hoechst
Hoe 16410 H **isoproturon** Hoechst
Hoe 17411 F **carbendazim** Hoechst
Hoe 22870 H **clofop-isobutyl**
Hoe 23408 H **diclofop-methyl** Hoechst
Hoe 25682 I **hyquincarb**
Hoe 30374 H **anilofos** Hoechst
Hoe 33171 H **fenoxaprop-ethyl** Hoechst
Hoe 35609 H **fenthiaprop-ethyl**
Hoe 39866 H **glufosinate-ammonium** Hoechst
Hoefluran H **trifluralin** Hoechst
Hoegrass H **diclofop-methyl** Hoechst
Hoelon H **diclofop-methyl** Hoechst
HOK-7501 H **MCPA-thioethyl** Hokko
Hokko Suzu FLM **fentin acetate** Hokko
Hoko M **metaldehyde** Hokochemie
Holdfast D G **dicamba + paclobutrazol** ICI
Hollratox CPN R **chlorophacinone** Holler
Holtox H **atrazine + cyanazine** Burts & Harvey, Shell
Homai F **thiophanate-methyl + thiram**
Hong Nien F **phenylmercury acetate**
Hongal F **captan** Agriplan
Hopcin I **fenobucarb**
HORA-Combi H **2,4-D + MCPA** HORA
HORA-Curan H **chlorotoluron** HORA
HORA D H **2,4-D** HORA
HORA DP H **dichlorprop** HORA
HORA Fenoxim H **bromofenoxim** HORA
HORA-Fenoxim spezial H **bromofenoxim + 2,4-D + MCPA** HORA
HORA-Flor F **imazalil** HORA
HORA Fluron H **thiazafluron** HORA
HORA Fluron plus H **karbutilate + thiazafluron** HORA
HORA Grasstop G **maleic hydrazide** HORA
HORA Karbutil H **karbutilate** HORA
HORA-Kupferspritzmittel F **copper oxychloride** Ciba-Geigy
HORA KV H **mecoprop** HORA
HORA KV Combi H **2,4-D + mecoprop** HORA
HORA M H **MCPA** HORA
HORA-Mazin H **simazine** HORA
HORA Oel S **petroleum oils** HORA
HORA Prim H **atrazine** HORA
HORA-Saatgutpuder B I **bromophos** Ciba-Geigy
HORA-Trazin H **atrazine** HORA
HORA-Tryn H **terbutryn** HORA
HORA-Turon H **isoproturon** HORA
Horco H **2,4-DB + dinoseb** Sandoz
Horlan F **copper oxychloride** Nordisk Alkali
Hormin H **2,4-D** OHIS
Hormo-Cornox H **mecoprop** Gullviks
Hormo-DP H **dichlorprop** Gullviks
Hormoblend H **dicamba + MCPA** Gullviks
Hormoblend Trippel H **dicamba + MCPA + mecoprop** Gullviks
Hormon-Mix H **dichlorprop + MCPA** Nordisk Alkali
Hormone TP G **fenoprop** Du Pont
Hormoneste H **MCPA** Kemira
Hormonex H **MCPA** Protex
Hormonil G **2,4-D** Argos
Hormonyl-D H **2,4-D** Bourgeois
Hormopear G **1-naphthylacetic acid** Biolchim
Hormoprin G **1-naphthylacetamide + 1-naphthylacetic acid** Probelte
Hormopron G **2,4-D** Probelte
Hormoprop H **MCPA + mecoprop** Kemira
Hormosan H **MCPA** Bourgeois
Hormosap G **4-indol-3-ylbutyric acid** Franco-Belge
Hormostar H **2,4-D** Gullviks
Hormotex H **MCPA** Gullviks
Hormotuho H **MCPA** BASF, Kemira
Hornoska-Golf mit Unkrautvernichter H **2,4-D + dicamba** Guenther
Hortag Aquasulf F **sulphur** Avon Packers
Hortag Hexaflow I **lindane** Avon Packers
Hortag Tec Ten Granules FG **tecnazene** Avon Packers
Hortag Tecnacarb FG **carbendazim + tecnazene** Avon Packers
Hortag Tecnazene Plus FG **tecnazene + thiabendazole** Avon Packers
Hortamon P I **carbaryl + lindane** KenoGard

Hortamon S.R. p. IA **carbaryl + dimethoate** KenoGard
Hortex I **lindane** Agrolinz, Felleskjoepet, Shell
Hortex RP FI **lindane + thiram** Shell
Horti Rasenduenger mit Moosvertilger H **iron sulphate** Kwizda
Horti Rasenduenger mit Unkrautvertilger H **2,4-D + dicamba** Kwizda
Hortichem Spraying Oil IA **petroleum oils** Hortichem
Horto-Rose F **dodemorph acetate + dodine** Agro
Hortosan F **captafol + mancozeb + sulphur** Maag
Hoslima M **metaldehyde** Hoechst
Hostafume G **propham** Hoechst
Hostamonda D H **2,4-D** Hoechst
Hostamonda M H **MCPA** Hoechst
Hostaquick I **heptenophos** Argos, Hoechst, Pluess-Staufer, Procida, Roussel Hoechst, Siegfried, Zupa
Hostathion IA **triazophos** Argos, Condor, Hoechst, Roussel Hoechst, Zupa
Hotspur H **clopyralid + fluroxypyr + ioxynil** Farm Protection
House Plant Leaf Shine Plus Pest Killer I **permethrin** Synchemicals
House Plant Pest Killer H **pyrethrins + resmethrin** Synchemicals
HOX 1901 I **ethiofencarb** Bayer
HRS 924 A **fluorbenside**
HS 186 A **dicofol + tetradifon** Tecniterra
HTG 10 H **chlorthiamid** C.F.P.I.
Huile 970 S **petroleum oils** BASF
Huile blanche IA **petroleum oils** Gaillard
Humectante Bayer S **polyglycolic ethers** Bayer
Humectol S **polyglycolic ethers** Afrasa
Humextra H **metolachlor** Ciba-Geigy
Hungaria L-7 I **lindane** Budapesti Vegyimuvek
Hungazin H **atrazine** Budapesti Vegyimuvek, Chemolimpex
Hus og have insektspray I **piperonyl butoxide + pyrethrins** Kirk
HWG 1608 F **terbuconazole**
Hy-TL F **thiabendazole + thiram** Agrichem
Hy-TL Turbo FI **bendiocarb + thiabendazole + thiram** Agrichem
Hy-Vic F **thiabendazole + thiram** Agrichem
Hyban H **dicamba + mecoprop** Agrichem
Hydon H **bromacil + picloram** Chipman
Hydraguard IF **lindane + thiram** Agrichem
hydramethylnon I [67485-29-4] 5,5-dimethylperhydropyrimidin-2-one 4-trifluoromethyl -α-(4-trifluoromethylstyryl) cinnamylidenehydrazone, pyramdron, amidinohydrazone, AC 217,300, Amdro, Combat, Cyaforgel, Maxforce, Wipeout
hydrazide maleique G **maleic hydrazide**
hydrogen cyanide IR [74-90-8] hydrogen cyanide, acide cyanhydrique, prussic acid, Cyanosil, Cyclon, Cymag (sodium cyanide), Zyklon
Hydrol I **allyxycarb**
hydroprene I [41096-46-2] ethyl (*E*)-3,7,11-trimethyldodeca-2,4-dienoate, OMS 1696, Altozar, Gencor, ZR-512
Hydrosekt I **butocarboxim** Luxan
Hydroter F **fentin hydroxide** Eurofyto
Hydrothol H **endothal** Decco-Roda, Desarrollo Quimico, Pennwalt
Hydrotin F **fentin hydroxide** Liro
hydroxyisoxazole F **hymexazol**
8-hydroxyquinoline sulphate FB [134-31-6] bis(8-hydroxyquinolinium) sulphate, Albisal, Chinosol, Cryptonol
Hyflier FI **bendiocarb + thiram** Agrichem
Hygaea algefjerner L **benzalkonium chloride + tributyltin naphthenate** Hygaea
Hygaea desinfektion L **benzalkonium chloride + tributyltin naphthenate** Hygaea
Hygiegazon H **2,4-D + MCPA** Hygiena
Hygrass H **dicamba + mecoprop** Agrichem
Hylemox AI **ethion** Rhodiagri-Littorale, Rhone-Poulenc
Hylogam I **lindane** Chimac-Agriphar
Hymec H **mecoprop** Agrichem
hymexazol F [10004-44-1] 5-methylisoxazol-3-ol, hydroxyisoxazole, F-319, SF-6505, Tachigaren
Hymush I **thiabendazole** Agrichem
Hypercuivre F **copper oxychloride + zineb** Vital
Hyperkil R **calciferol** Antec
Hypnoline F **cuprammonium** Franco-Belge
Hyprone H **dicamba + MCPA + mecoprop** Agrichem
Hyquat G **chlormequat chloride** Agrichem

hyquincarb I [56716-21-3] 5,6,7,8-tetrahydro-2-methyl-4-quinolyl dimethylcarbamate, Hoe 25682
Hyspray S **tallow amine ethoxylate** Sanac
Hystore FG **tecnazene** Agrichem
Hysward H **dicamba + MCPA + mecoprop** Agrichem
Hytane H **isoproturon** Ciba-Geigy
Hytec FG **tecnazene** Agrichem
Hytec Super FG **tecnazene + thiabendazole** Agrichem
Hytin F **fentin acetate + maneb** Agrichem
Hytox M **metaldehyde** Agrichem
Hytox I **isoprocarb** Planters Products
Hytrol H **amitrole + 2,4-D + diuron + simazine** FCC
Hyvar H **bromacil** Bayer, Du Pont, OHIS, Rhone-Poulenc, Schering, Siapa
Hyvar General Weedkiller H **isocil** Du Pont
Hyzon H **chloridazon** Agrichem

IAA G **indol-3-ylacetic acid**
Ibertox IAF **DNOC** Siapa
IBP F **iprobenfos**
Ica I **lindane** Sedagri
Icaphos IA **dimethoate** Sedagri
Icaplouse H **2,4-D + mecoprop** Sedagri
Icedin H **2,4-D + dicamba** Dudesti
ICI Anti-Mos H **iron sulphate** ICI
ICI Slug Pellets M **metaldehyde** ICI
ICI Wetter S **polyglycolic ethers** ICI
Icon I **lambda-cyhalothrin** ICI
Idrolene F **maneb** Chimiberg
Idrolin IA **endosulfan** Margesin
Idrorame F **Bordeaux mixture** Chimiberg, Sivam
Idunesto G **chlorpropham + propham** Kemira
IFC HG **propham**
Igepa-Unkrautjaeger H **sodium chlorate** IGEPA
Igran H **terbutryn** Ciba-Geigy, Ligtermoet, Pinus, Scam, Viopharm
Igran combi H **metolachlor + terbutryn** Ciba-Geigy
Igran Special H **chlorotoluron + terbutryn** VCH
Igrater H **metobromuron + terbutryn** Ciba-Geigy, Ligtermoet, Liro
IH 773B H **fluazifop-butyl** Ishihara Sangyo
Ijzerammonium sulfaat H **iron sulphate** Graham
IKI-7899 I **chlorfluazuron** Ishihara Sangyo
Ikkokuso H **chlomethoxyfen** Ishihara Sangyo
Ikurin H **ammonium sulphamate** Hodogaya
Ildenal I **dichlorvos + propoxur** Bayer
Illoxan H **diclofop-methyl** Chromos, Hoechst, Nitrokemia, Organika-Sarzyna, Pluess-Staufer, Procida, Rhone-Poulenc, Roussel Hoechst
Illoxan Combi H **bromoxynil + diclofop-methyl + ioxynil** Procida
Iloxan H **diclofop-methyl** Argos, Hoechst
imazalil F [35554-44-0] 1-(*β*-allyloxy-2,4-dichlorophenylethyl)imidazole, enilconazole, chloramizol, Bromazil, Deccozil, Fecundal, Freshgard, Fungaflor, Fungazil, R23979
imazamethabenz-methyl H [81405-85-8] methyl 6-(4-isopropyl-4-methyl-5-oxo-2-imidazolin-2-yl)-*m*-toluate/methyl 2-(4-isopropyl-4-methyl-5-oxo-2-imidazolin-2-yl)-*p*-toluate mixture, AC222,293, AC 293, Assert, Dagger
imazapyr H [81334-34-1] 2-(4-isopropyl-4-methyl-5-oxo-2-imidazolin-2-yl)nicotinic acid, AC 252,925, Arsenal, Assault, Chopper, CL 252,925
imazaquin H [81335-37-7] 2-(4-isopropyl-4-methyl-5-oxo-2-imidazolin-2-yl)-3-quinolinecarboxylic acid, imazaquine, AC 252,214, CL 252,214, Scepter
imazaquine H **imazaquin**
imazethapyr H [81335-77-5] (*RS*)-5-ethyl-2-(4-isopropyl-4-methyl-5-oxo-2-imidazolin-2-yl)nicotinic acid, AC 263,499, CL 263,499, Pivot, Pursuit
Imidan I **phosmet** Agrishell, Efthymiadis, Kwizda, Ligtermoet, OHIS, Rhone-Poulenc, RousselHoechst, Serpiol, Siegfried, Stauffer
Imidan-Captan FI **captan + phosmet** Efthymiadis
Impact F **flutriafol** ICI, Siegfried, Spolana, Zupa
Impact CL F **chlorothalonil + flutriafol** ICI
Impact CLT F **chlorothalonil + flutriafol** ICI
Impact Excel F **chlorothalonil + flutriafol** ICI

Impact R F **carbendazim + flutriafol** ICI, Sopra
Impact RM F **carbendazim + flutriafol** Sopra
Impact T F **captafol + flutriafol** ICI, Sopra
Impact TX F **chlorothalonil + flutriafol** Sopra
Imperator I **permethrin** ICI
Imperator I **cypermethrin** ICI
Impraline Koper FW **copper naphthenate** Hermadix
Improved Golden Malrin Vliegenkorrels I **methomyl** Sandoz
Imugan F **chloraniformethan**
inabenfide G [82211-24-3] 4'-chloro-2'-(α-hydroxylbenzyl)isonicotinanilide, CGR-811, Seritard
Inacin F **zineb** Inagra
Inacop F **copper oxychloride** Inagra
Inagron IA **monocrotophos** Inagra
Inakor H **atrazine + prometryn** Radonja
Inakor H **atrazine**
Inakor extra H **atrazine + cyanazine + prometryn** Radonja
Inakor-T H **atrazine + prometryn** Radonja
Inaman F **maneb** Inagra
Inavid F **copper oxychloride + folpet** Inagra
Incisekt I **chlorpyrifos** Henkel
Incisekt I **permethrin** Henkel
Increcel G **chlormequat chloride** Sarabhai
Indar F **triazbutil** Rohm & Haas
indol-3-ylacetic acid G [87-51-4] indol-3-ylacetic acid, IAA, AIA, heteroauxin, Rhizopon A
4-indol-3-ylbutyric acid G [133-32-4] 4-indol-3-ylbutyric acid, IBA, AIB, Chryzoplus, Chryzopon,Hormodin, Rhizopon AA, Seradix
Indrex F **dodine** Baslini
Inex H **linuron + pendimethalin** Du Pont
Inexit I **lindane** Celamerck
Infernal R **thallium sulphate** Fabel-Maxenri
Inhibiteur 360 G **maleic hydrazide** Procida
Inject a cide I **oxydemeton-methyl** Steinbauer
Insation IA **malathion** Safor
Insectalac IA **diazinon** Sorex
Insecticida Oro IA **petroleum oils** Quimicas Oro
Insecticida Oro amarillo IA **DNOC + petroleum oils** Quimicas Oro
Insectisol I **azinphos-ethyl** Reis
Insecto Strip H **dichlorvos** Secto
Insecto-Solo I **lindane** Permutadora
Insectophene IA **endosulfan** Pepro
Insectrol I **diazinon + piperonyl butoxide + pyrethrins** Rentokil
Insectrol P.O. I **propoxur** Rentokil
Insegar I **fenoxycarb** La Quinoleine, Maag
Insekt chok I **piperonyl butoxide + pyrethrins** Aeropak
Insekt I IA **pyrethrins** Ledona
Insekten spuitmiddel Roxion IA **dimethoate** Shell
Insekten-Killer I **permethrin + pyrethrins** Perycut
Insekten-Spritzpulver Hortex I **lindane** Shell
Insekten-Staeubemittel Hortex neu IA **endosulfan** Shell
Insekten-Staeubemittel Hortex I **lindane** Shell
Insekten-Staeubemittel Nexion I **bromophos** Shell
Insekten-Streumittel Nexion I **bromophos** Shell
Insekten-strooimiddel Nexion I **bromophos** Shell
Insektenbestrijdingsmiddel Nexion I **bromophos** Shell
Insektenil-DCV-Spray I **dichlorvos + piperonyl butoxide + pyrethrins** Hentschke & Sawatzki
Insektenil-fluessig-M I **malathion + piperonyl butoxide + pyrethrins** Hentschke & Sawatzki
Insektenil-fluessig-N-'HA' I **lindane** Hentschke & Sawatzki
Insektenil-fluessig-N-'HS'-forte I **dichlorvos + lindane** Hentschke & Sawatzki
Insektenil-fluessig-V-'Spezial' I **lindane + piperonyl butoxide + pyrethrins** Hentschke & Sawatzki
Insektenil-Konzentrat-M IA **malathion** Hentschke & Sawatzki
Insektenil-Raumnebel I **pyrethrins** Hentschke & Sawatzki
Insektenil-Raumnebel I **dichlorvos** Hentschke & Sawatzki
Insektenil Raumnebel forte I **piperonyl butoxide + pyrethrins** Hentschke & Sawatzki
Insektenil Raumnebel forte trocken DDVP I **dichlorvos + piperonyl butoxide + pyrethrins** Hentschke & Sawatzki

Insektenil-Strip I **dichlorvos** Hentschke & Sawatzki
Insektenil-Tetrat I **tetrachlorvinphos** Hentschke & Sawatzki
Insektenkiller Tropical Floracid I **bioallethrin + permethrin** Perycut
Insektenstaeubemittel Hortex Neu I **endosulfan** Shell
Insektizid fluessig/liquide IA **diazinon** Spedro
Insektizid/insecticide Gesal IA **diazinon** Reckitt & Colman
Insektizid/insecticide Pirox I **chlorpyrifos + pirimicarb** Maag
Insektizides Spritzpulver Mioplant IA **diazinon** Migros
Insektizides Staeubemittel/poudrage insecticide mio-plant I **chlorpyrifos + pirimicarb** Migros
Insetticida Due I **formothion** Chemia
Insex DN I **chlorpyrifos + dichlorvos** Kwizda
Instop pas I **dichlorvos** Chemika
Integral Granules H **amitrole + diuron + sodium thiocyanate** Desinfection Integrale
Integral Herbazol H **amitrole + diuron** Desinfection Integrale
Integral Imper Rat R **chlorophacinone** Desinfection Integrale
Integral Muskrat R **chlorophacinone** Desinfection Integrale
Integral Prefix H **dichlobenil** Shell
Integral Rat R **coumatetralyl** Desinfection Integrale
Integral Rattox R **chlorophacinone** Desinfection Integrale
Intox I **chlordane** Sandoz
Intrasol IN **carbofuran** Burri
Intration IA **thiometon** Dimitrova
Invert 155 H **picloram + 2,4,5-T** C.F.P.I.
Invert DP H **dichlorprop + picloram** C.F.P.I.
Invicta H **bifenox + isoproturon** Farm Protection
iodofenphos IA [18181-70-9] *O*-2,5-dichloro-4-iodophenyl *O,O*-dimethyl phosphorothioate, jodfenphos, OMS 1211, C 9491, Elocril, Nuvanol N
Ioniz VR H **diflufenican + ioxynil + isoproturon + mecoprop** Rhodiagri-Littorale
Iorowit S **sodium sulphosuccinate** IPO-Sarzyna
Iotox H **ioxynil + mecoprop** Rhone-Poulenc
Iotril H **ioxynil** Makhteshim-Agan
Iotrilex H **ioxynil** COPCI
ioxynil H [1689-83-4] 4-hydroxy-3,5-diiodobenzonitrile, 15380 RP, ACP 63-303, Actril, Actrilawn, Bentrol, Certrol, Iotril, M&B 11641, M&B 8873, Mate, Topper, Totril, Trevespan
IP 50 H **isoproturon** Agriben, Rhone-Poulenc
IP Flo H **isoproturon** Agriben, Budapesti Vegyimuvek, Condor, Rhodiagri-Littorale, Rhone-Poulenc
Ipam FIHN **metham-sodium** PVV
ipazine H [1912-25-0] 6-chloro-N^2, N^2-diethyl-N^4-isopropyl-1,3,5-triazine-2,4-diamine, G30031, Gesabal
IPC HG **chlorpropham**
IPC HG **propham** Protex, Quimigal
IPC G **chlorpropham + propham** ICI, Sanac
Ipecap F **captan** Interphyto
Ipersan H **trifluralin** Quimica Estrella
Ipiclor H **alachlor** KenoGard
Ipitox I **permethrin** Gullviks
Ipocide FIHN **metham-sodium** Scam
Ipofos I **bromfenvinfos** Organika-Azot
iprobenfos F [26087-47-8] *S*-benzyl *O,O*-di-isopropyl phosphorothioate, IBP, Kitazin
Iprodial F **iprodione** Ciba-Geigy
iprodione F [36734-19-7] 3-(3,5-dichlorophenyl)-*N*-isopropyl-2,4-dioxoimidazolidine-1-carboxamide, 26019 RP, Chipco 26019, Kidan, LFA 2043, NRC 910, ROP 500F, Rovral, Verisan
Iprolate F **carbendazim + iprodione** Ravit
iprymidam H [30182-24-2] 6-chloro-N^4-isopropylpyrimidine-2,4-diamine, SAN 52123H
Ipslure T **pheromones** Borregaard
IPSP I [5827-05-4] *S*-ethylsulphinylmethyl *O,O*-di-isopropyl phosphorodithioate, Aphidan, PSP-204
IPT FI **isoprothiolane**
IPX H **proxan-sodium**
Irokar F **dinocap** Ciba-Geigy
Irol S **alkyl phenol ethoxylate** Agrochemiki, Siapa
IS I **bendiocarb** Schering
Isa FI **fenchlorphos + 8-hydroxyquinoline sulphate** Sedagri
Isathrin I **resmethrin** Procida

Isathrine I **resmethrin** Hoechst, Procida, Roussel Uclaf
isazofos NI [42509-80-8] *O*-5-chloro-1-isopropyl-1*H*-1,2,4-triazol-3-yl *O,O*-diethyl phosphorothioate, CGA 12223, Miral, Triumph
Iso-Cornox H **mecoprop** AAsepta, Bjoernrud, Schering, Schering AAgrunol
isobenzan I [297-78-9] 1,3,4,5,6,7,8,8-octachloro-1,3,3a,4,7,7a-hexahydro-4,7-methanoisobenzofuran, telodrin, OMS 206, ENT 25545, SD 4402, Telodrin
Isocal F **calcium chloride** ISO
isocarbamid H [30979-48-7] *N*-isobutyl-2-oxoimidazolidine-1-carboxamide, isocarbamide
isocarbamide H **isocarbamid**
isocil H [314-42-1] 5-bromo-3-isopropyl-6-methyluracil, isoprocil, Du Pont Herbicide 82, Hyvar
isodrin I [465-73-6] (1*R*,4*S*,5*R*,8*S*)-1,2,3,4,10,10-hexachloro-1,4,4a ,5,8,8a-hexahydro-1,4:5,8-dimethanonaphthalene, isodrine, Compound 711
isodrine I **isodrin**
isofenphos I [25311-71-1] *O*-ethyl *O*-2-isopropoxycarbonylphenyl isopropylphosphoramidothioate, isophenphos, Amaze, Bay SRA 12869, Oftanol
Isomer I **lindane** Terranalisi
isomethiozin H [57052-04-7] 6-*tert*-butyl-4-isobutylideneamino-3-methylthio-1,2,4-triazin-5- one, Tantizon
Isomil I **permethrin** Ital-Agro
isonoruron H [28805-78-9] 1,1-dimethyl-3-(perhydro-4,7-methanoinden-1-yl)urea mixture with 1,1-dimethyl-3-(perhydro-4,7-methanoinden-2-yl)urea
Isopan H **chlorpropham** Van Wesemael
Isopestox IA **mipafox**
isophenphos I **isofenphos**
isoprocarb I [2631-40-5] *o*-cumenyl methylcarbamate, isoprocarbe, MIPC, OMS 32, ENT 25670, Bay 105807, Etrofolan, Hytox, Mipcin
isoprocarbe I **isoprocarb**
isoprocil H **isocil**
isopropalin H [33820-53-0] 4-isopropyl-2,6-dinitro-*N,N*-dipropylaniline, isopropaline, EL-179, Paarlan
isopropaline H **isopropalin**
isoprothiolane FI [50512-35-1] di-isopropyl 1,3-dithiolan-2-ylidenemalonate, IPT, Fudiolan, Fuji-one, Isoran, NNF-109, SS 11946
Isoproturee H **isoproturon** Interphyto
Isoproturee M H **isoproturon + mecoprop** Interphyto
Isoproturee MD H **dicamba + isoproturon + mecoprop** Interphyto
isoproturon H [34123-59-6] 3-(4-isopropylphenyl)-1,1-dimethylurea, Alon, Arelon, CGA 18731, Graminon, Hoe 16410, Hytane, IP 50, IP Flo, Tolkan
Isoran FI **isoprothiolane** Jin Hung
Isortal G **propham** Quimagro
isothiocyanate de methyle NFIH **methyl isothiocyanate**
Isotox I **lindane** Chevron
Isotril H **ioxynil + isoproturon + mecoprop** Ravit
isouron H [55861-78-4] 3-(5-*tert*-butylisoxazol-3-yl)-1,1-dimethylurea, isuron, Conserve, Isoxyl, SSH 43
isoxaben H [82558-53-7] *N*-[3-(1-ethyl-1-methylpropyl)isoxazol-5-yl]-2,6-dimethoxybenzamide, Cent-7, EL-107, Flexidor
isoxathion I [18854-01-8] *O,O*-diethyl *O*-5-phenylisoxazol-3-yl phosphorothioate, E-48, Karphos, SI-6711
Isoxyl H **isouron** Shionogi
isuron H **isouron**
IT-931 F **dithianon** Celamerck
IT 3233 H **flurecol-butyl** Celamerck
IT 3456 GH **chlorflurecol-methyl** Celamerck
IT-5732 G **chlorfluren**
IT-5733 G **dichlorflurecol-methyl**
Ital-Fen I **fenitrothion** Ital-Agro
I.T. Schneckenkorn M **metaldehyde** Drogenhansa
I.T. Wuehlmausfertigkoeder R **warfarin** Drogenhansa
Ivorin Super H **dinoseb acetate + monolinuron** Hoechst
Ivosit H **dinoseb acetate** Hoechst
Ivosta anti-bladluis I **pirimicarb** Klasmann
Ixo-7 H **isoproturon + isoxaben** Eli Lilly
Izal germicide FBV **phenol** Sterling

J-38 IA **formothion** Sandoz
Jackyl S HI **DNOC** La Littorale
Jacutin I **lindane** Agrolinz, Shell
Jaguar H **benazolin + bromoxynil + ioxynil + mecoprop** Schering

Jarbel antimousse gazon H **chloroxuron + iron sulphate** Ciba-Geigy
Jarbel engrais desherbant gazon H **dicamba + mecoprop** Ciba-Geigy
Jarbel gazon liquide H **2,4-D + mecoprop** Ciba-Geigy
Jarbel Special liseron H **2,4-D + MCPA** Ciba-Geigy
jasmolin I IA **pyrethrins**
jasmolin II IA **pyrethrins**
Javelin I **Bacillus thuringiensis** Sandoz
Javelin H **diflufenican + isoproturon** Rhone-Poulenc
Jebodimethoaat IA **dimethoate** Luxan
Jebolindaan IA **lindane** Luxan
Jebolinpar I **lindane + parathion** Luxan
Jeboloofdood-DNBP H **dinoseb** Luxan
Jeboscab IA **lindane** Luxan
Jeboterra-korrels I **parathion** Luxan
Jebotuban H **dicamba + mecoprop** Luxan
Jepolinex H **2,4-D + dicamba** Luxan
Jernvitriol H **iron sulphate** Broeste
Jet flugnaeifur I **piperonyl butoxide + pyrethrins** Islenzk Ameriska
Jetfix-Ampfer-Streumittel CMPP H **mecoprop** Marktredwitz
Jetfix-MPD H **2,4-D + mecoprop** Hokochemie
Jiffy Grow G **4-indol-3-ylbutyric acid + 1-naphthylacetic acid** Klijn
jodfenphos IA **iodofenphos**
Joker H **fenthiaprop-ethyl**
Juolavehnantuho H **TCA** Kemira
Jupiter I **chlorfluazuron** Ciba-Geigy
Juurikasteho IA **dimethoate** Kemikaalimiehet

K-15 H **amitrole + atrazine + simazine** Spyridakis
K-15 Extra H **amitrole + diuron + sodium thiocyanate** Spyridakis
K 223 H **dymron** SDS Biotech
K 1441 H **methyldymron** SDS Biotech
K 6451 A **chlorfenson** Dow
K 22023 H **DMPA**
Kabaprim H **amitrole + atrazine** Viopharm
Kabat I **methoprene** KenoGard, Sandoz
Kabre H **dinoseb** Leu & Gygax
Kabrol H **dinoseb** Pluess-Staufer
Kack H **dimethylarsinic acid** Drexel
kadethrin I [58769-20-3] [1R-[1α,3α(*E*)]]-[5-(phenylmethyl)-3-furanyl]methyl 3-[(dihydro-2-oxo-3(2*H*)-thienylidene)methyl]-2,2-dimethylcyclopropanecarboxylate, ENT 29117, Kadethrin, RU 15525, Spray-Tox
Kadizol A **dicofol + tetradifon** Agrocros
Kadizol Triple AI **dicofol + ethion + tetradifon** Agrocros
Kafil I **permethrin** ICI
Kafil Super I **cypermethrin** ICI, La Quinoleine
Kailan H **MCPA + propanil** Serpiol
Kakengel F **polyoxin D zinc salt**
Kali-Kane M **metaldehyde** Sandoz
Kaliram F **copper oxychloride** Sandoz
Kalkosan F **calcium chloride** Chema
Kalkstickstoff gemahlen FH **calcium cyanamide** SKW Trostberg
Kama Sanguano Spezial Rasenduenger mit UnkrautvernichterH **2,4-D + MCPA** Kama-Organ-Duenger, Mahle Duenger
Kama Sanguano Spezial Rasenduenger mit Moosvernichter H **iron sulphate** Kama-Organ-Duenger
Kamilon D H **clopyralid + dicamba + dichlorprop + MCPA** KVK
Kan-Alje L **benzalkonium chloride + tributyltin naphthenate** Dyrup
Kanakymppi I **piperonyl butoxide + pyrethrins** Farmos
Kankerdood WF **mercuric oxide** Bayer, Kemira, Leu & Gygax
Kankerex FW **mercuric oxide** UCP
Kankerfix FW **carbendazim** Leu & Gygax
Kankersept FBW **copper oxychloride** Asepta
Kankertox FW **mercuric oxide** Ciba-Geigy
Kankerwering FBW **copper oxychloride** Zerpa
Kap-bio-Spray IA **piperonyl butoxide + pyrethrins** Materna
Kapitol F **captan + nuarimol** Elanco
Kaptan F **captan** OHIS, Organika-Azot
Kaptazor F **captan** Zorka Subotica
Kaptogal F **captan** Galenika
Kaput R **chlorophacinone** Desinfection Integrale
Kar F **carbendazim** KenoGard
Karamate N F **mancozeb** Rohm & Haas
Karate IA **lambda-cyhalothrin** ICI, Maag, Sopra, Spolana
Karathane FA **dinocap** Agrocros, Aragonesas, AVG, Condor, Du Pont,

Farmos, ICI-Valagro, KVK, LaLittorale, Maag, Pepro, Permutadora, Pinus, Protex, Ravit, Rhodiagri-Littorale, Rohm & Haas, Sandoz, Speiss, Staehler, Urania
Karathane combi F **dinocap + mancozeb** KVK
Karbaril I **carbaryl** Pinus, Zorka Sabac, Zupa
Karbaspray IG **carbaryl**
karbation FNHI **metham-sodium**
Karbatox I **carbaryl** Organika-Azot
Karbatox-Extra I **carbaryl + chlorfenvinphos** Organika-Azot
Karbofos I **malathion** SEGE
Karbol 3 IA **anthracene oil + DNOC** Burri
Karbolina DNK I **DNOC + tar oils** Organika-Fregata
karbutilate H [4849-32-5] 3-(3,3-dimethylureido)phenyl *tert*-butylcarbamate, karbutylate, CGA 61837, FMC 11092, Tandex
karbutylate H **karbutilate**
Karcide H **diuron** ERT
Kariben H **linuron** Burri
Kariver A **dicofol** KenoGard
Kariver doble TK A **dicofol + tetradifon** KenoGard
Karmex H **diuron** Du Pont, Duphar, Franken-Chemie, Gullviks, Nordisk Alkali, Pepro, Procida, Quimigal, Rhone-Poulenc, Schering
Karnak F **captan** Lainco
Karpasbakteeri I **Bacillus thuringiensis** Farmos
Karpesin AD H **amitrole + diuron** Bredologos
Karphos I **isoxathion** Sankyo
KarRATe R **chlorophacinone** Dexstar
Kartiram F **thiram** Agro-Kemi
Kartril EV H **amitrole + atrazine + diuron** Rhodic
Kasim-G H **dichlobenil + simazine** Galenika
Kasmiron F **kasugamycin + phospdiphen**
Kasser H **diphenamid** Schering
Kastrix-Myyransyotti R **crimidine** Berner
kasugamycin FB [6980-18-3] [5-amino-2-methyl-6-(2,3,4,5,6-pentahydroxycyclohexyloxy)tetrahydropyran-3-yl]amino-α-iminoacetic acid, KSM, Kasugamin, Kasumin
Kasumin FB **kasugamycin** Hokko, Lainco
Kasumin Cobre FB **copper oxychloride + kasugamycin** Lainco
Kasurabcide F **fthalide + kasugamycin**
Kasuran F **copper oxychloride + kasugamycin**
Kasurbaron F **kasugamycin + phosdiphen**
Katan F **dinocap** Eli Lilly
Katben H **amitrole + diuron** ERT
Kauritil F **copper oxychloride** BASF, Sapec
Kayabest F **methasulfocarb** Nippon Kayaku
Kayametone H **methoxyphenone** Nippon Kayaku
Kayanex R **bisthiosemi** Nippon Kayaku
Kayaphos I **propaphos** Nippon Kayaku
Kayazinon IA **diazinon** Nippon Kayaku
Kayazol IA **diazinon** Nippon Kayaku
KB Anti-Mousse H **iron sulphate** Agriben, Rhodic
KB Araignee Rouge A **dicofol** Agriben, Rhodic
KB Bloc Appat R **chlorophacinone** Agriben, Rhodic
KB Bombe Polish I **petroleum oils + rotenone** Agriben, Rhodic
KB Bombe Totale FIA **dicofol + dinocap + fenitrothion + maneb +piperonyl butoxide + pyrethrins** Agriben
KB Bouillie Bordelaise F **Bordeaux mixture** Rhodic
KB Carox H **linuron** Agriben, Rhodic
KB Chlorose F **iron compounds** Rhodic
KB Cloque du Pecher F **thiram** Rhodic
KB Cochenille I **malathion + petroleum oils** Agriben, Rhodic
KB Courtilliere-ver Gris I **lindane** Rhodic
KB Desherbant Gazon H **2,4-D + mecoprop** Agriben, Rhodic
KB Desherbant Selectif H **2,4-D** Rhodic
KB Fourmis IA **diazinon** Agriben, Rhodic
KB Granex H **carbetamide + oxadiazon** Rhodic
KB Herbonex Desherbant Total H **amitrole + simazine** Agriben, Rhodic
KB Herbonex Granule H **amitrole + atrazine** Agriben
KB Herbonex Liquide H **amitrole + simazine** Agriben, Rhodic
KB Insecte Sol I **lindane** Agriben, Rhodic
KB Insecte-liquide IA **malathion** Agriben
KB Insectes IA **phosalone** Agriben
KB Jardim Batateira IF **carbaryl + lindane + maneb** Rhone-Poulenc
KB Jardim Herbicida Relva H **2,4-D + mecoprop** Rhone-Poulenc
KB Jardim Hortic H **linuron** Rhone-Poulenc

KB Jardim Insecticida para plantas I **dimethoate** Rhone-Poulenc
KB Jardim Insectos do solo I **lindane** Rhone-Poulenc
KB Jardim Limace M **metaldehyde** Rhone-Poulenc
KB Limaces S M **metaldehyde** Agriben, Rhodic
KB Maladies F **iprodione + maneb + sulphur** Agriben
KB Mastic Spray FW **bitumen** Agriben
KB Mauvaises Herbes H **amitrole + ammonium thiocyanate** Agriben, Rhodic
KB Mildiou F **copper oxychloride + maneb + zineb** Agriben
KB Mouche des Legumes I **ethion** Agriben
KB Pucerons I **dimethoate** Agriben, Rhodic
KB Rats R **chlorophacinone** Rhodic
KB Stop Germe G **chlorpropham** Rhodic
KB Total Poudre FI **folpet + phosalone + sulphur** Agriben
KB Traitement d'Hiver IA **malathion + petroleum oils** Agriben, Rhodic
KCO-3001 A **fenothiocarb** Kumiai
Kedifon FA **dinocap** Agrocros
Kefo-varsisto- ja rikkahavite H **phosphoric acid** Mauri Takala
Keim-Stop G **chlorpropham** Fahlberg-List
Keim-Stop-Fumigant 83 G **chlorpropham + propham** Fahlberg-List
Keimhemmer 'Marktredwitz' G **propham** Marktredwitz
Kelaran A **propargite** Eli Lilly
kelevan I [4234-79-1] ethyl 1,1a,3,3a,4,5,5,5a,5b,6-decachlorooctahydro-2-hydroxy-γ-oxo-1,3,4-metheno-1*H*-cyclobuta[*cd*]pentalene-2-pentanoate, Despirol
Kelmor S A **chlorbenside + dicofol + fenson** Aziende Agrarie Trento
Kelted A **dicofol + tetradifon** Agrimont
Kelteran A **dicofol + tetradifon** Aragonesas
Kelthane A **dicofol** Agrotec, Broeste, Condor, Farmos, Hoechst, ICI-Valagro, KVK, LaLittorale, Maag, Permutadora, Plantivern-Kjemi, Protex, Ravit, Rohm & Haas, Sandoz, Schering AAgrunol, Speiss, Staehler, Urania
Kelthane mixte IA **dicofol + parathion-methyl** La Littorale
Kelthane mixte P FIA **dicofol + parathion-methyl + sulphur** La Littorale
Kelthanoide A **dicofol + tetradifon** Condor
Kelthion A **dicofol + tetradifon** Agrochemiki, Hellenic Chemical, La Quinoleine
Kemate F **anilazine** Bayer
Kemazina H **simazine** Argos
Kemdazin F **carbendazim** Argos, Kemichrom
Kemicid FIHN **metham-sodium** Kemichrom
Kemifam H **phenmedipham** Kemira, Nordisk Alkali
Kemikar F **carboxin** Kemira
Kemntazin F **carbendazim** Filocrop
Kemolate IA **phosmet**
Ken F **sulphur** Szovetkezet
Kenitral IA **endosulfan** Key
Kenofuran IAN **carbofuran** KenoGard
Kenopel FX **guazatine acetates** KenoGard, Pennwalt
Kentac G **fatty alcohols** BP
Kenyatox I **piperonyl butoxide + pyrethrins** Drilexco
Kenyatox Grain Protectant I **piperonyl butoxide + pyrethrins** Copyr
Kenyatox Verde I **piperonyl butoxide + pyrethrins** Copyr, Margesin
Kepone IF **chlordecone**
Keraton A **dicofol + fenson** Agronova
Keraunos Bait R **zinc phosphide** Agrotechnica
Kerb H **propyzamide** EniChem, KVK, Maag, Pan Britannica, Pinus, Procida, Protex, Ravit, Rohm& Haas, Serpiol, Spiess, Urania
Kerb Mix H **diuron + propyzamide** Rohm & Haas, Serpiol, Siapa
Kerb ultra H **diuron + propyzamide** Procida, Rohm & Haas
Kerex Fungicide F **maneb + zineb** Temana
Kerfex I **lindane** Agrolinz
Keriguards I **dimethoate** ICI
Keriroot F **captan + 1-naphthylacetic acid** ICI
Kerispray I **pirimiphos-methyl + pyrethrins** ICI
Keropur H **benazolin** Schering
Keropur H **benazolin + dicamba + MCPA** Agrolinz
Kexels Unkrautvertilger H **sodium chlorate** Kexel Drogen

Key Amarillo IA **DNOC + petroleum oils** Key
Key Azufre F **sulphur** Key
Key Semillas FI **lindane + maneb** Key
Keyazil IA **azinphos-methyl** Key
Keyazil IA **azinphos-methyl** Key
Keyazil IA **azinphos-methyl** Key
Keycorc I **carbon disulphide + carbon tetrachloride** Key
Keydane I **lindane** Key
Keyram F **ziram** Key
Keythion IA **malathion** Key
Keytrol H **amitrole + atrazine + 2,4-D** Burts & Harvey
Keyvin I **carbaryl** Key
Keyzin F **zineb** Key
KF-32 F **fthalide** Kureha
Kibrill spray mod bladlus I **piperonyl butoxide + pyrethrins** Praestrud & Kjeldsmark
Kibron plantenspray I **piperonyl butoxide + pyrethrins** Van Nielandt
Kicker I **piperonyl butoxide + pyrethrins** Fairfield
Kidan F **iprodione** Rhone-Poulenc, Zorka Sabac
Kiemremmer Extra G **chlorpropham + propham** Agriben
Kiemremmer-SF G **chlorpropham + propham** Agriben
Kilakar A **proclonol** Diana
Kilerb H **sodium chlorate** Research Development
Kill-it I **piperonyl butoxide + pyrethrins + resmethrin** Teknosan
Kill-it blomsterspray S I **resmethrin** Agro-Kemi
Kill-it fluespray D I **piperonyl butoxide + resmethrin** Agro-Kemi
Kill-it flugnaeitur I **piperonyl butoxide + pyrethrins + resmethrin** Teknosan
Kill-it havepudder FI **captan + lindane + sulphur** Agro-Kemi
Kill-it insektspray S I **permethrin + piperonyl butoxide + pyrethrins** Agro-Kemi
Kill-it myrepudder I **lindane** Agro-Kemi
Kill-it rosenspray S I **resmethrin** Agro-Kemi
Kill-it staldspray N I **piperonyl butoxide + pyrethrins** Agro-Kemi
Kill-it vinter og foraars sproejtemiddel I **petroleum oils** Agro-Kemi
Kill-Net H **chloroxuron** Chimac-Agriphar
Kill-Star 1 I **deltamethrin** Gisga
Kill-Star 2 I **chlorpyrifos** Gisga
Kill-Star 3 I **permethrin + piperonyl butoxide** Gisga
Killgerm Fly & Pest Strip I **dichlorvos** Killgerm
Killgerm Lindacide I **lindane + pyrethrins** Killgerm
Killgerm Tetracide I **fenitrothion + lindane + tetramethrin** Killgerm
Kilper Blau F **copper oxychloride + mancozeb** Du Pont
Kilprop H **mecoprop**
Kiltin Fosforbrinte R **aluminium phosphide** Kiltin
Kiltin W.3 R **warfarin** Kiltin
Kilumal AI **fenpropathrin** Shell
Kilval IA **vamidothion** Rhone-Poulenc
Kinalux IA **quinalphos** United Phosphorus
Kinol F **8-hydroxyquinoline sulphate** De Weerdt
Kinolat V-4-X F **carboxin + oxine-copper** Galenika
Kinolat-15 F **oxine-copper** La Quinoleine
kinoprene I [42588-37-4] 2-propynyl (*E*)-3,7,11-trimethyl-2,4-dodecadienoate, Enstar
Kiporon H **cyanazine + dichlorprop** R.S.R.
Kirk myremiddel I **diazinon** Ciba-Geigy
Kisvax F **carboxin** Jin Hung
Kitabasitac F **iprobenfos + mepronil**
Kitazin F **iprobenfos** Kumiai
Kition I **azinphos-methyl** Caffaro, Ergex
Kitron I **acephate** Jin Hung
Kivax F **carboxin + thiram** Sepran
Kiwi Lustr F **dicloran**
KKB R **calcium carbide** Sasse
Kladex F **dodine** Rhone-Poulenc
Klartan IA **fluvalinate** Sandoz
Klerat R **brodifacoum** Chromos, ICI, ICI-Zeltia, Maag
Klevamol H **mecoprop** Plantevern-Kjemi
Klinopalm H **ametryn + 2,4-D + MSMA** Ciba-Geigy
Kloben H **neburon** Chimac-Agriphar, Du Pont
Kloratul H **sodium chlorate** Neuber
Klorex H **sodium chlorate** Efthymiadis, KemaNord, KenoGard
Klorpikrin FIHN **chloropicrin**
Klorprofam H **chlorpropham**
KM H **sodium chlorate** Kerr-McGee
KM 72 G **chlorpropham** CTA

Knave I **disulfoton + quinalphos** Hortichem
Knockmate F **ferbam** Nihon Nohyaku
Knox out 100 IA **diazinon** Desarrollo Quimico, Pennwalt, Pennwalt France, Umupro
KO Mouche I **dimethoate + fenitrothion** Procida
Koban F **etridiazole** Mallinckrodt
Kobra Rax Difenacoum Raattgift R **difenacoum** ICI
Kobra-Rax Majsbete R **warfarin** Sanerings
Kobu F **quintozene** Takeda
Kobutol F **quintozene** Hokko
Kocide F **copper hydroxide** Chiltern, Chimac-Agriphar, Efthymiadis, Fattinger, Grima, Kennecott, Rhodiagri-Littorale, Sipcam
Kofumin I **dichlorvos** Pliva
K.O. Dorine I **malathion** Procida
Kolfugo F **carbendazim** Chinoin
Kolodust FA **sulphur** FMC
Kolofog FA **sulphur** FMC
Kolospray FA **sulphur** FMC
Kolosul F **sulphur** Helinco, Zorka Sabac
Kolsol AF **sulphur** Sivam
Koltar H **oxyfluorfen** Rohm & Haas
Kolthior F **sulphur** La Quinoleine
Kombat F **carbendazim + mancozeb** La Littorale
Kombat S F **carbendazim + mancozeb + sulphur** Hoechst
Kombi 25-25 H **2,4-D + MCPA** Dow
Kombi Rosenspritzpulver FI **dinocap + lindane + zineb** Shell
Kombicid G-5 I **fenitrothion + lindane** Zorka Sabac
Kombifix-D H **dalapon-sodium + dichlobenil** Shell
Kombinal TO I **tributyltin oxide** Bitterfeld
Kombyrone-P1 H **linuron + monolinuron** Imex-Hulst
K.O. Moss H **iron sulphate** Chimac-Agriphar
Koneprox FB **copper oxychloride** Shell
Konesta H **TCA** Akzo Zout, Aragonesas
Konker F **carbendazim + vinclozolin** BASF
Kontra Schneckex Kruemel M **metaldehyde** Vogger
Koper F **copper oxychloride** Bayer, Eurofyto, Sanac, Van Wesemael
Kopersept FB **copper oxychloride** Asepta
Koplan FA **dinocap** Sandoz
Kopmite A **chlorobenzilate**
Kopperkalk Bayer F **copper oxychloride** Bayer
Kor 80 F **mancozeb** Du Pont, Hoechst, Rohm & Haas
Koral H **amitrole + ammonium thiocyanate + simazine** Etisa
Koril H **bromoxynil + dicamba + mecoprop** La Quinoleine
Korit X **ziram** Kwizda
Korlan I **fenchlorphos**
Kornesta H **TCA** Akzo
Kornitol X **natural fats** Kornitol
Kornitol X **neutral hydrocarbons + organic bases** Carchim
Kornitol-RSM-Rosenspritzmittel F **sulphur + zineb** Kornitol
Kornitol-Unex-Spezial H **borax + bromacil** Kornitol
Kornkaefer-Cit-Staub I **pyrethrins** Cit
Korovicid H **2,4-D** Zorka Sabac
Korovicid combi H **2,4-D + mecoprop** Zorka Sabac
Korrensaade G **chlormequat chloride** BASF
Korrenvahvistaja CCC G **chlormequat chloride** Kemira
Kortal H **amitrole + diuron + petroleum oils + simazine** La Quinoleine
Korthane F **dinocap** Du Pont
Korynex F **thiram + ziram** SEGE
Korzebe F **mancozeb** Tradi-agri
K.O. Souris R **difenacoum** Valmi
K-Othrin I **deltamethrin** Hoechst, Pluess-Staufer, Procida, Roussel Hoechst, Siapa
Kothrin Sovilo I **deltamethrin** Sovilo
Kouman EL F **ziram** Viopharm
Kozan F **sulphur** Viopharm
Kraakalos R **chloralose** Gotlands
Kraft Rasenduenger + Moosvernichter H **iron sulphate** Kraft
Kraft Rasenduenger + Unkrautvernichter mit Langzeitwirkung H **2,4-D + dicamba** Kraft
Kraft Rasenduenger mit Unkrautvernichter H **2,4-D + MCPA** Kraft
Krater H **asulam + diuron** Rhone-Poulenc
Kregan I **chlorpyrifos + lindane** Schering
Krenite H **fosamine-ammonium** Burts & Harvey, C.F.P.I., Du Pont, Procida, Siapa
Kreosan I **DNOC** Zorka
Kreozan FHI **DNOC** Zorka Sabac
Krezotol I **DNOC** Syntetyka
Kripid R **antu**
Kriss IA **parathion** La Littorale

Kriss M IA **parathion-methyl** La Littorale
Kriss M soufre FIA **parathion-methyl + sulphur** La Littorale
Krootazone H **chloridazon** Colvoo
Krovar H **bromacil + diuron** Du Pont, Lauff, Sapec
Krumkil R **coumafuryl**
Kryocide I **cryolite** Pennwalt
krysid R **antu**
Ksilogal I **fenitrothion** Galenika
KSM FB **kasugamycin**
KT 22 A **dicofol + tetradifon** Sipcam
KT 48 A **dicofol** Quimicas Oro
Ku 13-032-c F **dichlofluanid** Bayer
Kuenstliche Rinde Lac Balsam WF **captafol** Scheidler
Kumatin popras R **coumachlor** Druchema
Kumatox R **warfarin** Druchema
Kumirol H **linuron** Kumiai, Nichimen
Kumulan F **nitrothal-isopropyl + sulphur** Agrolinz, BASF, Collett, Intrachem, Maag, Margesin, Siegfried, Spiess, Urania
Kumulus FA **sulphur** Agrolinz, BASF, Collett, Sapec
Kunilent X **fish oil** Chromos
Kuoriaistuho-ruiskute I **methoxychlor** Kemira
Kupfer F **copper oxychloride** Agroplant, Burri, CTA, Hokochemie, Kwizda, Leu & Gygax, Staehler
Kupfer Sandoz F **cuprous oxide** Sandoz
Kupfer-Folpet F **copper oxychloride + folpet** Leu & Gygax
Kupfer-Fusilan F **copper oxychloride + cymoxanil** Kwizda
Kupfer-Phaltan F **copper oxychloride + folpet** Chemia
Kupfer-Schwefel Pulver D F **copper oxychloride + sulphur + zineb** Agrotex
Kupferkalk F **copper oxychloride** Bayer, Ciba-Geigy, Hoechst, Neudorff, Schering, Spiess, Urania, Wacker
Kupferoxid F **cuprous oxide**
Kupferoxychlorid F **copper oxychloride**
Kupfersol F **copper oxychloride** Eli Lilly
Kupferspritzmittel F **copper oxychloride** Austria Metall, Schacht
Kupfervitriol F **copper sulphate** Austria Metall
Kuprijauhe F **copper oxychloride** Kemira
Kuprikol F **copper oxychloride** Spolana
Kupriksalin F **copper oxychloride + cymoxanil + zineb** OHIS
Kuprokalk F **copper oxychloride** Agronova
Kupроneb F **copper oxychloride + propineb** Leu & Gygax
Kupropin F **copper oxychloride** Pinus
Kupropur F **copper oxychloride** Land-Forst
Kuprotox F **copper oxychloride** KVK
Kuril H **bromoxynil + dicamba + mecoprop** Pluess-Staufer
Kuron HG **fenoprop** Dow
Kurosal HG **fenoprop** Dow
Kusagard H **alloxydim-sodium** Kemira, Nippon Soda, Plantevern-Kjemi, VCH
Kusatol H **sodium chlorate** Hodogaya
KV-Kombi-Getreideherbizid H **2,4-D + mecoprop** Land-Forst
Kvit mod krybende og flyvende insekter I **allethrin + permethrin** Agro-Kemi
KVK Buskmedel H **MCPA** KVK
KVK-Flue-Aerosol I **piperonyl butoxide + pyrethrins** KVK
KVK fluemiddel I **piperonyl butoxide + pyrethrins + resmethrin** KVK
KVK Herbatox-BV plaenemiddel H **2,4-D + dicamba + mecoprop** KVK
KVK-Herbatox-D H **2,4-D** KVK
KVK Plus H **dicamba + MCPA + mecoprop** KVK
KVK sproejtesvovl F **sulphur** KVK
KVK Svovl-Thiram F **sulphur + thiram** KVK
KVK-Vesakkoruiskute B H **MCPA** Bang
Kwik-Kil R **strychnine**
KWP 61 IA **carbaryl + tetradifon** Schering
Kylar G **daminozide** Uniroyal
Kypchlor I **chlordane**
Kypfarin R **warfarin**
Kypfos IA **malathion**
Kypman F **maneb**
Kypzin F **zineb**
Kytrol H **amitrole + ammonium thiocyanate + diuron + simazine** C.F.P.I.

L-2 I **lindane** Argos
L 16/184 IA **oxydisulfoton**
L 395 IA **dimethoate** Montedison
L-561 IA **phenthoate** Montedison
L 1058 IA **azothoate**
L-34314 H **diphenamid** Eli Lilly
L-36352 H **trifluralin** Elanco
L-flamprop-isopropyl H **flamprop-M-isopropyl**
Labiazole TA H **amitrole + ammonium thiocyanate** L.F.P.C.

Labilite F **maneb + thiophanate-methyl** Nippon Soda
Labo'sol H **amitrole + atrazine** L.F.P.C.
Labrax H **benazolin + clopyralid** Lachema
Labuctril H **bromoxynil** Lachema
Lacbalsam FW **captafol + poly(vinyl propionate)** Coppyn Boomchirugen, Scheidler, Turba Projar
Laddok H **atrazine + bentazone** Zorka Sabac
Ladob H **dinoseb** Lachema
Lafarex N I **methoprene** Lachema
Lagran H **alachlor + atrazine + pyridate** Monsanto
Laicon F **polyoxins** Lainco
Laidan IA **diazinon** Lainco
Laiguant G **2,4-D** Lainco
Laikuaj G **gibberellic acid** Lainco
Laincobre F **copper oxychloride + zineb** Lainco
Laincoil agrios IA **petroleum oils** Lainco
Lainsect IA **naled** Lainco
Lainzufre FA **sulphur** Lainco
Laipar M I **parathion-methyl** Lainco
Lairam F **ziram** Lainco
Lairana Adulticida A **dicofol** Lainco
Lairana Total A **dicofol + tetradifon** Lainco
Laisol FIHN **metham-sodium** Lainco
Laitane Fuerte F **dinocap** Lainco
Laitane Normal F **dinocap + sulphur** Lainco
Laiterra H **amitrole + diuron** Lainco
Laition IA **dimethoate** Lainco
Laitom IA **azinphos-methyl** Lainco
Laivin I **carbaryl** Lainco
Lambast H **butachlor** Monsanto
lambda-cyhalothrin I [91465-08-6] α-cyano-3-phenoxybenzyl 3-(2-chloro-3,3,3-trifluoropropenyl)-2,2-dimethylcyclopropanecarboxylate, a 1:1mixture of the (*Z*)-(1*R*, 3*R*), S-ester and (*Z*)-(1*S*, 3*S*), R-ester, Icon, Karate, PP 321
Lambrol I **fluenetil**
Lance IN **cloethocarb** BASF
Lancer H **flamprop-methyl** ICI
Lancer plus H **flamprop-M-isopropyl** ICI
Lancord I **cypermethrin + methomyl** BASF
Landmaster H **2,4-D + glyphosate** Monsanto
Landrin IM **trimethacarb** Shell
Langzeit Rasenduenger mit Unkrautvernichter H **dicamba + mecoprop** Hoechst
Lanirat R **bromadiolone** Liro
Lannate IN **methomyl** Du Pont, OHIS, Permutadora, Ravit, R.S.R., Spiess, Szovetkezet, Urania
Lanox IA **methomyl** Crystal
Lanray H **orbencarb** Kumiai, Kwizda
Lanray H **linuron + orbencarb** Siegfried
Lanray L H **linuron + orbencarb** Kumiai, Kwizda
Lanslide H **lenacil + linuron** Pan Britannica
Lantironce H **2,4-D + 2,4,5-T** LHN
Laptran F **ditalimfos** Dow
Larsen R **coumachlor** Schering
Larvadex I **cyromazine** Ciba-Geigy
Larvakil I **diflubenzuron** Antec
Larvatox I **dimethoate + endosulfan** Tecniterra
Larvex I **polybutene** Tuhamij
Larvin I **thiodicarb** Ravit, Rhodiagri-Littorale, Union Carbide
Laser H **cycloxydim** BASF
Lasso H **alachlor** Hellafarm, Leu & Gygax, Monsanto, Quimigal, Sandoz, Shell, Siapa, Sipcam
Lasso + Atrazina H **alachlor + atrazine** Monsanto, Quimigal, Sandoz
Lasso AT H **alachlor + atrazine** Hellafarm
Lasso combi-tekuci H **alachlor + atrazine** Pinus
Lasso GD H **alachlor + atrazine** Condor, Monsanto, Sipcam
Lasso N H **alachlor** Spolana
Lasso/Atrazin H **alachlor + atrazine** Pinus, Zorka Sabac
Lasso/Linopin H **alachlor + linuron** Zorka Sabac
Lastanox F F **bis(tributyltin) oxide + formaldehyde** Lachema
Latex Forestier X **odorous substances** Sema Vinyl
Lathion I **azinphos-ethyl** Geopharm
Lathion Combi I **azinphos-ethyl + azinphos-methyl** Geopharm
Lathion Metil IA **azinphos-methyl** Sipcam
Latox I **malathion** Geophyt
Laubrex II H **dalapon-sodium + diuron + MCPA** Lauff
Laubrex III H **diuron + hexazinone** Lauff
Lawa gaspatroner R **sulphur** Lassen & Wedel
Lawa mod lus pa planter, spray med glans

I **piperonyl butoxide + pyrethrins + rotenone** Lassen & Wedel
Lawa mod lus paa planter 87 I **piperonyl butoxide + pyrethrins** Lassen & Wedel
Lawa mod lus paa planter, med glans 87 I **piperonyl butoxide + pyrethrins** Lassen & Wedel
Lawa plantevask FI **lindane + petroleum oils + piperonyl butoxide+ pyrethrins + rotenone** Lassen & Wedel
Lawi-Oel S **petroleum oils** Dow
Lawn Fertilizer with Weedkiller H **2,4-D + mecoprop** Asef Fison
Lawn Mosskiller and Fertiliser H **dichlorophen** May & Baker
Lawn Mosskiller and Fertiliser H **iron sulphate** Boots
Lawn Plus H **2,4-D + dicamba** ICI
Lawn Plus H **2,4-D + mecoprop** Agrolinz
Lawn sand vervelours H **iron sulphate** Truffaut
Lawnsman Mosskiller I **chloroxuron + dichlorophen + iron sulphate** ICI
Lawnsman Weed and Feed H **2,4-D + dicamba** ICI
Lazeril H **diflufenican + ioxynil + mecoprop** Rhodiagri-Littorale
Lazo H **alachlor** Condor, Monsanto
LE 79-519 I **permethrin**
LE 79600 I **cypermethrin** Rhone-Poulenc
Leafex H **sodium chlorate** Simplot
Lebaycid I **fenthion** Bayer, Pinus
Ledax I **piperonyl butoxide + pyrethrins** Ledona
Ledax-san F **sulphur** Ledona, Schaette
Legor H **benzoylprop-ethyl** Siapa
Legumex Extra H **benazolin + 2,4-DB + MCPA** Schering
Legumex M H **MCPB** Schering
Legurame H **carbetamide** Agriben, Condor, Rhodiagri-Littorale, Rhone-Poulenc
Legurame PM + Ronstar H **carbetamide + oxadiazon** Rhodiagri-Littorale
Lekinol-Cu F **oxine-copper** Lek
Lemonene F **biphenyl**
Lenac H **lenacil** Chemia
lenacil H [2164-08-1] 3-cyclohexyl-1,5,6,7-tetrahydrocyclopentapyrimidine-2,4(3*H*)-dione, lenacile, Adol, Du Pont 634, Elbatan, Venzar, Vizor
lenacile H **lenacil**
Lenacilo H **lenacil** Aragonesas
Lenamon H **lenacil** Du Pont
Lentacol X **thiram** Margesin, Shell
Lentagran plus H **dichlorprop + pyridate** Ruse
Lentagran WP H **pyridate** Agrichem, Agrishell, Agrolinz, Agroplant, KenoGard, Margesin, Protex, Ruse, Schering AAgrunol, Shell
Lentazin total H **atrazine** Agrolinz
Lentemul D H **2,4-D** Agrolinz
Lentrix H **cyanazine + pyridate** Shell
Lepidos I **dichlorvos** Visplant
Lepit R **chlorophacinone** Kwizda, Schering
Lepit Gifkorrels R **zinc phosphide** Denka
Lepit Konzentrat R **chlorophacinone + sulphaquinoxaline** Schering
Lepracin F **ziram** Afrasa
leptophos I [21609-90-5] *O*-4-bromo-2,5-dichlorophenyl *O*-methyl phenylphosphonothioate, MBCP, OMS 1438, Abar, Phosvel, VCS 506
Leptox I **chlorfenvinphos** Sapec
Lermol H **clopyralid** Siapa
Lermol 3 H **2,4-D + dichlorprop + triclopyr** C.F.P.I.
Lesan F **fenaminosulf** Bayer
Lester G **1-naphthylacetic acid** Farmer
Letal-Rat R **chlorophacinone** Agriplan
Leucon F **ditalimfos** Rustica
Leutox I **ethylene oxide** Leuna-Werke
Levanox F **copper oxychloride + cymoxanil + zineb** Argos
Leven F **zineb** Baslini
Lexone H **metribuzin** Du Pont
LFA 2043 F **iprodione** Rhone-Poulenc
LH 30/Z F **propineb** Bayer
Licor G **4-CPA + (2-naphthyloxy)acetic acid** Scam
Lidastop I **diazinon + lindane** Hermoo
Lidax I **lindane** Chimac-Agriphar
Lidazon I **diazinon + lindane** Ligtermoet, Liro
Lider I **fenitrothion + trichlorfon** Ital-Agro
Lig 20 I **lindane** Chemia
Lignasan F **carbendazim** Du Pont
Likiplouse H **2,4-D + dicamba** Sedagri
Limacir M **metaldehyde** EniChem
Limaclor M **metaldehyde** Fisons, Pfizer
Limacol M **metaldehyde** Reis
Limaflor Ex M **metaldehyde** Leu & Gygax
Limagran M **metaldehyde** Inagra
Limaldehyde M **metaldehyde** Bourgeois
Limalo M **metaldehyde** Maag
Limargos M **metaldehyde** Argos
Limasivam M **metaldehyde** Sivam

Limaslak M **metaldehyde** Eurofyto
Limasol Super M **metaldehyde** Salomez
Limastop M **metaldehyde** C.N.C.A.T.A.
Limatex M **metaldehyde** Sapec
Limatic M **metaldehyde** C.N.C.A.T.A.
Limatox M **metaldehyde** Chimiberg, Kwizda
Limatrin M **metaldehyde** Enotria
Limavit M **metaldehyde** Agriben
Limax M **metaldehyde** Franco-Belge, Maag, Ruse
Limbolid G **heptopargil** EGIS, EGYT
lime sulphur FIA [1344-81-6] calcium polysulphide, calcium polysulphide, polysulfure de calcium, Orthorix, Security Lime Sulphur
Limit H *N*-[(acetylamino)methyl]-2-chloro-*N*-(2,6-diethylphenyl)acetamide Monsanto
Limort M **metaldehyde** Bayer
Limoter H **linuron + monolinuron** Bourgeois
Linamex H **butralin + linuron** Sopra
Linarol H **trifluralin** EniChem
Linazol H **MCPA** Procida
Linda-Solo I **lindane** ICI-Valagro
lindaan I **lindane**
Lindacoop I **lindane** Ficoop
Lindacot I **lindane** Lapapharm
Lindaflor I **lindane** Agriben
Lindafor I **lindane** Agriben, Pepro, Rhone-Poulenc, Sedagri
Lindagranox I **lindane** Rhone-Poulenc
Lindal I **lindane** Eurofyto, Mylonas
Lindaline I **lindane** Bourgeois
Lindamul I **lindane** Agriben, Condor, Rhone-Poulenc
Lindan I **lindane** Agrolinz, Bjoernrud, Chromos, Epro, OHIS, Pinus, Propfe, Radonja, Shell, Staehler, Zorka Sabac, Zorka Subotica, Zupa
lindane I [58-89-9] 1α,2α,3β,4α,5α,6β-hexachlorocyclohexane, gamma-HCH, gamma-BHC, gamma benzene hexachloride, gamma-HKhTsH, OMS 17, ENT7796, Agronexa, Agronexit, Exagama, Gallogama, Gamaphex, Gammalin, Gammex, Gammexane, Inexit, Isotox, Lindafor, Lindagranox, Lindamul, Lindaterra, Lintox, Nexit, Novigam, Silvanol
Lindanex I **lindane** Agrofarm, Protex
Lindanil I **lindane** Sopepor
Lindano I **lindane** CAG
Lindanol I **lindane** Geochem
Lindasect I **lindane** Agro-Kemi
Lindaterra I **lindane** Condor, Rhone-Poulenc
Lindatox I **lindane** Borzesti, Dudesti
Lindavis I **lindane** Visplant
Lindex H **cyanazine + linuron** Shell
Lindex I **lindane** Agriphyt, Agriplan, Anticimex, La Quinoleine, Schwebda, Shell
Lindex-Plus IF **fenpropimorph + lindane + thiram** Dow
Lindexan I **lindane** Gefex
Lindit I **lindane** Ellagret
Lindol IA **lindane** Agrotechnica, Van Wesemael
Lindozal I **lindane** Hellenic Chemical
Lindram IF **lindane + thiram** Lucebni zavody
Linex H **linuron** Griffin
Linnet H **linuron + trifluralin** Pan Britannica
Linocin H **linuron + simazine** CTA
Linol H **MCPA** La Littorale
Linopan H **linuron** Maag
Linormona H **MCPA** Rhone-Poulenc
Linorox H **linuron** Du Pont
Linorto H **linuron** Caffaro
Linosil H **diuron** Platecsa
Linoxone H **MCPA** Aragonesas, La Quinoleine
Linoxone extra H **MCPA + mecoprop** La Quinoleine
Linozerba H **linuron** Sapec
Lintaxin A **cyhexatin** Lintaplant
Lintox I **lindane** Agrocros, Siapa
Linuben H **linuron** ERT
Linukey H **linuron** Key
Linurac H **linuron** Agriphyt
Linuragrex H **linuron** Sadisa
Linural H **linuron** Tradi-agri
Linuran Super H **linuron + monolinuron** Eurofyto
Linuree H **linuron** Interphyto
Linuree special H **linuron + monolinuron** Interphyto
Linurex H **linuron** Makhteshim-Agan, Permutadora
linuron H [330-55-2] 3-(3,4-dichlorophenyl)-1-methoxy-1-methylurea, Afalon, Du Pont Herbicide 326, Hoe 02810, Linex, Linorox, Linurex, Lorox, Rotalin, Sarclex
Linusint H **linuron** Sintagro
Linvur I I **lindane** I.C.M.M., I.C.P.P.

Liosol S **polyglycolic ethers** Sepran, Terranalisi
Liphadione R **chlorophacinone** Lipha
Lipomel F **lecithin** Meyer
Liqua-Tox R **warfarin** Bell Labs
Liquid Club Root Control F **thiophanate-methyl** May & Baker
Liquid Copper Fungicide F **cuprammonium** Murphy
Liquid Derris I **rotenone** Pan Britannica
Liquifos I **parathion** Caffaro
Liquizol AF **sulphur** EniChem, Mormino
lirimfos I [38260-63-8] *O*-6-ethoxy-2-isopropylpyrimidin-4-yl *O,O*-dimethyl phosphorothioate, SAN 201I
Liro Antilimaces M **metaldehyde** Liro
Liro Carmazin F **maneb + zineb** Liro
Liro Croton F **dinocap** Liro
Liro Faneron H **bromofenoxim** Liro
Lirogam IA **lindane** Ligtermoet
Lirogazon H **2,4-D + dicamba + MCPA + mecoprop** Ligtermoet
Liro Gazon H **2,4-D + dicamba + MCPA + mecoprop** Liro
Liro Grassol N G **maleic hydrazide** Ligtermoet
Liro Macotin F **fentin acetate + mancozeb** Ligtermoet
Liro Macotin N F **fentin acetate + mancozeb** Ligtermoet
Liro Manzeb F **maneb + zineb** Ligtermoet
Liro Matin F **fentin acetate + maneb** Ciba-Geigy, Ligtermoet, Liro
Liromort H **dinoseb** Ligtermoet
Liron H **linuron** Zupa
Liro Nefal S **glycols + methylene chloride** Ligtermoet
Lironion H **difenoxuron** Ciba-Geigy, Ligtermoet
Liro Nogos I **dichlorvos** Ligtermoet
Lironox Extra H **dicamba + MCPA + mecoprop** Ligtermoet
Liropon H **dalapon-sodium** Ligtermoet
Liro-Stanol F **fentin acetate** Ligtermoet
Lirotan F **zineb** Ligtermoet
Lirotect F **thiabendazole** Ligtermoet, Liro
Lirotect Extra F **imazalil + thiabendazole** Ligtermoet, Liro
Lirotect M F **maneb + thiabendazole** Liro
Lirotectim F **imazalil + thiabendazole** Ligtermoet
Lirothion I **parathion** Ligtermoet
Liro-Tin FLM **fentin acetate** Liro
Liro-Trithion Spuitbus N A **carbophenothion** Ligtermoet
Liro Vurex F **folpet + maneb** Ligtermoet
Lisamon H **alachlor** Du Pont
Lisolan H **linuron** Schering
Litarol H **bromoxynil** La Littorale, Rhodiagri-Littorale
Litexa I **lindane** La Littorale
Lithofin algex L **quaternary ammonium** Stingel
Litocide IA **endosulfan + parathion** Tecniterra
Livin H **clopyralid + MCPA + mecoprop** Schering
Lizetan IA **methiocarb + propoxur** Bayer
Lizetan-Zierpflanzenspray IA **omethoate** Bayer
LM 91 R **chlorophacinone** Lipha
LM 637 R **bromadiolone** Lipha
Lo-Drift S **polyacrylamide** Amchem, Union Carbide
Logic I **fenoxycarb** Maag
Loginet R **chlorophacinone** Hygiene Service
Logran H **triasulfuron** Ciba-Geigy
Lonacol F **zineb** Bayer, Berner
Lonacol M F **maneb** Bayer
Lonacol Ramato F **copper oxychloride + zineb** Bayer
Londax H **bensulfuron-methyl** Du Pont
Lonpar H **clopyralid + 2,4-D + MCPA** Prochimagro
Lontranil H **clopyralid + cyanazine** Dow
Lontrel H **clopyralid** Dow, Kwizda, Lachema, Schering, Zorka Sabac
Lontrel CM H **clopyralid + ioxynil + mecoprop** Shell
Lontrel DP H **clopyralid + dichlorprop** Esbjerg
Lontrel Kombi H **clopyralid + MCPA + mecoprop** Dow
Lontrel Nuovo H **clopyralid + MCPA + mecoprop** ICI
Lontrel P H **clopyralid + mecoprop** Chimac-Agriphar, Dow
Lontrel Plus H **clopyralid + dichlorprop + MCPA** ICI
Lontryx H **clopyralid + ioxynil + mecoprop** Dow
Lonza Rasenduenger mit Unkrautvernichter H **chlorflurecol-methyl + MCPA** Lonza-Werke
Lonza Schneckenkorn M **metaldehyde** Lonza-Werke
Loovex H **dinoseb** L.A.P.A.
Lop-tox FI **lindane + thiram** Esbjerg

Lord Rasenduenger mit Moosvernichter H **iron sulphate** SKW Trostberg
Lord Rasenduenger mit Unkrautvernichter H **chlorflurecol-methyl + MCPA** SKW Trostberg
Lorox H **linuron** Du Pont, Farmos
Lorox Plus H **chlorimuron-ethyl + linuron** Du Pont
Lorsban I **chlorpyrifos** Dow, Siapa
Lorsban C I **carbaryl + chlorpyrifos** Agrocros
Lorsban L I **chlorpyrifos + lindane** Dow, Prochimagro
Lorvek F **pyroxychlor**
Lostal F **fentin acetate + propiconazole** Schering
Lotetu Arvalin R **zinc phosphide** Budapesti Vegyimuvek
Lovozal A **fenazaflor**
Loxiran I **chlorpyrifos** Neudorff
Loxytril H **bromoxynil + dichlorprop + ioxynil** Lachema
LS 74783 F **fosetyl-aluminium** Rhone-Poulenc
Lucel F **chlorquinox**
Lucenit H **diuron** Nitrokemia
Ludocyan H **difenzoquat methyl sulphate** Siapa
Lumachene M **metaldehyde** Bimex
Lumachicida M **metaldehyde** Ciba-Geigy
Lumacid M **metaldehyde** Fivat
Lumakidin M **metaldehyde** Industrialchimica
Lumakill M **metaldehyde** Agronova
Lumakorn M **metaldehyde** Gobbi
Lumascam M **metaldehyde** Scam
Lumeton Forte H **chlorotoluron + mecoprop** VCH
Luqdiuron H **diuron** Luqsa
Luqmullant S **polyglycolic ethers** Luqsa
Luqsacel G **chlormequat chloride** Luqsa
Luqsathion IA **malathion** Luqsa
Luqsazufre FA **sulphur** Luqsa
Luqsol IA **petroleum oils** Luqsa
Luqsol invierno IA **DNOC + petroleum oils** Luqsa
Luqzinon I **diazinon** Luqsa
Lurat R **coumafuryl**
Luron H **linuron** Chemia
Lurontil H **linuron + trifluralin** Visplant
Lusadon IA **endosulfan** Caffaro
Lusagran I **azinphos-ethyl** Quimigal
Luserb H **simazine** Siapa
Lutin-Neu-Winterspritzmittel IA **DNOC** Marktredwitz
Lution I **fenitrothion** Luqsa
Lutiram F **copper oxychloride + metiram** Sapec
Luxamix F H **2,4-D + dicamba + mecoprop** Luxan
Luxan Algendood L **quaternary ammonium** Luxan
Luxan Anti Spruit G **chlorpropham + propham** Luxan
Luxan Anti-Germe G **chlorpropham + propham** Luxan
Luxan Anti-Schuim S **silicones** Luxan
Luxan azidro F **carbendazim** Luxan
Luxan Captan-M F **captan + maneb** Luxan
Luxan Captan-Zwavel F **captan + sulphur** Luxan
Luxan Carbendazim-M F **carbendazim + maneb** Luxan
Luxan Chloor-IPC H **chlorpropham** Luxan
Luxan Corbel F **fenpropimorph** Luxan
Luxan corbel star F **chlorothalonil + fenpropimorph** Luxan
Luxan Dicamix H **2,4-D + dicamba + MCPA** Luxan
Luxan Dicamix F H **2,4-D + dicamba + mecoprop** Luxan
Luxan Dizalin I **diazinon + lindane** Luxan
Luxan Emeltenkorrels IA **lindane** Luxan
Luxan Flowerspray I **fenitrothion** Luxan
Luxan Gro-Stop G **chlorpropham + propham** FSA
Luxan Hydramin G **maleic hydrazide** Luxan
Luxan Insegar I **fenoxycarb** Luxan
Luxan insekten dood-P I **permethrin** Luxan
Luxan luizendood I **pirimicarb** Luxan
Luxan Mecomix H **MCPA + mecoprop** Luxan
Luxan Mollenpatroon R **sulphur** Luxan
Luxan Mollentabletten R **aluminium phosphide** Luxan
Luxan myrelokkedase I **trichlorfon** Agro-Kemi
Luxan naturen IA **piperonyl butoxide + pyrethrins** Luxan
Luxan naturen-P I **permethrin** Luxan
Luxan Nevocol S **ethylene glycols** Luxan
Luxan Nevolin N S **glycols + methylene chloride** Luxan
Luxan Onkruidkorrels Extra H **dichlobenil** Luxan

Luxan Parasect I **dimethoate + fenitrothion** Luxan
Luxan plantenspray I **deltamethrin** Luxan
Luxan quinolate Pro F **carbendazim + oxine-copper** Luxan
Luxan Repulsif Contre Gibier X **ziram** Luxan
Luxan Rootone-F G **4-indol-3-ylbutyric acid + 2-methyl-1-naphthylacetamide + 2-methyl-1-naphthylacetic acid + 1-naphthylacetamide + thiram** Luxan
Luxan Scabisix IA **lindane** Luxan
Luxan Sitrol H **amitrole + simazine** Luxan
Luxan Slakkenkorrels M **metaldehyde** Luxan
Luxan Spuitzwavel F **sulphur** Luxan
Luxan Stalvliegendood I **dimethoate + fenitrothion** Luxan
Luxan Teceal-vloeibaar H **chloral hydrate** Luxan
Luxan Uitvloeier H S **polyglycolic ethers** Luxan
Luxan Uracom H **amitrole + bromacil + diuron** Luxan
Luxan Uratan H **bromacil + diuron** Luxan
Luxan Wildafweermiddel X **ziram** Luxan
Luxan Zimanaat F **maneb + zineb** Luxan
Luxarin Rattenkorrels R **warfarin** Luxan
Luxatox DP H **dichlorprop** Luxan
Luxatox Kombi H **2,4-D + MCPA** Luxan
Lyatox I **piperonyl butoxide + pyrethrins** Hiven
Lypor I **temephos** Cyanamid
Lyso-champ IN **phenol** Sterling
Lysol no 2 I **phenol** Sterling
Lysozid-fluens doed I **trichlorfon** PLK
lythidathion I [2669-32-1] *S*-5-ethoxy-2,3-dihydro-2-oxo-1,3,4-thiadiazol-3-ylmethyl *O,O*-dimethyl phosphorodithioate, GS 12968, NC 2962
Lyzatex T **hydrolysed proteins** Rhone-Poulenc

2M-4Kh H **MCPA**
2M-4Kh-M H **MCPB**
371 M H **atrazine + diuron** Urania
M-50 IA **malathion** Argos
M 52 H **MCPA** Schering
M 52 Kombi H **2,4-D + MCPA** Schering
M-70 F **mancozeb** Du Pont
M-74 IA **disulfoton**
M-81 IA **thiometon**
M 96 olieemulsion A **petroleum oils** Midol
M 1568 IA **cyanthoate**
M 2060 I **fluenetil**
M 3432 H **tiocarbazil** Montedison
M 8164 F **chlozolinate** Montedison
M 9834 F **benalaxyl** Montedison
M Special F **captan + zineb** ICI, Maag, Viopharm
M&B 8873 H **ioxynil** May & Baker
M&B 9057 H **asulam** May & Baker
M&B 10064 H **bromoxynil** May & Baker
M&B 11641 H **ioxynil** May & Baker
M&B 25-105 G **propyl 3-tert-butylphenoxyacetate** May & Baker
M&B 38544 H **diflufenican** May & Baker
M3/158 IA **demeton-S-methyl sulphone** Bayer
Maben F **maneb** ERT
Mac F **sulphur** Koch & Reis
Mac 2 F **sulphur + triadimenol** R.S.R.
Macbal I **XMC** Hodogaya
Machete H **butachlor** Monsanto
Macuprax F **Bordeaux mixture + cufraneb** Agribus, Comac
Maestro II H **ioxynil + mecoprop** Ciba-Geigy
Maeusegiftweizen R **zinc phosphide** Schacht
Maeusetod R **potassium nitrate + sulphur** Hauri
Mafu IA **dichlorvos** Bayer
Mafu Creepy Crawly Spray I **dichlorvos + propoxur** Bayer
Magarzel H **dicamba + MCPA** ICI-Zeltia
Magfos R **magnesium phosphide** Nicholas-Mepros
Magic F **fenpropimorph + prochloraz** La Quinoleine
Magnetic 6 FA **sulphur** OHIS, Shell, Stauffer
Magnetic 72 FA **sulphur** Serpiol
Magnum H **chloridazon + ethofumesate** BASF
Magtoxin IR **magnesium phosphide** Breymesser, Colkim, Degesch
Mahle Rasenduenger Sanguano + UV H **chlorflurecol-methyl + MCPA** Mahle Duenger, Rossgatterer
Mahle Rasenduenger Sanguano + MV H **iron sulphate** Rossgatterer
maiblue Ameisen-Spray I **bendiocarb** Samen-Maier
maiblue Ameisenstaub I **bendiocarb** Samen-Maier

maiblue Baum- und Wundbehandlungsmittel FW **captafol + thiabendazole** Samen-Maier
maiblue Blattlaus- und Pflanzenspray IA **dimethoate** Samen-Maier
maiblue Insekten-Frei I **pirimiphos-methyl** Samen-Maier
maiblue Insekten-Staub I **pirimiphos-methyl** Samen-Maier
maiblue Mehltau-Frei F **bupirimate** Samen-Maier
maiblue Rasenduenger mit Unkrautvernichter H **2,4-D + dicamba** Samen-Maier
Maiblue Rasenduenger mit Moosvernichter + Super MosskilH **iron sulphate** Samen-Maier
maiblue Rasenduenger mit Moosvernichter H **iron sulphate** Samen-Maier
maiblue Schneckenkorn M **metaldehyde** Samen-Maier
maiblue Total-Unkrautfrei gegen Unkraeuter auf Wegen und PlaetzenH **amitrole + MCPA + simazine** Samen-Maier
maiblue Unkrautfrei fuer den Rasen H **dicamba + MCPA** Samen-Maier
maiblue Wuehlmausbrocken R **zinc phosphide** Samen-Maier
Maidis H **atrazine** Chemia
Maintain GH **chlorflurecol-methyl**
Mais-Banvel H **dicamba** Agrolinz
Mais-Bentrol H **bromoxynil + simazine** Spiess, Urania
Mais-Certrol H **atrazine + bromoxynil** Pluess-Staufer, Spiess, Union Carbide, Urania
Maisdiserb H **atrazine** Chemia
Maitac AI **amitraz** Agrishell, Schering
Maitrine H **ametryn** Sipcam-Phyteurop
Maizim H **atrazine** Burri
Maizina H **atrazine** Sipcam
Maizol H **atrazine** Candilidis
Maizor H **atrazine + ethalfluralin** Grima
Makaber R **bromadiolone** De Rauw
Maki R **bromadiolone** Lipha
Mala I **malathion** C.G.I.
Malacyde IA **prothoate** Agronova
Maladan I **malathion** Agro-Kemi
Maladrex I **malathion** Shell
Maladust I **malathion** Gefex, EniChem
Malafex I **malathion** Gefex
Malafin IA **malathion** Agrocros
Malafos formulazione inodore I **malathion** EniChem
Malagrain I **malathion** Procida
Malagrex IA **malathion** Sadisa
Malamar IA **malathion**
Malan-polyte I **malathion** Berner
Malan-ruiskute I **malathion** Berner
Malaphele IA **malathion** Rhone-Poulenc
Malasan I **malathion** Eli Lilly
Malasiini-polyte I **malathion** Kemira
Malasiini-ruiskute I **malathion** Kemira
Malaspray IA **malathion**
Malatex I **malathion** Anticimex
Malathane I **malathion** Chimac-Agriphar, Permutadora
Malathex IA **malathion** Protex
Malathin IA **malathion** Fattinger
malathion IA [121-75-5] *S*-1,2-bis(ethoxycarbonyl)ethyl *O,O*-dimethyl phosphorodithioate, malathon, carbofos, mercaptothion, maldison, mercaptotion, OMS 1, ENT 17034, Calmathion, Celthion, Cythion, EI 4049, Emmatos, Fyfanon, Hilthion, Karbofos, Kypfos, Malamar, Malaphele, Malaspray, Malatol, Malmed, MLT, Sumitox, Vegfru Malatox, Zithiol
malathon IA **malathion**
Malathyne IA **malathion** Chimac-Agriphar
Malation I **malathion** Collett, Nordisk Alkali, Zorka Sabac, Zupa
Malatival I **malathion** ICI-Valagro
Malatol IA **malathion** Cyanamid
Malaton I **malathion** Quimigal
Malatox I **malathion** Agrochemiki, Bourgeois, Siapa
Malatox P I **malathion + parathion** Siapa
Malavia IA **malathion** Frere
Malavis I **malathion** Visplant
Malaxone I **malathion** Quimigal
maldison IA **malathion**
Malehid G **maleic hydrazide** Zupa
maleic hydrazide G [123-33-1; 10071-13-3] 6-hydroxy-2*H*-pyridazin-3-one, hydrazide maleique, MH, Burtolin, De-Cut, Fair-2, Fair-Plus, Malzid, Mazide, MH-30, OMH-K, Regulox, Retard, Slo-Gro, Sprout-Stop, Stunt-Man, Sucker-Stuff, Super Sucker-Stuff, Super-De-Sprout, Vondalhyd
Malerbane Cereali H **2,4-D** Chimiberg
Malerbane Giavoni H **molinate** Chimiberg
Malerbane Prati H **2,4-DB** Chimiberg
Malermais H **atrazine** Chimiberg
Malermais E H **atrazine + simazine** Chimiberg

Malertox Bietomin H **chloridazon** Sivam
Malertox D.P.Na H **dalapon-sodium** Sivam
Malertox DMU H **diuron** Sivam
Malertox Giavonil H **propanil** Sivam
Malertox GM Combi H **2,4-D + MCPA** Sivam
Malertox Grano Complex H **dicamba + MCPA** Sivam
Malertox Grano Estere H **2,4-D** Sivam
Malertox Grano Giallo HIAF **DNOC** Sivam
Malertox Grano Riso H **MCPA** Sivam
Malertox Luron H **linuron** Sivam
Malertox M.S. H **simazine** Sivam
Malertox Mais A H **atrazine** Sivam
Malertox Mais A.S. H **atrazine + simazine** Sivam
Malertox Mais L H **alachlor** Sivam
Malertox Medica S H **diuron + propyzamide** Sivam
Malertox Prati S H **2,4-DB** Sivam
Malertox Premerg TL H **linuron + trifluralin** Sivam
Malertox Riso Fluid H **molinate** Sivam
Malertox Riso Granuli H **molinate** Sivam
Malertox T.C.Na H **TCA** Sivam
Malertox Trialin H **trifluralin** Sivam
Malinebe F **maneb** Agrishell
Malipur F **captan** Berlin Chemie
Malital I **malathion** Ital-Agro
Malix IA **endosulfan** Hoechst
Malix-Combi I **dimethoate + endosulfan** Roussel Hoechst
Malixol I **malathion** Sedagri
Malmed IA **malathion** Montedison
Maloran H **chlorbromuron** Ciba-Geigy, Ligtermoet, Nitrokemia
Malpineks G **maleic hydrazide** Pinus
Maltex I **malathion** Agrotechnica
Malyphos I **malathion** Sipcam-Phyteurop
Malzid G **maleic hydrazide** Spezialchemie Leipzig
Malzid combi HG **2,4-D + maleic hydrazide** Spezialchemie Leipzig
Man Power F **copper hydroxide + maneb** Agtrol
Man-Zox F **maneb** Cumberland
Manacol F **maneb** Agro-Kemi
Managrex F **maneb** Sadisa
Manasim F **maneb** BASF
Manatam F **maneb** Eurofyto
Manatane F **mancozeb** Eli Lilly
Manate F **maneb** EMTEA, Hoechst, Procida
Mancatene F **maneb + metiram** Bourgeois
Mancobleu F **copper oxychloride + mancozeb** La Quinoleine
Mancocide F **mancozeb** Ciba-Geigy
Mancofol F **folpet + mancozeb** Condor, Pepro, Rohm & Haas, Visplant
Mancokar F **dinocap + mancozeb** Rohm & Haas
Mancolan F **mancozeb** Hellenic Chemical
Mancomil F **dithianon + mancozeb** Epro
Mancomix F **mancozeb** Eurofyto
Manconyl F **mancozeb** Gen. Rep., Prochimagro
Mancoplus F **mancozeb** J.S.B., Tradi-agri
Mancospor F **mancozeb** Enotria
Mancothane F **mancozeb** Candilidis
Mancovin F **mancozeb** Hoechst
Mancozan F **maneb + zineb** Rhone-Poulenc, SPE
mancozeb F [8018-01-7] manganese ethylenebis(dithiocarbamate) (polymeric) complex with zinc salt, mancozebe, manzeb, Dithane 945, Dithane M-45, Dithane Ultra, Fore, Formec, Mancozin, Manzate 200, Manzin, Nemispor, Penncozeb, Policar MZ, Sandozebe, Vondozeb Plus
mancozebe F **mancozeb**
Mancozin F **mancozeb** Crystal
Mancur F **cymoxanil + mancozeb** Fattinger
Mandane F **maneb** Sipcam-Phyteurop
Manderol F **chlozolinate** Montedison
Mandorlin FA **dinocap** Platecsa
maneb F [12427-38-2] manganese ethylenebis(dithiocarbamate) (polymeric), manebe, MEB, ENT 14875, Dithane M-22, Farmaneb, Kypman, Lonacol M, Man-Zox, Manebgan, Manesan, Mangavis, Manzate, Nespor, Plantineb, Polyram M, Remasan, Rhodianebe, TersanLSR, Trimangol
Maneb Brestan F **fentin acetate + maneb** Hoechst
Maneb Combi F **fentin acetate + maneb** Leu & Gygax
Maneb Forte F **fentin hydroxide + maneb** Sandoz
Maneb-tin F **fentin acetate + maneb** Agriben, Van Wesemael
Maneba F **maneb** Kemira
manebe F **maneb**
Manebgan F **maneb** Makhteshim-Agan
Manebina F **maneb** Caffaro, Ergex
Maneor F **maneb** BP, R.S.R.

Manesan F **maneb** Agriben, Rhone-Poulenc
Manesan F **fentin acetate + maneb** Pluess-Staufer
Manesan Flo IF **lindane + maneb** Condor
Manex F **maneb** Chimac-Agriphar, Schwebda
Manex F **maneb + zinc** Chiltern
Manezine F **maneb + zineb** Agronova
Mangaline F **maneb** Aveve
Manganex F **maneb** Protex
Manganil F **maneb** Bourgeois
Mangastan F **fentin acetate + maneb** Van Wesemael
Mangatex F **maneb** Van Wesemael
Mangavis F **maneb** Visplant
Mankuprox F **copper oxychloride + mancozeb** Organika-Azot
Manocupryl F **copper sulphate + maneb** Prochimagro, Visplant
Manogil F **maneb** Agriphyt
Manolate F **maneb** Prochimagro
Manolate corbeaux FX **anthraquinone + maneb** Prochimagro
Manolate triple FIX **anthraquinone + lindane + maneb** Prochimagro
Manoran F **maneb** Procida
Manovit F **maneb** Vital
Manoxyl F **copper oxychloride + maneb** Vital
Manteb F **maneb** Scam
Mantrebe F **maneb** La Littorale
Manzagrex F **mancozeb** Sadisa
Manzate F **maneb** Du Pont, Zootechniki
Manzate 200 F **mancozeb** Du Pont
Manzeb F **mancozeb** Chemia
Manzeco F **mancozeb** Vieira
Manzib K F **maneb + zineb** Chemia
Manzicarb F **maneb + zineb** BASF
Manzin F **mancozeb** Crystal
Manzocure F **mancozeb** Proval
Mape H **MCPA + MCPB** Leu & Gygax
Mapin S **petroleum oils** Pinus
Maposol FNHI **metham-sodium** EMTEA, Procida, Roussel-Hoechst
Maqbal I **XMC** Hodogaya
Mar-Frin R **warfarin**
Marathon H **prodiamine** Sandoz
Marba K-T A **dicofol + tetradifon** Marba
Marba Naranjos IA **petroleum oils** Marba
Marbaset G **gibberellic acid** Marba
Marbazina D H **atrazine + simazine** Marba
Marbazineb F **zineb** Marba
Marcan F **mancozeb** Margesin
Marisan F **dicloran** Siapa
Marks DPD H **2,4-D + dichlorprop** PLK
Marksman H **linuron + trifluralin** FCC
Marksman H **atrazine + dicamba** Sandoz
Marlate I **methoxychlor** Kincaid
Marmer H **diuron**
Marnis Ratten- und Maeusekoeder R **warfarin** Marni
Marshal I **carbosulfan** Ewos, FMC, Grima, ICI, La Quinoleine, Rhone-Poulenc, Sandoz, Schering, Wacker
Marsoline I **diazinon + lindane** Protex
Marstan Fly Spray I **lindane + pyrethrins** Killgerm
Marvel H **diphenamid** Visplant
Marvex IA **dichlorvos**
Masolon F **carbendazim + pyrazophos** Hoechst
Massorrat Plus R **difenacoum** Masso
MasterSpray H **bromoxynil + ioxynil + trifluralin** Pan Britannica
Mastor I **cypermethrin** R.S.R.
Masulfan IA **endosulfan** Marba
Matacar A **hexythiazox** Sipcam
Matacil I **aminocarb** Bayer
Matahormigas I **carbaryl** KenoGard
Mataranha Total A **dicofol + tetradifon** Sopepor
Matas Algerens L **benzalkonium chloride** Perfektion
Matas insekt spray universal I **permethrin + piperonyl butoxide + pyrethrins** Agro-Kemi
Matas Insekta-lak I **chlorpyrifos** Agro-Kemi
Matas insektlak I **dieldrin** Agro-Kemi
Matas kobber sproejtemiddel F **copper oxychloride** Agro-Kemi
Matas mosfjerner H **chloroxuron** Ciba-Geigy
Matas musekorn R **bromadiolone** Agro-Kemi
Matas musekorn D R **difenacoum** ICI
Matas myrelokkedase I **phoxim** Agro-Kemi
Matas myremiddel I **lindane** Agro-Kemi
Matas myremiddel til vanding I **phoxim** Agro-Kemi
Matas myrespray I **permethrin + piperonyl butoxide + pyrethrins** Agro-Kemi
Matas plaenerens til vanding H **2,4-D + mecoprop** Agro-Kemi
Matas plantespray I **resmethrin** Agro-Kemi
Matas rosesproejtemiddel FI **dinocap + dodine + monocrotophos** Ciba-Geigy

Matas sproejtemiddel mod krybende insekter I **allethrin + permethrin** Agro-Kemi
Matas Total ukrudtsfjerner H **atrazine** Ciba-Geigy
Mataven H **flamprop-methyl** Shell
Mate H **ioxynil** Rhone-Poulenc
Matikus R **brodifacoum** ICI
Matin F **fentin acetate + maneb** Bayer
Matox R **potassium nitrate + sulphur** Urech
Matrak R **difenacoum** Sopra
Matrigon H **clopyralid** Dow, KVK
MatriKerb H **clopyralid + propyzamide** Rohm & Haas
Mausan-Giftweizen R **zinc phosphide** Lettenmayer
Mausex-Giftkoerner R **zinc phosphide** Nagel
Mausex-Koeder R **bromadiolone** Frowein
Mavrik IA **fluvalinate** Chimac-Agriphar, Inagra, Sandoz, Mitsubishi
Mavrik B I **fluvalinate + thiometon** Sandoz
Maxforce I **hydramethylnon** Cyanamid
Maxi-Sovitox R **scilliroside** Sovilo Fertiligene
Maxim F **carbendazim** FCC
maxima Pflanzenschutz IA **dimethoate** Dehag
MaxiMate F **carbendazim + maneb** FCC
Maytril H **bromoxynil + ioxynil + mecoprop** Rhone-Poulenc
Mazaline H **simazine** Hoechst, Roussel Hoechst
Mazide G **maleic hydrazide** Synchemicals
Mazimix F **maneb + zineb** Van Wesemael
Mazin F **maneb + zinc oxide** UCP
Mazinol H **simazine** Scam
Mazipron H **atrazine + petroleum oils** BP, R.S.R.
Mazolan F **mancozeb** Ellagret
MB 25-105 G **propyl 3-tert-butylphenoxyacetate** Agriben
MB 3046 H **MCPB** May & Baker
MB M 750 H **MCPA** PLK
MB MPD 575 H **2,4-D + mecoprop** PLK
MBC F **carbendazim**
MBCP I **leptophos**
MBPMC H **terbucarb**
MBR-8251 H **perfluidone**
MBR 12325 GH **mefluidide** 3M
MC 25 FX **guazatine acetates**
MC 0035 H **benzadox**
MC 474 IA **mecarbam**
MC 833 F **carbamorph**
MC 1053 AF **dinobuton**
MC 1488 H **medinoterb acetate**
MC 1945 AF **dinocton**
MC 1947 AF **dinocton**
MC 2188 I **chlormephos**
MC 2420 I **mecarphon**
MC-4379 H **bifenox** Mobil
MC 10109 H **acifluorfen-sodium**
MCC H **swep**
McN-1025 R **norbormide** McNeil Laboratories
MCP H **MCPA**
MCPA H [94-74-6] (4-chloro-2-methylphenoxy)acetic acid, MCP, 2,4-MCPA, 2M-4Kh, Agritox, Agroxone, Bordermaster, Chiptox, Empal, Hedonal M, Mephanac, Phenoxylene 50, Rhomene, Rhonox, Shamrox, U 46 M Fluid, Weed-Rhap, Zelan
MCPA-thioethyl H [25319-90-8] *S*-ethyl 4-chloro-*o*-tolyloxythioacetate, phenothiol, Fenobit, Herbit, HOK-7501, Zero One
MCPB H [94-81-5] 4-(4-chloro-*o*-tolyloxy)butyric acid, 2,4-MCPB, 2M-4Kh-M, Can-Trol, MB 3046, Thistrol, Tropotox
MCPP H **mecoprop**
MDBA H **dicamba**
ME 605 Spritzpulver IA **parathion-methyl** Bayer
MEB F **maneb**
Mebatryne H **ametryn** Rhone-Poulenc
Mebazine H **atrazine** Rhone-Poulenc
mebenil F [7055-03-0] *o*-toluanilide, BAS 3050F, BAS 3053F, BAS 305F
Mebrom FNHI **methyl bromide** Mebrom
mecarbam IA [2595-54-2] *S*-(*N*-ethoxycarbonyl-*N*-methylcarbamoylmethyl) *O,O*-diethyl phosphorodithioate, mecarbame, MC 474, Murfotox, P 474, Pestan
mecarbame IA **mecarbam**
mecarbinzid F [27386-64-7] methyl 1-(2-methylthioethylcarbamoyl)benzimidazol-2-ylcarbamate, mecarbinzide, BAS 3201F
mecarbinzide F **mecarbinzid**
mecarphon I [29173-31-7] methyl [methoxy(methyl) phosphinothioylthio]acetyl (methyl)carbamate, MC 2420
mechlorprop H **mecoprop**
Mecomec H **mecoprop** PBI Gordon
Mecopex H **mecoprop**
mecoprop H [7085-19-0] (*RS*)-2-(4-chloro-*o*-tolyloxy)propionic acid, MCPP, mechlorprop, CMPP, Chipco Turf

Herbicide MCPP, Clovotox, Compitox, Duplosan, Hedonal MCPP, Iso-Cornox, Kilprop, Mecomec, Mecopex, Mepro, Methoxone, Propal, RD 4593, U 46 KV Fluid
Mecormona H **mecoprop** Rhone-Poulenc
Mecotex H **mecoprop** L.A.P.A., Protex
Mectril H **bromoxynil + ioxynil + MCPA + mecoprop** Agronorden
Medeclorex F **quintozene** Agrocros
Medelinon H **linuron** Agrocros
Medex H **diuron + linuron + TCA** Nitrokemia
Mediben H **dicamba** Sandoz
Medicadino H **dinoseb** L.A.P.A.
Medinebe Z F **copper sulphate + zineb** Sedagri
Medinex H **dicamba + mecoprop** L.A.P.A.
medinoterb acetate H [2487-01-6] 6-*tert*-butyl-2,4-dinitro-*m*-cresyl acetate, medinoterbe acetate, MC 1488, P 1488
medinoterbe acetate H **medinoterb acetate**
Medipham H **phenmedipham** Benagro
Medixone H **2,4-D + dicamba + MCPA** Zerpa
Medopaz I **petroleum oils** Agrochemiki
Medrin IA **thiometon** Shell
mefenacet H [73250-68-7] 2-(1,3-benzothiazol-2-yloxy)-*N*-methylacetanilide, FOE 1976, Hinochloa, NTN 801, Rancho
mefluidide GH [53780-34-0] 5'-(1,1,1-trifluoromethanesulphonamido)acet-2',4'-xylidide, Embark, MBR 12325, Mowchem
Mefos I **parathion-methyl** Shell
Mega-D H **2,4-D** Akzo Zout
Mega-DP H **dichlorprop** Akzo
Mega-M H **MCPA** Akzo, Akzo Zout
Mega-MD H **2,4-D + MCPA** Akzo
Meganet H **difenzoquat methyl sulphate + imazamethabenz-methyl** Cyanamid
Mega-P H **mecoprop** Akzo Zout
Mega-PD H **2,4-D + mecoprop** Akzo
Megaplus H **imazamethabenz-methyl + pendimethalin** Cyanamid
Mehltau EX F **pyrazophos** Ratzenberger
Mehltaumittel F **dodemorph acetate** BASF
Mekohormo H **MCPA + mecoprop** Maatalouspalvelu
Melipax IRA **camphechlor** Fahlberg-List
Melophen IA **endosulfan** Maag
Melprex F **dodine** Collett, Cyanamid, Lapapharm, Rhodiagri-Littorale, Zorka Sabac
Melprex Combi F **dodine + thiophanate-methyl** Lapapharm
Meltatox F **dodemorph acetate** Agrolinz, Badilin, BASF, Collett, Intrachem
Meltatox Combi F **dodemorph acetate + nitrothal-isopropyl** BASF
MEMA F **methoxyethylmercury acetate**
Mendene IA **endosulfan + parathion-methyl** Visplant
Mendiplex H **chlorotoluron + diuron** Diana
Mengmeststof 12 + 6 + 6 met onkruidverdelger Lawn Plus H **2,4-D + mecoprop** ICI
Menno ter forte L **quaternary ammonium** Tuhamij
Mennozid I **trichlorfon** Reese
Meobal I **xylylcarb** Sumitomo
Meoc M **metaldehyde** Meoc
Meol IA **parathion + petroleum oils** Meoc
Meothrin AI **fenpropathrin** Sumitomo, Shell
MEP I **fenitrothion**
Mep 20 I **parathion-methyl** ICI
Mepatox I **parathion-methyl** Agrocros
Mephanac H **MCPA**
Mephetol Extra H **dicamba + dichlorprop + MCPA** Chafer
mephosfolan IA [950-10-7] diethyl 4-methyl-1,3-dithiolan-2-ylidenephosphoramidate, mephospholan, ENT 25991, Cytro-Lane, Cytrolane, EI 47470
mephospholan IA **mephosfolan**
mepiquat chloride G [24307-26-4] 1,1-dimethylpiperidinium chloride, BAS 08306W, Pix
Mepro H **mecoprop**
Mepro Special H **dicamba + MCPA + mecoprop** Kemira
mepronil F [55814-41-0] 3'-isopropoxy-*o*-toluanilide, Basitac
mercaptodimethur MIAX **methiocarb**
mercaptofostion + mercaptostiol IA **demeton**
mercaptophos I **fenthion**
mercaptothion IA **malathion**
mercaptotion IA **malathion**
mercuric oxide FW [21908-53-2] mercury oxide, oxyde mercurique, yellow oxide of mercury, Kankerdood, Santar
mercurous chloride FI [7546-30-7]

mercury(I) chloride, chlorure mercureux, calomel, Cyclosan
Merfusan F **mercurous chloride + mercuric chloride** Rhone-Poulenc
Mergamma FI **lindane + phenylmercury acetate** ICI
Meri Muls I **dichlorvos + malathion** Meriel
Merikill K Othrine I **deltamethrin** Meriel
Merit H **bromoxynil** Sopra
Merkazin H **prometryn** Budapesti Vegyimuvek
Merpafol F **captafol** COPCI, Makhteshim-Agan, Masso
Merpan F **captan** Anorgachim, Edefi, Makhteshim-Agan, Sapec
Merpelan H **benzthiazuron + lenacil** Bayer
Merpelan AZ H **isocarbamid + lenacil** Bayer, Organika-Zarow, Shell
Merphitheio I **azinphos-ethyl** Efthymiadis
merphos G [150-50-5] tributyl phosphorotrithioite, Folex, Profal
Mersil F **mercurous chloride + mercuric chloride** Rhone-Poulenc
Mersolite F **phenylmercury acetate**
Mertect F **thiabendazole** MSD AGVET
Merz-Cumarin-Fertigkoeder R **warfarin** Merz
Mesamate H **MSMA** Vertac
Mesodrin RL IA **oxydemeton-methyl** Shell
mesoprazine H [1824-09-5] 6-chloro-N^2-isopropyl-N^4-(3-methoxypropyl)-1,3,5-triazine-2,4-diamine, CGA 4999, G34698
Mesoranil H **aziprotryne** Ciba-Geigy
Mesthan IA **parathion-methyl** Du Pont
Mestro IA **dimethoate** Embrica
Mesurol MIX **methiocarb** Agro-Kemi, Agroplant, Bayer, Berner, Pinus
Mesurol-Combi FI **methiocarb + thiram** Bayer
MET 1486 H **ethidimuron** Bayer
Meta M **metaldehyde** ICI-Zeltia
Metabloc M **metaldehyde** CIFO
Metabrom FIHN **chloropicrin + methyl bromide** Bromine Compounds, Edefi, Rasmussen
metacetaldehyde M **metaldehyde**
Metacid M **metaldehyde** Terranalisi
Metacide IA **parathion-methyl** Bayer
Metacidine I **methidathion** Sapec
Metaclor-R Antilimacos M **metaldehyde** Ciba-Geigy
Metacrate I **metolcarb** Sumitomo
Metadelphene X **diethyltoluamide**
Meta-Dipterex I **demeton-S-methyl + trichlorfon** Bayer
Metafos IA **azinphos-methyl** Inagra
Meta-Fum FIHN **metham-sodium** Sivam
Meta-Isosystox I **demeton-S-methyl** Bayer
Meta Isosystoxsulfon IA **demeton-S-methyl sulphone** Bayer
Metakey M **metaldehyde** Key
Metal I **malathion** Baslini
metalaxyl F [57837-19-1] methyl *N*-(2-methoxyacetyl)-*N*-(2,6-xylyl)-DL-alaninate, Apron, CGA48988, Ridomil, Subdue
Metaldehid M **metaldehyde** Sojuzchimexport
Metaldehida M **metaldehyde** Hispagro, Sojuzchimexport
Metaldehido M **metaldehyde** Shell
Metaldehyd M **metaldehyde** ACP, Sojuzchimexport
metaldehyde M [108-62-3] (tetramer); [9002-91-9] (homopolymer) *r*-2,*c*-4,*c*-6,*c*-8-tetramethyl-1,3,5,7-tetroxocane, metacetaldehyde, Ariotox, Cekumeta, Halizan, Helarion, Metason, Mifaslug, Namekil
Metaldene M **metaldehyde** Visplant
Metam FIHN **metham-sodium** Chemia, Nordisk Alkali
Metam Sodio FNHI **metham-sodium**
Metambane H **dicamba + MCPA** Chimiberg
metamitron H [41394-05-2] 4-amino-4,5-dihydro-3-methyl-6-phenyl-1,2,4-triazin-5-one, metamitrone, DRW 1139, Goltix, Herbrak
metamitrone H **metamitron**
Metamix H **metamitron** RACROC
Metanex FNHI **metham-sodium** Shell
Metanil H **methabenzthiazuron** Quimigal
metaphos IA **parathion-methyl**
Metaran A **cyhexatin** Chemia
Metaran FX **thiram** Inagra
Metarex M **metaldehyde** de Sangosse
Metaros M **metaldehyde** ERT
Meta-Schneckenkorn M **metaldehyde** Agro
Meta-Schneckentod M **metaldehyde** Agro
Metasol M **metaldehyde** Eli Lilly
Meta-Sol extra FIHN **metham-sodium** Sipcam-Phyteurop
Metason M **metaldehyde** Aveve, Jewnin-Joffe
Meta-Suspension M **metaldehyde** Lonza
metasystemox IA **oxydemeton-methyl**

Metasystox IA **demeton-S-methyl** Bayer
Metasystox (i) IA **demeton-S-methyl** Bayer
Metasystox 55 IA **demeton-S-methyl** Bayer
Metasystox R IA **oxydemeton-methyl** Agroplant, Bayer, Berner, Pinus, Schering
Metasystox S IA **oxydeprofos**
Metasystox S-O I **oxydemeton-methyl** Agro-Kemi
Metasystox spezial IA **oxydemeton-methyl + trichlorfon** Bayer
Metathion IA **azinphos-methyl** Sivam
Metation I **fenitrothion** Dimitrova, Radonja
Metazachloor H **metazachlor** Imex-Hulst
metazachlor H [67129-08-2] 2-chloro-*N*-(pyrazol-1-ylmethyl)acet-2',6'-xylidide, metazachlore, BAS 47900H, Butisan S
metazachlore H **metazachlor**
Metazin IA **azinphos-methyl** Safor
Metazintox IA **azinphos-methyl** Agrocros
metazoxolon F [5707-73-3] 4-(3-chlorophenylhydrazono)-3-methylisoxazol-5(4*H*)-one, PP 395
Metendox I **endosulfan + methomyl** Siapa
Metene I **parathion-methyl** Visplant
Meteor IA **dimethoate** Van Wesemael
Metex G **chlormequat chloride** Protex
metflurazon H [23576-23-0] 4-chloro-5-dimethylamino-2-(α,α,α-trifluoro-*m*-tolyl) pyridazin-3(2*H*)-one, metflurazone, SAN 6706H
metflurazone H **metflurazon**
Meth-O-Gas INFRHA **methyl bromide** Great Lakes
methabenzthiazuron H [18691-97-9] 1-benzothiazol-2-yl-1,3-dimethylurea, methibenzuron, Bay 74283, Tribunil
Methacid I **methidathion** CTA
methacrifos IA [62610-77-9] methyl (*E*)-3-(dimethoxyphosphinothioyloxy)-2-methacrylate, OMS 2005, CGA 20168, Damfin
metham FNHI **metham-sodium**
metham-sodium FNHI [137-42-8; 6734-80-1 (dihydrate)] sodium methyldithiocarbamate, metam-sodium, metham, carbam, karbation, SMDC, Arapam, Busan, Maposol, Monam, N-289, Nemasol, Sistan, Solasan, Sometam, Trimaton, Vapam, VPM
methamidophos IA [10265-92-6] *O,S*-dimethyl phosphoramidothioate, acephate-met, ENT 27396, Bay 71628, Filitox, Monitor, Ortho 9006, Patrole, SRA 5172, Tam, Tamanox, Tamaron
Methaphos I **methamidophos** Efthymiadis
Methar H **DSMA** Cleary
methasulfocarb F [66952-49-6] *S*-[4-[(methylsulfonyl)oxy]phenyl] methylcarbamothioate, Kayabest, NK 191
methazole H [20354-26-1] 2-(3,4-dichlorophenyl)-4-methyl-1,2,4-oxadiazolidine-3,5-dione, oxydiazol, Mezopur, Paxilon, Probe, Tunic, VCS-438
methfuroxam F [28730-17-8] 2,4,5-trimethyl-*N*-phenyl-3-furancarboxamide, furavax, Arbosan
methibenzuron H **methabenzthiazuron**
methidathion IA [950-37-8] *S*-2,3-dihydro-5-methoxy-2-oxo-1,3,4-thiadiazol-3-ylmethyl *O,O*-dimethyl phosphorodithioate, DMTP, OMS 844, ENT 27193, GS 13005, Supracide, Suprathion, Ultracide
Methil Cotnion IA **azinphos-methyl** Edefi
methiocarb MIAX [2032-65-7] 4-methylthio-3,5-xylyl methylcarbamate, mercaptodimethur, methiocarbe, metmercapturon, OMS 93, ENT 25726, Bay 37344, Draza, H 321, Mesurol, Ortho Slug-Geta, Slug-M
methiocarbe MIAX **methiocarb**
methiuron H [21540-35-2] 1,1-dimethyl-3-*m*-tolyl-2-thiourea, MPDT, MH 090
methocrotophos I [25601-84-7] (*E*)-2-(*N*-methoxy-*N*-methylcarbamoyl)-1-methylvinyl dimethyl phosphate, metocrotophos, C 2307
metholcarb I **metolcarb**
methometon H [1771-07-9] 6-methoxy-N^2,N^4-bis(3-methoxypropyl)-1,3,5-triazine-2,4-diamine, metometon, G 34690
methomyl IA [16752-77-5] *S*-methyl *N*-(methylcarbamoyloxy)thioacetimidate, OMS 1196, Du Pont 1179, Lannate, Lanox, Nudrin
methoprene I [40596-69-8] isopropyl (*E*)-(*RS*)-11-methoxy-3,7,11-trimethyldodeca-2,4-dienoate, OMS1697, Altosid, Apex, Diacon, Dianex, Kabat, Minex, Pharorid, Precor, ZR 515
methoprotryne H [841-06-5] 2-isopropylamino-4(3-methoxypropylamino)-6-methylthio-1,3,5-triazine, methotryne, metoprotryne, G 36393, Gesaran
Methormone H **MCPA** Interphyto
Methosan FIHN **metham-sodium** SPE

methotryne H **methoprotryne**
Methouram F **thiram** Diana
Methoxone H **mecoprop** ICI
methoxychlor I [72-43-5] 1,1,1-trichloro-2,2-bis(4-methoxyphenyl)ethane, methoxychlore, DMDT, OMS 466, ENT 1716, Higalmetox, Marlate
methoxychlore I **methoxychlor**
methoxyethylmercury acetate F [151-38-2] 2-methoxyethylmercury acetate, MEMA, Cekusil Universal A, Panogen M, Panogen Metox
2-methoxyethylmercury chloride F [123-88-6] chloro(2-methoxyethyl)mercury, MEMC, Bagalol, Emisan 6, Taysatto
methoxyphenone H [41295-28-7] 4-methoxy-3,3'-dimethylbenzophenone, Kayametone, NK-049
Methybrom FNHI **chloropicrin + methyl bromide** Grower
Methyl Bladan IA **parathion-methyl** Bayer
methyl bromide INFRHA [74-83-9] bromomethane, bromure de methyle, Haltox, Meth-O-Gas, Terr-O-Gas 100
methyl demeton IA **demeton-S-methyl**
methyl isothiocyanate NFIH [556-61-6] methyl isothiocyanate, isothiocyanate de methyle, MIT, Trapex
methyl niclate FB **nickel dimethyldithiocarbamate**
methyl parathion IA **parathion-methyl**
Methyl-Brom FNHI **chloropicrin + methyl bromide** De Ceuster
Methyl-Chloro FNHI **chloropicrin + methyl bromide** De Ceuster
Methyl-Paretox IA **parathion-methyl** Bourgeois
methyldimuron H **methyldymron**
methyldymron H [42609-73-4] *N*-methyl-*N'*-(1-methyl-1-phenylethyl)-*N*-phenylurea, dimelon-methyl, methyldimuron, K 1441, Stacker
Methyleuparene F **tolylfluanid** Bayer
methylmercaptofostiol IA **demeton-S-methyl**
methylmetiram FA [8064-35-5] tris[ammine[propylene-1,2-bis(dithiocarbamato)]zinc(2+)]-5-methyltetrahydro-1,2,4,7-dithiadiazocine-3,8-dithione, polymer, Basfungin
Metiamon I **parathion-methyl** Du Pont
Metidion I **methoxychlor** KVK
Metil Paraben I **parathion-methyl** ERT
Metil Parafene I **parathion-methyl** Condor
Metil Parathion I **parathion-methyl** Ficoop, Serpiol
Metil-Cotnion IA **azinphos-methyl** Budapesti Vegyimuvek
Metilan I **parathion-methyl** Safor, Tecniterra
Metilazinphos IA **azinphos-methyl** Roussel Hoechst
Metilbromid FIHN **chloropicrin + methyl bromide** Radonja
Metilfit IA **azinphos-methyl** Sofital
Metilmercaptofos IA **demeton-S-methyl** Sojuzchimexport
metilmercaptofosoksid IA **oxydemeton-methyl**
Metilparation I **parathion-methyl** Budapesti Vegyimuvek
metiltriazotion IA **azinphos-methyl**
metiram F [9006-42-2] zinc ammoniate ethylenebis(dithiocarbamate) - poly(ethylenethiuram disulphide), metirame zinc, Carbatene, FMC 9102, Polyram, Polyram-Combi
metirame zinc F **metiram**
metmercapturon MIAX **methiocarb**
metobromuron H [3060-89-7] 3-(4-bromophenyl)-1-methoxy-1-methylurea, C 3126, Patoran, Pattonex
metocrotophos I **methocrotophos**
Metofan IA **endosulfan + methomyl** Aragonesas
Metofos I **methomyl + phosalone** Visplant
Metofos I **chlorfenvinphos + methoxychlor** Organika-Azot
metolachlor H [51218-45-2] 2-chloro-6'-ethyl-*N*-(2-methoxy-1-methylethyl)acet-*o*-toluidide, metolachlore, CGA 24705, Dual, Humextra, Pennant
metolachlore H **metolachlor**
metolcarb I [1129-41-5] *m*-tolyl methylcarbamate, metholcarb, MTMC, C-3, Metacrate, Ofunack M, Tsumacide
metometon H **methometon**
metoprotryne H **methoprotryne**
Metossi I **methoxychlor** Aziende Agrarie Trento
Metover I **methomyl** R.S.R.
Metox I **parathion-methyl** Sivam
Metox I **methoxychlor** Aragonesas, Organika-Azot
Metoxdem AF **binapacryl** Agrocros
Metoxiagrex I **methoxychlor** Sadisa
Metoxide I **azinphos-methyl + demeton-S-methyl sulphone** Gaillard

metoxuron H [19937-59-8] 3-(3-chloro-4-methoxyphenyl)-1,1-dimethylurea, Deftor, Dosaflo, Dosanex, Purivel, SAN 6915H, SAN 7102H, Sulerex
Metrazin H **atrazine + methabenzthiazuron** Leu & Gygax
metribuzin H [21087-64-9] 4-amino-6-*tert*-butyl-4,5-dihydro-3-methylthio-1,2,4-triazin-5-one, metribuzine, Bay 6159H, Bay 6443H, Bay 94337, Bay DIC 1468, Lexone, Sencor, Sencoral, Sencorex
metribuzine H **metribuzin**
metriphonate I **trichlorfon**
Metroc H **methabenzthiazuron** RACROC
Metrol I **carbophenothion + dimethoate + parathion** Hellenic Chemical
Metron H **2,4-D + MCPA** Chimac-Agriphar
metsulfuron-methyl H [74223-64-6] methyl 2-[[[[(4-methoxy-6- methyl-1,3,5-triazin-2-yl)amino] carbonyl]amino]sulfonyl] benzoate, Allie, Ally, Brushoff, DPX T6376, DPX-6376, Escort, Gropper
Metyphon I **parathion-methyl** Roussel Hoechst
Mevidrin IA **mevinphos** Hui Kwang
Mevinex I **mevinphos** Permutadora
Mevinox IA **mevinphos** Crystal
mevinphos IA [7786-34-7] 2-methoxycarbonyl-1-methylvinyl dimethyl phosphate, ENT 22374, Apavinphos, Duraphos, Mevidrin, Mevinox, OS-2046, PD 5, Phosdrin
mexacarbate IAM [315-18-4] 4-dimethylamino-3,5-xylyl methylcarbamate, Zectran
Mezene FX **ziram** Agrimont, Montedison
Mezopur H **methazole** Agrolinz, ICI-Zeltia, Sandoz
Mezotox H **nitrofen** Chinoin
MG-06 H **eglinazine-ethyl** Nitrokemia
MG-07 H **proglinazine-ethyl** Nitrokemia
MG2 F **pyrethrins** Killgerm
MG3 A **phenothrin + tetramethrin** Killgerm
Mglawik Extra I **methoxychlor + propoxur** Organika-Azot
MH G **maleic hydrazide**
MH 090 H **methiuron**
MH 30 G **maleic hydrazide** Kwizda, Uniroyal
Mibiol IA **petroleum oils** Baslini
Mical PS F **copper compounds (unspecified) + folpet + sulphur** Sofital
Miceb F **maneb + zineb** Geopharm
Micene F **mancozeb** Inagra, Sipcam
Miceram F **copper oxychloride + mancozeb** Agrotechnica
Micetox F **dodine** Eli Lilly
Micevit F **maneb + thiophanate-methyl + zineb** Sipcam
Micosen F **copper sulphate + folpet** La Littorale
Micosep F **mancozeb** Sepran
Micosin F **ziram** Chimiberg
Micospor F **captan** Sofital
Micoter F **captan + carboxin + 8-hydroxyquinoline sulphate** Sofital
Micozeb F **mancozeb** Terranalisi
Microbagnabile AF **sulphur** Solfotecnica
Microcap I **permethrin + piperonyl butoxide** SIMAR
Microcide I **methidathion** Chemia
Microline FB **copper oxychloride** La Littorale
Microlux F **sulphur** EMTEA, Procida, Roussel Hoechst
Micron F **sulphur** Agrochemiki
Microneb F **zineb** Prochimagro
Micronyl F **ziram** Prochimagro, Van Wesemael
Microrame F **calcium copper oxychloride** Chemia
Microsev I **carbaryl** Enotria
Microsol AF **sulphur** Protex, Sepran
Microsulfo F **sulphur** Bayer
Microter IN **carbofuran** Caffaro
Microthiol FA **sulphur** Afrasa, BP, Koch & Reis, R.S.R., Tarsoulis, UCAR
Microtox FA **sulphur** Agrocros
Microzidina F **dodine + zineb** Chemia
Microzineb F **zineb** Chemia
Microzineb SB F **sulphur + zineb** Chemia
Microzolfo AF **sulphur** EniChem
Microzufre F **sulphur** Aragonesas
Microzul R **chlorophacinone**
Midol blomsterspray I **resmethrin** Midol
Midol Feni I **fenitrothion** Midol
Midol fluespray I **piperonyl butoxide + pyrethrins** Midol
Midol frugtfinish F **petroleum oils** Midol
Midol meldugmiddel F **petroleum oils** Midol
Midol myregift I **lindane + piperonyl butoxide + pyrethrins** Midol
Midol nr. 21 I **lindane + piperonyl butoxide + pyrethrins** Midol
Midol nr. 70 I **piperonyl butoxide + pyrethrins** Midol

Midol olie emulsion AF **petroleum oils** Midol
Midol plaenerens H **2,4-D + dichlorprop** Midol
Midol rosenspray I **resmethrin** Midol
Midstream H **diquat dibromide** ICI
Miedzian F **copper oxychloride** Organika-Azot
Mierenmiddel I **phoxim** Bayer
Mifaslug M **metaldehyde** Farmers Crop Chemicals
Mifatox IA **demeton-S-methyl** Farmers Crop Chemicals
Mikal F **folpet + fosetyl-aluminium** Agriben, BVK, Condor, Pepro, Rhone-Poulenc, Zorka Sabac
Mikal M F **fosetyl-aluminium + mancozeb** Rhone-Poulenc
Mikal Z F **fosetyl-aluminium + zineb** Pepro
Mikantop I **dimethoate + fenvalerate**
Mikazol T F **thiabendazole** Pliva
Milan I **EPN + parathion-methyl** Helena
Milban F **dodemorph acetate** Mallinckrodt
Milbol A **dicofol** Delicia
Mil-Col F **drazoxolon** ICI
Milcozebe F **mancozeb** Sandoz
Milcurb F **dimethirimol** ICI
Milcurb Super F **ethirimol** ICI
Mildane F **dinocap** Chimiberg
Mildin F **folpet** EniChem
Mildothane FW **thiophanate-methyl** Rhone-Poulenc
Milfaron F **chloraniformethan**
Milfin F **mancozeb + oxadixyl** Atlansul
milfuram F **ofurace**
Milgo F **ethirimol** AVG, ICI, ICI-Zeltia
Milkil F **folpet + fosetyl-aluminium** ICI-Valagro
Miller-Aide S **di-1-p-menthene** Geopharm
Millie F **ditalimfos** Dow
Milmer F **oxine-copper**
milneb F [3773-49-7] 4,4',6,6'-tetramethyl-3,3'-ethylenedi-1,3,5-thiadiazinane-2-thione, thiadiazine, Banlate, Experimental Fungicide 328
Milo-Pro H **propazine** Griffin
Milocep H **propazine** Ciba-Geigy
Milogard H **propazine** Ciba-Geigy
Milraz F **cymoxanil + propineb** Bayer
Milraz Cobre F **copper oxychloride + cymoxanil + propineb** Bayer
Milraz Duplo F **cymoxanil + propineb + triadimefon** Bayer
Milstem F **ethirimol** ICI
Miltox F **copper oxychloride + zineb** PVV, Sandoz
Miltoxan F **maneb + zineb** Sandoz
Miltoxan Blau F **iron compounds (unspecified) + maneb + zineb** Sandoz
Miltoxan Spezial F **copper compounds (unspecified) + iron compounds (unspecified) +maneb + zineb** Sandoz
Milzan F **cymoxanil + zineb** Aragonesas
Minacid I **methidathion** Visplant
Minarix G **mefluidide** La Quinoleine
Minazol H **amitrole + ammonium thiocyanate** Interphyto
Mindex H **2,4-DB** Diana
Mindi I **dichlorvos + malathion** Industrialchimica
mineral oils AIH **petroleum oils**
Mineraloel IA **petroleum oils** Pluess-Staufer, Sandoz
Minerol IA **petroleum oils** Burri
Minex I **methoprene** Sandoz
Minodrin forte IA **petroleum oils** Avenarius
Minosina F **maneb** Platecsa
Minotor mixte T FI **carbaryl + copper oxychloride + endosulfan + zineb** La Quinoleine
Minphos I **dichlorvos** Aziende Agrarie Trento
Mio-plant M **metaldehyde** Migros
mipafox IA [371-86-8] *N,N'*-di-isopropylphosphorodiamidic fluoride, Isopestox, Pestox 15
MIPC I **isoprocarb**
Mipcin I **isoprocarb** Mitsubishi
Mipzinon I **diazinon + isoprocarb**
Miraflor-Spray I **malathion + piperonyl butoxide + pyrethrins** Varo
Miral IN **isazofos** Ciba-Geigy
Mirazyl P I **permethrin** Chimac-Agriphar
Mirbel G **chlormequat chloride + choline chloride** La Littorale
Mirotin F **fentin acetate + maneb** Eurofyto
Mirowet S **polyglycolic ethers** Eurofyto
Mirvale G **chlorpropham** Ciba-Geigy
Misan F **captan** Agrimont
Miscela Solfocalcica FIA **lime sulphur** Aziende Agrarie Trento
Miseram Extra Blu F **copper oxychloride + zineb** Geopharm
Missile F **pyrazophos** Hoechst
Mistral F **fenpropimorph** Rhone-Poulenc

MIT NFIH **methyl isothiocyanate**
Mitac A **amitraz** Agrolinz, Budapesti Vegyimuvek, Ciba-Geigy, ICI-Valagro, NOR-AM, Ruse, Schering, Schering AAgrunol
Mitacid A **cyhexatin** Gaillard, Leu & Gygax
Mitacid T A **cyhexatin + tetradifon** Inagra
Mitarex A **dicofol** Interphyto
Mitaxan A **cyhexatin + tetradifon** Leu & Gygax
Mitchell 360 H **amitrole + diuron + MCPA + TCA** Sico
Mitchell 550 H **bromacil + 2,4-D + dalapon-sodium + diuron + TCA** Sico
Mitekil A **dicofol + tetradifon** Platecsa
Mitekil R AI **dicofol + dimethoate + tetradifon** Platecsa
Mitex A **amitraz** Shell
Mithal A **fenson + propargite** Tecniterra
Mitifon A **tetradifon** Hellenic Chemical
Mitigan A **dicofol** Anorgachim, Makhteshim-Agan
Mition A **dicofol + tetradifon** Lapapharm
Mitiveex A **cyhexatin + tetradifon** Agriplan
Mito FOG G **chlorpropham** Frowein
Mitox A **chlorbenside**
Mitox A **dicofol** Diana, Hispagro
Mitran A **cyhexatin** Caffaro
Mitran A **chlorfenethol + chlorfenson** Nippon Soda
Mitrazon A **benzoximate + cyhexatin** Siegfried
Mitron H **amitrole + simazine + sodium thiocyanate** Spyridakis
Mix Top Schering S **petroleum oils** Schering
Mixan I **dichlorvos + tetradifon** Agro-Kanesho
Mixi-Tok H **neburon + nitrofen** Rohm & Haas
Mixol S **petroleum oils** Burri
Mixone Super H **dichlorprop + MCPA + mecoprop** Eurofyto
MK-23 F **fluoromide** Mitsubishi
MK-90 F **copper oxychloride + mancozeb** Chemia
MK-936 IA **abamectin** MSD AGVET
ML 50 H **linuron** Avenarius
MLT IA **malathion** Sumitomo
MO H **chlornitrofen** Mitsui Toatsu
MO 338 H **chlornitrofen** Mitsui Toatsu
Mocap NI **ethoprophos** Agriben, Agrotec, Argos, Duphar, Hellafarm, Procida, Protex, Radonja, Ravit, Rhone-Poulenc, Sandoz, Schering, Sipcam
Mocap Super NI **disulfoton + ethoprophos** Rhone-Poulenc
Modown H **bifenox** Rhone-Poulenc, Mobil
Modra galica F **copper sulphate** Radonja
Mogebron H **ACN + MCPB + simetryn** Agro-Kanesho
Mogeton HL **ACN** Agro-Kanesho
Mogiol S **polyglycolic ethers** Masso
Mogol I **malathion** Agrimont
Mogran I **parathion** Agrimont
Mohojuurensuoja F **mercurous chloride** Berner
Mojante S **polyglycolic ethers** Agrocros, Argos, Inagra, Probelte, Quimicas Oro
Mojasaf S **polyglycolic ethers** Safor
Mojaver S **polyglycolic ethers** KenoGard
Mol 35 IA **endosulfan** Inagra
Mole Ban R **lindane + tecnazene + thiourea** Rhone-Poulenc
Mole Death R **strychnine**
Molidram H **molinate** Hoechst
Molinam H **molinate** Sipcam, Sipcam-Phyteurop
Molinan H **molinate** Inagra
molinate H [2212-67-1] *S*-ethyl *N,N*-hexamethylenethiocarbamate, Higalnate, Ordram, R-4572, Sakkimol
Molinex H **molinate** Vieira
Molipan H **linuron + monolinuron** Maag
Moliram H **molinate** Ellagret
Mollendood R **strychnine** Bogena
Mollona F **folpet** Agrishell
Mollona Super C F **copper sulphate + folpet** Agrishell
Mollux M **metaldehyde** Siegfried
Moloss F **Bordeaux mixture + maneb + metiram** EMTEA
Moloss M F **copper oxysulphate + maneb** Roussel Hoechst
Moltran H **molinate** Hellenic Chemical
Moltranil H **molinate + propanil** Hellenic Chemical
Mon-097 H **acetochlor** Monsanto
MON 0358 H **terbuchlor**
MON-0573 H **glyphosate** Monsanto
monalide H [7287-36-7] 4'-chloro-2,2-dimethylvaleranilide, Potablan, SN 35830
Monalox H **alloxydim-sodium** Kwizda
Monam FNHI **metham-sodium** BASF
Monamex H **butralin + monolinuron** Union Carbide
Moncereen F **pencycuron** Bayer

Monceren F **pencycuron** Agroplant, Bayer
Moncut F **flutolanil** Nihon Nohyaku
Mondak H **dicamba + MCPA** Sandoz
Mondariso H **molinate** Caffaro
Mondepor H **MCPA** Sopepor
Monilate F **carbendazim** Farmon Agrovia
Monisarmiovirus I **Heliothis polyhedrosis virus** Kemira
Monitor IA **methamidophos** Chevron, Geopharm, Ruse
Monocron IA **monocrotophos** Alpha, Ciba-Geigy, Makhteshim-Agan
monocrotophos IA [6923-22-4] dimethyl (*E*)-1-methyl-2-(methylcarbamoyl)vinyl phosphate, OMS 834, ENT 27129, Apadrin, Azodrin, Bilobran, C 1414, Crisodrin, Hazodrin, Monocron, Monodrin, Nuvacron, Pandar, Pillardrin, Plantdrin, SD 9129, Susvin
Monodrin IA **monocrotophos** Hui Kwang
Monofos IA **monocrotophos** Chemia
Monokrotofos I **monocrotophos** Zorka Subotica
monolinuron H [1746-81-2] 3-(4-chlorophenyl)-1-methoxy-1-methylurea, Afesin, Aresin, Arresin, Hoe 02747
Monosan herbi H **2,4-D** Galenika
Monosan herbi specijal H **2,4-D + MCPA + mecoprop** Galenika
Monosan kombi super H **2,4-D + mecoprop** Galenika
Monosan-S H **2,4-D + MCPA** Galenika
Monosan super-DP H **dichlorprop + MCPA + mecoprop** Galenika
Monotrel-DP H **clopyralid + dichlorprop** Galenika
Monotrel kombi H **clopyralid + MCPA + mecoprop** Galenika
Monotrel-M H **clopyralid + MCPA** Galenika
Monsur IG **carbaryl** Nordisk Alkali
Monur H **monuron** Chemia
monuron H [150-68-5]; [140-41-0] (trichloroacetate) 3-(4-chlorophenyl)-1,1-dimethylurea, CMU, chlorfenidim, Rosuran, Urox (trichloroacetate)
Monuryl H **monolinuron** Interphyto
Moos-Frei H **iron sulphate** A.C.I.E.R.
Moos-Killer H **iron sulphate** Schomaker
Moos KO Neu H **iron sulphate** Neudorff
Moos-Stop H **iron sulphate** Staehler
Moosuran H **iron sulphate** Spiess, Urania
Moos Vernichter H **iron sulphate** Bermel
Moosvertilger H **iron sulphate** Schacht
Moos-Vertilger Schola H **iron sulphate** Schomaker
Mop fluid F **sulphur** Marty et Parazols
Moratox R **chloralose** Salomez
Morestan FAI **quinomethionate** Bayer, Berner
morfamquat dichloride H [4636-83-3] 1,1'-bis(3,5-dimethylmorpholinocarbonylmethyl)-4,4'-bipyridylium dichloride, Morfoxone, PP 745
Morfon F **mancozeb + trimorphamide** Organika-Azot
Morfos Methyl I **parathion-methyl** Efthymiadis
Morfoxone H **morfamquat dichloride**
Morgan F **sulphur** Prochimagro
Moris H **molinate + tiocarbazil** Inagra
Moritor R **chlorophacinone** Ital-Agro
Morkit X **anthraquinone** Bayer
Morocide AF **binapacryl** Argos, Hoechst, Pluess-Staufer
Morogal H **mecoprop** Zupa
Morozin IA **dinoseb** Zupa
Morphactin GH **chlorflurecol-methyl**
Morphos I **parathion** Efthymiadis
morphothion I [144-41-2] *O,O*-dimethyl *S*-morpholinocarbonylmethyl phosphorodithioate, Ekatin M
Morrocid AF **binapacryl** Hoechst
Morsuvin X **natural resins + pine oil** Spolana
Mortalin gift mod kakerlakker I **dieldrin** Mortalin
Mortalin giftvand R **thallium sulphate** Mortalin
Mortalin kakerlakgift I **chlorpyrifos** Mortalin
Mortalin muldvarpe- og mosegrise-gas R **aluminium phosphide** Mortalin
Mortalin roeggaspatron R **sulphur** Mortalin
Mortegg Emulsion IHFX **tar oils** Dow
Mortherbe H **sodium chlorate** Franco-Belge
Mortin-Gaspatrone R **sulphur** Tschapek
Mortitaupe R **chloralose** Sovilo
Mos vaek H **iron sulphate** Midol
Mosdood H **iron sulphate** Melchemie
Mosefjerner H **iron sulphate** Vadheim
Moskill H **iron sulphate** Lejeune-Jardirama
Mosmiddel H **iron sulphate** Bayer
Moss Gun H **dichlorophen** ICI

Moss Vaeck H **iron sulphate** MV-Produkter
Mossfritt H **ammonium sulphate + iron sulphate** Svaloef
Mosskil-A H **iron sulphate** Fisons
Mosskill Plus H **iron sulphate** Fisons
Mosskiller for Lawns H **iron sulphate** ICI
Mosstox LFB **dichlorophen** May & Baker
Mostop H **iron sulphate** Eurofyto
Mostox LFB **dichlorophen** May & Baker
Mosweeder H **iron sulphate** Sanac
Mota-Kvalster A **dicofol** Gullviks
Mota-Vilt X **thiram** Nordisk Alkali
Motecide C-50 F **captan** Ciba-Geigy
Motedin F **dodine** Afrasa
Motox IRA **camphechlor**
Mous-Rid R **strychnine**
Mouse-Pak R **warfarin**
Mouser R **brodifacoum** ICI
Mouxine mural I **permethrin** C.N.C.A.T.A.
Mowchem GH **mefluidide** Rhone-Poulenc
Moxiline S **polyglycolic ethers** Agriben
MP 58 H **mecoprop** Schering
MP-Combi H **2,4-D + mecoprop** Burri, Hokochemie
MP-D H **2,4-D + mecoprop** Dow
MP-Kombi H **2,4-D + mecoprop** Schering
MPDT H **methiuron**
MPMC I **xylylcarb**
MPP I **fenthion**
M-propionat NAB H **mecoprop** Nordisk Alkali
MSMA H [2163-80-6] sodium hydrogen methylarsonate, Ansar 170, Ansar 529, Arsonate, Bueno, Daconate, Dal-E-Rad, Drexar, Mesamate, Superarsonate, Trans-Vert, Weed-E-Rad, Weed-Hoe
M-Special F **captan + zineb** Ruse
MSS Mircam H **dicamba + mecoprop** Mirfield
MSS Mircam Plus H **dicamba + MCPA + mecoprop** Mirfield
MSS Sugar Beet Herbicide H **chlorpropham + fenuron + propham** Mirfield
MT-14 F **maneb + thiram** Bourgeois
MT-101 H **naproanilide** Mitsui Toatsu
MTI 500 I **ethofenprox** Mitsui Toatsu
MTMC I **metolcarb**
MTO-460 F **phosdiphen** Mitsui Toatsu
Mudekan H **linuron + trifluralin** Shell
Mullant S **polyglycolic ethers** Marba
Multamat I **bendiocarb** Schering, Schering AAgrunol
Multamat K.O. I **captan + fosetyl-aluminium + thiabendazole** Schering
Multapon IA **azinphos-methyl + demeton-S-methyl sulphone** Schering
Multar F **maneb + sulphur + thiabendazole** Van Wesemael
Multexol IA **dimethoate** Neudorff
Multi-goal H **dalapon-sodium + oxyfluorfen** Rohm & Haas
Multi-W F **carbendazim + maneb** Pan Britannica
Multicide Concentrate F-2271 I **phenothrin** Sumitomo
Multidis H **dichlorprop + MCPA + mecoprop** Agrimont
Multiflor-Rapid H **dicamba + MCPA** Gerhardtz
Multiflora-Moosvernichter H **iron sulphate** Gerhardtz
Multiflora-Rosenduenger mit Unkrautstop H **diuron + methabenzthiazuron** Gerhardtz
Multiflora-Supergruen Rasenduenger mit Moosvernichter H **iron sulphate** Gerhardtz
Multiprop GH **chlorflurecol-methyl** Celamerck
Multivall IA **dimethoate** Eli Lilly
Murbenine F **guazatine acetates** Dow
Murbenine Plus F **guazatine acetates + imazalil** Dow
Murfotox I **mecarbam** Dow, Efthymiadis, Grima
Murfotox Oil I **mecarbam + petroleum oils** Efthymiadis
Murganic RPB Liquid Seed Treatment F **carboxin + phenylmercury acetate** Rothwell Plant Health
Muriol R **chlorophacinone** Condor
Muris Esca R **warfarin** Visplant
Muronit H **acetochlor + chlorbromuron** Nitrokemia
Murox F **nuarimol** Dow
Murphy Ant Killer Powder I **lindane** Murphy
Murphy Combined Seed Dressing F **captan + lindane** Murphy
Murphy Covershield Weed Preventer H **propachlor** Fisons
Murphy Fentro I **fenitrothion** Murphy
Murphy Greenhouse Aerosol I **malathion** Murphy

Murphy Hormone Rooting Powder F **captan + 1-naphthylacetic acid** Murphy
Murphy Lawn Wormkiller IG **carbaryl** Murphy
Murphy Path Weedkiller H **amitrole + atrazine** Murphy
Murphy Pest and Disease Smoke FI **lindane + tecnazene** Murphy
Murphy Slugit Liquid M **metaldehyde** Murphy
Murphy Slugits M **metaldehyde** Murphy
Murphy Systemic Insecticide IA **dimethoate** Murphy
Murphy Tumblebug F **heptenophos + permethrin** Murphy
Murphy Tumblemoss H **chloroxuron + iron sulphate** Murphy
Murphy Tumbleslug M **metaldehyde** Murphy
Murphy Tumbleweed F **glyphosate** Murphy
Murphy Wasp Destroyer IG **carbaryl** Murphy
Murphys mosaeydir H **chloroxuron + iron sulphate** Murphy
Murvesco A **fenson** Murphy
Murvin IG **carbaryl** Dow
Murvis A **fenson** Visplant
Murvite IA **fenson + prothoate** Eli Lilly
MUS H **chlorthal-dimethyl** Grima
Muscaton I **dimethoate + malathion** Zucchet
Muscatox-Smeermiddel I **cyfluthrin + phoxim** Bayer
Muscid-Giftweizen R **zinc phosphide** Kwizda
Muscin I **dichlorvos** Shell
Muscodel L **quaternary ammonium** Vosimex
Musketeer H **ioxynil + isoproturon + mecoprop** Hoechst
Muskitol I **dimethoate + fenitrothion** Luxan
Muster H **glyphosate** ICI
Mustoxin R **coumachlor** Fivat
Mustrin I **cypermethrin** Maag
Mutan H **triclopyr** Chimac-Agriphar
Mutox I **dichlorvos** Bitterfeld
Muurahais-Baition I **phoxim** Berner
MUW 1193 H **dimepiperate**
Mux-N I **dichlorvos + trichlorfon** Wolfen
MV 4 F **maneb** Akzo, Shell
MV 119A F **dithianon** Celamerck
MY 93 H **dimepiperate**
myclobutanil F [88671-89-0] *α*-butyl-*α*-(4-chlorophenyl)-1*H*-1,2,4-triazole-1-propanen itrile, RH-3866, Systhane
myclozolin F [54864-61-8] (*RS*)-3-(3,5-dichlorophenyl)-5-methoxymethyl-5-methyl-1,3-oxazol idine-2,4-dione, myclozoline, BAS 436F
myclozoline F **myclozolin**
Mycocid F **copper sulphate + folpet** Interphyto
Mycodifol F **captafol + folpet** Chemia, Chevron, Geopharm, Quimigal, Shell, Sopra, Zupa
Mycofax F **thiabendazole** Coophavet
Mycotal I **Verticillium lecanii** Tate & Lyle, Microbial Resources
Mycotox F **copper sulphate + folpet** Gaillard, Sipcam-Phyteurop
Mycozol F **thiabendazole** MSD AGVET
Myfusan Rookkaars F **tecnazene** Duphar
Mylone H **ioxynil + mecoprop** Agronorden
Mylone NFHI **dazomet** Union Carbide
Myocarb F **carbendazim** Tuhamij
myprozine F **pimaricin**
Myrr I **malathion + piperonyl butoxide + pyrethrins** Ewos, Finnewos
Myrr C I **cypermethrin** Ewos
Myzafan IA **endosulfan** Afrasa
MZ Chrom F **mancozeb** Phytorgan
MZ-80 F **mancozeb** Visplant

N 252 HG **dimethipin** Uniroyal
N-289 FNHI **metham-sodium** Stauffer
N-521 NFHI **dazomet** Stauffer
N-2790 I **fonofos** Stauffer
NA-53M A **benzoximate** Nippon Soda
NA-73 A **hexythiazox** Nippon Soda
NA Betafam H **phenmedipham** Nordisk Alkali
NA dicamba mix H **dicamba + dichlorprop + MCPA** Nordisk alkali
NA MotOgraes H **2,4-D + dichlorprop** Nordisk Alkali
NA sproejtesvovl F **sulphur** Nordisk Alkali
NAB Kombi 667 H **2,4-D + dichlorprop** Nordisk Alkali
Nabac F **hexachlorophene** Efthymiadis
nabam FL [142-59-6] disodium ethylenebis(dithiocarbamate), nabame, Dithane A-40, Nabasan, X-Spor
nabame FL **nabam**

Nabasan FL **nabam** Pepro, Rhone-Poulenc
Nabu H **sethoxydim** Kemira, Nippon Soda, PVV, VCH
NAC IG **carbaryl**
Nacudivor F **copper naphthenate + dichlofluanid** I.C.P.P.
Nacuvor F **copper naphthenate** I.C.M.M., I.C.P.P.
Nafor F **copper oxychloride + mancozeb** Argos
Naftal G **1-naphthylacetic acid** Aifar
Naftane FI **carbaryl + lindane + maneb** Pepro
Naftene I **carbaryl** Agronova
Naftil I **carbaryl** Ravit
Naftil acaricide IA **carbaryl + chlorfenson** Pepro
Naftilo I **carbaryl + lindane** Pepro
naled IA [300-76-5] 1,2-dibromo-2,2-dichloroethyl dimethyl phosphate, BRP, bromchlophos, dibrom, OMS 75, ENT 24988, Bromex, Dibrom, Ortho Fly Killer, RE 4355
Nalkil H **bromacil**
Namate H **DSMA**
Namedit H **2,4-D + fenthiuron + nitrofen** Bitterfeld
Namekil M **metaldehyde**
Nankor I **fenchlorphos**
α-naphthaleneacetamide G **1-naphthylacetamide**
naphthenate de cuivre FX **copper naphthenate**
1-naphthylacetamide G [86-86-2] 2-(1-naphthyl)acetamide, α-naphthaleneacetamide, NAD, NAAm, Amid-Thin, Rootone
1-naphthylacetic acid G [86-87-3] 2-(1-naphthyl)acetic acid, acide naphtylacetique, NAA, Celmone, Floramon, Fruit Fix, Fruitone-N, NAA-800, Nafusaku, Phyomone, Planofix, Plucker, Rhizopon B, Rootone, Stik
naproanilide H [52570-16-8] 2-(2-naphthalenyloxy)-*N*-phenylpropanamide, MT-101, Uribest
napropamide H [15299-99-7] (*RS*)-*N,N*-diethyl-2-(1-naphthyloxy)propionamide, Devrinol, R-7465
Napser H **naptalam** Sepran
naptalam H [132-66-1] *N*-1-naphthylphthalamic acid, naptalame, NPA, alanap, Alanap
naptalame H **naptalam**
naramycin FG **cycloheximide**
Narsty X **aluminium ammonium sulphate + denatonium benzoate + sucroseoctaacetate + terpenes** Mandops
Nasiman T **hydrolysed proteins** EniChem
NaTA H **TCA** Argos, Hoechst, Pluess-Staufer, Roussel Hoechst, Sipcam-Phyteurop, Spiess, Urania
natamycin F **pimaricin**
Natrilon H **TCA** Aveve
Natriumchloraat H **sodium chlorate**
Natriumklorat H **sodium chlorate**
Naturen I **piperonyl butoxide + pyrethrins** Chimac-Agriphar
Naturinsektizid-Gesal IA **pyrethrins** Reckitt & Colman
Navetta IA **piperonyl butoxide + pyrethrins** Berner, Farmos
Navron R **fluoroacetamide**
Naworol H **alachlor** Bayer
NC 302 H **quizalofop-ethyl** Nissan
NC-311 H **pyrazosulfuron-ethyl** Nissan
NC 1667 H **trietazine**
NC 2962 I **lythidathion**
NC 3363 H **chlorflurazole**
NC 4780 H **fluoromidine**
NC 6897 I **bendiocarb** Schering
NC 8438 H **ethofumesate** Schering
NC 21314 A **clofentezine** Schering
NCI 96683 H **quizalofop-ethyl** Nissan
Nebrex F **maneb** Agriplan
Nebulin FG **tecnazene** Wheatley
Nebulo DDMA I **dichlorvos + malathion** C.G.I.
Neburane H **neburon** Bourgeois
neburea H **neburon**
Neburee H **neburon** Interphyto
Neburex H **neburon** Makhteshim-Agan, Sipcam-Phyteurop
neburon H [555-37-3] 1-butyl-3-(3,4-dichlorophenyl)-1-methylurea, neburea, Granurex, Kloben, Neburex
Neburyl H **neburon** Agriphyt
Necatol I **lindane** Agriphyt
Necro fourmis boite I **sodium dimethylarsinate** Sovilo
Nedvesitheto ken F **sulphur** Sojuzchimexport
Negal FW **maneb + zineb** Staehler
Negal-Extra WF **captan + carbendazim + 2,5-dichlorobenzoic acid** Staehler
Nekavin I **carbaryl** Ciba-Geigy
nekoe IA **rotenone**
Nellite N **diamidafos**
Nelpon H **tridiphane** Dow

Nelvek H **MCPA + propanil + triclopyr** Siapa
Nem-a-tak IN **fosthietan**
Nemacur N **fenamiphos** Bayer
Nemacur O IN **fenamiphos + isofenphos** Bayer
Nemafos IN **thionazin** Cyanamid
Nemafum FIHN **D-D** Visplant
Nemagon N **dibromochloropropane**
Nemal Grains New R **thallium sulphate** Chimac-Agriphar
Nemamort N **DCIP** SDS Biotech
Nemaphos NI **thionazin** Cyanamid
Nemasol FNHI **metham-sodium** United Agri Products
Nemasol IN **thionazin** Siapa
Nematin FN **metham-sodium** Dimitrova
Nematocide N **dibromochloropropane**
Nematosol IN **ethylene dibromide** Hellenic Chemical
Nematox II N **1,3-dichloropropene** Siapa
Neminfest H **linuron + trifluralin** Agrimont, CTA, Marktredwitz, Masso, Montedison
Nemispor F **mancozeb** Agrimont, Masso, Montedison
nendrin IR **endrin**
Nenningers Moostod H **iron sulphate** Nenninger
Neo-Combisan F **captan + dicloran** Asepta
Neo-Conserviet G **chlorpropham** Duthoit, Recter, Luxan
Neo-Conserviet mixte G **chlorpropham + propham** Duthoit
Neo Davlitox F **quintozene** Korleti
Neo-Jet Tox I **allethrin + piperonyl butoxide** Chimac-Agriphar
Neo Musol R **brodifacoum** Maag
Neo-Scabexaan IA **lindane** Gist-Brocades
Neo-Voronit F **fuberidazole + sodium dimethyldithiocarbamate** Bayer
Neobyne H **barban + MCPB + mecoprop** Schering
Neocid I **iodofenphos** Ciba-Geigy, Farmos
Neocid Insektenspray IA **dichlorvos** Ciba-Geigy
Neocidol IA **diazinon** Ciba-Geigy
Neodan F **captan + thiophanate-methyl** Lapapharm
Neodiprion sertifer NPV I Neodiprion sertifer Nuclear Polyhedrosis Virus, Virox
Neodiprion sertifer Nuclear Polyhedrosis Virus I **Neodiprion sertifer NPV**
Neodorm G **chlorpropham + propham** Sipcam-Phyteurop
Neofos I **malathion** Aziende Agrarie Trento
Neopec A **fenbutatin oxide** Sapec
Neopol FIA **barium polysulphide** Budapesti Vegyimuvek
Neopybuthrin I **bioallethrin + piperonyl butoxide + resmethrin** Wellcome
Neopynamin IA **piperonyl butoxide + tetramethrin** Sumitomo
Neoram F **copper oxychloride** Caffaro
Neoron A **bromopropylate** Ciba-Geigy, Ligtermoet, Liro, OHIS, Spolana
Neosan H **2,4-D + MCPA + mecoprop** Galenika
Neosorexa R **difenacoum** Sorex
Neostanox A **fenbutatin oxide**
Neotopsin FW **thiophanate-methyl** Lapapharm
Neotox IA **TEPP** Chevron, Tecniterra
Nephocarb IA **carbophenothion**
Neporex I **cyromazine** Ciba-Geigy, Liro
Neptan I **dichlorvos** Tecniterra
Nerkol IA **dichlorvos**
Neroxon F **copper oxychloride + zineb** Spolana
Nespor F **maneb** Montedison
Netagrone H **2,4-D** Rhodiagri-Littorale
Netard S H **bromacil + dalapon-sodium + diuron** Sipcam
Neterox H **amitrole + diuron + sodium thiocyanate** Galpro
Netosol H **sodium chlorate** Chimac-Agriphar
Netox I **carbaryl** Terranalisi
Nettafid I **piperonyl butoxide + pyrethrins** Siapa
Netzmittel Liquido S **alkyl phenol ethoxylate + polyglycolic ethers** Bitterfeld, Chimiberg, Ciba-Geigy
Netzschwefel FA **sulphur** Agrotec, Avenarius, BASF, Bayer, Burri, Ciba-Geigy, CTA, Dow, Du Pont, Hoko-Chemie, Kwizda, Leu + Gygax, Schacht, Sideco-Chemie, Staehler, VAW
Netz-Schwefelit FA **sulphur** Neudorff
Neu Substral Rasen-Duenger mit Unkrautvernichter H **2,4-D + dicamba** Barnaengen
Neudo 2 R **calcium phosphide** Leu & Gygax
Neudo-Fungan F **maneb** Neudorff, Staehler

Neudo-Phosphid R **calcium phosphide** Neudorff, Pluess-Staufer
Neudo-Phosphid S R **aluminium phosphide** Neudorff
Neudorff's Raupenspritzmittel I **Bacillus thuringiensis** Neudorff
Neudosan IA **potassium soap** Neudorff
Neurasen-Unkraut-Ex H **bromofenoxim** Stolte & Charlier
Neutra-Weissteer X **tar oils** Staehler
Neutrion IA **parathion-methyl + tetradifon** La Quinoleine
Neviken FIA **barium polysulphide + petroleum oils** PVV
Nevirol G **N-phenylphthalamic acid** Nehezvegyipari
Nevisox I **carbaryl** Ciba-Geigy
Nevugon I **trichlorfon** Bayer
New 5C Cycocel G **chlormequat chloride** Cyanamid
New Atraflow Plus H **amitrole + atrazine** Burts & Harvey
New Chlorea H **atrazine** Chipman
New Estermone H **2,4-D + dicamba** Synchemicals
New Formulation SBK Brushwood Killer H **2,4-D + dicamba + mecoprop** Synchemicals
New Formulation Weed & Brushkiller H **2,4-D + dicamba + mecoprop** Synchemicals
New Hysede FL FI **lindane + thiabendazole + thiram** Agrichem
New Hystore FG **tecnazene** Agrichem
New Kotol I **lindane** Shell
New Murbetex H **chloridazon** Dow
Nex IN **carbofuran** Tripart
Nexa moelhaenger I **lindane** Lopex
Nexa-Rat R **warfarin** Permutadora
Nexagan IA **bromophos-ethyl** BASF, Celamerck, Chimac-Agriphar, Shell, Sovilo, Zupa
Nexalotte I **empenthrin**
Nexen I **piperonyl butoxide + pyrethrins** Maag
Nexion I **bromophos** Celamerck, Chimac-Agriphar, Epro, Geopharm, Margesin, OHIS, Shell, Sovilo, Szovetkezet, Zupa
Nexion L FI **bromophos + captan** Shell
Nexion-aerosoli I **bromophos + piperonyl butoxide + pyrethrins** Kemira
Nexit I **lindane** Celamerck, Shell
NF-35 F **thiophanate** Nippon Soda
NF-44 FW **thiophanate-methyl** Nippon Soda
NF-114 F **triflumizole** Nippon Soda
nickel dimethyldithiocarbamate FB [15521-65-0] nickel bis(dimethyldithiocarbamate), methyl niclate, Sankel
niclosamide M [50-65-7] 2',5-dichloro-4'-nitrosalicylanilide, clonitralid, Bay 25648, Bayluscid, Bayluscide, SR-73, Yomesan
Nico Soap I **nicotine** Campbell
Nicogen I **nicotine** Aziende Agrarie Trento
Nicol I **nicotine** Caffaro
nicotine I [54-11-5] (*S*)-3-(1-methylpyrrolidin-2-yl)pyridine, Black Leaf 40, Nico Soap, XL All Insecticide
Nictol H **molinate** ERT
Nifam F **thiram** Agrimont
Nikal-Fix X **nicotine** Chemia
Nikonal H **phenmedipham** Montenegroexport
Nikotiinikarytenauha I **nicotine** Farmos
Nikotoxin I **nicotine** Ewos
Nilirex H **linuron + nitrofen** Aragonesas
Nimitex I **temephos** Cyanamid
Nimrod F **bupirimate** AVG, Galenika, ICI, ICI-Valagro, ICI-Zeltia, Maag, Schering, Sopra
Nimrod T F **bupirimate + triforine** ICI
Niomil I **bendiocarb** Schering
Niovol A **fenson** Fivat
Nioxyl F **ziram** Chimac-Agriphar
NIP H **nitrofen**
Nippon Ant & Crawling Insect Killer I **permethrin + tetramethrin** Synchemicals
Nippon Ant Killer Liquid I **borax** Synchemicals
Nippon Ant Killer Powder I **chlordane** Synchemicals
Nippon Insektsmedel I **piperonyl butoxide + pyrethrins** Gripen
Nippon myrmedel I **borax** Gripen
Nipsan IA **diazinon**
Niptan H **EPTC** Chemolimpex, Nitrokemia
Niran IA **parathion** Monsanto
Nissorun A **hexythiazox** BASF, Nippon Soda, Zorka Subotica
Niticid H **propachlor** Chemolimpex, Nitrokemia
Nitiran H **chlorbromuron + propachlor** Nitrokemia
Nitrador IAHF **DNOC**
nitralin H [4726-14-1] 4-methylsulphonyl-2,6-dinitro-*N*,*N*-dipropylaniline, Planavin, SD 11831

Nitricide FHI **DNOC** Sedagri
nitrilacarb I [29672-19-3] 4,4-dimethyl-5-(methylcarbamoyloxyimino)valeronitrile, nitrilacarbe, AC 82258, Accotril
nitrilacarbe I **nitrilacarb**
Nitro-Mop FI **DNOC + petroleum oils** Marty et Parazols
Nitrofan H **dinoseb** Burri
nitrofen H [1836-75-5] 2,4-dichlorophenyl 4-nitrophenyl ether, nitrofene, NIP, nitrophen, FW-925, Tok E-25, Tok WP-50, Tokkorn, Trizilin
nitrofene H **nitrofen**
nitrofluorfen H [42874-01-1] 2-chloro-α,α,α-trifluoro-*p*-tolyl 4-nitrophenyl ether, nitrofluorfene, RH-2512
nitrofluorfene H **nitrofluorfen**
nitrolime HF **calcium cyanamide**
nitrophen H **nitrofen**
Nitrosan I **DNOC** Dimitrova, Spolana
nitrothal-isopropyl F [10552-74-6] di-isopropyl 5-nitroisophthalate, nitrothale-isopropyl, BAS 30000F
nitrothale-isopropyl F **nitrothal-isopropyl**
Nitrox IA **parathion-methyl**
Niva Zrna R **zinc phosphide** JZD Chovatel
Nix-Scald G **ethoxyquin**
Nixol DA H **dinoseb** Agriben
NK-049 H **methoxyphenone** Nippon Kayaku
NK 191 F **methasulfocarb** Nippon Kayaku
NK 483 F **trichlamide** Nippon Kayaku
NK-1158 I **propaphos** Nippon Kayaku
NK 8116 I **cycloprothrin** Nippon Kayaku
NK-15561 R **bisthiosemi** Nippon Kayaku
NMC IG **carbaryl**
NNF-109 FI **isoprothiolane** Nihon Nohyaku
NNF-136 F **flutolanil** Nihon Nohyaku
NNI-750 IA **buprofezin** Nihon Nohyaku
No-Brot G **fatty alcohols** Sadisa
No Bunt F **hexachlorobenzene**
No-Rats R **difenacoum** De Rauw
No Scald G **diphenylamine** Sipcam
No-Seed G **(2-naphthyloxy)acetic acid** Bruinsma, Zaadhandel Hollandia
No Sekt I **piperonyl butoxide + pyrethrins + resmethrin** KenoGard
No-Sprout G **chlorpropham + propham** Sepran, Van den Broeke, Visplant
nobormide R **norbormide**
NOC I **methomyl** Sivam
Nociolex F **carbendazim** Platecsa
Nodust I **cypermethrin + fenitrothion + malathion** La Quinoleine
Noetrine I **deltamethrin** C.N.C.A.T.A.
Noexine I **propetamphos** C.N.C.A.T.A.
Nogerma G **chlorpropham + propham** Agriphyt, Chimac-Agriphar, Van Wesemael
Nogos IA **dichlorvos** Chromos, Ciba-Geigy, Spolana, Viopharm
Noita I **bromophos + piperonyl butoxide + pyrethrins** Kemira
Noita-karpasruiskute I **dimethoate + fenitrothion** Kemira
Noita-koitabletti I **hexachloroethane + naphthalene** Kemira
Nokad G **1-naphthylacetic acid** EniChem
Nolinex H **linuron + monolinuron** Protex
Nomersan FX **thiram** ICI
Nomolt I **teflubenzuron** Celamerck, Shell, Sovilo
Nopalmate H **hexaflurate**
Nopon S **petroleum oils** Agrolinz, Ruse
norbormide R [991-42-4] 5-(α-hydroxy-α-2-pyridylbenzyl)-7-(α-2-pyridylbenzylidene) bicyclo[2.2.1]hept-5-ene-2,3-dicarboximide, nobormide, ENT 51762, McN-1025, Raticate, Shoxin
Nordox F **copper hydroxide** Efthymiadis
norea H **noruron**
norflurazon H [27314-13-2] 4-chloro-5-methylamino-2-(α,α,α-trifluoro-*m*-tolyl) pyridazin-3(2*H*)-one, norflurazone, Evital, H 52143, H 9789, Solicam, Telok, Zorial
norflurazone H **norflurazon**
Norlan T H **neburon + terbutryn** Agriben
Norosac H **dichlobenil** PBI Gordon
Nortron H **ethofumesate** Efthymiadis, Fisons, Kobanyai, Maag, NOR-AM, Organika-Sarzyna, Schering, Schering AAgrunol, VCH, Zupa
Nortron Combi H **ethofumesate + lenacil** Siapa
Nortron Kombi H **ethofumesate + lenacil** Schering
Nortron Leyclene H **bromoxynil + ethofumesate + ioxynil** Schering
Nortron Tandem H **ethofumesate + phenmedipham** Maag
noruron H [2163-79-3] 1,1-dimethyl-3-(perhydro-4,7-methanoinden-5-yl)urea, norea, Herban, Hercules 7531
Norvan A **fenbutatin oxide** Shell
Nos Punt G **chlorpropham + propham** KenoGard
Noslak M **metaldehyde** De Weerdt

Nospor F **copper oxychloride + zineb** Siegfried
Nospor F F **copper oxychloride + folpet** Prometheus
Nospor R F **copper oxychloride + mancozeb** Solfotecnica
Nospor S F **mancozeb** Solfotecnica
Nospor-Schwefel AC FA **copper oxychloride + cyhexatin + folpet + sulphur** Siegfried
Notar F **chlorothalonil** Eli Lilly
Notaret F **chlorothalonil + folpet** Eli Lilly
Notrotox D F **zineb** Nitrofarm
Novacorn H **bromoxynil + ioxynil** Farmers Crop Chemicals
Novagrin IA **monocrotophos** Agronova
Novam FIHN **metham-sodium** Agronova
Novanox Plus H **atrazine + diuron + simazine** Schering
Novathion I **fenitrothion** Cheminova
Novazina H **atrazine** Agronova
Novege H **erbon**
Noven I **piperonyl butoxide + pyrethrins** KenoGard
Novenda IAF **DNOC** Nitrokemia
Novermone H **2,4-D** C.F.P.I., Etisa
Novermone Especial H **MCPA** Etisa
Novermone extra H **2,4-D + MCPA + mecoprop** C.F.P.I.
Novermone G H **dichlorprop + ioxynil** C.F.P.I.
Novermone gazon H **2,4-D + mecoprop** C.F.P.I.
Novermone gazons H H **dicamba + ioxynil + mecoprop** C.F.P.I.
Novermone gazons P H **2,4-D + dicamba + ioxynil** C.F.P.I.
Novermone Special H **MCPA** C.F.P.I.
Novigam I **lindane**
Novo-Tak IA **diazinon** Siegfried
Novocar IA **fenson + prothoate** Tecniterra
Novofix F active F **cymoxanil + folpet + zineb** R.S.R.
Novorail H **amitrole + atrazine + diuron + ethidimuron** C.F.P.I.
Novothion I **parathion** Aveve
Novozir F **mancozeb** Duslo
Nox Moos H **iron sulphate** Kinkel & Gebel
Nox Mos H **iron sulphate** Protex
Noxfire IA **rotenone** Penick
Noxfish IA **rotenone** Penick
NP-48 Na H **alloxydim-sodium** Nippon Soda
NP-55 H **sethoxydim** Nippon Soda
NPA H **naptalam**
NPD I **O,O,O',O'-tetrapropyl dithiopyrophosphate**
NPH 83 IA **vamidothion**
NRC 910 F **iprodione** Rhone-Poulenc
NRDC 104 I **resmethrin**
NRDC 143 I **permethrin**
NRDC 149 I **cypermethrin**
NRDC 161 I **deltamethrin**
NSC 14083 B **streptomycin**
NTN 801 H **mefenacet** Bayer
NTN 9306 I **sulprofos** Bayer
Nu Film S **di-1-p-menthene** Agrichem, Chimiberg, Intracrop
Nu-Sol H **amitrole + diuron + simazine** Koch & Reis, R.S.R.
nuarimol F [63284-71-9] (±)-2-chloro-4'-fluoro-α-(pyrimidin-5-yl)benzhydryl alcohol, EL-228, Gauntlet, Murox, Trimidal, Triminol
Nucidol IA **diazinon** Ciba-Geigy
Nudor H **alachlor** ERT
Nudor Extra H **alachlor + atrazine** ERT
Nudrin I **methomyl** Shell, Sipcam
Nunstar F **flusilazole** Du Pont
Nurelle F **pyroxychlor**
Nurelle I **cypermethrin** Bianchedi, Dow
Nurelle D I **chlorpyrifos + cypermethrin** Dow
Nurmikko-Hedonal H **2,4-D** Berner
Nurmikon Rikkaruohontuho H **2,4-D** Kemira
Nustar F **flusilazol** Du Pont
Nuvacron IA **monocrotophos** Chromos, Ciba-Geigy, Nitrokemia, Viopharm
Nuvagrain I **chlorpyrifos-methyl** Arkovet, Ciba-Geigy
Nuvan IA **dichlorvos** Chromos, Ciba-Geigy, Ligtermoet
Nuvan bitotal I **chlorpyrifos-methyl + dichlorvos** Ciba-Geigy
Nuvan duree I **chlorpyrifos-methyl** Arkovet, Ciba-Geigy
Nuvan Top I **dichlorvos + fenitrothion** Ciba-Geigy
Nuvanex I **dichlorvos + iodofenphos** Ciba-Geigy, Ligtermoet
Nuvanol IA **iodofenphos** Arkovet, Ciba-Geigy, Ligtermoet, Liro
Nuvaton Larvicida Polvere I **carbaryl + fenitrothion + malathion** Zucchet
Ny chok I **piperonyl butoxide + pyrethrins** Aeropak
Nytek F **oxine-copper** Maag

Nyvol I **parathion + petroleum oils** Agrishell

OB 21 F **copper oxychloride** Agro-Kemi, Bayer, Berner
Oben AF **sulphur** Agronova
Oborex FW **copper naphthenate**
Obsthormon G **1-naphthylacetic acid** Gobbi
Obstspritzmittel Orthocid F **captan** Shell
Occi 308 R **warfarin** Logissain
Occi 308 grains R **chlorophacinone** Logissain
Occi 308 H R **chlorophacinone** Logissain
Occi 318 R **chlorophacinone** Logissain
Occi 408 mulots et campagnols R **chlorophacinone** Logissain
Occi 608 limaces M **metaldehyde** Logissain
Occi 928 poudre de piste R **chlorophacinone** Logissain
Occi 938 desherbant granule H **atrazine + TCA** Logissain
Occi 958 debroussaillant special H **2,4-D + 2,4,5-T** Logissain
Occi 1018 desherbant total H **amitrole + atrazine** Logissain
Occi courtilieres 708 I **lindane** Logissain
Occi fourmis I **sodium dimethylarsinate** Logissain
Occi limaces M **metaldehyde** Logissain
Occi Mouches etables K Othrine I **deltamethrin** Logissain
Occi souris R **difenacoum** Logissain
Occi taupes RFX **chloralose** Logissain
Occi total herbes poudre mouillable H **amitrole + atrazine** Logissain
Occigram H **TCA** La Littorale
Occysol H **sodium chlorate** Agriben
OCD "Macclesfield" F **copper oxychloride + zineb** La Cornubia
Octa-Klor I **chlordane**
Octachlor I **chlordane** Sandoz
Octalene I **aldrin** Sandoz
Octalox I **dieldrin** Velsicol
Octamyl F **thiram** Aveve
Octave F **prochloraz-manganese complex** Fisons, Schering
octhilinone WFB [26530-20-1] 2-octylisothiazol-3(2*H*)-one, Pancil-T, RH-893
Octylan H **2,4-D** O.P.
Octylan KV-spezial H **2,4-D + mecoprop** Sandoz
Octylan MP-M H **MCPA + mecoprop** Staehler
Ofal H **isoproturon** Ciba-Geigy
Ofal Extra H **isoproturon + triasulfuron** Ciba-Geigy
Off X **diethyltoluamide**
Off-Shoot O G **fatty acid esters** Keyser & Mackay
Off-Shoot T G **fatty alcohols** Du Pont, Seppic, Verchim Asterias
Off-Shoot T Super G **chlorpropham + fatty alcohols** Du Pont
Ofnack IA **pyridafenthion** Mitsui Toatsu
Oftan HI **DNOC** Pluess-Staufer
Oftanol I **isofenphos** Bayer, Berner
Oftanol Combi I **isofenphos + phoxim** Bayer
Oftanol T F **isofenphos + thiram** Agro-Kemi, Bayer, Berner
Oftanol-Zaadbehandeling F **isofenphos + thiram** Bayer
Ofunack IA **pyridafenthion** Candilidis, Inagra, Mitsui Toatsu, Radonja, Sipcam
Ofunack M I **metolcarb**
ofurace F [58810-48-3] (±)-2-chloro-*N*-(2,6-dimethylphenyl)-*N*-(tetrahydro-2-oxo -3-furanyl)acetamide, milfuram, Ortho 20615
Ohza H **butachlor + naproanilide**
Oidiase F **sulphur** Ciba-Geigy
Oidifin F **triforine** Epro
Oidin F **copper oxychloride + sulphur** Zorka Sabac
Oidiol AF **sulphur** Tecniterra
Oidox F **sulphur** Geopharm
OK-174 I **benfuracarb** Otsuka
Oko IA **dichlorvos** Bayer
Oktagam Neu I **lindane + methoxychlor** Staehler
Okultin CMPP H **mecoprop** Marktredwitz
Okultin Combi H **2,4-D + MCPA** Marktredwitz
Okultin D H **2,4-D** Marktredwitz
Okultin DP H **dichlorprop** Marktredwitz
Okultin DP-M H **dichlorprop + MCPA** Marktredwitz
Okultin M H **MCPA** Marktredwitz
Okultin MP H **mecoprop** Marktredwitz
Ole F **chlorothalonil** Chiltern
Oleane IA **chlorfenvinphos + petroleum oils** Shell
Oleanol IA **petroleum oils** Schering

Oleanol S IA **fenitrothion + petroleum oils** Schering
Oleo SI **petroleum oils** ACA, Du Pont, HORA, ICI
Oleoagrex Blanco IA **petroleum oils** Sadisa
Oleo Atratex H **atrazine + petroleum oils** Protex
Oleo Basudin IA **diazinon + petroleum oils** Ciba-Geigy
Oleo Bladan I **parathion + petroleum oils** Bayer
Oleoc IA **petroleum oils** Meoc
Oleocine IA **petroleum oils** Agroplant
Oleocobre F **cuprous oxide + petroleum oils** Procida
Oleocuivre F **cuprous oxide + petroleum oils** Roussel Hoechst
Oleo-Danathion-Hoko IA **fenitrothion + petroleum oils** Hokochemie
Oleo-Diazinon IA **diazinon + petroleum oils** Agroplant, Burri, Ciba-Geigy, Leu & Gygax, Pinus, Pluess-Staufer, Sandoz, Siegfried
Oleo-ekalux IA **petroleum oils + quinalphos** Sandoz
Oleo-Ekamet I **etrimfos + petroleum oils** Bjoernrud, Sandoz
Oleo-Endosulfan IA **endosulfan + petroleum oils** Leu & Gygax
Oleoethiargos IA **ethion + petroleum oils** Argos
Oleo Ethion IA **ethion + petroleum oils** Quimicas Oro
Oleofanan H **dinoseb** Sipcam-Phyteurop
Oleofen IA **fenitrothion + petroleum oils** Burri
Oleo folidol IA **parathion + petroleum oils** Bayer
Oleofos IA **fenitrothion + petroleum oils** Agriplan
Oleofos I **parathion + petroleum oils** Agrimont
Oleofos Par I **parathion-methyl + petroleum oils** Agriplan
Oleo Gesaprim H **atrazine** Ciba-Geigy
Oleo Kriss IA **parathion + petroleum oils** Gaillard
Oleometation IA **fenitrothion + fenson + petroleum oils** Dimitrova
Oleoparaben IA **parathion-methyl + petroleum oils** ERT
Oleoparafene I **parathion-methyl + petroleum oils** Condor
Oleoparaphene I **parathion + petroleum oils** Rhone-Poulenc
Oleoparathion IA **parathion + petroleum oils** Burri, Hokochemie, Sandoz
Oleoparation IA **parathion-methyl + petroleum oils** Afrasa, Key
Oleo Sovi-Tox R **warfarin** Sovilo
Oleotan IA **petroleum oils** Bimex
Oleoter IA **petroleum oils** Terranalisi
Oleo-Thiodan IA **endosulfan + petroleum oils** Pluess-Staufer
Oleothion I **parathion + petroleum oils** Sandoz
Oleo Ultracid IA **methidathion + petroleum oils** Ciba-Geigy, Viopharm
Oleoverdecion IA **fenitrothion + petroleum oils** KenoGard
Oleovis IA **petroleum oils** Visplant
Oleovis Attivato I **parathion + petroleum oils** Visplant
Oleo-Wofatox IA **parathion-methyl** Bitterfeld
Oleozolon I **petroleum oils + phosalone** Hellafarm
Olicron I **phosphamidon** Ciba-Geigy
Olidoc IA **DNOC + petroleum oils** Chemia
Oliocin IA **petroleum oils** Bayer
Olitref H **trifluralin** Budapesti Vegyimuvek
Olivar H **amitrole + simazine** Andreopoulos
Olivin ulje I **petroleum oils + phenthoate** Chromos
Olmex G H **amitrole + ammonium thiocyanate + diuron** Chimac-Agriphar
Olmex liquid H **amitrole + diuron** Chimac-Agriphar
Olmex liquide H **amitrole + diuron** Agriphyt
Olmex S-T H **dichlobenil** Chimac-Agriphar
Olmex Super H **amitrole + diuron** Chimac-Agriphar
Olparin IA **parathion + petroleum oils** Avenarius
Olpisan F **trichlorodinitrobenzene** Bitterfeld
Olymp F **flusilazol** Du Pont
OM 2424 F **etridiazole** Uniroyal
Omega F **prochloraz** Schering
omethoate IA [1113-02-6] *O,O*-dimethyl *S*-methylcarbamoylmethyl phosphorothioate, dimethoate-met, Bay 45432, Folimat, S-6876

Omexan I **bromophos** Delicia
OMH-K G **maleic hydrazide** Otsuka
Omite A **propargite** Argos, Ciba-Geigy, Hellafarm, ICI, Kemi-Intressin, Kwizda, Lucebnizavody, Sipcam, Uniroyal, Zorka Sabac
Omite-Fenson A **fenson + propargite** Uniroyal
Omnidal H **dalapon-sodium + simazine** Agronova
OMPA I **schradan**
OMS 1 IA **malathion**
OMS 2 I **fenthion**
OMS 14 IA **dichlorvos**
OMS 17 I **lindane**
OMS 18 I **dieldrin**
OMS 19 IA **parathion**
OMS 32 I **isoprocarb**
OMS 33 I **propoxur**
OMS 37 NI **fensulfothion**
OMS 43 I **fenitrothion**
OMS 75 IA **naled**
OMS 93 MIAX **methiocarb**
OMS 94 IA **dimethoate**
OMS 111 IA **dimethoate**
OMS 170 I **aminocarb**
OMS 186 IA **azinphos-methyl**
OMS 193 I **heptachlor**
OMS 194 I **aldrin**
OMS 206 I **isobenzan**
OMS 213 IA **parathion-methyl**
OMS 214 I **dicapthon**
OMS 219 IA **EPN**
OMS 226 I **cyanophos**
OMS 239 I **crotoxyphos**
OMS 244 IA **carbophenothion**
OMS 252 IA **ethoate-methyl**
OMS 253 IA **dicrotophos**
OMS 410 I **fonofos**
OMS 412 I **trichloronat**
OMS 466 I **methoxychlor**
OMS 468 I **allethrin**
OMS 469 IA **diazinon**
OMS 570 IA **endosulfan**
OMS 571 AF **binapacryl**
OMS 578 I **trichloronat**
OMS 595 IA **tetrachlorvinphos**
OMS 646 IA **phosfolan**
OMS 658 I **bromophos**
OMS 659 IA **bromophos-ethyl**
OMS 698 IA **formothion**
OMS 716 I **promecarb**
OMS 755 A **tetrasul**
OMS 771 IAN **aldicarb**
OMS 773 I **allyxycarb**
OMS 786 I **temephos**
OMS 800 I **trichlorfon**
OMS 834 IA **monocrotophos**
OMS 844 IA **methidathion**
OMS 864 IAN **carbofuran**
OMS 869 I **cyanophos**
OMS 971 I **chlorpyrifos**
OMS 1011 I **tetramethrin**
OMS 1017 F **fentin hydroxide**
OMS 1020 FLM **fentin acetate**
OMS 1056 AF **dinobuton**
OMS 1075 IA **phenthoate**
OMS 1078 I **TDE**
OMS 1089 IA **azothoate**
OMS 1102 I **dioxacarb**
OMS 1155 IA **chlorpyrifos-methyl**
OMS 1168 I **fospirate**
OMS 1170 I **phoxim**
OMS 1196 IA **methomyl**
OMS 1197 I **chlorphoxim**
OMS 1206 I **resmethrin**
OMS 1209 AI **chlordimeform**
OMS 1211 IA **iodofenphos**
OMS 1325 IA **phosphamidon**
OMS 1328 IA **chlorfenvinphos**
OMS 1394 I **bendiocarb**
OMS 1424 IA **pirimiphos-methyl**
OMS 1437 I **chlordane**
OMS 1438 I **leptophos**
OMS 1502 I **propetamphos**
OMS 1696 I **hydroprene**
OMS 1697 I **methoprene**
OMS 1804 I **diflubenzuron**
OMS 1806 IA **etrimfos**
OMS 1810 I **phenothrin**
OMS 1820 AI **amitraz**
OMS 1821 I **permethrin**
OMS 1825 I **azamethiphos**
OMS 1845 I **heptenophos**
OMS 1998 I **deltamethrin**
OMS 2000 IA **fenvalerate**
OMS 2002 I **cypermethrin**
OMS 2004 IA **profenofos**
OMS 2005 IA **methacrifos**
OMS 2007 I **flucythrinate**
OMU H **cycluron**
Oncol I **benfuracarb** Agrocros, Chimac-Agriphar, Du Pont, Farm Protection, Galenika, Otsuka, Sipcam
Ondatox C R **chlorophacinone** Salubrhygiene
One Shot H **bromoxymil + diclofop-methyl + MCPA** Hoechst
Onecide H **fluazifop-butyl** Ishihara Sangyo

Onkruidbestrijder + Gazonmest H **2,4-D + dicamba** Mommersteeg
Onkruidverdelger Scotts H **2,4-D + dicamba** Graham
Only A **cyhexatin** Eli Lilly
Onmex F **penconazole** Ciba-Geigy
Onslaught H **linuron + trifluralin** Quadrangle
Ontracic H **prometon** Ciba-Geigy
Onutex IAF **DNOC** Visplant
O.P. spray yellow I **diazinon + pyrethrins** Rentokil
Operats R **warfarin** C.N.C.A.T.A.
Operats 2000 R **coumatetralyl** C.N.C.A.T.A.
Opogard H **terbuthylazine + terbutryn** Ciba-Geigy
Oracle H **isoproturon + metsulfuron-methyl** Du Pont
Orangefix G **2,4-D** KenoGard
orbencarb H [97505-36-7] *S*-[2-(chlorophenyl)methyl] diethylcarbamothioate, orthobencarb, Lanray
Orbit F **fenpropimorph + prochloraz** Maag
Orblon F **carbendazim + maneb + pyrazophos** R.S.R.
Orchan S **petroleum oils** Du Pont, Ligtermoet
Orchard Herbicide H **amitrole + diuron** Promark
Orchex I **petroleum oils** Hoechst
Ordagrex H **molinate** Sadisa
Ordram H **molinate** BASF, Bayer, Caffaro, Condor, EniChem, ICI-Valagro, OHIS, Pepro, Ravit, Rhone-Poulenc, Sapec, Serpiol, SPE, Stauffer
Ordram S H **molinate + thiobencarb** Serpiol
Orfamone T **pheromones** Sipcam
Orfon I **trichlorfon** Agriplan
Organil 66 F **maneb + metiram** EMTEA, Procida
Organil 648 F **folpet + thiophanate-methyl** Procida
Orizan H **propanil** Zupa
Orizerba H **propanil** Sapec
Orizol H **molinate** Visplant
Ormo G **4-CPA + (2-naphthyloxy)acetic acid** Caffaro
Ormocaffaro G **gibberellic acid** Ergex
Ormogran H **MCPA** Sofital
Ormosep H **MCPA** Sepran
Ormosep Combi H **2,4-D + MCPA** Sepran
Ormotec G **4-CPA + (2-naphthyloxy)acetic acid** Tecniterra
Ornabel H **dichlobenil** Truffaut
Ornalin F **vinclozolin** Mallinckrodt
Oroap FA **dinocap** Quimicas Oro
Orocaid G **2,4-D** Quimicas Oro
Orocobre F **copper oxychloride** Quimicas Oro
Orocobre Zineb F **copper oxychloride + zineb** Quimicas Oro
Orodan IA **endosulfan** Quimicas Oro
Orodip I **trichlorfon** Quimicas Oro
Orofos IA **azinphos-methyl** Quimicas Oro
Orolit I **fenitrothion** Quimicas Oro
Oromaneb F **maneb** Quimicas Oro
Oromyzus I **dichlorvos** Quimicas Oro
Oroseba F **maneb + thiram + zineb** Quimicas Oro
Orosit IA **dimethoate** Quimicas Oro
Orothion IA **malathion** Quimicas Oro
Orozam F **ziram** Quimicas Oro
Orozineb F **zineb** Quimicas Oro
Orozinon IA **diazinon** Quimicas Oro
Orthen I **acephate** Bayer, Chevron, Ortho, Spiess, Wacker
Orthene I **acephate** Agrocros, Chevron, Condor, Geopharm, Ligtermoet, Maag, Nordisk Alkali, Pepro, Protex, Rhone-Poulenc, Schering, Sipcam
Ortho 5865 F **captafol** Chevron
Ortho 8890 F **dichlozoline**
Ortho 9006 IA **methamidophos** Chevron
Ortho 12420 I **acephate** Chevron
Ortho 20615 F **ofurace** Chevron
Ortho Danitol AI **fenpropathrin** Chevron
Ortho Dibrom IA **naled** Agrocros, Chevron, Chromos, Du Pont, Farmos, Schering
Ortho Difolatan F **captafol** Agrocros, Chevron, Protex, Quimigal, Schering, Shell, Sopra
Ortho Dimecron I **phosphamidon** Vaultier
Ortho Fly Killer IA **naled** Chevron
Ortho Mix nieuw F **captan + maneb** Ligtermoet
Ortho Monitor IA **methamidophos** Schering
Ortho Phaltan F **folpet** Agrocros, Bayer, Budapesti Vegyimuvek, Chemia, Chevron, Liro, Ligtermoet, Maag, Nordisk Alkali, Protex, Quimigal, Schering, Shell, Zupa
Ortho Phosphate Defoliant G **S,S,S-tributyl phosphorotrithioate** Mobay

Ortho-Poudrol F **copper oxychloride + folpet** Maag
Ortho Prunit G **uniconazole** Chevron
Ortho Slug-Geta MIAX **methiocarb** Chevron
Ortho Spotless F **diniconazole** Chevron
Ortho Sumagic G **uniconazole** Chevron
orthobencarb H **orbencarb**
Orthocid F **captan** Bayer, Bjoernrud, Chemia, Nordisk Alkali, Shell
Orthocide F **captan** Agrarische Unie, Agrocros, Bayer, Chevron, Geopharm, Ligtermoet, Liro, Protex, Ravit, Rhodiagri-Littorale, Schering
Orthofat I **acephate** Shell
Orthofor F **captan** Caffaro
Orthorix FIA **lime sulphur** Chevron
Orthoscam F **captan** Scam
Orthotox I **methamidophos** Schering
Orthozid F **captan** Agroplant, Pluess-Staufer
Ortigran H **trifluralin** Scam
Orto Dibrom I **naled** Nordisk Alkali
Ortodiagrex IA **naled** Sadisa
Ortofenilfenolo F **2-phenylphenol** Decco-Roda
Ortomoni G **(2-naphthyloxy)acetic acid** Spyrou
Ortran I **acephate** Chevron
Ortril I **acephate** Chevron
Oryzaemate FB **probenazole** Meiji Seika
oryzalin H [19044-88-3] 3,5-dinitro-N^4,N^4-dipropylsulphanilamide, Dirimal, EL-119, Ryzelan, Surflan
Oryzemate FB **probenazole** Meiji Seika
OS-2046 IA **mevinphos** Shell
Osadan A **fenbutatin oxide** Shell
Osaquat H **paraquat dichloride** Productos
Osbac I **fenobucarb** Sumitomo
OSCORNA Insektenschutz IA **piperonyl butoxide + pyrethrins** Woelper
Ossiclor F **copper oxychloride** Tecniterra
Ossicloruro Caffaro F **copper oxychloride** Shell
Ossiram F **copper oxychloride** Sepran
Ossitan F **copper oxychloride + folpet** EniChem
Oterb H **amitrole + diuron + petroleum oils + simazine** R.S.R.
Otilan H **trifluralin** Hoechst
Oust H **sulfometuron-methyl** Du Pont
Outflank I **permethrin** Shell
Outfox H **cyprazine**
Ovacide A **dicofol + tetradifon** Chemia
Ovatox I **dinoseb** Efthymiadis
Ovatoxin AI **chlordimeform** Agro-Quimicas de Guatemala
Ovex A **chlorfenson** FMC
Ovicar T A **dioxathion + tetradifon** Du Pont
Ovifac IA **prothoate + tetradifon** Agrimont
Ovipron IA **petroleum oils** BP, R.S.R.
Ovirex IAS **petroleum oils** Shell
Ovomitex A **dicofol + fenson** Ravit
Ovoran F **Bordeaux mixture** Platecsa
Ovotek A **dicofol + tetradifon** Ravit
Ovotox A **fenson** Aziende Agrarie Trento
Ovotran A **chlorfenson** Dow
Owadofos plynny I **fenitrothion** Organika-Azot
Owadziak pylisty I **lindane** Organika-Azot
oxabetrinil S [74782-23-3] (*Z*)-1,3-dioxolan-2-ylmethoxyimino(phenyl)acetonitrile, CGA 92194, Concep II
oxadiazon H [19666-30-9] 5-*tert*-butyl-3-(2,4-dichloro-5-isopropoxyphenyl)-1,3,4-oxadiazo l-2(3*H*)-one, 17623 RP, Foresite, Ronstar
oxadixyl F [77732-09-3] 2-methoxy-*N*-(2-oxo-1,3-oxazolidin-3-yl)acet-2',6'-xylidide, Anchor, SAN 371F, Sandofan
oxamil IAN **oxamyl**
oxamyl IAN [23135-22-0] *N*,*N*-dimethyl-2-methylcarbamoyloxyimino-2-(methylthio)acetamide, oxamil, thioxamyl, DPX 1410, Vydate
oxapyrazon H [25316-57-8] 2-hydroxyethyldimethylammonium 5-bromo-1,6-dihydro-6-oxo-1-phenylpyridazin-4-yloxamate, oxapyrazone, BAS 350H
oxapyrazone H **oxapyrazon**
Oxatin F **carboxin + thiram** Visplant
Oxfor A **dicofol + fenson** Du Pont
Oxi-Cupro F **copper oxychloride** ICI-Valagro
Oxiclorura de Cupru F **copper oxychloride** Agria, Caffaro, Diana, Hispagro, ICI, Sandoz, Zorka
Oxicloruro F **copper oxychloride** Aporta
Oximan F **mancozeb + oxycarboxin** Sofital
Oximipiol F **captan** Serpiol
oxine-copper F [10380-28-6] bis(quinolin-8-olato)copper, oxine-Cu, oxine copper, oxine-cuivre, oxyquinoleate de cuivre, copper 8-hydroxyquinolate, copper oxinate, copper8-quinolinolate, Bioquin,

Cellu-Quin, Cunilate 2472, Dokirin, Fruitdo, Milmer, Nytek, Quinolate, Quinondo
oxine-Cu F **oxine-copper**
oxine-cuivre F **oxine-copper**
Oxinol H **bromoxynil + clopyralid + ioxynil + mecoprop** KVK
Oxiram F **cuprous oxide** Ciba-Geigy
Oxiser F **copper oxychloride** Serpiol
Oxitril H **bromoxynil + ioxynil** Agronorden, Rhone-Poulenc
Oxitril 4 H **bromoxynil + dichlorprop + ioxynil + MCPA** Ewos
Oxolon H **bromoxynil + clopyralid + mecoprop** Rhodiagri-Littorale
Oxotin A **cyhexatin** Oxon
oxycarboxin F [5259-88-1] 2,3-dihydro-6-methyl-5-phenylcarbamoyl-1,4-oxathi-in 4,4-dioxide, oxycarboxine, DCMOD, F-461, Oxykisvax, Plantvax, Ringmaster
oxycarboxine F **oxycarboxin**
oxychlorure de cuivre F **copper oxychloride**
Oxycuivre F **copper oxychloride** Siegfried
Oxycur F **copper oxychloride** Schering
oxyde cuivreux F **cuprous oxide**
oxyde mercurique FW **mercuric oxide**
oxydemeton-methyl IA [301-12-2] *S*-2-ethylsulphinylethyl *O,O*-dimethyl phosphorothioate, oxydemetonmethyl, oxydemeton-metyl, demeton-S-methyl sulphoxide, metasystemox, metilmercaptofosoksid, ENT 24964, Bay 21097, Metasystemox R, Metasystox R, R-2170
oxydemeton-metyl IA **oxydemeton-methyl**
Oxydemetox-M IA **oxydemeton-methyl** Phytorgan
oxydeprofos IA [2674-91-1] *S*-2-ethylsulphinyl-1-methylethyl *O,O*-dimethyl phosphorothioate, ESP, ENT 25647, Bay 23655, Estox, Metasystox S, S 410
oxydiazol H **methazole**
oxydimethin HG **dimethipin**
oxydisulfoton IA [2497-07-6] *O,O*-diethyl *S*-2-ethylsulphinylethyl phosphorodithioate, Bay 23323, Disyston-S, L 16/184, S 309
Oxyfane H **dinoseb** Agriphyt
oxyfluorfen H [42874-03-3] 2-chloro-α,α,α-trifluoro-*p*-tolyl 3-ethoxy-4-nitrophenyl ether, oxyfluorfene, Goal, Koltar, RH-2915
oxyfluorfene H **oxyfluorfen**
Oxyjonc H **dichlobenil** Aseptan
Oxykisvax F **oxycarboxin** Jin Hung
Oxykupfer 50 F **copper oxychloride** Siegfried
oxyquinoleate de cuivre F **oxine-copper**
Oxytane H **bromoxynil + fluroxypyr + ioxynil** Rhodiagri-Littorale
oxythioquinox FAI **quinomethionate**
Oxytril H **bromoxynil + ioxynil** Rhone-Poulenc
Oxytril CM H **bromoxynil octanoate + ioxynil octanoate** Rhone-Poulenc
Oxytril M H **bromoxynil + ioxynil + mecoprop** Agriben, Agrotec, Condor, Ravit, Rhodiagri-Littorale, Rhone-Poulenc, Shell
Oxytril MDP H **bromoxynil + dichlorprop + ioxynil + MCPA** Rhone-Poulenc
Oxytril P H **bromoxynil + dichlorprop + ioxynil** Dow
Oxyzenebe F **copper oxychloride + zineb** Tarsoulis

P 474 IA **mecarbam**
P 1108 HA **dinoterb acetate**
P 1488 H **medinoterb acetate**
2,4-PA H **2,4-D**
PA Conquest F **captafol** Portman
Paarlan H **isopropalin** Elanco, Eli Lilly
PAC H **chloridazon**
paclobutrazol G [76738-62-0] (2*RS*,3*RS*)-1-(4-chlorophenyl)-4,4-dimethyl-2-(1*H*-1,2,4-triazol-1-yl)pentan-3-ol, Bonzi, Clipper, Cultar, Parlay, PP 333
Pacol I **parathion + petroleum oils** Rhodiagri-Littorale, Rhone-Poulenc
Padan I **cartap hydrochloride** Takeda
Paicer H **pyrazoxyfen** Ishihara Sangyo
pallethrine I **allethrin**
Pallicap F **captan + nitrothal-isopropyl** BASF
Pallicap M F **captan + maneb + nitrothal-isopropyl** BASF
Palligold F **maneb + metiram + nitrothal-isopropyl + sulphur** BASF
Pallinal F **metiram + nitrothal-isopropyl** BASF, Intrachem
Pallinal C F **captan + metiram + nitrothal-isopropyl** Maag
Pallinal M F **maneb + metiram + nitrothal-isopropyl** BASF, Luxan
Pallitop F **metiram + nitrothal-isopropyl** BASF

Paloma 66 I **methoxychlor** Mariman
Palormone D H **2,4-D** UCP
Paluthion I **fenitrothion** Procida
Pamir I **azinphos-methyl + dimethoate** Protex
Pamisan F **phenylmercury acetate** Excel
Panacide LFB **dichlorophen**
Panam I **carbaryl** Agrimont
Panatac A **clofentezine** Schering
Pancid IA **azinphos-ethyl** Sandoz
Pancide IA **azinphos-methyl** Sandoz
Pancil T FW **octhilinone** Rohm & Haas
Pandar IA **monocrotophos** Rhone-Poulenc
Pandar F **fentin acetate** Eli Lilly
Pandox I **demeton-S-methyl + oxydemeton-methyl + parathion-methyl** Diana
Panil H **propanil** Rohm & Haas, Siapa
Panko 40 FHIN **metham-sodium** Rohm & Haas
Panko WP F **Bordeaux mixture + mancozeb** Rohm & Haas
Panocon A **fenothiocarb** Argos, Kumiai
Panoctin F **guazatine acetates** KenoGard, Shell
Panoctin GF F **fenfuram + guazatine acetates + imazalil** Shell
Panoctin Plus F **guazatine acetates + imazalil** KenoGard, Shell
Panoctin Spezial F **fenfuram + guazatine acetates** KenoGard, Shell
Panoctin Universal FX **fenfuram + guazatine acetates + imazalil** KenoGard, Shell
Panoctine FX **guazatine acetates** KenoGard, Kwizda, Ligtermoet, Liro, Shell, Sopra
Panoctine AT FI **guazatine acetates + lindane** Liro
Panoctine C F **carboxin + guazatine acetates** Kwizda
Panoctine extra F **guazatine acetates + imazalil** KenoGard
Panoctine Plus F **guazatine acetates + imazalil** Farmos, KenoGard, Kwizda, Ligtermoet, Liro, Shell
Panoctine Super FX **fenfuram + guazatine acetates** Maag
Panoctine universal F **fenfuram + guazatine acetates + imazalil** Maag
Panogen M F **methoxyethylmercury acetate** Farmos, Kwizda, Maag, Shell
Panogen Metox F **methoxyethylmercury acetate** KenoGard
Panogen-N F **methoxyethylmercury acetate** Liro
Panol-Gaspatrone R **potassium nitrate + sulphur** Delicia
Panolil FX **guazatine acetates** KenoGard
Panoram F **fenfuram** KenoGard, Ligtermoet, Quimigal, Shell
Panoram D-31 I **dieldrin** Shell
Pansercoll algefjerner L **benzalkonium chloride** KVK
Pansoil F **etridiazole** Uniroyal, Sankyo
Panter H **linuron + pendimethalin** Cyanamid
Panther H **diflufenican + isoproturon** Rhone-Poulenc
Panthion IA **parathion**
Pantrin I **carbaryl** Epro
PAP IA **phenthoate**
Papthion IA **phenthoate** Sumitomo
Parable H **diquat dibromide + paraquat dichloride** ICI
Paracoccidol oil I **parathion + petroleum oils** Ergex
Paradimal I **1,4-dichlorobenzene** Trans-Meri
Para-Diquat H **diquat dibromide + paraquat dichloride** Leu & Gygax
Parafenizon IA **fenson + parathion-methyl** Marty et Parazols
paraffin oils S **petroleum oils**
Paraffinicum S **petroleum oils** Schwebda
Parafitanol I **parathion + petroleum oils** Sapec
parafluron H [7159-99-1] 1,1-dimethyl-3-(α,α,α-trifluoro-*p*-tolyl)urea, C 15935
Paragrano N **parathion** CTA
Paragrex I **parathion-methyl** Sadisa
Paragrin I **parathion + petroleum oils** Chimiberg
Parakakes R **diphacinone** Motomco
Parakakes R **pindone** Motomco
Parakey Metil I **parathion-methyl** Key
Paral Garten-Spray I **resmethrin** Thompson
Paralight H **paraquat dichloride** Pluess-Staufer
Paralindex I **lindane + parathion-methyl** La Quinoleine
Paraluq M I **parathion-methyl** Luqsa
Paramaag-Sommer IA **petroleum oils** Fattinger
Paramar IA **parathion**
Parameth IA **parathion-methyl** L.A.P.A.
Paramethyl I **parathion-methyl** Gefex

Paramin P I **parathion + petroleum oils** Aziende Agrarie Trento
Paramon IA **petroleum oils** Du Pont
Paramouss H **dichlorophen** Truffaut
Paran-E I **parathion** Diana
Paran-M I **parathion-methyl** Diana
Paranol G **decan-1-ol**
Para-Pac R **pindone** Ceva
Paraphene I **parathion** Rhone-Poulenc
Paraphos I **parathion** Diana
Parapin IA **petroleum oils** Pinus
paraquat dichloride H [1910-42-5] 1,1'-dimethyl-4,4'-bipyridylium dichloride, Cekuquat, Crisquat, Dextrone X, Esgram, Gramoxone, Herboxone, Osaquat, Pillarxone, PP 148, Scythe, Simpar, Speedway, Toxer Total
Paraquin H **paraquat dichloride** Scam
Paraquone H **paraquat dichloride** Eli Lilly
Parasitex I **carbaryl** Tatsiramos
Parasol F **copper hydroxide** Comac
Para-sommer IA **petroleum oils** Propfe, Staehler
Parasoufre FIA **parathion-methyl + sulphur** R.S.R.
Parasoufre acaricide FIA **dicofol + parathion-methyl + sulphur** R.S.R.
Parasoufre F FIA **folpet + parathion-methyl + sulphur** R.S.R.
Parat R **warfarin** Colkim
Parater H **paraquat dichloride** Terranalisi
Parathene IA **parathion**
parathion IA [56-38-2] *O,O*-diethyl *O*-4-nitrophenyl phosphorothioate, parathion-ethyl, ethyl parathion, thiophos, OMS 19, ENT 15108, AC 3422, Alkron, Alleron, Bladan, Corothion, E-605, Eftol, Ekatox, Ethyl Parathion, Etilon, Folidol E 605, Fosferno, Niran, Panthion, Paramar, Parathene, Penncap-E, Phoskil, Pox, Rhodiatox, Soprathion, Stathion, Vitrex
parathion-ethyl IA **parathion**
parathion-methyl IA [298-00-0] *O,O*-dimethyl *O*-4-nitrophenyl phosphorothioate , methyl parathion, metaphos, OMS 213, ENT 17292, Bladan M, Cekumethion, Folidol-M, Fulkil, Metacide, Nitrox, Paratox, Partron M, Penncap-M, Tekwaisa, Wofatox
Parathionex I **parathion** Permutadora
Parathol I **piperonyl butoxide + pyrethrins** Rentokil
Parathox-E I **parathion** Diana
Parathox-M I **parathion-methyl** Diana
Paratiao Reis I **parathion** Reis
Paratidol I **parathion + petroleum oils** ICI-Valagro
Paratil I **parathion** Sopepor
Paration IA **parathion** Pliva
Paratoil I **parathion + petroleum oils** Siapa
Paratoleo I **parathion + petroleum oils** Quimigal
Paratox IA **parathion-methyl** All-India Medical, Chimac-Agriphar, Van Wesemael, Visplant
Para-Weiss I **lindane + methoxychlor + petroleum oils** Staehler
Parax I **parathion-methyl** Argos
Parazin H **paraquat dichloride + simazine** Terranalisi
Parazone H **paraquat dichloride** Ellagret
Pardi H **diquat dibromide + paraquat dichloride** ICI
Pardner H **bromoxynil** Rhone Poulenc
Pared H **paraquat dichloride** Agrimont, Agrocros
Parenil metile I **parathion-methyl** Enotria
Parethyl I **parathion** Hellenic Chemical, Viopharm
Paretox IA **parathion** Bourgeois
Parexan IA **piperonyl butoxide + pyrethrins + rotenone** Hoechst, Pluess-Staufer
Parexan F IA **piperonyl butoxide + pyrethrins + S421** Hoechst
Paridol IA **parathion** Ciba-Geigy
Paridol T 20 I **parathion-methyl** Ciba-Geigy
Park + Herbicide H **2,4-D + dicamba** Barenbrug
Park Antimousse H **iron sulphate** Barenbrug
park extra H **chlorflurecol-methyl + MCPA + mecoprop** Du Pont
Park gazonmest met onkruidverdelger H **2,4-D + dicamba** Barenbrug
Park koolvlieg-weg I **chlorfenvinphos** Grondmix
Park met mosbestrijdingsmiddel H **iron sulphate** Barenbrug
Park onkruid-weg H **dichlobenil** Grondmix
park Rasenduenger + Unkrautvernichter H **chlorflurecol-methyl + MCPA** Du Pont
park Rasenduenger + UV neu H **2,4-D + dicamba** Du Pont
park Super H **dicamba + MCPA** Du Pont
Park uienvlieg-weg I **chlorfenvinphos** Grondmix

Park wortelvlieg-weg I **chlorfenvinphos** Grondmix
Parlay G **paclobutrazol** ICI, Sopra
Parmathion I **fenitrothion + fenvalerate**
Parmethyl I **parathion-methyl** Viopharm
Paroil IA **parathion-methyl + petroleum oils** Avenarius, Luqsa
Partner F **captan + maneb + zineb** ICI
Partron M IA **parathion-methyl**
Parzate F **zineb** Du Pont
Parzon I **cypermethrin + phosalone** Rhone-Poulenc
Pasta Caffaro F **copper oxychloride** Caffaro, Ergex
Pasta Lumachicida in Grani M **metaldehyde** Ital-Agro
Pasta Rameica F **copper oxychloride** Chimiberg
Pasturol H **dicamba + MCPA + mecoprop** FCC
Patafol F **mancozeb + ofurace** Sopra
Patafol Plus F **maneb + ofurace + zineb** ICI
Patap I **cartap hydrochloride** Takeda
Patatol Activado I **carbaryl + malathion** Agrocros
Patatol Croszintox IA **azinphos-methyl** Agrocros
Patatol Metazintox IA **azinphos-methyl** Agrocros
Pathclear H **amitrole + diquat dibromide + paraquatdichloride + simazine** ICI
Patolin H **linuron + monolinuron** Protex
Patoran H **metobromuron** Agrolinz, BASF, Budapesti Vegyimuvek, Chromos, Ciba-Geigy, Imex-Hulst, Ligtermoet, Liro, Sapec
Patoran Special H **metobromuron + metolachlor** Budapesti Vegyimuvek
Patrin IG **carbaryl**
Patrol F **fenpropidin** ICI, Maag
Patrole IA **methamidophos** Productos
Pattonex H **metobromuron** Makhteshim-Agan
Paxilon H **methazole** Sandoz
Pay-Off I **flucythrinate** Cyanamid
PBI Slug Pellets M **metaldehyde** Pan Britannica
PBI Spreader S **alkyl phenol ethoxylate** Pan Britannica
PCA H **chloridazon**
PCBPCPS A **chlorbenside**
PCNB F **quintozene**
PCP IFH **pentachlorophenol**
PCPA-Marks G **4-CPA** Anorgachim
PCPBS A **fenson**
PCQ R **diphacinone** Bell Labs
PD 5 IA **mevinphos** Celamerck, Schering, Shell
PDD I **diflubenzuron** Duphar
PDU H **fenuron**
Peaweed H **prometryn + terbutryn** Pan Britannica
PEBC H **pebulate**
pebulate H [1114-71-2] *S*-propyl butyl(ethyl)thiocarbamate, PEBC, R-2061, Tillam
Pecotot Schneckenkorn Feingranulat mit VPA M **metaldehyde** Pfeiffer Glanzit
pedinex AIM **dinex**
Pedroxina F **oxine-copper** Sopepor
Pekeldrin A **dicofol + tetradifon** Ciba-Geigy
Pel-sel H **2,4-D** CARAL
Pell-Pac R **pindone** Ceva
Pellacol X **thiram** Agrolinz, Plantevern-Kjemi
Pelousonet H **2,4-D + mecoprop** Colvoo
Pelt F **thiophanate-methyl** Procida
Pelt Sol F **thiophanate** Procida
Peltar F **maneb + thiophanate-methyl** Procida
Peltis F **thiophanate-methyl** Procida
Peltisan F **maneb + sulphur + thiophanate-methyl** Procida
Pencal IH **calcium arsenate**
Penchloral IFH **pentachlorophenol**
penconazole F [66246-88-6] 1-(2,4-dichloro-β-propylphenethyl)-1*H*-1,2,4-triazole, Award, CGA 71818, Onmex, Topas, Topaz, Topaze
pencycuron F [66063-05-6] 1-(4-chlorobenzyl)-1-cyclopentyl-3-phenylurea, Bay NTN 19701, Monceren
Pendic H **chlorotoluron + pendimethalin** Cyanamid, Liro
pendimethalin H [40487-42-1] *N*-(1-ethylpropyl)-2,6-dinitro-3,4-xylidine, pendimethaline, penoxalin, AC 92553, Accotab, Go-Go-San, Herbadox, Prowl, Sipaxol, Stomp, WayUp
pendimethaline H **pendimethalin**
Pendiron H **chlorotoluron + pendimethalin** Ciba-Geigy, Cyanamid
Pennant H **metolachlor** Ciba-Geigy
Penncap E IA **parathion** Pennwalt
Penncap M IA **parathion-methyl** Decco-Roda, Grima, Kwizda, Pennwalt, Sandoz, Sopra, SPE

Penncapthrin IA **permethrin** Desarrollo Quimico

Penncapthrine I **permethrin** Pennwalt France

Penncozeb F **mancozeb** Elanco, Pennwalt, Pennwalt France, Pennwalt Holland, Sanac, Shell

Pennflo F **mancozeb** Agrishell

Penngar lad I **diazinon** Pennwalt France

Pennmanzone F **ferbam + mancozeb** Pennwalt

Pennout HLG **endothal** Pennwalt

Pennside IA **parathion** Pennwalt France

Pennthal H **endothal** Siapa

Penntox MS I **parathion-methyl** Siapa

penoxalin H **pendimethalin**

penta IFH **pentachlorophenol**

Pentac A **dienochlor** Argos, Broeste, Dow, Farmos, Hellafarm, La Quinoleine, Ligtermoet, Liro, Plantevern-Kjemi Sandoz, Schering, Stinnes

pentachlorophenol IFH [87-86-5] pentachlorophenol, PCP, penta, Penchloral, Pentacon, Penwar, Sinituho

Pentaclor Z I **bioallethrin + methoxychlor + piperonyl butoxide + trichlorfon** Zucchet

Pentacon IFH **pentachlorophenol**

Pentagan G **chlormequat chloride + choline chloride** COPCI

Pentagen F **quintozene** Mitsui Toatsu, Sankyo

pentanochlor H [2307-68-8] 3'-chloro-2-methylvaler-*p*-toluidide, solan, CMMP, CMA, Dakuron, Dutom, FMC 4512, Solan

Pentasol NF A **dicofol + tetradifon** Eli Lilly

Penter I **parathion-methyl + petroleum oils** Baslini

Penwar IFH **pentachlorophenol**

Peprol I **trichlorfon** Condor

Peptoil AIH **petroleum oils** Drexel

Per-Sintol F **bis(ethoxydihydroxydiethylamino) sulphate** Morera

Peral vigne H **diuron + simazine** Sipcam-Phyteurop

Peran F **zineb** Efthymiadis

perchlorobenzene F **hexachlorobenzene**

Percide I **perthane** Aziende Agrarie Trento

Perenot Vite F **calcium copper oxychloride + zineb** Enotria

Perenox F **cuprous oxide** ICI

Perfect IA **malathion + piperonyl butoxide + pyrethrins** Johansson, Volcke

Perfektan I **lindane** BASF

Perfekthion IA **dimethoate** Agrolinz, BASF, Compo, Intrachem, Maag, Sapec

Perflan H **tebuthiuron** Elanco

perfluidone H [37924-13-3] 1,1,1-trifluoro-*N*-(4-phenylsulphonyl-*o*-tolyl)methanesulphonamide, Destun, MBR-8251

Pericide R **difenacoum** Diffusion

Perigen I **permethrin** Wellcome

Perizin pour abeilles A **coumaphos** Bayer

Perlka FH **calcium cyanamide** SKW Trostberg

Permanent I **permethrin + pyrethrins** Europlant, Kwizda

Permasect I **permethrin** Afrasa, Brinkman, Kemira, Megafarm, Mitchell Cotts, Sandoz, Tuhamij

Permax I **permethrin** Edialux

permethrin I [52645-53-1] 3-phenoxybenzyl (1*RS*)-*cis,trans*-3-(2,2-dichlorovinyl)-2,2-dimethylcyclopropanecarboxylate, permethrine, OMS 1821, Ambush, Atroban, Coopex, Corsair, Dragon, Ectiban, Eksmin, FMC 33297, Imperator, Kafil, LE 79-519, NRDC 143, Outflank, Perigen, Permasect, Permit, Perthrine, Picket, Pounce, PP 557, Pramex, Pynosect, Quamlin, Stomoxin, Talcord, Torpedo

permethrine I **permethrin**

Permetrin I **permethrin** Mostyn

Permilan F **zineb** Agrolinz

Permit I **permethrin** Pan Britannica

Permussenito F **sodium arsenite** Permutadora

Permutex I **carbaryl** Permutadora

Permutine I **thiometon** Permutadora

Permuzan F **quintozene** Permutadora

Perocin F **zineb** Agria

Perocur F **cymoxanil + mancozeb** Epro

Perolan-Super F **copper oxychloride + folpet** Pluess-Staufer

Peronal F **copper oxychloride + zineb** Meoc

Peronal Super F **copper oxychloride + folpet** Meoc

Perontan F **zineb** Kwizda

Perontan ZMF F **ferbam + maneb + zineb** Kwizda

Peropal A **azocyclotin** Agroplant, Bayer

Perosporine Super Blue F **copper oxychloride + zineb** Hellenic Chemical

Perozin F **zineb** Agria, Nitrofarm
Persevtox H **dinoseb** La Quinoleine
Persulon F **fluotrimazole** Bayer
perthane I [72-56-0] 1,1-dichloro-2,2-bis(4-ethylphenyl)ethane, ethylan, Perthane
Perthrine I **permethrin** ICI, Siegfried, Sopra
Pertubir G **chlorpropham + propham** Pliva
Peruran H **atrazine + diuron + simazine** Spiess, Urania
Pervit Bleu F **copper oxychloride + zineb** Agronova
Pesco H **MCPA + 2,3,6-TBA** Schering
Pescolan FX **thiram** Afrasa
Pesconex H **MCPA + mecoprop + 2,3,6-TBA** Schering
Pesguard I **phenothrin + tetramethrin** Sumitomo
Pest Master FHIN **chloropicrin + methyl bromide** Kondovas
Pest pel R **difenacoum** Chemsearch
Pest-Oil IS **petroleum oils** Du Pont
Pestan IA **mecarbam** Takeda
Pestox 3 I **schradan**
Pestox 15 IA **mipafox**
Pestox III I **schradan**
Pestox XIV IA **dimefox**
Petafol F **captafol + folpet** Caffaro
Petran H **trifluralin** Montenegroexport
Petrisan IA **petroleum oils** Bayer
petroleum oils AIHS mineral oils, white oils, paraffin oils, adjuvant oils, spray oils, Actipron, Agrol Plus, Fyzol, Peptoil, Volck
PF-50 I **parathion** Sapec
Pflanzen-Paral fuer Gartenpflanzen IF **cetoctaelate + resmethrin** Thompson
Pflanzen-Paral fuer Rosen IF **cetoctaelate + resmethrin** Thompson
Pflanzen-Paral fuer Topfpflanzen IA **piperonyl butoxide + pyrethrins** Thompson
Pflanzen-Paral gegen Blattlaeuse I **butocarboxim** Thompson, Wacker
Pflanzen-Paral gegen Blattlaeuse und Pilzkrankheiten FI **butocarboxim + fenarimol** Thompson
Pflanzen-Paral gegen Pilzkrankheiten F **fenarimol** Thompson
Pflanzen-Paral Pflanzenschutz-Zaepfchen gegen Blattlaeuse und Spinnmilben IA **butoxycarboxim** Thompson
Pflanzen-Paral Spritzmittel gegen Pflanzenschaedlinge I **permethrin** Thompson
Pflanzen-Paral Spruehschutz gegen Blattlaeuse I **butocarboxim** Thompson
Pflanzen-Paral Spruehschutz gegen Pilzkrankheiten F **fenarimol** Thompson
Pflanzen-Paral Universal Spritzmittel gegen Pflanzenschaedlinge I **permethrin** Thompson
Pflanzen-Schaedlings-frei I **bromophos** Flora-Frey
Pflanzen-Spray FIA **lecithin + piperonyl butoxide + pyrethrins** Flora-Frey
Pflanzenfreund Rasenduenger mit Unkrautvernichter H **2,4-D + MCPA** Scipio
Pflanzenspray IA **dimethoate** GABI-Biochemie
Pflanzenspray Hortex IA **lindane + petroleum oils** Shell
Pflanzol-Kaltnebel I **dicofol + lindane + piperonyl butoxide +pyrethrins** Delicia
Pflanzol-Rosenspray IA **dicofol + lindane + piperonyl butoxide +pyrethrins** Delicia
Pflanzol-Spray IA **dicofol + lindane + piperonyl butoxide +pyrethrins** Delicia
PG 53 H **dinoseb** EniChem
PH 60-40 I **diflubenzuron** Duphar
Phacira R **chlorophacinone** Valmi
Phaltan F **folpet** Agroplant, Chevron, Geopharm, Pluess-Staufer, Sandoz, Siegfried
Phaltane F **folpet** Bayer
Phaltocide F **folpet** Ciba-Geigy, Rhodiagri-Littorale
Phaltocuivre BX F **copper sulphate + folpet** Rhodiagri-Littorale
Phaltozid F **folpet** Ciba-Geigy
Pharorid I **methoprene** Sandoz
PHC I **propoxur**
Phenacide IRA **camphechlor**
Phenador-X F **biphenyl**
phenaminosulf F **fenaminosulf**
phenamiphos N **fenamiphos**
Phenatol A **phenkapton**
Phenatox IRA **camphechlor**
phencapton A **phenkapton**
phencyclate I **cycloprothrin**
phenetacarbe I **fenethacarb**
phenisobromolate A **bromopropylate**
phenisopham H [57375-63-0] isopropyl 3-(*N*-ethyl-*N*-phenylcarbamoyloxy)phenylcarbamate, phenisophame, Diconal, SN 58132, Verdinal

phenisophame H **phenisopham**
phenkapton A [2275-14-1] *S*-(2,5-dichlorophenylthiomethyl) *O,O*-diethyl phosphorodithioate, phencapton, Phenatol, Phenudin
phenmedipham H [13684-63-4] methyl 3-(3-methylcarbaniloyloxy)carbanilate, phenmediphame, Alegro, Beetomax, Beetup, Betanal, Betanal E, Betosip, Fender, Goliath, Gusto, Kemifam, Pistol, Protrum K, SN 38584, Spin-aid, Vangard
phenmedipham-ethyl H [13684-44-1] 3-ethoxycarbonylaminophenyl 3'-methylcarbanilate, phenmediphame-ethyl, SN 38574
phenmediphame H **phenmedipham**
phenmediphame-ethyl H **phenmedipham-ethyl**
phenobenzuron H [3134-12-1] 1-benzoyl-1-(3,4-dichlorophenyl)-3,3-dimethylurea, Benzomarc, PP 65-25
Phenopal H **fenoprop** Anagnostopoulos, UCP
Phenoseptyl special champignonnieres F **2-phenylphenol** Sicca
Phenostat-A FLM **fentin acetate** Nitto Kasei
Phenostat-H F **fentin hydroxide** Nitto Kasei
Phenotan H **dinoseb acetate** Pepro
Phenoterb H **dinoterb + nitrofen** Pepro
phenothiol H **MCPA-thioethyl**
Phenothion I **carbophenothion** Hellenic Chemical
phenothrin I [26002-80-2] 3-phenoxybenzyl 2,2-dimethyl-3-(2-methylprop-1-enyl)cyclopropanecarboxylate, phenothrine, d-phenothrin, OMS 1810, ENT 27972, Multicide Concentrate F-2271, S-2539, Sumithrin
phenothrine I **phenothrin**
Phenoxylene H **MCPA** Efthymiadis, Schering
phenthoate IA [2597-03-7] *S*-α-ethoxycarbonylbenzyl *O,O*-dimethyl phosphorodithioate, PAP,. dimephenthoate, OMS 1075, ENT 27386, Aimsan, Cidial, Elsan, L-561, Papthion, S-2940, Tanone
Phenudin A **phenkapton**
Phenylbenzene F **biphenyl**
phenylmercury acetate F [62-38-4] phenylmercury acetate, acetate de phenylmercure, PMA, Agrosan, Cekusil, Celmer, Ceresol, Hong Nien, Mersolite, Pamisan, Phix, PMAS, Seedtox, Unisan
phenylmercury chloride F [100-56-1] chlorophenylmercury, Stopspot
2-phenylphenol F [90-43-7] biphenyl-2-ol, phenyl-2 phenol, Dowicide 1, Nectryl
Pherocon T **pheromones** Zoecon
Pheroprax T **pheromones** Sovilo
Philocap F **captan** Filocrop
Phior P W **Trichoderma harzanium** Orsan
Phix F **phenylmercury acetate**
Pholozim F **carbendazim + maneb** Hellenic Chemical
Phomopsin F **folpet** Efthymiadis
phorate IAN [298-02-2] *O,O*-diethyl *S*-ethylthiomethyl phosphorodithioate, timet, ENT 24042, AC 3911, Agrimet, Geomet, Granutox, Rampart, Thimenox, Thimet, Vegfru Foratox
phosacetim R [4104-14-7] *O,O*-bis(4-chlorophenyl) *N*-acetimidoylphosphoramidothioate, phosacetime, Bay 38819, Gophacide
phosacetime R **phosacetim**
phosalone IA [2310-17-0] *S*-6-chloro-2,3-dihydro-2-oxobenzoxazol-3-ylmethyl *O,O*-diethyl phosphorodithioate, ENT 27163, 11974 RP, Azofene, Rubitox, Zolone
Phosan Plus I **dimethoate + malathion + methoxychlor** Chimac-Agriphar
phosdiphen F [36519-00-3] bis(2,4-dichlorophenyl) ethyl phosphate, MTO-460
Phosdrin I **mevinphos** Agrishell, Agroplant, Sandoz, Shell, Siegfried, Spolana, Szovetkezet
phosethyl Al F **fosetyl-aluminium**
Phosfleur G **chlorphonium chloride** Hortus, Perifleur, Schering
phosfolan IA [947-02-4] diethyl 1,3-dithiolan-2-ylidenephosphoramidate, phospholan, OMS 646, ENT 25830, Cyalane, Cylan, Cyolan, Cyolane, EI 47031
Phosfon G **chlorphonium chloride** Meier
Phoskil IA **parathion**
Phoslit IA **mevinphos** La Littorale
phosmet IA [732-11-6] *O,O*-dimethyl *S*-phthalimidomethyl phosphorodithioate, PMP, phtalofos, ENT 25705, Appa, Imidan, Kemolate, Prolate, R-1504
phosphamidon IA [13171-21-6] 2-chloro-2-diethylcarbamoyl-1-methylvinyl dimethyl phosphate, OMS 1325, ENT 25515, Apamidon, C 570, Dimecron, Phosron

phosphine IR [7803-51-2] phosphine, Al-Phos (aluminium phosphide), Alutal, Celphide (aluminium phosphide), Celphine (aluminium phosphide), Celphos (aluminium phosphide), Delicia-Gastoxin (aluminium phosphide), Detia-Gas-Ex (aluminium phosphide), Detiaphos (magnesium phosphide), Fumi-Cel (magnesium phosphide), Fumi-Strip(magnesium phosphide), Fumitoxin (aluminium phosphide), Magtoxin (magnesiumphosphide), Phostoxin (aluminium phosphide), Polytanol (calcium phosphide), Quickphos (aluminium phosphide)
Phosphit IA **dichlorvos** Nippon Soda
phospholan IA **phosfolan**
phosphure de zinc R **zinc phosphide**
Phosron IA **phosphamidon** Hui Kwang
Phostek I **aluminium phosphide** Agrar-Speicher
Phostoxin IR **aluminium phosphide** Anticimex, Breymesser, Colkim, Degesch, DGS, Geopharm, Heerdt-Lingler, Heydt Bona, Rentokil, Skadedyr
Phostoxin Mg I **magnesium phosphide** Heydt Bona
Phostoxin-Plates I **magnesium phosphide** DGS
Phosulphon IA **demeton-S-methyl sulphone**
Phosvel I **leptophos**
Phosvin R **zinc phosphide**
Phosvit IA **dichlorvos** Nippon Soda
Phoundzex F **cuprammonium** Efthymiadis
phoxim I [14816-18-3] *O,O*-diethyl *α*-cyanobenzylideneamino-oxyphosphonothioate, phoxime, OMS 1170, Bay 5621, Bay 77488, Baythion, Volaton
phoxime I **phoxim**
Phroutolan F **copper oxychloride + sulphur + zineb** Agrotechnica
phtalofos IA **phosmet**
phthalide F **fthalide**
phthalthrin I **tetramethrin**
Phygon F **dichlone** Aveve, Uniroyal
Phynazol G **chlormequat chloride + ethephon** Bitterfeld
Phyomone G **1-naphthylacetic acid** ICI, ICI-Zeltia
Phytactine G **kinetin** Timac
Phytar H **dimethylarsinic acid** Crystal, Vertac
Phyto-Medipham H **phenmedipham** Phyto
Phyto-Pyrazol H **chloridazon** Phyto
Phyto-toluron H **chlorotoluron** Phyto
Phytocape F **captan** Bayer
Phytochlor F **quintozene** Sipcam-Phyteurop
Phytocide H **ammonium sulphamate** Truffaut
Phytocuivre F **copper hydroxide** Ciba-Geigy
Phytocur V IA **dimethoate** Bayer
Phytokupfer F **copper hydroxide** Ciba-Geigy
Phytolan F **copper oxychloride + zineb** Tsilis
Phytomalatox I **malathion** Phytopharmaceutiki
Phyton-27 LF **copper sulphate** Source Technology Biologicals
Phytonic G **1-naphthylacetamide** Leu & Gygax
Phytosol I **trichloronat** Bayer
Phytosoufre F **sulphur** Bayer
Phytosyl Broussaille H **2,4-D + 2,4,5-T** Protex
Phytox F **zineb** Staehler
Phytox + Ultraschwefel F **sulphur + zineb** Staehler
Phytox-M F **maneb + zineb** Staehler
Phytox Mz 80 F **mancozeb** Chimiberg
Phytox Rame F **copper oxychloride + zineb** Chimiberg
Phytox-Super F **metiram** Staehler
Phytoxone H **2,4-D + mecoprop** Agroplant
Phytoxone DNBP H **dinoseb** Van Wesemael
PI 63 F **copper oxychloride + zineb** ICI
Pibutox I **kadethrin + permethrin** Organika-Fregata
Pibutrin Insecticida No. 44 I **piperonyl butoxide + pyrethrins** Cassels
Pic-Clor NIFRH **chloropicrin**
Picket I **permethrin** ICI
picloram H [1918-02-1] 4-amino-3,5,6-trichloropyridine-2-carboxylic acid, piclorame, Grazon, Tordon
piclorame H **picloram**
Picrin NIFRH **chloropicrin** Mitsui Toatsu
Pictyl I **fenoxycarb** Maag
Piedor F **carbendazim** R.S.R.
Pielik H **2,4-D** Organika-Rokita
Pielisam H **linuron + methabenzthiazuron + terbutryn** Organika-Zarow

Pika-Radar I **perthane + piperonyl butoxide + pyrethrins** Farmos
Pillarcap F **captan** Pillar
Pillardrin IA **monocrotophos** Pillar
Pillarfuran IAN **carbofuran** Pillar
Pillargon I **propoxur** Pillar
Pillarsete H **butachlor** Pillar
Pillarstin F **carbendazim** Pillar
Pillartan F **captafol** Pillar
Pillarxone H **paraquat dichloride** Pillar
Pillarzo H **alachlor** Pillar
Pilot H **quizalofop-ethyl** Schering
pimaricin F [7681-93-8] (8*E*,14*E*,16*E*,18*E*,20*E*)-(1*S*,3*R*,5*S*,7*S*,12*R*,24*R*,25*S*,26*R*)-22-(3-amino-3,6-dideoxy-*β*-D-mannopyranosyloxy)-1,3,26-trihydroxy-12-methyl-10-oxo-6,11,28-trioxatricyclo[22.3.1.0^{5,7}]octacosa-8,14,16,18,20-pentaene-25-carboxylicacid, tennecetin, natamycin, myprozine, Delvolan
Pinazon I **diazinon** Pinus
pindone R [83-26-1] 2-pivaloylindan-1,3-dione, pivaldione, Contrax-P, Duocide, Para-Pac, Parakakes, Pell-Pac, Pivacin, Pival, Pivalyn, Tri-Ban
Pinodrin R **endrin** Pinus
Pinofon A **tetradifon** Pinus
Pinofos I **trichlorfon** Pinus
Pinozeb F **mancozeb** Pinus
Pinulin F **vinclozolin** Pinus
Pio F **polyoxins**
Piomy F **polyoxins**
Piomycin F **polyoxin B**
piperalin F 3-(2-methylpiperidino)propyl 3,4-dichlorobenzoate, Pipron
piperonyl butoxide S [51-03-6] 5-[2-(2-butoxyethoxy)ethoxymethyl]-6-propyl-1,3-benzodioxole, piperonyl butoxyde, ENT 14250, Butacide, Prentox, Pybuthrin
piperonyl butoxyde S **piperonyl butoxide**
piperophos H [24151-93-7] *S*-2-methylpiperidinocarbonylmethyl *O,O*-dipropyl phosphorodithioate, C 19490, Rilof
piproctanyl bromide G [56717-11-4] 1-allyl-1-(3,7-dimethyloctyl)piperidinium bromide, piproctanylium bromide, Alden, Ro 06-0761, Stemtrol
piproctanylium bromide G **piproctanyl bromide**
Pipron F **piperalin** Elanco
piprotal S [5281-13-0] 5-[bis[2-(2-butoxyethoxy)ethoxy]methyl]-1,3-benzodioxole, heliotropin acetal, Tropital
Pirazone H **chloridazon** Chemia
Pirenyl E I **piperonyl butoxide + pyrethrins** Ravit
Piresintal I **permethrin** Chimiberg, Sepran
Pirigrain I **pirimiphos-methyl** C.G.I.
Pirigrain choc I **dichlorvos** C.G.I.
Pirigrain plus IA **dichlorvos + pirimiphos-methyl** C.G.I.
Pirigrain port IA **dichlorvos + pirimiphos-methyl** C.G.I.
pirimicarb I [23103-98-2] 2-dimethylamino-5,6-dimethylpyrimidin-4-yl dimethylcarbamate, pyrimicarbe, ENT 27766, Abol, Aficida, Aphox, Fernos, Pirimor, PP 062, Rapid
pirimiphos-ethyl I [23505-41-1] *O*-2-diethylamino-6-methylpyrimidin-4-yl *O,O*-diethyl phosphorothioate, pyrimiphos-ethyl, Fernex, PP 211, Primicid, Primotec, Solgard
pirimiphos-methyl IA [29232-93-7] *O*-2-diethylamino-6-methylpyrimidin-4-yl *O,O*-dimethyl phosphorothioate, pyrimiphos-methyl, OMS1424, Actellic, Actellifog, Blex, Pirigrain, PP 511, Silosan, Sybol 2
Pirimol I **pirimicarb** ICI
Pirimor I **pirimicarb** Berner, Deletia, ICI, ICI-Valagro, Lucebni zavody, Maag, Organika-Azot, Pinus, Plantevern-Kjemi, Schering, Sopra, Spiess, SPU, Urania
Pirimouche I **cypermethrin** C.G.I.
Pirimouche I **pirimiphos-methyl** C.G.I.
Piripin H **propanil** Pinus
Pirital I **piperonyl butoxide + pyrethrins** Ital-Agro
Pirox fluid FIA **diazinon + tetradifon + triforine** Ruse
Pirox neu FI **endosulfan + sulphur + zineb** Agro
Pirox-Spray FI **chlorbenside + dinocap + malathion + mancozeb + methoxychlor** Agro
Piruvel H **metoxuron** Hanson & Moehring
Pisec H **diuron + propham** Chemia
Pist'Operats R **warfarin** C.N.C.A.T.A.
Pistol H **phenmedipham** ABM Chemicals
Pitezin H **atrazine** Pitesti
Pivacin R **pindone**
Pival R **pindone** Motomco
pivaldione R **pindone**
Pivalyn R **pindone** Motomco
Pivot H **imazethapyr** Cyanamid
Pix G **mepiquat chloride** BASF, Intrachem

PKhNB F **quintozene**
PL-80 I **lindane** Inagra
Placusan F **copper oxychloride + copper sulphate + maneb + zineb** Platecsa
Plakin H **asulam** Chinoin
Plamazina H **simazine** Platecsa
Planavin H **nitralin** Shell
Plant Pin IA **butoxycarboxim** Dow, KenoGard, Luxan, Wacker
Plant-O Aerosol FI **dinocap + fenitrothion + maneb + pyrethrins** Efthymiadis
Plantdrin IA **monocrotophos** Planters Products
Plantect F **thiabendazole** Ewos
Plantenspray L 77/1217 IA **piperonyl butoxide + pyrethrins** Enna
Plantenspray L 80/1647 I **phenothrin + tetramethrin** Enna
Plantex H **sodium chlorate** Schacht
Plantineb F **maneb** Hoechst, Roussel, Roussel Uclaf
Plantinebe F **maneb** Roussel Hoechst
Plantisoufre F **sulphur** Procida
Plantomycin B **streptomycin** ICI
Plantonit H **terbutryn** Chemolimpex
Plantox IA **piperonyl butoxide + pyrethrins** Eurofill
Plantulin H **ametryn + nitrofen** Bitterfeld
Plantvax F **oxycarboxin** Cillus, Hellafarm, ICI-Valagro, ICI-Zeltia, Kemi-Intressen, Kwizda, La Quinoleine, Siapa, Uniroyal
Plantvax C F **captan + oxycarboxin** Uniroyal
Plantvax M F **mancozeb + oxycarboxin** Uniroyal
Plavi kamen F **copper sulphate** Zupa
Plictet A **cyhexatin** Dow
Plictran A **cyhexatin** Agro, Bjoernrud, Chromos, Condor, Dow, Duphar, ICI-Zeltia, Inca, Maag, Nitrokemia, Nordisk Alkali, Prochimagro, Rhodiagri-Littorale, Rhone-Poulenc, Sandoz, Schering AAgrunol, Siapa, Umupro
Plidion A **cyhexatin + tetradifon** Chemia
plifenate I [21757-82-4] 2,2,2-trichloro-1-(3,4-dichlorophenyl)ethyl acetate, benzethazet, BAY MEB 6046, Baygon MEB
PLK-Betafam H **phenmedipham** PLK
PLK-D 500 H **2,4-D** PLK
PLK-DP 640 H **dichlorprop** PLK
PLK-DPD 667 H **2,4-D + dichlorprop** PLK
PLK-DPM 750 H **dichlorprop + MCPA** PLK
PLK flue-aerosol til stalde I **piperonyl butoxide + resmethrin** PLK
PLK-fluedraeber I **piperonyl butoxide + pyrethrins** PLK
PLK-Kobberoxychlorid F **copper oxychloride** PLK
PLK-M 750 H **MCPA** PLK
PLK-MH 300 HG **maleic hydrazide** PLK
PLK-MP 500 H **mecoprop** PLK
PLK-MPD 575 H **2,4-D + mecoprop** PLK
PLK-Pyrethrum mod fluer I **piperonyl butoxide + pyrethrins** PLK
PLK-Sproejtesvovl F **sulphur** PLK
PLK-Trifocid H **DNOC** PLK
PLK-Trimangol F **maneb** PLK
PLK-Vondocarb F **carbendazim + maneb + zineb** PLK
PLK-Vondozeb F **maneb + zineb** PLK
Plondrador F **carbendazim + ditalimfos** Prochimagro
Plondrel F **ditalimfos** Dow, Kwizda
Plonemil F **benomyl + ditalimfos** Zorka Sabac
Pludgerm G **chlorpropham** SCAC-Fisons
Pludiserb H **simazine** Montedison
Pluevel H **dicamba + MCPA** Pluess-Staufer
Plural PS F **copper compounds (unspecified) + folpet +sulphur** Sofital
Pluraline H **trifluralin** Sedagri
Plurisol H **atrazine + simazine** Sedagri
Plydal-E H **chlorpropham + fenuron + propham** Grima
PMA F **phenylmercury acetate**
PMAS F **phenylmercury acetate** Cleary
PMP IA **phosmet**
Poast H **sethoxydim** BASF
Podquat G **chlormequat chloride + di-1-p-menthene** Mandops
Pokon anti-mos H **iron sulphate** Bendien
Pokon anti-slakken M **metaldehyde** Bendien
Pokon Gazonmest extra met onkruidverdelger H **2,4-D + dicamba** Bendien
Pokon Meeldauw Spray F **pyrazophos** Bendien
Pokon Mehltau Spray F **pyrazophos** Braun
Pokon Mildew Spray F **pyrazophos**
Pokon mosaeydir H **ammonium sulphate + iron sulphate** Bendien
Pokon Plantenspray IA **piperonyl butoxide + pyrethrins** Bendien, Braun

Pokon plantenspray plus I **deltamethrin** Bendien
Pokon plantespray I **pyrethrins + rotenone** Praestrud & Kjeldsmark
Pokon plontuudi I **piperonyl butoxide + pyrethrins** Bendien
Pokon stekpoeder G **1-naphthylacetic acid** Bendien
Pokon Vaextspray I **piperonyl butoxide + pyrethrins** Stuifbergen
Pol-Acaritox A **tetradifon** Azot
Pol-Akaritox A **tetradifon** Organika-Azot
Pol-Enolofos I **chlorfenvinphos** Organika-Azot
Pol-Foschlor I **trichlorfon** Azot
Pol-Kupraman F **copper oxychloride + maneb** Organika-Azot
Pol-Kupritox F **copper oxychloride** Organika-Azot
Pol-Metox I **methoxychlor** Organika-Azot
Pol-Owadofos I **fenitrothion** Azot
Pol-Pielik H **2,4-D** Organika-Rokita
Pol-Sulcol Extra F **sulphur** Sajina
Pol-Sulkol Extra F **sulphur** Organika-Sarzyna
Pol-Thiuram F **thiram** Organika-Azot
Poladan H **dalapon-sodium** EniChem
Polado H **glyphosate** Monsanto
Polaris G **glyphosine**
Polbor F **Bordeaux mixture** Eli Lilly
Poliagua S **polyacrylamide** Meristem
Poliba FI **barium polysulphide** Scam
Polibar FI **barium polysulphide** Bario
Policar MZ F **mancozeb** Cequisa, Interfertil, Phytorgan, Pluess-Staufer, Visplant
Policritt F **carbendazim** Siapa
Policritt C F **maneb + tridemorph** Siapa
Policritt M F **carbendazim + maneb** Siapa
Polimal I **malathion** Agrotechnica
Polirend GI G **gibberellic acid** KenoGard
Polisar MT I **parathion-methyl** EniChem
Polisolfuro di bario FI **barium polysulphide**
Polisolfuro di calcio FIA **lime sulphur**
Polisulfura de Bariu IAF **barium polysulphide**
Polisulfura de Calciu F **lime sulphur**
Polisulfuro de Bario IAF **barium polysulphide**
Politrin I **cypermethrin** Ciba-Geigy
Politrin N IA **cypermethrin + monocrotophos** Ciba-Geigy
Politrin P IA **cypermethrin + profenofos** Ciba-Geigy
Pollux R **zinc phosphide** O.P., Wuelfel
Polsol FI **barium polysulphide** Sivam
Poltan F **Bordeaux mixture** Bimex
Poltiglia F **Bordeaux mixture** Agrimont, Manica, Scam, Terranalisi
Poltiglia Z F **copper sulphate + zineb** Manica
Polvere F **copper oxychloride** Caffaro, EniChem, Ergex
Polvo Cuprico F **copper oxychloride** Agrocros, Probelte, Sadisa
Polvosol Cupro F **copper oxychloride + sulphur** Sivam
Polvosol F F **folpet + sulphur** Sivam
Polybarit FA **barium polysulphide** Spolana
Polycarbazin F **metiram** MCU
polychlorcamphene IRA **camphechlor**
Polycron IA **profenofos** Ciba-Geigy
Polycur F **cymoxanil + metiram** Sapec
Polydial S I **cypermethrin + diazinon** Ciba-Geigy
Polykarbacin F **metiram** Sojuzchimexport
Polymone HG **dichlorprop** Universal Crop Protection
Polymone 60 H **2,4-D + mecoprop** Sopra
Polymone Supra H **2,4-D + mecoprop** ICI
Polymone X H **2,4-D + dichlorprop** UCP
Polyoxin AL F **polyoxin B** Nihon Nohyaku
polyoxin B F **polyoxins**
polyoxin D F **polyoxins**
Polyoxin Z F **polyoxin D zinc salt**
polyoxins F [19396-06-6 (polyoxin B); 22976-86-9 (polyoxin D)] 5-(2-amino-5-*O*-carbamoyl-2-deoxy-L-xylonamido)-5-deoxy-1-(1,2,3,4-tetrahydro-5-hydroxymethyl-2,4-dioxopyrimidinyl)-*β*-D-allofuranuronic acid (polyoxin B), polyoxin B, polyoxin D, Kakengel (polyoxin D zincsalt), Pio, Piomy, Piomycin (polyoxin B), Polyoxin AL (polyoxin B), Polyoxin Z(polyoxin D zinc salt)
Polyram F **metiram** Agrolinz, BASF, Compo, Siegfried, Shell
Polyram Combi F **metiram** BASF, Intrachem, Nitrokemia, Sapec
Polyram Combi-M F **maneb + metiram** Luxan
Polyram Combi-MS F **maneb + metiram + sulphur** Luxan
Polyram-Kupfer F **copper oxychloride + metiram** BASF, Siegfried
Polyram M F **maneb** BASF
Polyram-Nospor F **copper oxychloride + metiram** Siegfried

Polyram Ultra FX **thiram** BASF
Polyram-Z F **zineb** BASF
polysulfure de calcium FIA **lime sulphur**
Polytanol R **calcium phosphide** Chemia, CTA, Wuelfel
Polythanol R **calcium phosphide** Wuelfel
Polytox H **bromacil + 2,4-D + diuron + simazine** Schering
Polytril H **amitrole + atrazine** Ciba-Geigy
Polytrin I **cypermethrin** Ciba-Geigy, Ligtermoet
Polyxane L IA **parathion** Agriben
Polzopin M **metaldehyde** Pinus
Pomarsol FX **thiram** Agroplant, Bayer, Berner
Pomarsol Z F **ziram** Bayer
Pomex I **carbaryl** Agrochemiki, Siapa
Pommetrol G **chlorpropham + propham** Decco-Roda, Fletcher
Pomodorin F **captan** Platecsa
Pomona Floks IA **diazinon** Organika-Fregata
Pomonit G **1-naphthylacetic acid** IPO-Sarzyna
Pomorol I **petroleum oils** Permutadora
Pomoroleo I **parathion + petroleum oils** Permutadora
Pomoxon G **1-naphthylacetic acid** Novotrade
Pomuran F **captan + mancozeb** Spiess
Ponnax F FG **carbendazim + chlormequat chloride + cholinechloride** BASF
Portman Isotop H **isoproturon** Portman
Portman Superquat G **chlormequat chloride + choline chloride** Portman
Portman Tri-Lin H **linuron + trifluralin** Portman
Posidor I **dimethoate + endosulfan** Roussel Hoechst
Posse I **carbosulfan** FMC, Zupa
Post-Kite H **ioxynil + isoproturon + mecoprop** Schering
Potablan H **monalide** Schering
potassium cyanate H [590-28-3] Aero Cyanate
Potivin H **cyanazine + linuron** Shell
Poudrage nexion I **bromophos** Sovilo
Poudramor R **coumachlor** Salomez
Poudre C Mortis R **chlorophacinone** Sovilo
Poudre siccative F **copper sulphate + sulphur** R.S.R.
Pounce I **permethrin** FMC
Power Manplex F **mancozeb** Power
Power Platoon G **ethephon + mepiquat chloride** Power
Power Spray-Save S **tallow amine ethoxylate** Power
Pox IA **parathion** Chimiberg
Pox M I **parathion-methyl** Chimiberg
PP 005 H **fluazifop-P-butyl** ICI
PP 009 H **fluazifop-butyl** ICI
PP 021 H **fomesafen** ICI
PP 062 I **pirimicarb** ICI
PP 140 F W **oxine-copper + petroleum oils** Chauvin
PP 140 F FW **8-hydroxyquinoline sulphate** Staehler
PP 148 H **paraquat dichloride** ICI
PP 149 F **ethirimol** ICI
PP 199 A **fentrifanil**
PP 211 I **pirimiphos-ethyl** ICI
PP 296 F **diclobutrazol** ICI
PP 321 I **lambda-cyhalothrin** ICI
PP 333 G **paclobutrazol** ICI
PP 383 I **cypermethrin** ICI
PP 395 F **metazoxolon**
PP 447 F **azithiram**
PP 450 F **flutriafol** ICI
PP 484 I **primidophos**
PP 493 H **haloxydine**
PP 511 IA **pirimiphos-methyl** ICI
PP 523 F **hexaconazole** ICI
PP 557 I **permethrin** ICI
PP 563 I **cyhalothrin** ICI
PP 581 R **brodifacoum** ICI
PP 588 F **bupirimate** ICI
PP 604 H **tralkoxydim** ICI
PP 618 IA **buprofezin** ICI
PP 65-25 H **phenobenzuron**
PP 675 F **dimethirimol** ICI
PP 745 H **morfamquat dichloride**
PP 781 F **drazoxolon** ICI
PP 831 H **diethamquat**
PP 993 I **tefluthrin** ICI
PP-Kombi H **2,4-D + mecoprop** Pluess-Staufer
PP-Kombi-Plus H **2,4-D + mecoprop** Pluess-Staufer
PP Mini Leatherjacket Pellets I **lindane** ICI
P.R. 20 I **Parathion** EniChem
Prado H **atrazine + pyridate** Agrolinz
Pradone H **carbetamide + dimefuron** Agrotec, Rhone-Poulenc
Prairil H **dicamba + MCPA** La Quinoleine
Praixone H **dicamba + MCPA** La Quinoleine

Praixone 300 H **2,4-D + mecoprop** La Quinoleine
prallethrin I [23031-36-9] (*S*)-2-methyl-4-oxo-3-(2-propenyl)cyclopent-2-enyl (1*R*)-*cis,trans*-chrysanthemate, Etoc, S 4068
Pramex I **permethrin** Penick
Pramitol H **prometon** Ciba-Geigy
Pramitol AT H **atrazine** Ciba-Geigy, KVK
Pre-Empt H **linuron + trietazine + trifluralin** Schering
Pre-San H **bensulide** Mallinckrodt
Prebane H **terbutryn** Ciba-Geigy
Precor I **methoprene** Sandoz
Precuran H **chlorotoluron + terbutryn** Ciba-Geigy
Preeglone H **diquat dibromide + paraquat dichloride** ICI, ICI-Zeltia, Plantevern-Kjemi
Prefar H **bensulide** Geopharm, Serpiol, Stauffer
Prefix H **chlorthiamid** Shell
Prefix C H **dichlobenil** Shell
Prefix D H **dichlobenil** Burts & Harvey, Shell
Prefix G H **dichlobenil** Agrishell, Shell
Prefix-Kombi H **chlorthiamid + dalapon-sodium** Shell
Prefongil F **carbendazim + chlorothalonil** Sipcam-Phyteurop
Preforan H **fluorodifen** Ciba-Geigy, Liro
Prefox H **ethiolate**
Pregard H **profluralin** Ciba-Geigy
Pregard Combi H **profluralin + prometryn** Viopharm
Pregolfnet H **siduron** Truffaut
Prelude F **prochloraz** Schering, Schering AAgrunol
Prelude FE F **mancozeb + prochloraz-manganese complex** Schering
Premalox H **chlorpropham + fenuron + propham** Hortichem
Premazin H **simazine** Protex
Premerge H **dinoseb** Siapa, Vertac
Premilan H **dinoseb** KVK
Premium H **neburon + terbutryn** Pepro
Premix G **chlormequat chloride** Edefi
Premol H **2,4-D + MCPA** La Littorale
Prenap H **cycloate + phenmedipham** Agriben
Prentox S **piperonyl butoxide** Prentiss
Prep G **ethephon** Union Carbide
Prephon I **parathion** Ellagret
Preroc H **chlorthiamid** RACROC
Presan F **copper oxychloride + zineb** Baslini
Preskill H **ioxynil** Sipcam-Phyteurop
pretilachlor H [51218-49-6] 2-chloro-2',6'-diethyl-*N*-(2-propoxyethyl)acetanilide, pretilachlore, CGA 26423, Rifit, Sofit, Solnet
pretilachlore H **pretilachlor**
Pretox Royal H **siduron** Graines Loras
Prevalan F **captan + nuarimol** Schering
Prevenol H **chlorpropham** Hoechst, Schering, Schering AAgrunol
Prevex F **propamocarb hydrochloride** NOR-AM
Previcur F **prothiocarb**
Previcur N F **propamocarb hydrochloride** Diana, Gullviks, Hoechst, Huhtamaki, ICI, Kwizda, Maag, Pinus, Plantevern-Kjemi, Schering, Schering AAgrunol
Preview H **chlorimuron-methyl + metribuzin** Du Pont
Prexene F **dodine** Visplant
Pride H **fluridone** Elanco
Priglone H **diquat dibromide + paraquat dichloride** Maag
Primagram H **atrazine + metolachlor** Ciba-Geigy, Ruse
Primagram Tz H **metolachlor + terbuthylazine** Ciba-Geigy
Primal R **difenacoum** Esoform
Primatol H **amitrole + 2,4-D + simazine** Ciba-Geigy
Primatol 21 H **amitrole + atrazine + simazine + thiazafluron** Ciba-Geigy
Primatol 220 H **amitrole + atrazine + picloram + simazine** Ciba-Geigy
Primatol A H **atrazine** Ciba-Geigy, Ligtermoet, Liro, Plantevern-Kjemi
Primatol AD H **amitrole + atrazine + 2,4-D** Ciba-Geigy
Primatol ATA H **amitrole + atrazine** Ligtermoet, Liro
Primatol M H **terbuthylazine** Ciba-Geigy
Primatol O H **prometon** Ciba-Geigy
Primatol P H **propazine** Drexel
Primatol Q H **prometryn** Ciba-Geigy
Primatol S H **simazine** Ciba-Geigy
Primatol SE H **amitrole + simazine** Ciba-Geigy
Primatol TD H **amitrole + atrazine + diuron + karbutilate** Ciba-Geigy
Primatope H **atrazine + simazine** Ciba-Geigy
Primawett S **polyglycolic ethers** Sanac
Primdal H **alachlor + atrazine** Agrocros
Prime G **flumetralin** Ciba-Geigy
Primextra H **atrazine + metolachlor**

BASF, Ciba-Geigy, Ligtermoet, Liro, Ruse, Viopharm
Primicid I **pirimiphos-ethyl** ICI, Sopra
primidophos I [39247-96-6] *O,O*-diethyl *O*-(2-*N*-ethylacetamido-6-methylpyrimidin-4-yl) phosphorothioate, prymidophos, PP 484
Primma H **2,4-D** Agrocros
Primma Pen H **MCPA** Agrocros
Primmatrel H **clopyralid + 2,4-D** Agrocros
Primol H **prometryn** Peppas
Primosol AF **sulphur** Eurozolfi
Primotec I **pirimiphos-ethyl** ICI
Princep H **simazine** Ciba-Geigy
Princotyl H **MCPA** Agrishell
Printagal H **dichlorprop + fluroxypyr + MCPA** Procida
Printan H **chlorotoluron + mecoprop** Procida
Printazol H **2,4-D + MCPA** Procida, Roussel Hoechst
Printazol n H **2,4-D + MCPA + picloram** Procida
Printazol total H **2,4-D + MCPA + mecoprop + picloram** Procida
Printop H **simazine** Ciba-Geigy
Printormona H **MCPA** Rhone-Poulenc
Printrex H **atrazine + simazine** Amvac
Prior-Brom R **bromadiolone** Esoform
Prior-Dione R **chlorophacinone** Esoform
PRM 12 G **ethephon** Ciba-Geigy
PRM fruit G **ethephon + fenoprop** C.F.P.I.
Pro-Drone I **1-(8-methoxy-4,8-dimethylnonyl)-4-(1-methylethyl)benzene** Stauffer
Pro-Limax M **metaldehyde** Staehler
Pro-Tex F **fentin hydroxide + maneb** Griffin
Proacido G **gibberellic acid** Probelte
Probanil H **chlorpropham + propazine** Fahlberg-List
Probatox H **dicamba + dichlorprop + MCPA** Agro-Kemi
Probe H **methazole** Plantevern-Kjemi, Sandoz
Probel Doble A **dicofol + tetradifon** Probelte
Probel G IA **azinphos-methyl** Probelte
Probel MP I **parathion-methyl** Probelte
Probel R IA **dimethoate** Probelte
Probel S I **fenitrothion** Probelte
Probeltane B F **dinocap** Probelte
Probelte Cobre F **copper oxychloride** Probelte
Probelthion Especial I **lindane + malathion** Probelte
probenazole FB [27605-76-1] 3-allyloxy-1,2-benzisothiazole 1,1-dioxide, Oryzaemate, Oryzemate
Procarbine H **chlorpropham** Aveve
Procarpil G **4-CPA + (2-naphthyloxy)acetic acid** Condor, Rhodiagri-Littorale
Procer H **2,4-D** Probelte
Procer M H **MCPA** Probelte
prochloraz F [67747-09-5] *N*-propyl-*N*-[2-(2,4,6-trichlorophenoxy)ethyl]imidazole-1-carboxamide, BTS 40542, Omega, Prelude, Sporgon, Sportak
Procibellinne G **gibberellic acid** Roussel Hoechst
Procina H **simazine** Probelte
Procina doble H **atrazine + simazine** Probelte
Procithio F **thiram** Hoechst
Prococel G **chlormequat chloride** Probelte
Procop F **Bordeaux mixture** Prochimagro
procymidone F [32809-16-8] *N*-(3,5-dichlorophenyl)-1,2-dimethylcyclopropane-1,2-dicarboximide, S-7131, Sialex, Sumiboto, Sumilex, Sumisclex
Prodacap F **captan** Aprodas
Prodactif Flo I **lindane** Aprodas
Prodaline H **trifluralin** Aprodas
Prodalon H **atrazine** Aprodas
Prodaminol H **MCPA** Aprodas
Prodan I **sodium fluosilicate** Tamogan
Prodanate F **maneb** Procida
Prodanax H **TCA** Aprodas
Prodapel F **folpet** Procida
Prodaram FX **ziram** Procida
Prodathio F **thiram** Aprodas
Prodavapam extra FIHN **metham-sodium** Aprodas
Prodazine H **simazine** Aprodas
Prodazol H **2,4-D** Aprodas
prodiamine H [29091-21-2] 5-dipropylamino-α,α,α-trifluoro-4,6-dinitro-*o*-tolui dine, CN-11-2936, Endurance, Marathon, Rydex, USB-3153
Prodiaz I **diazinon** Hoechst
Prodipte I **trichlorfon** Probelte
Prodix H **isoproturon + neburon** Rhodiagri-Littorale
Profa Algefjerner L **sodium hypochlorite** Profa

Profal G **merphos**
Profalon H **chlorpropham + linuron** Hoechst
Profan F **fentin acetate + maneb** Burri
Profen A **fenson + propargite** Sofital
profenofos IA [41198-08-7] *O*-4-bromo-2-chlorophenyl *O*-ethyl *S*-propyl phosphorothioate, OMS 2004, CGA 15324, Curacron, Polycron, Selecron
Proflan H **trifluralin** Safor
profluralin H [26399-36-0] *N*-(cyclopropylmethyl)-α,α,α-trifluoro-2,6-dinitro-*N*-propyl-*p*-toluidine, profluraline, CGA 10832, Pregard, Tolban
profluraline H **profluralin**
Profos IA **monocrotophos** Probelte
Progazon H **clopyralid + mecoprop** Chimac-Agriphar
Progazon Plus H **dicamba + dichlorprop + MCPA + mecoprop** Chimac-Agriphar
Progibb G **gibberellic acid** Abbott, Ceva, Schering
proglinazine-ethyl H [68228-18-2] *N*-(4-chloro-6-isopropylamino-1,3,5-triazin-2-yl)glycine ethyl ester, MG-07
Prograss H **ethofumesate** NOR-AM
Prohelan H **prometryn** Radonja
Prokamix-DPD H **2,4-D + dichlorprop** Agro-Kemi
Prolan H **trifluralin** Probelte
Prolan doble H **amitrole + diuron** Probelte
Prolapon H **dalapon-sodium** Probelte
Prolate IA **phosmet** Stauffer
Prolex H **propachlor** Aako, Makhteshim-Agan
Prolin F **ziram** Baslini
Prolin R **sulfaquinoxaline + warfarin** Hopkins
Promalin G **6-benzyladenine + gibberellins** Abbott, Schering AAgrunol
Promat-50 H **atrazine + prometryn** Pliva
Promazin H **simazine** L.A.P.A.
promecarb I [2631-37-0] 5-methyl *m*-cumenyl methylcarbamate, promecarbe, OMS 716, ENT 27300, Carbamult, SN 34615
promecarbe I **promecarb**
Promephos I **parathion** Prometheus
Promepin H **prometryn** Pinus
Promet H **prometryn** Chemia
Promet I **furathiocarb** Ciba-Geigy
Prometen I **endosulfan + methoxychlor + propoxur** Organika-Azot
prometon H [1610-18-0] 2,4-bis(isopropylamino)-6-methoxy-1,3,5-triazine, prometone, G 31435, Gesafram, Ontracic, Pramitol, Primatol
prometone H **prometon**
Prometran H **prometryn** Diana
Prometrex H **prometryn** Anorgachim, Makhteshim-Agan
Prometrin H **prometryn** Diana, Ellagret, Pliva, Zorka, Zorka Sabac
Prometron H **prometryn** Ellagret
Prometryl H **prometryn** Hellenic Chemical
prometryn H [7287-19-6] 2,4-bis(isopropylamino)-6-methylthio-1,3,5-triazine, prometryne, Caparol, Cotton-Pro, G 34161, Gesagard, Primatol Q, Prometrex, Uvon
prometryne H **prometryn**
Promexil H **prometryn** Hellafarm
Promildor FIA **carbaryl + maneb + metiram + parathion** Procida
Pronalid H **propanil** Hellenic Chemical
pronamide H **propyzamide**
Prondane I **lindane** Probelte
Prondatox I **parathion-methyl** SPE
Pronone H **hexazinone** Proserve
PRONTOX-Wuehlmausgas R **phosphine** Wuelfel
Prop Job H **propanil** Drexel
Propa H **propanil**
propachlor H [1918-16-7] 2-chloro-*N*-isopropylacetanilide, propachlore, Albrass, Bexton, CP 31393, Niticid, Prolex, Ramrod
propachlore H **propachlor**
Propacip H **chlorpropham + propazine** Liro
Propaclor Doble H **alachlor + atrazine** Probelte
Propacor H **propanil** Du Pont
Propaflo H **propachlor** Agriben
propafos I **propaphos**
Propagrex H **propanil** Sadisa
Propal H **mecoprop** Universal Crop Protection
Propal riso H **propanil** Caffaro
Propalane H **propachlor** Top
propamocarb hydrochloride F [25606-41-1] propyl 3-(dimethylamino)propylcarbamate hydrochloride, chlorhydrate du propamocarbe, Banol, Filex, Prevex, Previcur N, SN 66752
Propanex H **propanil** Cumberland, Crystal

propanil H [709-98-8] 3',4'-dichloropropionanilide, DCPA, Bay 30130, Chem-Rice, Erban, Erbanil, FW-734, Herbax, Prop Job, Propa, Propanex, Riselect, Stam, Stampede, Strel, Supernox, Surcopur, Surpur, SynpranN

propaphos I [7292-16-2] 4-(methylthio)phenyl dipropyl phosphate, propafos, Kayaphos, NK-1158

propargite A [2312-35-8] 2-(4-*tert*-butylphenoxy)cyclohexyl prop-2-ynyl sulphite, BPPS, ENT 27226, Comite, DO 14, Omite

Propariz H **propanil** Rhone-Poulenc

Propazin F **propineb** Terranalisi

propazine H [139-40-2] 2-chloro-4,6-bis(isopropylamino)-1,3,5-triazine, G 30028, Gesamil, Milo-Pro, Milocep, Milogard, Primatol P, Prozinex

Propendive H **chlorpropham** L.A.P.A.

propetamphos I [31218-83-4] (*E*)-*O*-2-isopropoxycarbonyl-1-methylvinyl *O*-methyl ethylphosphoramidothioate, OMS 1502, Safrotin, SAN 52139 I, Seraphos

propham HG [122-42-9] isopropyl phenylcarbamate, prophame, IFC, IPC, Birgin, Chem-Hoe, Triherbide-IPC

prophame HG **propham**

Propiamba H **dicamba + mecoprop** Interphyto

propiconazole F [60207-90-1] (±)-1-[2-(2,4-dichlorophenyl)-4-propyl-1,3-dioxolan-2-ylmethyl]-1 *H*-1,2,4-triazole, Banner, CGA 64250, Desmel, Radar, Tilt

Propimix H **dichlorprop + MCPA** Agro-Kemi

propineb F [12071-83-9] polymeric zinc propylenebis(dithiocarbamate), propinebe, Airone, Antracol, Bay 46131, LH 30/Z, Taifen

propinebe F **propineb**

Propinex H **dichlorprop** Visplant

Propinox-D H **dichlorprop** Agro-Kemi

Propinox-M H **mecoprop** Agro-Kemi

Propinox-MD H **2,4-D + mecoprop** Agro-Kemi

Propion-DP H **dichlorprop** Bjoernrud

Propiormone H **mecoprop** Interphyto

Propixine H **atrazine** Sedagri

Propogon I **propoxur** Crystal

Proponex H **mecoprop** Shell

Proponex KV Kombi H **2,4-D + mecoprop** Shell

Propotox M I **methoxychlor + propoxur** Organika-Azot

Propoxone H **mecoprop** Sedagri

propoxur I [114-26-1] 2-isopropoxyphenyl methylcarbamate, PHC, arprocarb, OMS 33, ENT 25671, 58 12 315, Bay 39007, Baygon, Blattanex, Pillargon, Propogon, Rhoden, Sendran, Suncide, Unden, Undene

proprop H **dalapon-sodium**

Propson H **propanil** Rohm & Haas

Propuron H **neburon** Sedagri

propyl 3-tert-butylphenoxyacetate G [66227-09-6] propyl 3-*tert*-butylphenoxyacetate, M&B 25-105, M&B 25105

propyzamide H [23950-58-5] 3,5-dichloro-*N*-(1,1-dimethylpropynyl)benzamide, pronamide, Clanex, Kerb, RH-315

Prosem IF **lindane + maneb** ERT

Prosem C FX **anthraquinone + maneb** Procida

Prosem R IF **lindane + maneb** ERT

Prosevit FI **carbaryl + sulphur** Roussel Hoechst

Prosevor IG **carbaryl** Roussel Hoechst

prosulfocarb H *S*-benzyldipropylthiocarbamate, Boxer, SC-0574

Protector R IF **lindane + maneb** Afrasa

Protekt F **thiabendazole** Kemira

Protektal L **quaternary ammonium** Protekta

Protekton 7 I **dichlorvos** Protekta

Protekton CP I **chlorpyrifos** Protekta

Protekton DCV I **dichlorvos** Protekta

Protekton MA I **malathion** Protekta

Protekton Pyr I **piperonyl butoxide + pyrethrins** Protekta

prothidathion A *S*-2,3-dihydro-5-isopropoxy-2-oxo-1,3,4-thiadiazol-3-ylmethyl *O,O*-diethyl phosphorodithioate, GS 13010

prothiocarb F [19622-19-6] *S*-ethyl *N*-(3-dimethylaminopropyl)thiocarbamate hydrochloride, Dynone, Previcur

prothiofos I [34643-46-4] *O*-2,4-dichlorophenyl *O*-ethyl *S*-propyl phosphorodithioate, prothiophos, Bay NTN8629, Bideron, Tokuthion

prothiophos I **prothiofos**

prothoate AI [2275-18-5] *O,O*-diethyl *S*-isopropylcarbamoylmethyl phosphorodithioate, ENT 24652, EI 18682, Fac, Fostion

Protodan AI **endosulfan** Probelte

Protrazin H **atrazine + propachlor** Pliva
Protrum K H **phenmedipham** Atlas-Interlates
Proturf H **bromoxynil + dichlorprop + MCPA** May & Baker
Providex H **flamprop-M-isopropyl** Siapa
Provin F **chlorothalonil** Kwizda
Prowl H **pendimethalin** Cyanamid, Quimigal
proxan-sodium H [140-93-2] sodium *O*-isopropyl dithiocarbonate, proxane-sodium, IPX, Good-rite n.i.x.
proxane-sodium H **proxan-sodium**
proximpham H [2828-42-4] 2-propanone *O*-[(phenylamino)carbonyl]oxime
Proxol I **trichlorfon** NOR-AM
Proxon H **paraquat dichloride** Prometheus
Proxtat H **2,4-D + mecoprop** Shell
Prozinex H **propazine** Makhteshim-Agan
Prozinon IA **diazinon** Probelte
Prunit G **uniconazole** Sumitomo
prussic acid IR **hydrogen cyanide**
prymidophos I **primidophos**
prynachlor H [21267-72-1] 2-chloro-*N*-(1-methylprop-2-ynyl)acetanilide, prynachlore, BAS 290H, Basamaize
prynachlore H **prynachlor**
Psilan I **parathion-methyl** Scam
PSP-204 I **IPSP** Hokko
Psyllion I **fenitrothion** Agronova
Pucerons L I **dimethoate** Umupro
Puliris H **MCPA + mecoprop + propanil** Agrimont
Pulsan F **cymoxanil + mancozeb + oxadixyl** Sandoz
Pulsan TS F **cymoxanil + oxadixyl** Sandoz
Pulsar H **bentazone + MCPB** BASF
Pulsfog K G **chlorpropham** Stahl
Pulsfog-Draagstof S **methylene chloride + petroleum oils** Brinkman
Puma H **fenoxaprop-ethyl** Hoechst, Procida
Punch F **flusilazol** Du Pont
Punch C F **carbendazim + flusilazol** Du Pont
Purgarol H **MCPA + terbacil** Urania
Purit H **sodium chlorate** Nagel
Purivel H **metoxuron** Sandoz, VCH
Pursuit H **imazethapyr** Cyanamid
Puutarha-aerosoli IA **piperonyl butoxide + pyrethrins** Farmos
Puutarhan Rikkahavite H **glyphosate** Monsanto
Puutarharuiskute IA **piperonyl butoxide + pyrethrins** Farmos
Puzomor M **metaldehyde** Chromos
Py IA **pyrethrins**
Py Garden Insecticide IA **piperonyl butoxide + pyrethrins** Synchemicals
Py-Kill I **tetramethrin** Killgerm
Py Kill Plus I **phenothrin + tetramethrin** Killgerm
Py Kill-W I **pyrethrins** Killgerm
Py Powder IA **piperonyl butoxide + pyrethrins** Synchemicals
Py-sekt I **pyrethrins** KenoGard
Pybuthrin I **piperonyl butoxide + pyrethrins** ICI, Wellcome
Pybutrin I **piperonyl butoxide + pyrethrins** Industrialchimica
pydanon G [22571-07-9] (±)-hexahydro-4-hydroxy-3,6-dioxopyridazin-4-ylacetic acid, H 1244
Pydon I **pyrethrins + S421** Protex
Pydrin IA **fenvalerate** Shell
Pymax I **malathion + piperonyl butoxide + pyrethrins** Anticimex
Pymaxol I **malathion + piperonyl butoxide + pyrethrins** Anticimex
Pynamin I **allethrin** Sumitomo
Pynosect I **resmethrin** Mitchell Cotts
Pynosect 2 I **resmethrin + tetramethrin** Mitchell Cotts
Pynosect 6 I **resmethrin + tetramethrin** Mitchell Cotts
Pynosect 30 I **pyrethrins + resmethrin** Mitchell Cotts
Pynosect 40 I **piperonyl butoxide + resmethrin + tetramethrin** Mitchell Cotts
Pynosect PCO I **permethrin** Mitchell Cotts
Pyomix IA **naled** Bayer
pyracarbolid F [24691-76-7] 3,4-dihydro-6-methyl-2*H*-pyran-5-carboxanilide, Sicarol
Pyracur H **chloridazon + metolachlor** BASF
Pyradex H **chloridazon + di-allate** BASF, Rhodiagri-Littorale
Pyradex TF H **chloridazon + tri-allate** Shell
Pyradur H **chloridazon + metolachlor** BASF, Chromos
Pyral Rep double F **sodium arsenite + ziram** Marty et Parazols
Pyral Rep Fort FI **sodium arsenite** Marty et Parazols

Pyralesca RS FI **sodium arsenite** La Littorale
Pyralumnol FI **sodium arsenite** Procida
pyramdron I **hydramethylnon**
Pyramin H **chloridazon** Agrolinz, BASF, Budapesti Vegyimuvek, Chinoin, Chromos, Collett, Dimitrova,Imex-Hulst, Intrachem, Maag, Organika-Azot, Sapec, Shell
Pyramine H **chloridazon** BASF, Rhodiagri-Littorale
pyranocoumarin R [518-20-7] 3,4-dihydro-2-methoxy-2-methyl-4-phenyl-2*H*,5*H*-pyrano[3,2- *c*][1]benzopyran-5-one, Actosin
Pyrarsene F **sodium arsenite** Gen. Rep.
Pyrasan H **chloridazon** Sanac
Pyraser I **malathion + pyrethrins** Agriphyt
Pyrasol H **chloridazon** Top
Pyrasur H **chloridazon + lenacil** BASF
Pyrazon H **chloridazon** Eurofyto, KVK
pyrazophos F [13457-18-6] *O*-6-ethoxycarbonyl-5-methylpyrazolo[1,5-*a*]pyrimidin-2-yl *O,O*-diethyl phosphorothioate, Afugan, Curamil, Hoe 02873, Missile, Pokon Mildew Spray
pyrazosulfuron-ethyl H [93697-74-6] ethyl 5-[[[[(4,6-dimethoxy-2-pyrimidinyl) amino]carbonyl]amino]sulfonyl] -1-methyl-1*H*-pyrazole-4-carboxylate, Daioh, NC-311
pyrazoxyfen H [71561-11-0] 2-[4-(2,4-dichlorobenzoyl)-1,3-dimethylpyrazole-5-yloxy]acetophenone, Paicer
Pyrel I **piperonyl butoxide + pyrethrins** De Weerdt
Pyrem I **piperonyl butoxide + pyrethrins** KenoGard, Norsk Pyrethrum
Pyremex I **piperonyl butoxide + pyrethrins** Anticimex
Pyremist I **pyrethrins** Anticimex
Pyrenone I **piperonyl butoxide + pyrethrins** ATE, Fairfield
Pyresin I **allethrin**
Pyresoufre FI **cypermethrin + sulphur** R.S.R.
Pyreth IA **piperonyl butoxide + pyrethrins** Staehler
pyrethrins IA [121-21-1; 4466-14-2; 25402-06-6; 121-29-9; 1172-63-0; 121-20-0 respectively], pyrethrin I, jasmolin I, cinerin I, pyrethrin II, jasmolin II, cinerin II, Py, Pyrethrum
Pyrethrum IA **piperonyl butoxide + pyrethrins** Bayer, Rentokil, Synchemicals
Pyretox I **piperonyl butoxide + pyrethrins** Maatalouspalvelu
Pyretrex Special I **piperonyl butoxide + pyrethrins** Formulex
Pyrex I **piperonyl butoxide + pyrethrins + rotenone** Wikholm
Pyrex insektsspray I **malathion + piperonyl butoxide + pyrethrins** Wikholm
Pyrex Pumpspray I **piperonyl butoxide + pyrethrins** Wikholm
Pyrexcel I **allethrin**
pyriclor H [1970-40-7] 2,3,5-trichloropyridin-4-ol, Daxtron
pyridafenthion IA [119-12-0] *O, O*-diethyl-*O*-[2-phenyl-3(2*H*)-pyridazinone-6-yl] phosphorothioate, pyridaphenthion, CL 12,503, Ofnack, Ofnak, Ofunack
pyridaphenthion IA **pyridafenthion**
pyridate H [55512-33-9] 6-chloro-3-phenylpyridazin-4-yl *S*-octyl thiocarbonate, CL 11344, Lentagran, Tough
Pyridazin H **chloridazon** Visplant
pyridinitril F [1086-02-8] 2,6-dichloro-4-phenylpyridine-3,5-dicarbonitrile
pyrifenox F [88283-41-4] 2',4'-dichloro-2-(3-pyridyl)acetophenone *O*-methyloxime, ACR-3651 A, Dorado, Ro 15-1297
pyrimicarbe I **pirimicarb**
pyrimiphos-ethyl I **pirimiphos-ethyl**
pyrimiphos-methyl IA **pirimiphos-methyl**
Pyrinex I **chlorpyrifos** Aragonesas, Makhteshim-Agan
pyrinuron R [53558-25-1] 1-(4-nitrophenyl)-3-(3-pyridylmethyl)urea, RH-787, Vacor
Pyriquat special H **paraquat dichloride** Interphyto
Pyrit I **piperonyl butoxide + pyrethrins** Staehler
Pyrocide I **allethrin** McLaughlin Gormley King
Pyron H **clopyralid + pyridate** Prochimagro
Pyron H **chloridazon + cycloate** BASF
pyroquilon F [57369-32-1] 1,2,5,6-tetrahydropyrrolo[3,2,1-*ij*]quinolin-4-one, pyroquilone, CGA 49104, Coratop, Fongoren, Fongorene
pyroquilone F **pyroquilon**
Pyrotex I **allethrin + piperonyl butoxide** Protex
Pyrotox I **piperonyl butoxide + tetramethrin** Chinoin
pyroxychlor F [7159-34-4] 2-chloro-6-methoxy-4-trichloromethylpyridine, pyroxychlore, Dowco 269, Lorvek, Nurelle
pyroxychlore F **pyroxychlor**

Pyrsol I **piperonyl butoxide + pyrethrins** Nordisk Alkali
Pytoxan I **piperonyl butoxide + pyrethrins** Agro-Kemi

Qikron A **chlorfenethol** Nippon Soda
Quad-Keep F **tecnazene** Quadrangle
Quad-Store F **tecnazene** Quadrangle
Quadban H **decamba + MCPA + mecoprop** Quadrangle
Quamlin I **permethrin** Wellcome
Quartz GT H **diflufenican + isoproturon** Rhodiagri-Littorale
quaternary ammonium H Gloquat
Quatrol H **paraquat dichloride** SEGE
Queckenvertilger H **TCA** CTA
Queletox I **fenthion** Bayer
Quex H **sodium chlorate** Kwizda
Quick R **chlorophacinone** Rhodiagri-Littorale, Rhone-Poulenc, Sapec
Quick-Dip G **4-indol-3-ylbutyric acid + 1-naphthylacetic acid** Benfried
Quickfixer H **2,4-D + dalapon + maleic hydrazide + TCA** Kemira
Quickphos I **aluminium phosphide** Anorgachim, United Phosphorus
Quickra R **chlorophacinone** Sereg
Quikcide Concentrate Insecticide I **piperonyl butoxide + pyrethrins** Kem
Quilan H **benfluralin** Elanco, Grima
Quimato IA **dimethoate** Masso
Quimol H **molinate** Quimigal
quinacetol sulphate F [57130-91-3] bis(5-acetyl-8-hydroxyquinolinium) sulphate, quinacetol sulfate, Fongoren, G 20072, Risoter
quinalphos IA [13593-03-8] *O,O*-diethyl *O*-quinoxalin-2-yl phosphorothioate, chinalphos, ENT 27394, Bay 77049, Bayrusil, Ekalux, Kinalux, Sandoz 6538, Savall
Quinaris F **carbendazim + folpet** La Quinoleine
quinazamid F *p*-benzoquinone monosemicarbazone, quinazamide, BTS 8684, RD 8684
quinazamide F **quinazamid**
Quino Blanc D I **chlorpyrifos** La Quinoleine
Quinochancre F **oxine-copper** La Quinoleine
quinofop-ethyl H **quizalofop-ethyl**
Quinogam FI **lindane + oxine-copper** La Quinoleine
Quinolate F **oxine-copper** Argos, Budapesti Vegyimuvek, Hellafarm, La Quinoleine
Quinolate AC FX **anthraquinone + oxine-copper** La Quinoleine
Quinolate F F **fuberidazole + oxine-copper** Lucebni zavody
Quinolate FL FIX **anthraquinone + lindane + oxine-copper** La Quinoleine
Quinolate Mais FX **anthraquinone + captan** La Quinoleine
Quinolate MG FIX **anthraquinone + endosulfan + lindane + oxine-copper** La Quinoleine
Quinolate MG SAFI FIX **endosulfan + lindane + oxine-copper** La Quinoleine
Quinolate MG SAFLO FIX **endosulfan + lindane + oxine-copper** La Quinoleine
Quinolate plus AC FX **anthraquinone + oxine-copper** La Quinoleine
Quinolate plus anticorbeaux Eco FX **anthraquinone + oxine-copper** La Quinoleine
Quinolate plus antitaupins Eco FI **lindane + oxine-copper** La Quinoleine
Quinolate plus HI FX **anthraquinone + flutriafol + oxine-copper** La Quinoleine
Quinolate plus MG FIX **anthraquinone + endosulfan + lindane + oxine-copper** La Quinoleine
Quinolate plus MG SAFI FIX **endosulfan + lindane + oxine-copper** La Quinoleine
Quinolate plus MG SAFLO FIX **endosulfan + lindane + oxine-copper** La Quinoleine
Quinolate plus Semences F **oxine-copper** La Quinoleine
Quinolate plus tripl'eco FIX **anthraquinone + lindane + oxine-copper** La Quinoleine
Quinolate plus triple FIX **anthraquinone + lindane + oxine-copper** La Quinoleine
Quinolate plus V-4-X AC FX **anthraquinone + carboxin + oxine-copper** La Quinoleine
Quinolate plus V-4-X Semences F **carboxin + oxine-copper** La Quinoleine
Quinolate plus V-4-X triple FIX **anthraquinone + carboxin + lindane + oxine-copper** La Quinoleine
Quinolate Pro F **carbendazim + oxine-copper** La Quinoleine
Quinolate Pro AC FX **anthraquinone +**

carbendazim + oxine-copper La Quinoleine
Quinolate semences F **oxine-copper** La Quinoleine
Quinolate triple FIX **anthraquinone + lindane + oxine-copper** Carchim, La Quinoleine
Quinolate V-4-X F **carboxin + oxine-copper** Budapesti Vegyimuvek, La Quinoleine, Ligtermoet
Quinolate V-4-X AC FX **anthraquinone + carboxin + oxine-copper** La Quinoleine
Quinolate V-4-X Semences F **carboxin + oxine-copper** Lucebni zavody
Quinolate V-4-X triple FIX **anthraquinone + carboxin + lindane + oxine-copper** La Quinoleine
quinomethionate FAI [2439-01-2] 6-methyl-1,3-dithiolo[4,5-*b*]quinoxalin-2-one, chinomethionat, chinomethionate, oxythioquinox, quinoxalines, ENT 25606, Bay 36205, Morestan, SS 2074
quinonamid LH [27541-88-4] 2,2-dichloro-*N*-(3-chloro-1,4-naphthoquinon-2-yl)acetamide, quinonamide, chinonamid, Alginex
quinonamide LH **quinonamid**
Quinondo F **oxine-copper** Kanesho
Quinophos huileux IA **parathion-methyl** La Quinoleine
Quinorexone H **dicamba + mecoprop** La Quinoleine
Quinosol H **diuron + simazine** La Quinoleine
Quinoter H **linuron + monolinuron** La Quinoleine
quinothionate AF **thioquinox**
quinoxalines FAI **quinomethionate**
Quinoxone H **2,4-D** La Quinoleine
Quintalic F **carbendazim + iprodione** Agriben
Quintar FL **dichlone** Hopkins
Quintex H **chlorpropham + fenuron + propham** Truchem
Quintox R **cholecalciferol** Bell Labs
quintozene F [82-68-8] pentachloronitrobenzene, PCNB, PKhNB, Avicol, Botrilex, Brassicol, Earthcide, Folosan, Kobu, Kobutol, Kobutol, Pentagen, Saniclor, Terraclor, Tritisan, Tubergran, Turfcide
Quintozyl F **quintozene** Prochimagro
Quiritox R **warfarin** Neudorff, Polanz
Quirotex I **lindane** Permutadora
quizalofop-ethyl H [76578-14-8] ethyl 2-[4-(6-chloro-2-quinoxalinyloxy)phenoxy]propionate, quinofop-ethyl, xylofop-ethyl, Assure, DPX Y6202, EXP 3864, FBC-32197, NC 302, NCI 96683, Pilot, Targa
Quomadin I **cypermethrin** Bayer

R-3 IA **dimethoate** Argos
R 6 Erresei M-50 F **mancozeb** Ravit
R 6 Erresei Oroblu F **copper oxychloride + cymoxanil + mancozeb** Ravit
R 6 Erresei P.B. F **copper oxychloride + zineb** Ravit
R 6 Erresei S.B. M-10 F **copper oxychloride + mancozeb** Ravit
R 6 Erresei Stop R F **copper oxychloride + cymoxanil** Ravit
R 6 Erresei Triplo F **cymoxanil + fosetyl-aluminium + mancozeb** Ravit
R-40 IA **dimethoate** Afrasa, Argos
R-50 IA **dimethoate** Afrasa
R-1303 IA **carbophenothion** Stauffer
R-1504 IA **phosmet** Stauffer
R 1513 IA **azinphos-ethyl** Bayer
R 1582 IA **azinphos-methyl** Bayer
R-1607 H **vernolate** Stauffer
R-1608 H **EPTC**
R-1910 H **butylate** Stauffer
R-2061 H **pebulate** Stauffer
R-2063 H **cycloate** Stauffer
R-2170 IA **oxydemeton-methyl** Bayer
R-4461 H **bensulide** Stauffer
R-4572 H **molinate** Stauffer
R 5158 IA **amiton**
R-7465 H **napropamide** Stauffer
R 23979 F **imazalil** Janssen
R-25788 S **dichlormid** Stauffer
R-40244 H **flurochloridone** Stauffer
R-Bix H **paraquat dichloride** Sopra
R-Dimetoaatti IA **dimethoate** BASF, Farmos
RA 15-Neu H **diuron** Hentschke & Sawatzki
RA 17-Neu H **bromacil + diuron** Hentschke & Sawatzki
RA-2000 H **diuron + methabenzthiazuron** Hentschke & Sawatzki
RA-C-9 H **sodium chlorate** Hentschke & Sawatzki
Raatt-o-mat R **warfarin** Mellansvenska
Rabbe A **propargite** Grima
Rabcide F **fthalide** Kureha

rabenzazole F [40341-04-6] 2-(3,5-dimethylpyrazol-1-yl)-1*H*-benzimidazole, GAU 1356
Rabon IA **tetrachlorvinphos** Shell
Rabond IA **tetrachlorvinphos** Shell
Rabpoly F **fthalide + polyoxins**
Race H **mecoprop** Akzo
Racer H **flurochloridone** Duphar, OHIS, Sandoz, Schering, Serpiol, Stauffer
Racer L H **flurochloridone + linuron** Schering
Racer L spuitpoeder H **gibberellic acid + MCPA-thioethyl** Stauffer
Racidin G **4-indol-3-ylbutyric acid** Phytorgan
Racumin R **coumatetralyl** Bayer, Pan Britannica, Pinus
Racumin D R **cholecalciferol** Bayer
Racumin rat des champs appat R **chlorophacinone** Bayer
Racusan IA **dimethoate** KenoGard
Rad-E-Cate H **dimethylarsinic acid** Vineland
Radam FX **guazatine acetates** KenoGard
Radapon H **dalapon-sodium** Dow
Radar F **propiconazole** Farm Protection, ICI
Radar Dos I **piperonyl butoxide + pyrethrins** KenoGard
Radar insektsspray I **bioallethrin + methoxychlor + piperonyl butoxide + pyrethrins** Nobel
Radar universal I **methoxychlor + piperonyl butoxide + pyrethrins** Barnaengen
Radazin H **atrazine** Radonja
Radeks H **cyanazine** Radonja
Radeks-plus H **atrazine + cyanazine** Radonja
Radian H **napropamide + simazine** Pepro
Radical G **4-indol-3-ylbutyric acid** Tecniterra
Radicante G **1-naphthylacetic acid** Biolchim
Radinex F **thiabendazole** BASF
Radisol H **sodium chlorate** Borgofranco
Radixon G **2,4-D + 4-indol-3-ylbutyric acid + 1-naphthylacetic acid** Farmer
Radocid I **methidathion** RACROC
Radocineb F **zineb** Radonja
Radociram F **ziram** Radonja
Radofon I **trichlorfon** Radonja
Radokaptan F **captan** Radonja
Radokor H **simazine** Radonja
Radosan F **methoxyethylmercury acetate** Radonja
Radosan 1,2 F **phenylmercury acetate** Radonja
Radosan-M F **methoxyethylmercury acetate** Radonja
Radotion IA **malathion** Radonja
Radotiram F **thiram** Radonja
Radovit-M S **polyglycolic ethers** Radonja
Radox-Ameisenfresslack D1 I **sodium arsenate** Chemierax Radebeul
Radox-Ameisenfresslack P I **trichlorfon** Chemierax Radebeul
Radoxone TL H **amitrole + ammonium thiocyanate** Sopra
Radspor F **dodine** Truchem
Raeuchermittel Jacutin I **lindane** Shell
Rafene R **warfarin** C.F.P.I.
Rafix R **bromadiolone** Agrotec, Rhone-Poulenc
Raflo mit UV H **2,4-D + dicamba** BASF
Ragumon IA **azinphos-methyl** Agrimont
Raid House & Garden Insect Killer IA **piperonyl butoxide + pyrethrins** Trans-Meri
Raiffeisen Rasenduenger mit Moosvernichter H **iron sulphate** Raiffeisen
Raiffeisen-Spezial Rasenduenger mit Unkrautvernichter H **2,4-D + MCPA** Raiffeisen
Raisan FIHN **metham-sodium** Lainco
Raisan K FIHN **metham-potassium** Lainco
RAK 1 Pheromon Einbindiger Traubenwickler I **pheromones** BASF
Rakampf-Spezial-Frischkoeder R **warfarin** Detia Freyberg, Vorratsschutz
Rakatop I **dialifos** Leu & Gygax
Rakumin R **coumatetralyl** Berner
Ramag FW **thiram** Maag
Raman F **copper oxychloride + mancozeb** Sofital
Ramato F **copper oxychloride + mancozeb** Agrimont
Ramato P F **copper oxychloride + zineb** Agrimont
Rambo H **alachlor + atrazine** Siapa
Rame Caffaro F **calcium copper oxychloride** Caffaro
Rame Ossicloruro F **copper oxychloride** Mormino
Rame Sariaf F **calcium copper oxychloride** EniChem

Rame Zolfo F **copper oxychloride + mancozeb + sulphur** Chimiberg
Ramecalce F **Bordeaux mixture** Sepran
Ramedit Combi F **copper oxychloride + cymoxanil** Siapa
Ramedit MC F **copper oxychloride + mancozeb** Siapa
Rametrin PB F **calcium copper oxychloride** Enotria
Rametrin WP F **copper oxychloride** Enotria
Ramezin F **copper oxychloride + zineb** Caffaro, Ergex
Ramezin K F **calcium copper oxychloride + copper oxychloride + maneb** Caffaro
Ramezin KC F **calcium copper oxychloride + maneb** Caffaro
Ramik R **diphacinone**
Ramin F **copper oxychloride** Chemia
Ramolio F **copper oxychloride + petroleum oils** Chemia
Rampage R **cholecalciferol** Motomco, Sanofi
Rampart IAN **phorate**
Ramrod H **propachlor** Bjoernrud, Inagra, Leu & Gygax, Monsanto, Nordisk Alkali, Pinus, Shell, Sipcam
Ramucide R **chlorophacinone**
Rancho H **mefenacet** Bayer
Randal IA **fenpropathrin** Shell
Randox H **allidochlor**
Ranfor G **chlormequat chloride + ethephon** Ciba-Geigy
Rangado I **dimethylvinphos** Shell Kagaku
Rapid I **pirimicarb** ICI
Rapid Root G **4-indol-3-ylbutyric acid** Tuhamij
Rapid'Tox souris R **crimidine** Sovilo
Rapid-Ex H **sodium chlorate** Staehler
Rasana + M Supergazon Meststof met Mosverdelger H **iron sulphate** Graham
Rasana + U H **2,4-D + dicamba** Graham
Rasana gazonmest met mosbestrijder H **iron sulphate** Wolf
Rasana Plus U H **2,4-D + dicamba** Wolf
Rasen-Banvel H **dicamba + MCPA** Schering
Rasen-Certrol H **dichlorprop + ioxynil** Urania
Rasenduenger Mannadur mit Moosvernichter H **iron sulphate** Manna-Duengerwerk
Rasenduenger mit Moosvernichter H **iron sulphate** Spiess, Wolf
Rasenduenger mit Unkrautvernichter H **2,4-D + dicamba** Wolf
Rasenduenger mit UV H **dicamba + mecoprop** Hoechst
Rasenduenger + Moosvernichter H **iron sulphate** Du Pont
Rasenduenger plus Moosvernichter H **iron sulphate** Flora-Frey
Rasenduenger plus Unkrautvernichter H **2,4-D + dicamba** Flora-Frey
Rasenduenger Rasokur mit Unkrautvernichter H **chlorflurecol-methyl + MCPA** Shell
Rasenduenger spezial mit Unkrautvernichter H **2,4-D + dicamba** Terrasan
Rasenfloranid mit Moosvernichter H **iron sulphate** BASF
Rasenfloranid mit Unkrautvernichter H **2,4-D + dicamba** BASF, Compo
Rasenfloranid Rasenduenger mit Moosvernichter H **iron sulphate** Compo
Rasengruen mit Unkrautvernichter H **dicamba + MCPA** Schering
Rasenhedomat H **dicamba + MCPA** Bayer
Rasen Kap-Horn mit Moosvernichter H **iron sulphate** Leyffer & Nellen
Rasen Kap-Horn mit Unkrautvernichter H **2,4-D + dicamba** Leyffer & Nellen
Rasenkorn Rasenduenger mit Moosvernichter H **iron sulphate** GFG
Rasen-Moos-Ex H **iron sulphate** Shell
Rasen Neudotox S H **dicamba + MCPA** Neudorff
Rasen-RA-5 H **2,4-D + mecoprop** Hentschke & Sawatzki
Rasenstolz H **iron sulphate** Urania
Rasen-Terlavan H **2,4-D + MCPA** Schacht
Rasen Unkraut-frei Spritz- und Giessmittel H **dicamba + MCPA** Flora-Frey
Rasenunkraut-Vernichter H **2,4-D + dicamba** Wolf
Rasen-Unkrautvernichter Banvel H **dicamba + mecoprop** Epro
Rasen-Unkrautvernichter Banvel M H **dicamba + MCPA** Shell
Rasen-Utox H **dicamba + MCPA** Spiess, Urania
Rasikal H **sodium chlorate** A.C.I.E.R.
Rassapron H **amitrole + atrazine + diuron** BP
Rastop "Bloc C" R **chlorophacinone** Rastop
Rastop "Bloc D" R **difenacoum** Rastop

Rastop carottes lyophilisees R **chlorophacinone** Rastop
Rastop "Cereales C" R **chlorophacinone** Rastop
Rastop Special Rats Musques R **chlorophacinone** Rastop, Steininger
Rastop "Super Flocons" R **difenacoum** Rastop
Rasunex H **MCPA + mecoprop** Staehler
Ratafin R **coumafuryl**
Ratak R **difenacoum** Agro-Kemi, Berner, ICI, ICI-Valagro, ICI-Zeltia
Ratan E R **bromadiolone** Sepran
Rataplan R **warfarin** CIFO
Rat-A-Way R **coumafuryl**
Rat-billen R **warfarin** Vitafarm
Ratenit R **warfarin** De Rauw
Rat-Free Special R **chlorophacinone** Shell
Raticate R **norbormide** McNeil Laboratories
Raticida R **sulphaquinoxaline + warfarin** Probelte
Raticide A R **warfarin** Aufra
Raticide appat D R **difenacoum** SCAC-Fisons
Raticide D.I. 5 R **difenacoum** Aufra
Raticide EV R **bromadiolone** Rhodic
Raticide tout puissant R **chlorophacinone** SCAC-Fisons
Raticide tout puissant B R **bromadiolone** SCAC-Fisons
Raticide Umupro R **bromadiolone** Umupro
Ratilan R **coumachlor** Ciba-Geigy
Ratimus R **bromadiolone**
Ratiwarf R **warfarin** Rentokil
Ratkill R **warfarin** Colkim
Rat Killer CF R **chlorophacinone** Zucchet
Rat Killer Super R **warfarin** Zucchet
Ratocid R **warfarin** Aporta
Ratol R **zinc phosphide** United Phosphorus
Ratomet R **chlorophacinone**
Ratomide R **norbormide** Colkim
Ratona R **warfarin** Colkim
Ratorex R **warfarin**
Ratox R **thallium sulphate**
Ratox R **difenacoum** Quimigal
Ratox-mamak R **chlorophacinone** Radonja
Rat-Pak R **warfarin**
Ratrick R **difenacoum** Farm Protection
Ratrin R **chlorophacinone** Diana
Rat's R **warfarin** Arkovet
Ratten Vergif "Destructa" R **warfarin** Destructa
Rattenbekaempfungsmittel Gruen-Rot R **coumatetralyl** Angelkort
Rattenex R **warfarin** Bergophor
Rattengifttropfen R **thallium sulphate** Eckert
Rattenriegel Knax R **warfarin** Marktredwitz
Rattentod R **warfarin** USHBG
Rattex R **coumatetralyl** Perycut-Schrammel, Topical
Rattomix-Fertigkoeder R **warfarin** Breiler
Rattosin R **warfarin** Colkim
Rattosin Pellet R **sulphaquinoxaline + warfarin** Colkim
Rattox R **warfarin** Fivat
Rattrack R **antu**
Ravage H **buthidazole**
Ravatox R **scilliroside** Chemika
Ravel H **chloridazon** Burri
Ravelan G **chlormequat chloride** Avenarius
Raviac R **chlorophacinone** Agrotec, Lipha
Ravion I **carbaryl** Makhteshim-Agan
Ravyon IG **carbaryl** Alpha, Edefi, Makhteshim-Agan
RAX R **warfarin** Prentiss
Raxil F **terbuconazole**
RCR Multi-Tip Dressing IA **malathion** Rodent Control
RCR No. 25 Concentrate I **lindane + pyrethrins** Rodent Control
RCR No. 27 Ready for Use IA **piperonyl butoxide + pyrethrins** Rodent Control
RCR No. 28 Concentrate I **lindane + pyrethrins** Rodent Control
RCR No. 33 Ready for Use I **lindane + pyrethrins** Rodent Control
RCR No. 36 Aerosol I **lindane + pyrethrins** Rodent Control
RCR No. 40 Concentrate A **fenitrothion** Rodent Control
RCR No. 41 Insect Killer H **dichlorvos** Rodent Control
RCR No. 48 Ready for Use H **pyrethrins** Rodent Control
RD 406 HG **dichlorprop**
RD 2454 A **fluorbenside**
RD 4593 H **mecoprop**
RD 6584 F **dicloran**
RD 7693 H **benazolin**
RD 7901 F **sultropen**
RD 8684 F **quinazamid**
RD 14639 A **butacarb**

RE 4355 IA **naled** Chevron
RE-36290 H **cloproxydim**
Rebel H **glyphosate** Monsanto
Rebel F **copper oxychloride + mancozeb** KenoGard
Rebelate IA **dimethoate** BASF
Rebenveredlungswachs 'Riedel' FW **8-hydroxyquinoline sulphate** Hoechst
Rebwachs WF W **2,5-dichlorobenzoic acid + 8-hydroxyquinoline sulphate** Chauvin, Leu & Gygax, Staehler
Recin Super F **copper oxychloride + zineb** PVV
Recoil F **mancozeb + oxadixyl** Schering
Recop F **copper oxychloride** Sandoz
recozit Pflanzenspray IA **dimethoate** Reckhaus
recozit Rattentod R **warfarin** Reckhaus
red copper oxide F **cuprous oxide**
red squill R **scilliroside**
Redentin R **chlorophacinone** Chemolimpex, Reanal
Reducymol G **ancymidol** Elanco, Schering
Reflex H **fomesafen** ICI
Regazol H **diquat dibromide** Burri
Regidina F **dodine** Reis
Regisdrin I **mevinphos** Reis
Reglex H **diquat dibromide** ICI, Siapa, Sopra
Reglone H **diquat dibromide** Agrolinz, Agroplant, AVG, Berner, Ciba-Geigy, CTA, Delicia, Hokochemie, ICI, ICI-Valagro, ICI-Zeltia, Maag, Organika-Sarzyna, Pinus, Plantevern-Kjemi, Siegfried, Sinteza, Sopra, Spolana
Reglox H **diquat dibromide** ICI
Regran H **MCPA** Caffaro
Regufol G **chlormequat chloride** Inagra
Regulex G **gibberellic acid** ICI, Sopra
Regulor G **chlormequat chloride** Sedagri
Regulox G **maleic hydrazide** Burts & Harvey, Chipman
Reiniger van groene aanslag L **quaternary ammonium** Simus
Reldan I **chlorpyrifos-methyl** Bianchedi, Dow, Prochimagro, Ravit, Serpiol, Zorka-Sabac
Remacid A **cyhexatin** Burri
Remadion A **dicofol** Sedagri
Remanex A **tetradifon** Burri
Remanol A **dicofol** Burri
Remaphos I **azinphos-methyl + demeton-S-methyl sulphone** Burri
Remasan F **maneb** Rhone-Poulenc, Sedagri
Remiltine F **cymoxanil + mancozeb** Sandoz
Remiltine C F **copper oxychloride + copper sulphate + cymoxanil + mancozeb** Sandoz
Remiltine cuivre F **copper oxychloride + copper sulphate + cymoxanil + mancozeb** Sandoz
Remiltine F F **cymoxanil + folpet + mancozeb** Sandoz
Remiltine S F **cymoxanil + mancozeb** Sandoz
Remolex H **chloridazon + lenacil** Condor
Remtal H **simazine + trietazine** Schering
Remuron H **bromacil + diuron + simazine** Aseptan
Remuron TX H **amitrole + ammonium thiocyanate + diuron + simazine** Aseptan
Ren K S **hydroxyethyl cellulose + lauryl alcohol ethoxylate** Visplant
Renatox R **sulphur** Orastie
Renofluid royal H **glyphosate** Graines Loras
Renox H **MCPA** Ellagret
Rentokil Alphakil R **chloralose** Rentokil
Rentokil Ant & Crawling Insect Powder I **carbaryl** Rentokil
Rentokil Aquaspray ruiskute IA **fenitrothion** Rentokil
Rentokil Aquatox ruiskutejauhe IA **fenitrothion** Rentokil
Rentokil Biotrol R **warfarin** Rentokil
Rentokil Bromard R **bromadiolone** Rentokil
Rentokil Bromatrol R **bromadiolone** Rentokil
Rentokil Bromex R **bromadiolone** Rentokil
Rentokil Carpet Beetle Killer and Moth Proofer I **lindane + pyrethrins** Rentokil
Rentokil Cockroach Bait H **iodofenphos** Rentokil
Rentokil Deadline Rodenticide R **bromadiolone** Rentokil
Rentokil dispray white fogging fluid I **piperonyl butoxide + pyrethrins** Rentokil
Rentokil Fentrex R **difenacoum** Rentokil
Rentokil Fentrol R **difenacoum** Rentokil
Rentokil Fluorakil 3 X **fluoroacetamide** Rentokil
Rentokil Fly & Wasp Killer I **permethrin + tetramethrin** Rentokil
Rentokil Flytrol IA **diazinon** Rentokil

Rentokil FW R **warfarin** Rentokil
Rentokil Graanfumigant I **dichlorvos** Rentokil
Rentokil Insectrol Aerosol IA **diazinon + pyrethrins** Rentokil
Rentokil Insektrine I **deltamethrin** Rentokil
Rentokil Knox-Out 2FM IA **diazinon** Rentokil
Rentokil Liquid Bromatrol R **bromadiolone** Rentokil
Rentokil Malatox I **malathion** Rentokil
Rentokil Muscatrol I **permethrin** Rentokil
Rentokil Phostoxin F **aluminium phosphide** Rentokil
Rentokil Scarecrow strib X **polybutene + polyethylene wax** Rentokil
Rentokil Tri-spot L **quaternary ammonium** Rentokil
Rentokil Wasp Nest Killer I **carbaryl** Rentokil
Rentokil Waspex H **iodofenphos** Rentokil
Rentokil's drikkegiftpraeparat R **thallium sulphate** Rentokil
Rentokil's insektdraeber I **piperonyl butoxide + pyrethrins** Rentokil
Rentospray White I **piperonyl butoxide + pyrethrins** Rentokil
Repelente Sapec X **fish oil** Sapec
Repentol 6 X **odorous substances** Chema
REPTO H **amitrole + 2,4-D + diuron** Bayer
Repulse F **chlorothalonil** ICI
Resbuthrin I **resmethrin** Cooper, Nordisk Alkali
Rescue H **2,4-DB + naptalam** Uniroyal
Resiben N H **linuron + trifluralin** Burri
Residox H **atrazine** Chipman
Residuren Extra H **chlorpropham + diuron** Farm Protection
Resisan F **dicloran** Nissan
Reslin I **bioallethrin + permethrin + piperonyl butoxide** Wellcome
resmethrin I [10453-86-8] 5-benzyl-3-furylmethyl (1*RS*)-*cis,trans*-chrysanthemate, resmethrine, OMS 1206, Chryson, FMC 17370, Isathrine, NRDC 104, Pynosect, Respond, SBP-1382, Scourge, Synthrin
resmethrine I **resmethrin**
Resolut-D H **2,4-D** Nordisk Alkali
Respond I **resmethrin** Penick
Responsar I **cyfluthrin** Bayer
Restanox A **fenbutatin oxide** Siapa
Retacel G **chlormequat chloride** Lucebni zavody, VCHZ Synthesia
Retacel Super G **chlormequat chloride + ethephon** Lucebni zavody
Retard G **maleic hydrazide** Drexel
Retenox G **propham** Siegfried
Retenox-combi G **chlorpropham + propham** Siegfried
Reuze Ratenit Blok R **warfarin** De Rauw
Revecel G **chlormequat chloride** Agriplan
Revenge H **dalapon + TCA** Hopkins
Revox H **isoproturon + trifluralin** Hoechst, Pluess-Staufer, Procida
Reward H **vernolate** Stauffer
Rexatiao I **malathion** Reis
Rexoquin G **ethoxyquin** Decco-Roda, Pennwalt France
Rezgalic F **copper oxychloride** Sojuzchimexport
Rezoxiklorid F **copper oxychloride** Sojuzchimexport
RH-0265 H **fluoroglycofen-ethyl** Rohm & Haas
RH-315 H **propyzamide** Rohm & Haas
RH73-2 F **mancozeb + sulphur** Rohm & Haas
RH-787 R **pyrinuron**
RH-893 WFB **octhilinone** Rohm & Haas
RH-2161 F **fenapanil**
RH-2512 H **nitrofluorfen**
RH-2915 H **oxyfluorfen** Rohm & Haas
RH-3866 F **myclobutanil** Rohm & Haas
RH-3928 F **furophanate**
RH-6201 H **acifluorfen-sodium** Rohm & Haas
Rhino H **atrazine + butylate** PPG
Rhizopon A G **indol-3-ylacetic acid** ACF, Aifar, Huebecker, Liro, Puteaux, Turba Projar
Rhizopon AA G **4-indol-3-ylbutyric acid** ACF, Huebecker, Liro, Puteaux, Turba Projar
Rhizopon B G **1-naphthylacetic acid** ACF, Aifar, Huebecker, Liro, Puteaux, Turba Projar
Rhizopon Plantenspray IA **piperonyl butoxide + pyrethrins** ACF
Rhizotox rookboom F **trioxymethylene** NOC
Rhodan I **endosulfan** Ellagret
Rhodan Super I **endosulfan + parathion-methyl** Ellagret
Rhodane I **TDE**
Rhodax F **folpet + fosetyl-aluminium** Hoechst

Rhodax F **fosetyl-aluminium + mancozeb** Rhone-Poulenc
Rhoden I **propoxur** Agrotec
Rhodiacide AI **ethion** Rhone-Poulenc
Rhodianebe F **maneb** Rhone-Poulenc
Rhodiasan F **thiram** Rhodiagri-Littorale
Rhodiasoufre F **sulphur** Rhodiagri-Littorale
Rhodiatox IA **parathion** Hellafarm, Rhodiagri-Littorale
Rhodiatox acaricide IA **ethion + parathion** Rhodiagri-Littorale
Rhodiatox Kombi IA **azinphos-methyl + demeton-S-methyl sulphone** Agrotec
Rhodocide IA **ethion** Hellafarm, Rhodiagri-Littorale, Rhone-Poulenc
Rhodofix G **1-naphthylacetic acid** Maag, Rhodiagri-Littorale, Rhone-Poulenc
Rhomene H **MCPA** Rhone-Poulenc
Rhonox H **MCPA** Rhone-Poulenc
Ribinol F **8-hydroxyquinoline sulphate** Riedel de Haen
Ribinol I **permethrin** Staehler
Ric 16 F **calcium copper oxychloride** Caffaro
Rico H **anilofos** Hoechst
Ricochet H **glyphosate + simazine** Monsanto
Rid Brush H **2,4-D + triclopyr** Certified Lab.
Ridak R **difenacoum** ICI
Ridall R **zinc phosphide** Lipha
Rideon H **diphenamid** Chemolimpex, Kobanyai
Ridomil F **metalaxyl** Ciba-Geigy, Ligtermoet, Liro, Viopharm
Ridomil Combi F **folpet + metalaxyl** Ciba-Geigy, Liro
Ridomil delta F **fentin acetate + maneb + metalaxyl** Ligtermoet
Ridomil Extra F **copper oxychloride + metalaxyl** Ciba-Geigy
Ridomil-Fitorex F **mancozeb + metalaxyl** Ciba-Geigy
Ridomil-Folpet F **folpet + metalaxyl** Ciba-Geigy
Ridomil-Kupfer F **copper oxychloride + metalaxyl** Ciba-Geigy
Ridomil MBC F **carbendazim + metalaxyl** Ciba-Geigy
Ridomil Multi F **chlorothalonil + metalaxyl** Ciba-Geigy
Ridomil MZ F **mancozeb + metalaxyl** BASF, Ciba-Geigy, Duslo, Ruse, Viopharm
Ridomil plus F **copper oxychloride + metalaxyl** Budapesti Vegyimuvek, Ciba-Geigy, Liro, Ruse, Spolana
Ridomil Special F **mancozeb + metalaxyl** Liro
Ridomil TK F **mancozeb + metalaxyl** Ciba-Geigy
Ridomil Triple F **copper oxychloride + folpet + metalaxyl** Ciba-Geigy
Ridomil Z F **metalaxyl + zineb** Ciba-Geigy, Spolana, Zupa
Ridomil zeta F **metalaxyl + zineb** Ligtermoet
Ridomil Zineb F **metalaxyl + zineb** Budapesti Vegyimuvek, Ciba-Geigy
Riem Antivermine I **pyrethrins + S421** Riem
Rifit H **pretilachlor** Ciba-Geigy
Rigenal G **1-naphthylacetic acid** CIFO
Rikkaruohontuho Prefix H **chlorthiamid** Kemira
Rilof H **piperophos** Ciba-Geigy
Rim-Killer R **warfarin** Mylonas
Rimidin F **fenarimol** Elanco, Eli Lilly
Rimidine Plus F **carbendazim + fenarimol + maneb** Eli Lilly
Rimitox R **chlorophacinone** Sovilo
Rinal-Giftkoerner R **zinc phosphide** Vorratsschutz
Rinal-Insekten-Strip I **dichlorvos** Vorratsschutz
Rinal-Maeusekoeder R **warfarin** Delitia, Detia Freyberg, Vorratsschutz
Rinditol I **malathion** Tsilis
Ringer F **tridemorph** Pan Britannica
Ringmaster F **oxycarboxin** Rhone-Poulenc
Ringo H **chlorthal-dimethyl + propachlor** Masso
Riozeb F **mancozeb** ERT
Riozeb Cobre F **copper sulphate + mancozeb** ERT
Riozeb fuerte F **carbendazim + mancozeb** ERT
Ripcord I **cypermethrin** Agrishell, Agroplant, Du Pont, Galenika, Grima, Kemira, Shell, Sipcam, Spiess, Spolana
Risagro V I **endosulfan** Ital-Agro
Riscald Control G **ethoxyquin** Caffaro
Riselect H **propanil** Agrimont, Montedison
Risolex F **tolclofos-methyl** Spiess, Sumitomo, Urania
Risolutiv H **glyphosate** Scam
Risonet STM H **propanil** Ravit
Risoter F **quinacetol sulphate**

Risoverd E I **endosulfan** Sivam
Rival H **glyphosate + simazine** Inagra
Rival F **fenpropimorph + prochloraz** Schering
Rizolex F **tolclofos-methyl** Kwizda, Nordisk Alkali, Schering, Schering AAgrunol, Siapa, Sopra, Sumitomo
Rizomonda H **propanil** Quimigal
Rizotril H **propanil** Condor
RL 20 IA **dimethoate** EniChem
RL 40 IA **dimethoate** EniChem
Ro 06-0761 G **piproctanyl bromide** Maag
Ro 07-6145 G **dikegulac-sodium**
Ro 10-3108 I **epofenonane**
Ro 12-3049 F **fenpropidin** Maag
Ro 15-1297 F **pyrifenox** Maag
Ro-Neet H **cycloate** Agriben, Bayer, Cillus, Hartsikemia, Kwizda, OHIS, Organika-Sarzyna, Procida, Rhone-Poulenc, Serpiol, Siapa, SPE, Spiess, Stauffer, Urania
Roach Prufe I **boric acid** Medley
Rodalon L **benzalkonium chloride** Ferrosan
Rodamon FF R **chlorophacinone + sulphaquinoxaline** Du Pont
Rodan R **warfarin** Billen
Rodan Overdose R **difenacoum** Billen
Rodeclor-Grani R **chlorophacinone** Sici
Roden Esca R **warfarin** Sofital
Rodene R **warfarin** Colkim, Farmer
Rodent Cake R **diphacinone** Bell Labs
Rodenticide AG R **zinc phosphide** Motomco
Rodentin R **warfarin** Nitor, Plantevaern
Rodentox R **chlorophacinone** Organika-Fregata
Rodeo H **glyphosate** Monsanto
Rodex R **warfarin** Hopkins, Hygan
Rodex R **fluoroacetamide** Jewnin-Joffe, Zootechniki
Rodinec R **calciferol**
Rodofix G **1-naphthylacetic acid** Condor
Rodontal R **bromadiolone** Ital-Agro
Rody AI **fenpropathrin** Sumitomo, Shell
Roegpatroner for Rotter R **sulphur** Zuschlag
Rogana TK I **dialifos + trifenofos** Rohm & Haas
Rogatox IA **dimethoate** Scam
Rogodan I **dimethoate + endosulfan** Agrimont
Rogor IA **dimethoate** Agrimont, Chromos, CTA, Diana, Hokochemie, ICI, Inca, Kwizda, Marktredwitz, Masso, Meoc, Montecatini, Montedison, Plantevern-Kjemi, Pluess-Staufer, Schering, Siapa, Sopra, Spiess, Urania
Rogor-Fly-Kombi I **dimethoate + fenitrothion** Pluess-Staufer
Rogoter IA **dimethoate** Terranalisi
Rokar X H **bromacil** Siapa
ROL IA **tetrachlorvinphos** SDS Biotech
Romal I **malathion + rotenone** Aziende Agrarie Trento
Romefos IA **dimethoate** Agrocros
Rometan IA **dimethoate** Aragonesas
Romin H **chloridazon + cycloate** Serpiol
Ronald S G **dichlorprop + 1-naphthylacetic acid** Farmer
Ronamid H **propyzamide** Visplant
Rondo F **captan + pyrifenox** Maag
Ronebeet H **chloridazon + cycloate** Siapa
Roneet H **cycloate** Stauffer
Ronex H **diuron** Protex
Ronicur F **cymoxanil + metiram + vinclozolin** BASF
Ronilan F **vinclozolin** Agrolinz, BASF, Ciba-Geigy, Collett, Compo, Intrachem, Lucebni zavody, Luxan, Mir, Nitrokemia, OHIS, Sapec, Siegfried, Spiess, Urania
Ronilan M F **maneb + vinclozolin** BASF
Ronilan S F **sulphur + vinclozolin** BASF, R.S.R.
Ronilan Speciaal F **chlorothalonil + vinclozolin** BASF
Ronilan T Combi F **thiram + vinclozolin** BASF
ronnel I **fenchlorphos**
Ronstar H **oxadiazon** Condor, Hortichem, Nitrokemia, Ravit, Rhodiagri-Littorale, Rhone-Poulenc, Zorka Subotica
Ronstar TX H **carbetamide + oxadiazon** Rhodic
Ronweed H **diuron** Aziende Agrarie Trento
Root Guard IA **diazinon** Murphy
Root Out H **ammonium sulphamate** Dax
Rooting Powder F **captan + 4-indol-3-ylbutyric acid + 1-naphthylacetic acid** Sorex
Rootone G **1-naphthylacetic acid** Agriben, C.F.P.I.
ROP F **iprodione** Rhone-Poulenc
Ros 1 I **piperonyl butoxide + pyrethrins** Toersleff
Rosabel FI **cypermethrin + propiconazole** Liro
Rosabel Extra FI **cypermethrin + penconazole** Liro

Roseclear FI **bupirimate + pirimicarb + triforine** ICI
Rosemox S **polyglycolic ethers** Rhodiagri-Littorale
Rosen Myctan kombiniert FIA **piperonyl butoxide + pyrethrins + sulphur** Neudorff
Rosen- und Zierpflanzenspray Spiess-Urania FI **butocarboxim + fenarimol** Spiess
Rosen-Spray FI **pirimicarb + triforine** Epro
Rosenspray FIA **piperonyl butoxide + pyrethrins + sulphur** Spiess
Rosenspray 119 F **dodemorph acetate** Thompson
Rosenspray Combi plus FIA **piperonyl butoxide + pyrethrins + sulphur** Shell
Rosen-Spritz S F **triforine** Flora-Frey
Rosenspritzmittel P F **propiconazole** Ciba-Geigy
Rosenspritzmittel Saprol F **triforine** Shell
Rosenspritzmittel Spezial FI **pirimicarb + triforine** Epro
Rosex R **warfarin**
Rosol Combi FI **dinocap + piperonyl butoxide + pyrethrins** Ewos
Rospan A **chloropropylate** Ciba-Geigy
Rospin A **chloropropylate** Ciba-Geigy, OHIS
Rospin IA **azinphos-methyl + demeton-S-methyl sulphone** Urania
Rosquiver I **cryolite + sodium fluosilicate** KenoGard
Rosuran H **monuron** Spiess, Urania
Rotacide IA **rotenone** Fairfield
Rotalin H **linuron** Farm Protection
Rotanmyrkky 42 Extra Syotti R **warfarin** Farmos
Rotanmyrkky "242" R **coumatetralyl** Farmos
Rotanmyrkky "342" R **bromadiolone** Farmos
Rotenol IA **piperonyl butoxide + pyrethrins** Schacht
rotenone IA [83-79-4] (2*R*,6a*S*,12a*S*)-1,2,6,6a,12,12a-hexahydro-2-isopropenyl-8,9-dimethoxychromeno[3,4-*b*]furo[2,3-*h*]chromene-6-one, derris, cube, timbo, barbasco, haiari, nekoe, ENT 133, Chem-Fish, Cuberol, Fish Tox, Noxfire, Noxfish, Rotacide, Sicid
Rotox special-opblandet rottemiddel R **warfarin** Rotox
Rotox stroepudder rottegift R **warfarin** Rotox
Rotta-Rakumin R **coumatetralyl** Berner
Rottal G H **borax + bromacil** Marktredwitz
Roundup H **glyphosate** Andersen, BASF, Bjoernrud, Condor, Esbjerg, Hellafarm, Hogervorst-Stokman, Imex-Hulst, Monsanto, Organika-Sarzyna, Pinus, Quimigal, Ravit, Sandoz, Schering, Shell, Sipcam, Sivam, Spolana
Rout H **bromacil**
Rout H **oryzalin + oxyfluorfen** Sierra
Rover F **chlorothalonil** Chiltern
Rovis IA **dimethoate** Visplant
Rovlinka I **dioxacarb** Nitrokemia
Rovral F **iprodione** Agriben, Agronorden, Agrotec, BVK, Chromos, Condor, Ewos, Farmos, Hoechst, Hortichem, Imex-Hulst, Lucebni zavody, Nitrokemia, Plantevern-Kjemi, Ravit, Rhodiagri-Littorale, Rhodic, Rhone-Poulenc
Rovral TS F **carbendazim + iprodione** Agriben, Rhone-Poulenc, Zorka Subotica
Rovral UFB F **carbendazim + iprodione** Agriben, Agrotec
Rovral UTB F **carbendazim + iprodione** Agrotec
Rowmate H **dichlormate**
Rowral F **iprodione** Rhone-Poulenc
Roxasect tegen luis en spint op planten I **phenothrin + tetramethrin** Duphar
Roxion IA **dimethoate** Bjoernrud, Celamerck, Chimac-Agriphar, Epro, Geopharm, Gullviks, Kemira, Margesin, Permutadora, Sandoz, Shell, Siegfried, Sovilo
Royal MH-30 G **maleic hydrazide** Ligtermoet, Uniroyal
Royal Slo-Gro HG **maleic hydrazide** Uniroyal
Royaltae G **1-decanol** Uniroyal
Rozen Fungicide Wolf-Gerate F **dodemorph acetate + dodine** Graham
Rozenspray Funginex FI **diazinon + tetradifon + triforine** Chimac-Agriphar
Rozol R **chlorophacinone** Lipha
Roztoczol extra plynny 8 A **tetradifon** Organika-Azot
7175 RP I **endothion**
10465 RP IA **vamidothion** Rhone-Poulenc
11561 RP H **carbetamide** Rhone-Poulenc
11974 RP IA **phosalone** Rhone-Poulenc
15380 RP H **ioxynil** Rhone-Poulenc
17623 RP H **oxadiazon** Rhone-Poulenc
23465 RP H **dimefuron**
26019 RP F **iprodione** Rhone-Poulenc

RP-Thion AI **ethion** Volrho
R.S. 2 Souris R **difenacoum** Steininger
RU 15525 I **kadethrin** Roussel Uclaf
RU 22974 I **deltamethrin** Roussel Uclaf
RU 25474 I **tralomethrin** Roussel Uclaf
Ruban I **bensultap** Takeda
Rubidor I **azamethiphos** Ciba-Geigy
Rubigan F **fenarimol** Budapesti Vegyimuvek, Elanco, Eli Lilly, Grima, Pepro, Radonja, Rhone-Poulenc, Sandoz, Siapa, Spiess, Urania, VCH
Rubigan Combi F **fenarimol + sulphur** Eli Lilly
Rubigan Plus F **dodine + fenarimol** Radonja
Rubinol I **parathion-methyl** Candilidis
Rubitox IA **phosalone** Agriben, Rhone-Poulenc, Urania
Rubrum G **1-naphthylacetic acid** Tecniterra
Rubson "anti-mos" L **quaternary ammonium** Roberts
Ruebenunkrautmittel H **chloridazon** Stinnes
Ruga H **dicamba + MCPA + mecoprop** Pluess-Staufer
Rugby NI **S,S-di-sec-butyl O-ethyl phosphorodithioate** FMC
Rugby N **ebufos** Pepro
Rumesan ulje I **DNOC + petroleum oils** Pinus
Rumetan R **zinc phosphide** Hoechst
Rumex-Jet-Blackengranulat H **2,4-D + mecoprop** Hokochemie
Rumexan H **dicamba + mecoprop** Kwizda
Runcatex-MCPP H **mecoprop** Van Wesemael
Runol X **animal meal + fatty acids** Forst-Chemie
Rustica Netzschwefel FA **sulphur** Du Pont
ryanodine I [15662-33-6] (2*S*,3*S*,4*R*,4a*S*,5*S*,5a*S*,8*S*,9*R*,9a*R*,9b*R*)-2,3,4a,5a,9,9b-hexahydro-3-isopropyl-2a,5,8-trimethylperhydro-2,5-methanobenzo[1,2]pentaleno[1,6-*bc*]furan-4-ylpyrrole-2-carboxylate, Ryno-tox
Rydex H **prodiamine**
Ryl F **folpet** Sedagri
Ryl cuivre F **copper oxychloride + zineb** Sedagri
Ryltex IG **carbaryl** Protex
Ryno-tox I **ryanodine**
Ryzelan H **oryzalin**

S-47 H **bromobutide** Sumitomo
S 276 IA **disulfoton** Bayer
S 309 IA **oxydisulfoton**
S 410 IA **oxydeprofos**
S421 S [127-90-2] 1,1'-oxybis[2,3,3,3-tetrachloropropane]
S 572 F **captan + dicloran** Siapa
S-767 NI **fensulfothion** Bayer
S-1046 I **xylylcarb** Sumitomo
S-1102A I **fenitrothion** Bayer
S-1358 F **buthiobate** Sumitomo
S-1605 F **diethofencarb** Sumitomo
S-1752 I **fenthion** Bayer
S-1844 I **esfenvalerate** Sumitomo
S-1942 I **bromophos** Celamerck
S-2225 IA **bromophos-ethyl** Celamerck
S-2539 I **phenothrin** Sumitomo
S-2852 Forte I **empenthrin** Sumitomo
S-2864 H **butamifos** Sumitomo
S-2940 IA **phenthoate** Sumitomo
S-3206 AI **fenpropathrin** Sumitomo
S-3307 D G **uniconazole** Sumitomo
S-3308L F **diniconazole** Sumitomo
S-3349 F **tolclofos-methyl** Sumitomo
S 4068 I **prallethrin** Sumitomo
S-4084 I **cyanophos** Sumitomo
S-4087 I **cyanofenphos**
S-4347 H **bromobutide** Sumitomo
S-5602 IA **fenvalerate** Sumitomo
S-5620Aα I **esfenvalerate** Sumitomo
S-5660 I **fenitrothion** Sumitomo
S 6000 H **cypromid**
S 6115 H **cyprazine**
S 6173 H **benzadox**
S-6876 IA **omethoate** Bayer
S-7131 F **procymidone** Sumitomo
S-15076 H **ethiolate**
S-18510 H **benzipram**
S 19073 H **cyprazole**
S 21634 H **cyperquat chloride**
S-32165 F **diethofencarb** Sumitomo
S-5620Aα I **esfenvalerate** Sumitomo
Sabet H **cycloate** Chemolimpex, Sajobabony
Sabithane F **dinocap + myclobutanil** Rohm & Haas
Sabre H **bromoxynil** La Quinoleine
SADH G **daminozide**
Sadofos I **malathion** Organika-Azot
Sadolin algefjerner L **benzalkonium chloride** Sadolin
Sadoplon F **thiram** Organika-Azot
Safethion I **malathion** Agronova
Safidon IA **phosmet** Budapesti Vegyimuvek, KenoGard, Safor, Sajobabony

Safroil IA **petroleum oils** Safor
Safrotin I **propetamphos** Esbjerg, Geopharm, Sandoz
Saibrom O FHIN **chloropicrin + methyl bromide** Aporta
Saindane I **lindane** Agrishell
SAIsan F **drazoxolon** ICI
Saitofos I **malathion + methoxychlor + parathion** Siapa
Sakarat R **warfarin** Killgerm
Sakarat Special H **chlorophacinone** Killgerm
Sakkimol H **molinate** Chemolimpex, Sajobabony
Sal amina 2,4-D H **2,4-D** Montecinca
Salama-sumutin IA **piperonyl butoxide + pyrethrins** Kemira
salclomide F **trichlamide**
Saldo S **alkyl phenol ethoxylate** SEGE
Salfor IA **dimethoate** Aporta
salicylanilide F [87-17-2] 2-hydroxy-*N*-phenylbenzamide, Shirlan
Salinkaroktono M **metaldehyde** Ellagret
Salithiex F **procymidone** ICI-Zeltia
Salithion I **dioxabenzofos** Sumitomo
Salut IA **chlorpyrifos + dimethoate** Shell
Salute H **metribuzin + trifluralin** Mobay
Salvagrano I **lindane** Scam
Salvo NFHI **dazomet** Stauffer
Salvo H **dimethylarsinic acid**
Saminol H **amitrole + simazine** Ciba-Geigy, Ligtermoet, Liro, Viopharm
Saminol Super H **amitrole + chlorbromuron + simazine** Liro
Samuron H **desmetryn**
San-75 H **MCPA** Plantevern-Kjemi
SAN 155I I **thiocyclam hydrogen oxalate** Sandoz
SAN 197I IA **etrimfos** Sandoz
SAN 201I I **lirimfos**
SAN 371F F **oxadixyl** Sandoz
SAN 6706H H **metflurazon**
SAN 6913I IA **formothion** Sandoz
SAN 6915H H **metoxuron** Sandoz
SAN 7102H H **metoxuron** Sandoz
SAN 52123H H **iprymidam**
SAN 52139I I **propetamphos** Sandoz
Sanac Plantspray IA **malathion + piperonyl butoxide + pyrethrins** Sanac
Sanac Totale Onkruiddoder H **sodium chlorate** Sanac
Sanagricola F **copper oxychloride + zineb** Agrocros
Sanam M F **maneb** Sanac
Sanam Z F **ziram** Sanac
Sanaseed R **strychnine**
Sanata H **TCA** Sanac
Sanatir F **fentin hydroxide** Sanac
Sanavit F **copper oxychloride + folpet + sulphur** Scam
Sanax Fluessigschwefel FA **sulphur** Kwizda
Sanazin H **simazine** Sanac
Sancap H **dipropetryn** Ciba-Geigy
Sancer F **copper oxychloride** KenoGard
Sancol F **oxadixyl + propineb** Bayer
Sandocar I **carbaryl** Sandoz
Sandofan F **oxadixyl** Sandoz
Sandofan C F **copper oxychloride + oxadixyl** Sandoz, Turda
Sandofan CM F **copper oxychloride + copper sulphate + mancozeb + oxadixyl** Sandoz
Sandofan F F **folpet + oxadixyl** Chromos, Sandoz
Sandofan M F **mancozeb + oxadixyl** Bjoernrud, Dow, Duslo, Sandoz, Spiess, Urania
Sandofan Manco F **mancozeb + oxadixyl** Sandoz
Sandofan Super F **copper oxychloride + oxadixyl** Sandoz
Sandofan YM F **cymoxanil + mancozeb + oxadixyl** Sandoz
Sandofan-Z F **folpet + oxadixyl** Chromos
Sandol F **oxadixyl + propineb** Bayer
Sandolin A IA **DNOC** Hortichem
Sandolina IA **DNOC** Sandoz
Sandoline HI **DNOC** Sandoz
Sandoline A FI **DNOC** Sandoz
Sandomil F **carbendazim** Sandoz
Sandomil C F **captan + carbendazim** Sandoz
Sandomil SZ F **carbendazim + mancozeb** Sandoz
Sandothion I **fenitrothion + formothion** Sandoz
Sandovit S **polyglycolic ethers** Chromos, Sandoz
Sandovit A S **hydroxyethyl cellulose + lauryl alcohol ethoxylate** Sandoz
Sandoz 6538 IA **quinalphos** Sandoz
Sandoz cuivre 407 F **cuprous oxide** Sandoz
Sandozeb F **mancozeb** Sandoz
Sanex BU F **TCMTB** KenoGard
Sanexter Nema FHIN **metham-potassium** KenoGard
Sanexter V FHIN **metham-sodium** KenoGard

Sanfol F **folpet** Sandoz
Sanguano-Spezial-Rasenduenger mit Moosvernichter H **iron sulphate** Mahle Duenger
Sani APM H **amitrole + atrazine** Sanigene
Sani ATL H **amitrole + atrazine** Sanigene
Saniclor F **quintozene** Rhone-Poulenc
Sanifum F **copper oxychloride + zineb** Argos
Sanigran Special IXF **anthraquinone + ethion + lindane + methoxyethylmercury silicate** Rhone-Poulenc
Saniter I **fenitrothion** Sepran
Saniterpen insecticide DK I **deltamethrin** DRT
Sanizole TA H **amitrole + ammonium thiocyanate** Sanigene
Sanjosan I **dinoseb** Bayer
Sankel FB **nickel dimethyldithiocarbamate** Mikasa
Sanmarton IA **fenvalerate**
Sanol I **carbaryl** Luqsa
Sanoplant Bio Spritzmittel IA **pyrethrins** Maag
Sanozoil S **petroleum oils** Sanac
Sanseal FW **captafol** Sandoz
Sanspor F **captafol** ICI
Sanspor Ramato F **captafol + copper oxychloride** ICI
Santar FW **mercuric oxide** Sandoz
Santar SM F **captafol** Sandoz, Urania
Santhane F **captan** Sipcam
Santoquin G **ethoxyquin**
Sanugec F **thiram** Sipcam-Phyteurop
Sanvex I **cartap hydrochloride** Sipcam, Takeda
Sanzerb H **sodium chlorate** LHN
SAP H **bensulide**
Sapecron IA **chlorfenvinphos** BASF, Ciba-Geigy, Ligtermoet, Liro
Saphiben I **fenitrothion** ERT
Sappiran A **chlorfenson** Nippon Soda
Saprol F **triforine** Bjoernrud, Celemewrck, Chimac-Agriphar, Epro, Geopharm, Kemira, KVK, Margesin, Nitrokemia, Permutadora, Prochimagro, Sandoz, Shell, Zupa
Sarapron H **amitrole + atrazine + dicamba** BP
Sarclex H **linuron** Pepro, Rhone-Poulenc
Sari F **maneb + vinclozolin** Procida
Sarmite A **fenson + propargite** EniChem
Sarolex IA **diazinon** Ciba-Geigy
Sartion IA **azinphos-methyl** EniChem
Satan M **metaldehyde** Interdrug
Satecid H **propachlor** Sajobabony
Saterb H **terbutryn** Sajobabony
Satisfar I **etrimfos** Nickerson, Sandoz
Satoklor H **alachlor** Sajobabony
Satrazinol H **atrazine** Sanac
Satunil H **propanil + thiobencarb**
Saturn H **thiobencarb** Agrotechnica, Argos, Eli Lilly, Kumiai, Sopepor
Saturn-D H **2,4-D + thiobencarb** Kumiai
Saturn-M H **chlornitrofen + thiobencarb**
Saturn-S H **molinate + thiobencarb** Argos, Sapec
Saturnal H **clopyralid + mecoprop** Schering
Saturnmate H **linivan + thiobencarb** Kumiai
Saturno H **thiobencarb** Kumiai
Saturyl H **bromoxynil + clopyralid + MCPA + mecoprop** Schering
Saunakukka Hedonal H **2,4-D + mecoprop** Berner
Saunakukka Herbotal H **2,4-D + mecoprop** Farmos
Savall I **quinalphos** Farm Protection
Saverit H **vernolate** Chemolimpex
Savey A **hexythiazox** Du Pont
Savirade H **chlorotoluron + metoxuron** Sandoz
Savirox H **vernolate** Chemolimpex
Savit IG **carbaryl** Griffin
SBP-1382 I **resmethrin** Penick
SC-0574 H **prosulfocarb**
Scabexol IA **lindane** Biopharm
Scabia IA **lindane** Feed Farm
Scabicidin IA **lindane** Apharma
Scabicurin IA **lindane** Duphar
Scabinol IA **lindane** Alfasan
Scaldex DPA G **diphenylamine** Margesin
Scasol I **lindane** SCAC-Fisons
Scepter H **imazaquin** Cyanamid, Zorka Sabac
Schaedlings-Vernichter Decis I **deltamethrin** Shell
Schaedlingsfrei Parexan IA **piperonyl butoxide + pyrethrins** Shell
Schaedlingsfrei Parexan N IA **piperonyl butoxide + pyrethrins** Shell
Schapenwasmiddel "Wolfederatie" IA **lindane** Kommer
Schaumann Fertigkoeder R **coumatetralyl** Bayer
Scherpa I **cypermethrin** Rhone-Poulenc
Schimmelbestrijdingsmiddel funginex FI **triforine** Shell
Schiumarat R **norbormide** Colkim

Schnecken-frei M **metaldehyde** Flora-Frey
Schneckenkoerner M **metaldehyde** COOP, CTA, Meoc, Pluess-Staufer, Reckitt & Colman, Wyss Samen & Pflanzen
Schneckenkorn M **metaldehyde** Baur, Dehner, Dow, Fisons, Shell, Spedro, Spiess, Urania, Wuelfel
Schneckenkorn Mesurol M **methiocarb** Bayer
Schneckenkorn Spezial M **metaldehyde** Pluess-Staufer
Schnecken-Loesung Limagard M **ethanol** Shell
Schneckenschutzband M **metaldehyde** Neomat
Schneckentod M **metaldehyde** A.C.I.E.R., Schacht
Schneckenvertilgungsmittel 'Radikal' M **metaldehyde** Nagel
Schneckex M **metaldehyde** Nagel
Schnecktex M **metaldehyde** Breiler
Schola Rasenduenger mit Unkrautvernichter H **2,4-D + MCPA** Schomaker
Schola-Unkrautvertilger H **sodium chlorate** Schomaker
schradan I [152-16-9] octamethylpyrophosphoric tetra-amide, schradane, OMPA, Pestox 3, Pestox III, Sytam
schradane I **schradan**
Schrozberger (WLZ) Giftweizen R **zinc phosphide** SBG
Schwefelsaeure H **sulphuric acid** Coswig
Sciandor H **linuron + trifluralin** Eli Lilly
scilliroside R [507-60-8] 3β-(β-D-glucopyranosyl)-17β-(2-oxo-2*H*-pyran-5-y l)-14β-androst-4-ene-6β,8,14-triol 6-acetate, red squill, Silmine, Silmurin
Scipio IA **cypermethrin + ethion** Rhodiagri-Littorale
Scleran F **dicloran** Scam
Sclerosan F **dicloran** Eli Lilly
Sclex F **dichlozoline**
Scoop H **metsulfuron-methyl + thiameturon-methyl** Procida
Scotts Gazonmest met Mosbestrijder H **iron sulphate** Wolf
Scotts-onkruidverdelger H **2,4-D + dicamba** Wolf
Scourge I **resmethrin** Penick
Scout H **acifluorfen-sodium** Ravit
Scout I **tralomethrin** Roussel Uclaf
Screen F **flurazole** Monsanto
Scrubmaster H **tebuthiuron** Burts & Harvey
Scythe H **paraquat dichloride** Cyanamid
SD 3562 IA **dicrotophos**
SD 4294 I **crotoxyphos** Shell
SD 4402 I **isobenzan**
SD 7859 IA **chlorfenvinphos** Shell
SD 8280 I **dimethylvinphos** Shell Kagaku
SD 8447 IA **tetrachlorvinphos** Shell
SD 9129 IA **monocrotophos** Shell
SD 11831 H **nitralin** Shell
SD 14114 A **fenbutatin oxide** Shell
SD 15418 H **cyanazine** Shell
Seal & Heal FW **thiophanate-methyl** May & Baker
Secantin H **paraquat dichloride** Du Pont
secbumeton H [26259-45-0] 2-*sec*-butylamino-4-ethylamino-6-methoxy-1,3,5-triazine, Etazine, GS 14254
Seccatutto H **diquat dibromide + paraquat dichloride** ICI
Secto Florgard Spike H **dichlorvos** Secto
Secto Fly & Wasp Killer IA **pyrethrins** Secto
Secto House & Garden Powder I **lindane + pyrethrins** Secto
Secto Insect Killer Powder I **lindane + pyrethrins** Secto
Secto Slow Release Fly Killer H **dichlorvos** Secto
Secto Wasp & Ant Killer Aerosol I **lindane + pyrethrins** Secto
Sectosol IA **lindane** Sectolin
Sectovap H **dichlorvos** Secto
Sectrol I **N-octylbicycloheptenedicarboximide + piperonylbutoxide + pyrethrins** 3M
Securex I **thiodicarb** Condor, Union Carbide
Security IH **calcium arsenate**
Securol I **malathion** Schering
Sediazol I **diazinon** Sepran
Sedit I **carbaryl** Chimiberg, Sivam
Sedlene G **(2-naphthyloxy)acetic acid** Aifar
Sedumil-Col S H **simazine** ICI-Zeltia
Sedumin-Col A H **atrazine** ICI-Zeltia
Sedumin doble AS H **atrazine + simazine** ICI-Zeltia
Seedox I **bendiocarb** Schering, Schering AAgrunol
Seedoxin I **bendiocarb** Schering, Staehler
Seedtox F **phenylmercury acetate** All-India Medical
Seftal F **captan** Sepran

Segetan Giftweizen R **zinc phosphide** Spiess, Urania
Segor IA **dimethoate** Sepran
Segrene F **maneb + thiram + zineb** Tecniterra
Seis-Tres I **parathion + parathion-methyl** Drexel
Sektivap I **dichlorvos** Shell
Sektivap Bio Green Arrow I **permethrin + pyrethrins** Blumoeller
Selecron IA **profenofos** Ciba-Geigy, Viopharm
Select ban H **dicamba + mecoprop** Certified Lab.
Selectaire I **dichlorvos** Wellcome
Selectane H **2,4-D + MCPA** C.G.I.
Selective Sandoz H **dinoseb** Sandoz
Selectone H **cloproxydim**
Selectone A H **2,4-D** Agriphyt
Selectone E H **2,4-D** Agriphyt
Selectox Royal H **dicamba + ioxynil + MCPA + mecoprop + 2,4,5-T** Graines Loras
Selectweed H **dinoseb** Protex
Selectyl H **MCPA** Agriphyt, Chimac-Agriphar
Selectyl MD H **2,4-D + MCPA** Agriphyt
Selektin Kombi H **prometryn + simazine** Dimitrova
Selepon H **dalapon-sodium** Sofital
Selevit F **sulphur + zineb** Sofital
Selinon IAHF **DNOC** Bayer
Sellapro FB **oxine-copper** Probelte
Selor Amine H **2,4-D** Agriben
Selor-Supra H **2,4-D + mecoprop** Agriben
Seloxone H **clopyralid + mecoprop** ICI
Seltoran H **amitrole + terbuthylazine** Ciba-Geigy
Semagrex I **carbaryl + malathion** Sadisa
Semakor F **ziram** Phytoprotect
Sembral Maneb Col F **maneb** ICI-Zeltia
Sembral Semillas M-L IF **lindane + maneb** ICI-Zeltia
Sembral Zelmais F **captan** ICI-Zeltia
Semefil G **gibberellic acid** Platecsa
Semer H **desmetryn** Chemia
Semeron H **desmetryn** Ciba-Geigy, Ligtermoet, Liro, Plantevern-Kjemi, Schering, Shell, Viopharm
Semesan I **malathion** Sepran
Semevin I **thiodicarb** Union Carbide
Semevital R F **copper oxychloride** Ital-Agro
Semul I **carbaryl** Condor
Sencor H **metribuzin** Agro-Kemi, Agroplant, Bayer, Imex-Hulst, Nitrokemia, Pinus
Sencoral H **metribuzin** Bayer
Sencorex H **metribuzin** Bayer
Sendran I **propoxur**
Sendrosil FA **dinocap** Luqsa
Senegil F **fenarimol + sulphur** Sandoz
Senkor H **metribuzin** Berner
Sentry IN **aldicarb + lindane** Embetec
Sepaphid IA **vamidothion** Du Pont
Sepicap F **captan** Du Pont
Sepilate F **ziram** Du Pont
Sepimate H **ammonium sulphamate** Seppic
Sepineb F **zineb** Du Pont
Sepisol N **D-D** Du Pont
Sepivam super FIHN **metham-sodium** Du Pont
Sepizin AC IA **azinphos-methyl + ethion** Du Pont
Sepizin L IA **azinphos-ethyl** Du Pont, Seppic
Sepizin M IA **azinphos-methyl** Du Pont, Seppic
Seppic 11 E S **petroleum oils** Du Pont, Seppic
Seppic ete IS **petroleum oils** Du Pont
Seppic lin H **lenacil + linuron** Du Pont
Seppic M.M.D. H **clopyralid + MCPA + mecoprop** Du Pont
Seppic verger FI **DNOC + petroleum oils** Du Pont
Seppic vigne FI **anthracene oil + DNOC + petroleum oils** Du Pont
Sepr-Oil IA **petroleum oils** Sepran
Sepraform IA **carbaryl + diazinon** Sepran
Sepraform PG I **malathion** Sepran
Sepralim M **metaldehyde** Sepran
Sepravax F **captan + oxycarboxin** Sepran
Sepravax P F **mancozeb + oxycarboxin** Sepran
Seprazina H **atrazine** Sepran
Septal F **carbendazim + maneb** Schering
Septene IG **carbaryl**
Serachlor F **quintozene** SEGE
Seradix G **4-indol-3-ylbutyric acid** Agriben, Agronorden, Fischer, Hortichem
Seraphos I **propetamphos** Sandoz
Serasol H **amitrole + simazine + sodium thiocyanate** Geochem
Serban F **dodine** Elanco
Sereno H **glyphosate** Sipcam
Serinal F **chlozolinate** Agroplant, Montedison
Seritard G **inabenfide** Chugai

Seritone H **dichlorprop + MCPA** Agriben
Seritone R H **dichlorprop + MCPA + mecoprop** Rhodiagri-Littorale
Seritox H **amitrole + ammonium thiocyanate** Rhodic
Seritox 50 H **dichlorprop + MCPA** Rhone-Poulenc
Seritox granule H **amitrole + diuron + sodium thiocyanate** Rhodic
Serk I **endosulfan + thiometon** Sandoz
Serpiol verano IA **petroleum oils** Serpiol
Sertrol Tetra H **bromoxynil + dichlorprop + ioxynil + MCPA** Berner
Sertrol Trippel H **dichlorprop + ioxynil + MCPA** Berner
Servorem G **chlorpropham + propham** Hermoo
sesamex S [51-14-9] 5-[1-[2-(2-ethoxyethoxy)ethoxy]ethoxy]-1,3-benzodioxole, ENT 20871, Sesoxane
Sesmetrin I **permethrin** Sepran
Sesoxane S **sesamex**
Sesum I **fenitrothion** Sepran
sethoxydim H [74051-80-2] (±)-(*ZE*)-2-(1-ethoxyiminobutyl)-5-[2-(ethylthio)propyl]-3-hydroxycyclohex-2-enone, sethoxydime, Alloxol-S, BAS 90520H, Checkmate, Expand, Fervinal, Grasidim, Nabu, NP-55, Poast, SN 81742
sethoxydime H **sethoxydim**
Setox I **carbaryl** Sepran
Seumin H **amitrole + atrazine + simazine** Bredologos
Sevilan I **carbaryl** Scam
Sevimol I **carbaryl** Serpiol
Sevin IG **carbaryl** Agriphyt, Chimac-Agriphar, La Littorale, Nitrokemia, Rodiagri-Littorale, Rhone-Poulenc, Serpiol, Union Carbide, Zorka Sabac
Sevin soufre FI **carbaryl + sulphur** La Littorale
Sevinil I **carbaryl** Fivat
Sevisol I **carbaryl** ICI
Sevitox I **carbaryl** Terranalisi
Sevton H **dinoseb acetate** Maag
Sewarin R **warfarin** Killgerm
Sextan H **isoxaben + simazine** Eli Lilly
Sezin F **zineb** Sepran
Sezin R F **copper oxychloride + zineb** Sepran
SF-1293 H **bialaphos** Meiji Seika
SF-6505 F **hymexazol** Sankyo
Shamrox H **MCPA** Diamond Shamrock
Shell Atrater H **atrazine** Shell
Shell-Austriebsspritzmittel A **petroleum oils** Shell
Shell Curam F **cymoxanil + fentin acetate + mancozeb** Shell
Shell D-50 H **2,4-D** Shell
Shell DPD H **2,4-D + dichlorprop** Shell
Shell Endrex I **endrin** Shell
Shell Grazafix H **2,4-D + dicamba + MCPA** Shell
Shell Kombi I H **2,4-D + MCPA** Shell
Shell koromix-N H **MCPA + mecoprop** Shell
Shell M-75 H **MCPA** Shell
Shell Malzide K G **maleic hydrazide** Shell
Shell Nitam F **fentin acetate + maneb** Shell
Shell NMC FIHN **metham-sodium** Shell
Shell Ovirex IAS **petroleum oils** Shell
Shell Past FW **oils** Shell
Shell Phosdrin I **mevinphos** Shell
Shell Prefix G H **dichlobenil** Shell
Shell Rizovin H **chloral hydrate** Shell
Shell Sproejtesvovl F **sulphur** Shell
Shell Torque A **fenbutatin oxide** Shell
Shell Tritivin H **methabenzthiazuron** Shell
Shell U-Forst H **atrazine + cyanazine** Shell
Shell Unkrauttod A H **allyl alcohol** Shell
Shell Unkrautvernichter W H **petroleum oils** Shell
Shell Vapona strip I **dichlorvos** Shell
Shell W H **petroleum oils** Shell
Shell Weedkiller W H **petroleum oils** Shell
Shell WN 250 H **dinoseb** Shell
Shell zimaneb F **maneb + zineb** Shell
Shellestol S **polyglycolic ethers** Shell
Shellfix H **benzoylprop-ethyl** Shell
Shellfix S H **flamprop-M-isopropyl** Shell
Shellphos I **parathion** Shell
Shellprox H **dichlorprop + MCPA** Shell
Shellprox super F H **2,4-D + dichlorprop + MCPA** Shell
Shelltox Vapona Pest Strip I **dichlorvos** Shell
Shelltox with Vaponer I **dichlorvos** Shell
Sheriff I **carbosulfan** FMC
Sherpa I **cypermethrin** Agriben, Pepro, Rhone-Poulenc
Sherpa DL IA **cypermethrin + dicofol** Rhone-Poulenc
Shield H **clopyralid** Dow
Shionox G **silicon dioxide** Shionogi
Shirlan F **salicylanilide**

Shourone M H **chlornitrofen + dymron** SDS Biotech
Showrone H **dymron** SDS Biotech
Shoxin R **norbormide** McNeil Laboratories
Shunt H **dicamba + MCPA** La Quinoleine
SI-6711 I **isoxathion** Sankyo
Siaazol H **amitrole + atrazine + simazine** Bredologos
Siacarb H **thiobencarb** Siapa
Siacarb M H **molinate + thiobencarb** Siapa
Siacourt G **chlormequat chloride** Protex
Siafos F **pyrazophos** Siapa
Sialan IA **endosulfan** Siapa
Sialex F **procymidone** Siapa, Sumitomo
Sialex T F **procymidone + thiram** Siapa
Sialite F **dinocap** Siapa
Siaprit F **etem + sulphur + zineb** Agrochemiki, Siapa
Siaram F **Bordeaux mixture** Siapa
Siarkol Extra F **sulphur** Organika-Sarzyna
Siarkol K F **carbendazim + sulphur** Organika-Sarzyna
Siarkol N F **nitrothal-isopropyl + sulphur** Organika-Sarzyna
Siatek F **thiabendazole** Siapa
Sibutol F **bitertanol + fuberidazole** Agroplant, Bayer, Mir
Sibutol Combi Slurry F **bitertanol + fuberidazole + triadimenol** Bayer
Sibutol FS FX **anthraquinone + bitertanol + fuberidazole** Bayer
Sibutol FS F **bitertanol + fuberidazole** Bayer
Sibutol MZ A FX **anthraquinone + bitertanol + mancozeb** Bayer
Sibutol-Morkit-Fluessigbeize FX **anthraquinone + bitertanol + fuberidazole** Bayer
Sicarol F **pyracarbolid** Hoechst
Sicid IA **rotenone** Siegfried
Sickle H **bromoxynil + fluroxypyr** Dow
Sickosul F **sulphur** Schwefelveredlung
Siden H **propyzamide + simazine** Siapa
siduron H [1982-49-6] 1-(2-methylcyclohexyl)-3-phenylurea, Du Pont 1318, Tupersan
Siege H **etiozin** Du Pont
Siganol S **petroleum oils** Siegfried
Sigma FW **thiophanate-methyl** Rhone-Poulenc
Sigmaton AI **chlorfenson + dicofol + tetradifon** Sandoz
Sil'Operats R **scilliroside** C.N.C.A.T.A.
Sil'souryl R **scilliroside** C.N.C.A.T.A.
Silbos F **thiram + vinclozolin** BASF
silicon dioxide G [7631-86-9] silicon dioxide, Shionox
Silmine R **scilliroside** Sandoz
Silmurin R **scilliroside** Chemika, Geopharm, Sandoz
Silo I **dichlorvos** Sipcam-Phyteurop
Silo mixte IA **dichlorvos + malathion** Sipcam-Phyteurop
Silodor F **maneb + sulphur** Koch & Reis
Silosan IA **pirimiphos-methyl** ICI
Silothion I **malathion** Sipcam-Phyteurop
Silox F **zineb** Gaillard
Siltrinul H **linuron + trifluralin** Siapa
Silvanol I **lindane**
Silvanol algefjerner L **benzalkonium chloride + tributyltin naphthenate** Hygaea
Silvapron D H **2,4-D + petroleum oils** Megafarm
Silvapron T H **2,4,5-T** BP
Silvetox 1 I **methoxychlor** Borzesti
Silvetox 2 I **lindane + methoxychlor** Borzesti
Silvetox 3 I **methoxychlor + trichlorfon** Borzesti
silvex HG **fenoprop**
Silvex Rikkasirote H **atrazine + dichlobenil** Kemira
Silvicide H **ammonium sulphamate** Sedagri
Silvitox H **ammonium sulphamate** Rhodic
Silvorex H **atrazine** Ciba-Geigy
Sim-Trol H **simazine** Griffin
Simadex H **simazine** Schering, Schering AAgrunol
Simafor H **simazine** Safor
Simagra H **simazine** Phytosan
Simakey H **simazine** Key
Simakor H **simazine** Gen. Rep.
Simalenox H **simazine** Schering
Simalon H **amitrole + atrazine + simazine** Hoechst
Simaneb F **maneb** Siegfried
Simanex H **simazine** Aako, Agrotechnica, Aragonesas, COPCI, Edefi, Maataloustpalvelu, Makhteshim-Agan
Simanix H **amitrole + 2,4-D + simazine** Hermoo
Simaphyt H **simazine** Sipcam-Phyteurop
Simapin H **simazine** Pinus
Simaquat H **paraquat dichloride + simazine** Caffaro
Simat Mais H **atrazine + simazine** Chemia
Simata H **amitrole + simazine** Bayer

Simatagrex H **amitrole + simazine** Sadisa
Simater H **simazine** Agrishell
Simatop H **simazine** Top
Simatraz H **atrazine + simazine** Interphyto
Simatred H **atrazine + simazine** Agrimont
Simatrin H **amitrole + simazine** Agrochemiki
Simatrol H **amitrole + simazine** Sapec, Sipcam-Phyteurop, Zorka Sabac
Simatrol kombi H **amitrole + MCPA + simazine** Zorka Sabac
Simatsin-neste H **simazine** Ciba-Geigy
Simatylone H **simazine** Agriphyt
Simavit H **simazine** Ravit
Simaz H **simazine** Chemia
Simazeno H **simazine** Enotria
Simazin H **simazine** Agroplant, Bayer, Berghoff, Burri, Ciba-Geigy, CTA, Du Pont, Fahlberg-List, Leu + Gygax, Montedison, Orgstieklo, Pluess-Staufer, Schering, Siegfried, Sojuzchimexport, Spiess, Staehler, Stefes, Terranalisi, Urania, Zorka, Zorka-Sabac
Simazina H **simazine** Bayer, Caffaro, EniChem, Ficoop, Permutadora
simazine H [122-34-9] 2-chloro-4,6-bis(ethylamino)-1,3,5-triazine, CAT, Aquazine, Caliber, G 27692, Gesatop, Primatol S, Princep, Sim-Trol, Simadex, Simanex
Simazip H **simazine** Interphyto
Simazol H **simazine** Ellagret, Visplant
Simazol H **amitrole + simazine** Edefi
Simazol doble H **amitrole + simazine** Probelte
Simazon H **diuron + simazine** Schering
simeton H [673-04-1] N^2,N^4-diethyl-6-methoxy-1,3,5-triazine-2,4-di amine, G 30044
Simetrax H **simazine** Peppas
simetryn H [1014-70-6] N^2,N^4-diethyl-6-methylthio-1,3,5-triazine-2,4 -diamine, simetryne, G 32911, Gy-bon
simetryne H **simetryn**
Simex H **amitrole + simazine** Gefex
Simflo Plus H **amitrole + ammonium thiocyanate + simazine** Agriben
Simflow H **simazine** Burts & Harvey
Simflow Plus H **amitrole + ammonium thiocyanate + simazine** Burts & Harvey
Simin H **simazine** Agrochemiki
Simonis Slakkenkorrels M **metaldehyde** Simonis
Simop H **simazine** Marty et Parazols
Simosol H **amitrole + simazine + sodium thiocyanate** Agrotechnica
Simpar H **paraquat dichloride** Visplant
Sinap DT H **2,4-D** Shell
Sinap MCPB H **MCPB** Shell
Sinasil H **paraquat dichloride** Hoechst
Sinbar H **terbacil** Du Pont, OHIS, Sedagri
Sinflouran H **trifluralin** Agro-Quimicas de Guatemala
Sinfonal I **chlorfenvinphos** Shell
Single Purpose F **phenylmercury acetate** Dow
Sinherban H **2,4-D** Ciba-Geigy
Sinituho F **pentachlorophenol** Kemira
Sinoratox IA **dimethoate** Sinteza
Sinox IAHF **DNOC** FMC
Sintofan IA **endosulfan** Platecsa
Sintovur I A **ethion** Sinteza
Siolcid H **linuron** Siapa
Sipaxol H **pendimethalin** Cyanamid
Sipcamol IA **petroleum oils** Sipcam
Sipcaplant F **captan + thiophanate-methyl** Sipcam
Sipcarix G **mefluidide** Sipcam-Phyteurop
Sipcavit F **folpet + thiophanate-methyl** Inagra, Sipcam
Siperin I **cypermethrin** Jewnin-Joffe
Siplen H **linuron + trifluralin** Sipcam
Sipulan-Neste H **chlorpropham** Berner
Sira F **sulphur** Sojuzchimexport, Spolana
Sirdate F **captafol + cymoxanil + folpet + oxadixyl** Du Pont
Sirdate P F **cymoxanil + maneb + oxadixyl** Du Pont
Sirocco F **fenpropimorph + iprodione** Rhone-Poulenc
Sistan FNHI **metham-sodium** Universal Crop Protection
Sistematon IA **dimethoate** Agrocros
Sistemin IA **dimethoate** Zupa
Sisthane F **fenapanil**
Sitophil IA **dichlorvos + malathion** C.N.C.A.T.A.
Sitophil total I **dichlorvos** Agrinet, C.N.C.A.T.A.
Sitradol H **pendimethalin** Siegfried
Sitrapon H **amitrole + dalapon-sodium + simazine** Eurofyto
Sitrazin H **atrazine + pendimethalin** Siegfried
Sitrazin H **atrazine + simazine** Bayer
Sitrol H **amitrole + 2,4-D + simazine** Eurofyto
Sivamcarb I **carbaryl** Sivam
Sivamdod F **dodine** Sivam

Sivamil F **benomyl** Sivam
Sivamil F F **benomyl + folpet** Sivam
Sivamlin P I **lindane** Sivam
Sivampar I **parathion** Sivam
Sivamtone S **ammonium sulphate** Sivam
Sivamvos I **dichlorvos** Sivam
Sivamwarf R **coumachlor** Sivam
Sixanol I **methoxychlor** Chimac-Agriphar
Skadedyrcentralens insektlak I **chlorpyrifos** Skadedyr
Skeetal I **Bacillus thuringiensis** Microbial Resources
Skill H **dicamba + DNOC + mecoprop** Sipcam-Phyteurop
Skipper M **thiodicarb** Pepro
Skovtjaere X **tar** Diana
Slam H **asulam + dalapon** Rhone-Poulenc
Slam C IA **azothoate**
Slaymor R **bromadiolone** Ciba-Geigy
Slim EG M **metaldehyde** Agrimont
Slo-Gro G **maleic hydrazide** Uniroyal
Slug Destroyer M **metaldehyde** Schering
Slug Gard M **methiocarb** Pan Britannica
Slug-M M **methiocarb**
Slugal M **metaldehyde** Ciba-Geigy
Slugit M **metaldehyde** Efthymiadis, Kemira
Slut effektiv insektdraeber I **methoxychlor + piperonyl butoxide + pyrethrins** Nielsen
Slut fluedraeber D I **piperonyl butoxide + resmethrin** Nielsen
Slut velduftende fluedraeber I **piperonyl butoxide + pyrethrins** Nielsen
SM-55 F **captan + sulphur** Ligtermoet
SM-55/Maneb F **captan + maneb + sulphur** Ligtermoet
SM 85 F **maneb + sulphur** Schering
SMA H **sodium monochloroacetate**
SMCA H **sodium monochloroacetate**
SMDC FNHI **metham-sodium**
Smedip IA **lindane** Langkamp
Smeesana R **antu**
SMT-emulsionable I **fenitrothion** ICI-Zeltia
SN 34615 I **promecarb** Schering
SN 35830 H **monalide** Schering
SN 36056 AI **formetanate hydrochloride** Schering
SN 38107 H **desmedipham** Schering
SN 38574 H **phenmedipham-ethyl**
SN 38584 H **phenmedipham** Schering
SN 49537 G **thidiazuron** Schering
SN 58132 H **phenisopham** Schering
SN 66752 F **propamocarb hydrochloride** Schering
SN 78314 F **cyprofuram** Schering
SN 81742 H **sethoxydim** Schering
Snapper H **amitrole + atrazine + 2,4-D** ICI
Snek-Vetyl Neu M **metaldehyde** Vetyl-Chemie
Snile-Kverk M **metaldehyde** Plantevern-Kjemi
Snip I **azamethiphos** Ciba-Geigy
Snip I **dimetilan**
Snip-Fliegenkoeder I **azamethiphos + tricosene** Ciba-Geigy
S.O. 50-10 I **carbaryl** ERT
Socatrine I **deltamethrin** Wellcome
Sodil B F **sulphur** La Littorale
sodium chlorate H [7775-09-9] sodium chlorate, chlorate de sodium, Atlacide, De-Fol-Ate, Defol, Dervan, Drop-Leaf, Fall, Harvest-Aid, Klorex, KM, Kusatol, Leafex, Tumble Leaf, Tumbleaf
sodium monochloroacetate H [3926-62-3] chloroacetic acid, sodium salt, SMA, SMCA, Croptex Steel, Somon
sodium pentaborate G [12007-92-0] boron sodium oxide
Soenflor Mosedreper H **iron sulphate** Soendenaa
Sofit H **pretilachlor** Ciba-Geigy
Sofit Sp H **molinate + pretilachlor** Ciba-Geigy
Sofol PS F **folpet + sulphur** Sofital
Sofril FA **sulphur** Pepro, Ravit, Rhone-Poulenc
Sofrital F **sulphur** Vital
Soil Fungicide 1823 F **chloroneb** Du Pont
Sok-Bt I **Bacillus thuringiensis** NOR-AM
Sol Micro-90 AF **sulphur** Sivam
Sol Net H **sodium chlorate** Chimac-Agriphar
Solacol F **validamycin A** AAgrunol, Schering AAgrunol
Solado H **glyphosate** Siapa
Solagro I **heptachlor** Hispagro
Solamort H **dinoseb** Agriben
Solamyl G **chlorpropham + propham** Aveve
solan H **pentanochlor**
Solarex H **linuron + terbacil** Seppic
Solasan FNHI **metham-sodium** Agriben, Condor, Rhone-Poulenc
Solbar S FI **barium polysulphide** Bayer
Solchim H **amitrole + atrazine** Chimique
Soldep IAN **warfarin** Spolana

Soldrex I **lindane** Shell
Soleol I **petroleum oils** Quimigal
Solethion IA **ethion** Afrasa
Solethion oil IA **ethion + petroleum oils** Afrasa
Solfa FA **sulphur** Farm Protection
Solfac I **cyfluthrin** Bayer
Solfac Combi I **cyfluthrin + fenthion** Bayer
Solfarin R **warfarin**
Solfato di Rame F **copper sulphate** Manica, Sepran
Solfiren AF **sulphur** Visplant
Solfo F **sulphur** Marty et Parazols
Solfobario FI **barium polysulphide** Agrochemiki, Siapa
Solfocol AF **sulphur** EniChem
Solforin Attivato TMTD F **sulphur + thiram** Sofital
Solfosan AF **sulphur** Eli Lilly
Solfosan Effe F **fenarimol + sulphur** Eli Lilly
Solfovit FA **sulphur** Agroplant
Solfum FHIN **metham-sodium** Tecniterra
Solgard I **pirimiphos-ethyl** ICI
Solicam H **norflurazon** Sandoz
Solnet H **pretilachlor** Ciba-Geigy
Solnet granule desherbant total H **amitrole + atrazine + simazine** Ciba-Geigy
Solo HH **chlorpropham + naptalam** Uniroyal
Soltair H **diquat dibromide + paraquat dichloride +simazine** ICI
Solthion IA **azinphos-methyl** ICI
Soltraz H **atrazine + simazine** Aseptan
Solution G **maleic hydrazide** C.F.P.I.
Solvenal IA **malathion** Platecsa
Solvigran IA **disulfoton** Farm Protection
Solvirex IA **disulfoton** Sandoz, Zupa
Sometam FNHI **metham-sodium** Visplant
Somio RX **chloralose**
Somol IA **petroleum oils** Margesin
Somon H **sodium monochloroacetate** Atlas-Interlates
Sonalan H **ethalfluralin** Elanco, Eli Lilly, Grima
Sonaptil I **carbaryl** Sopepor
Sonar H **fluridone** Elanco
Sonax F **etaconazole** Ciba-Geigy
Sonax C F **captan + etaconazole** Ciba-Geigy
Sonic H **glyphosate** Rigby Taylor
Sopragam IA **lindane + parathion** Sopra
Soprathion IA **parathion**
Sopratom S **polyethylene oxide** Sopra
Soraf PB F **copper compounds (unspecified) + folpet + sulphur** Sofital
Sorene PB 83 F **captan** Shell
Sorex Crawling Insect Bait H **iodofenphos** Sorex
Sorexa R **warfarin** Sorex
Sorexa C R I R **calciferol + warfarin** Sorex
Sorexa CD R **calciferol + difenacoum** Sorex
Sorexa CR R **calciferol + warfarin** Colkim
Sorexa Plus R **warfarin** Sorex
Sorgan H **propachlor + propazine** Makhteshim-Agan
Soria H **amitrole + ammonium thiocyanate + atrazine +hexazinone** Protex
Sorilan F **fenpropidin** La Quinoleine
Sorkax gaspatron R **barium nitrate + sulphur** Rentokil
Sorkil R **difenacoum** Edialux
Sotox I **permethrin + tetramethrin**
Soufravion F **sulphur** R.S.R.
soufre FA **sulphur**
Soufre aero acaricide FAI **carbophenothion + parathion-methyl + sulphur** Sedagri
Soufre aero methyl FIA **parathion-methyl + sulphur** Sedagri
Soufre charge cuprique BOB F **copper oxide + sulphur** Marty et Parazols
Soufrebe F **sulphur** Agrishell
Soufrugec F **sulphur** Sipcam-Phyteurop
Soultaphino G **fatty acid esters** Filocrop
Souricide R **difenacoum** Aufra, Servigeco, Steininger
Souricide R **bromadiolone** Rhodic, SCAC-Fisons, Umupro
Souricide R **chloralose** Gilbert
Souricide Rapid'Tox R **crimidine** Sovilo
Sovelix M **metaldehyde** Sovilo
Sovi-Tox R **warfarin** Sovilo
Sovicortil I **lindane** Sovilo
Sovifol A **dicofol** Sovilo
Sovilo liquide total I **diazinon + triforine** Sovilo
Sovinexit I **lindane** Sovilo
Sovion FI **DNOC + petroleum oils** Sovilo
Sovisol H **amitrole + atrazine** Sovilo
Sovisol H **amitrole + diuron + sodium thiocyanate** Sovilo
Sovitaup R **chloralose** Sovilo
Sovitox R **scilliroside** Sovilo
Sovitox poudre R **chlorophacinone** Sovilo
SP 1103 I **tetramethrin** Sumitomo
Spannit IA **chlorpyrifos** Pan Britannica

Spartcide F **fluoromide** Mitsubishi, Kumiai
Sparton H **amitrole + atrazine + simazine** Agrochemiki
Spasor H **glyphosate** Burts & Harvey, Monsanto, Procida, Rhone-Poulenc, Siapa
Special souris S R **bromadiolone** Lipha
Special Toxirat R **chlorophacinone** Steininger
Spectracide IA **diazinon** Ciba-Geigy
Spectron H **chloridazon + ethofumesate** Schering
Speedway H **paraquat dichloride** ICI
Spergon FX **chloranil**
Spersul FA **sulphur** ICI, ICI-Zeltia
Spezial-Inficin gegen Schnecken M **metaldehyde** Buchrucker
Spezial-Kalkstickstoff FH **calcium cyanamide** SKW Trostberg
Spezial-Kiepenkerl-Rasenduenger mit Unkrautvernichter H **2,4-D + MCPA** Nebelung
Spezial-Rasenduenger mit Unkrautvernichter VGC H **2,4-D + dicamba** Deutscher Garten-Center
Spezial-Rasenduenger mit Unkrautvernichter H **2,4-D + MCPA** Parco
Spezial-Unkrautvernichter Weedex H **glyphosate** Shell
Spica 10 H **dalapon-sodium** Procida
Spica 66 H **2,4-D + picloram** Procida
Spica 103 VPS H **amitrole + diuron + tebuthiuron** Procida
Spica 300 G H **picloram** Procida
Spica 320 S H **atrazine + TCA** Procida
Spica D H **2,4-D + triclopyr** Procida
Spica debroussaillant H **2,4-D + triclopyr** Procida
Spica liquide H **amitrole + 2,4-D + MCPA + TCA** Procida
Spica selectif H **2,4-D + MCPA** Procida
Spicacil H **bromacil + diuron + picloram** Procida
Spicadam H **amitrole + bromacil + diuron** Procida
Spicafor H **amitrole + bromacil** Procida
Spicagran H **bromacil + diuron** Procida
Spicagrass H **amitrole + diuron** Procida
Spicamat H **ammonium sulphamate** Procida
Spicatrak H **amitrole + bromacil** Procida
Spicatramp H **amitrole + diuron + simazine** Procida
Spicazine H **atrazine** Procida
Spider Spray IA **carbophenothion** Serpiol
Spike H **tebuthiuron** Elanco
Spin-aid H **phenmedipham** NOR-AM
Spinnaker F **triadimenol** Shell
Spira I **allethrin** Trenti
Spiringshaemmer-F G **chlorpropham** Cillus
Spirunarstyriefni F G **chlorpropham** Cillus
Spitfos I **parathion** Enotria
Splendor H **tralkoxydim** ICI
Splendor F **carbendazim + fenarimol + oxycarboxin** La Quinoleine
Spodnam DC GS **di-1-p-menthene** Mandops
Spomil A **bromopropylate** Maag
Spontal F **mancozeb + nuarimol** Sandoz
Spontan H **dicamba + dichlorprop + MCPA + mecoprop** Nordisk Alkali
Sporgon F **prochloraz-manganese complex** Pluess-Staufer, Schering, Schering AAgrunol
Sporgon delta F **chlorothalonil + prochloraz** Schering AAgrunol
Sporozal F **TCMTB** Diana
Sportak F **prochloraz** Bjoernrud, Pluess-Staufer, Schering, Schering AAgrunol
Sportak alpha F **carbendazim + prochloraz** Hoechst, Schering
Sportak Delta F **cyproconazole + prochloraz** Schering
Sportak FE F **mancozeb + prochloraz** Schering
Sportak FM F **fenpropimorph + prochloraz** Schering
Sportak MZ F **mancozeb + prochloraz** Schering
Sportak PF F **carbendazim + prochloraz** Schering
Spotless F **diniconazole** Sumitomo
Spotrete FX **thiram** Cleary
Spra-cal IH **calcium arsenate**
Spraid S **tallow amine ethoxylate** Avon Packers
Spray 214 S **petroleum oils** Lubrichim
Spray Antivermine I **propoxur** Natural Granen
Spray Oil 7-E IA **petroleum oils** Leu & Gygax
spray oils AIHS **petroleum oils**
Spray Ol IA **petroleum oils** Sivam
Spray-Tox I **kadethrin** Roussel Uclaf
Sprayban F **petroleum oils** R.S.R.
Sprayfast GS **di-1-p-menthene** Mandops
Sprayoel agro S **petroleum oils** Agroplant

Sprayprover S **petroleum oils** Fine Agrochemicals
Sprigone I **dichlorvos + permethrin + tetramethrin** Denka
Spring Bladluisspray IA **piperonyl butoxide + pyrethrins** Vernooy
Spring Bladluisspray N I **phenothrin + tetramethrin** Vernooy
Spring Spray R **bromophos** Synchemicals
Springcorn Extra H **dicamba + MCPA + mecoprop** FCC
Sprinkle Anti-Bladluis I **phenothrin + tetramethrin** Drift
Sprint F **fenpropimorph + prochloraz** Schering
Sprint FL F **carbendazim** Sipcam-Phyteurop
Sprintene G **vegetable extracts** Aifar
Sprion LG H **dichlobenil** Leu & Gygax
Spritex I **malathion** Lapapharm
Spritex Vloeibaar I **malathion + piperonyl butoxide + pyrethrins** Denka
Spritoxin A I **malathion** Denka
Spritoxin Super I **permethrin + tetramethrin** Denka
Spritz-Cit I **pyrethrins** Cit
Spritz-Cupral FB **copper oxychloride** Nickelhuette
Spritz-Hormin H **2,4-D** Bitterfeld
Spritz-Hormit H **2,4-D** Bitterfeld
Sprout Nip G **chlorpropham** Nordisk Alkali, PPG
Sprout-Stop G **maleic hydrazide** Drexel
Spruzit IA **piperonyl butoxide + pyrethrins** Europlant, Neudorff
Spud-Nic HG **chlorpropham**
Spuitzwavel F **sulphur** Aveve, Sanac
Spur IA **fluvalinate** Sandoz
Sputop I **deltamethrin** Coopers
Squadron H **imazaquin + pendimethalin** Cyanamid
SR-73 M **niclosamide** Bayer
SR 406 F **captan** Chevron
SRA 5172 IA **methamidophos** Bayer
S. Ramedit F **copper oxychloride + zineb** Siapa
SS 2074 FAI **quinomethionate** Bayer
SS 11946 FI **isoprothiolane** Nihon Nohyaku
SSH 43 H **isouron** Shionogi
Sta-Gon I **boric acid** Amvac
Staa-Free H **bromacil** Proserve
Stabilan G **chlormequat chloride** Agrolinz, Agroplant, Diana, Protex, Shell, Van Wesemael
Stabineb F **maneb** Pennwalt France
Stablix S I **chlorfenvinphos + phenol** Agrishell
Stacar A **cyhexatin** Solfotecnica
Stacker H **methyldymron** SDS Biotech
Staeubemittel Insektizid/poudrage insecticide Gesal I **chlorpyrifos + pirimicarb** Reckitt & Colman
Stake H **alachlor** Cardel
Stald chok I **piperonyl butoxide + resmethrin** Aeropak
Stallfly Combi I **dimethoate + fenitrothion** Leu & Gygax
Stallfly Permanent I **fenitrothion + permethrin** Leu & Gygax
Stallfly Zidil I **chlorpyrifos** Leu & Gygax
Stallspritzmittel Nexion I **bromophos** Shell
Stam H **propanil** EniChem, ICI-Valagro, Permutadora, Procida, Rohm & Haas, Sipcam
Stam Extra H **fenoprop + propanil** Rohm & Haas
Stamat I **piperonyl butoxide + pyrethrins** Drilexco
Stammschutzmittel Gamma I **lindane** Agrolinz
Stampede H **propanil** Rohm & Haas
Stampede CM H **MCPA + propanil** Rohm & Haas
Standak NIA **aldoxycarb** Union Carbide
Standup G **chlormequat chloride** Vass
Stanex FLM **fentin acetate** Protex
Stanex H F **fentin hydroxide** Protex
Stanoram F **decafentin**
Stantox H **2,4-D** Agriphyt, Chimac-Agriphar
Stanza F **fenpropimorph + prochloraz** Schering
Staphylini F **copper oxychloride + sulphur** Mylonas
Starane H **fluroxypyr** Dow, Schering
Starane Kombi H **clopyralid + fluroxypyr + ioxynil** Dow, Schering
Starane Super H **bromoxynil + fluroxypyr + ioxynil** Pluess-Staufer
Starox H **MCPA** Ciba-Geigy
Starter H **chloridazon** Truchem
Stathion IA **parathion**
statt-jaeten H **diuron + methabenzthiazuron** Neudorff
Stay Off X **aluminium ammonium sulphate** Synchemicals
Stayput G **fenoprop** Efthymiadis
STCA H **TCA**

Steatite cuprique MOP FB **copper oxychloride** Marty et Parazols
Steiner M **metaldehyde** CTA
Steladone IA **chlorfenvinphos** Ciba-Geigy
Stellon H **clopyralid + MCPA + mecoprop** KVK
Stellox H **bromoxynil + ioxynil** Ciba-Geigy
Stempor F **carbendazim** ICI
Stemtrol G **piproctanyl bromide** Maag
Steribox P F **2-phenylphenol** Fomesa
Sterilite IHFX **tar oils** Albright & Wilson, Tenneco
Sterisol CP I **lindane** Agriphyt
Sterzik-Laeuterungspatrone H **MCPA** Sterzik-Geraete-Versand
STI-Koll AF **sulphur** Solfotecnica
Sticereal 2,4 D-Pso H **2,4-D** Solfotecnica
Sticken S **alkyl phenol ethoxylate** Agrotechnica
Stigor IA **dimethoate** Solfotecnica
Stim-root G **4-indol-3-ylbutyric acid** Maasmond, Valimex
Stimal Cereali I **malathion** Solfotecnica
Stimroot G **4-indol-3-ylbutyric acid** Maasmond
Sting H **glyphosate** Monsanto
Stip-stop F **calcium chloride** Phyto
Stipanil H **propanil** Solfotecnica
Stipend I **chlorpyrifos** Dow
Stiphate I **acephate** Solfotecnica
Stiphos I **malathion** Solfotecnica
Stiralin H **trifluralin** Solfotecnica
Stiram F **copper oxychloride** Solfotecnica
stirofos IA **tetrachlorvinphos**
Stirpan Forte H **dinoterb** Maag
Stiryl I **carbaryl** Solfotecnica
Stiuron H **linuron** Solfotecnica
Stizinfos IA **azinphos-methyl** Solfotecnica
Stizir F **ziram** Solfotecnica
Stodiek Rasenduenger mit Moosvernichter fein H **iron sulphate** Stodiek
Stodiek-Spezial-Rasenduenger mit Unkrautvernichter H **2,4-D + MCPA** Stodiek
Stoko Unkrautvernichter H **sodium chlorate** Haniel
Stomophos I **pirimiphos-methyl** Coopers
Stomoxin I **permethrin** Wellcome
Stomp H **pendimethalin** AVG, BASF, Cyanamid, Lapapharm, Ligtermoet, Maag, Opopharma, Spiess, Urania, VCH, Zorka Sabac
Stomp-prometrin H **pendimethalin + prometryn** Zorka Sabac
Stop Schneckenkoerner M **metaldehyde** Wyss Samen & Pflanzen
Stop-Fly I **permethrin** Siegfried
Stop-Frut G **1-naphthylacetic acid** Luqsa
Stopgerme G **chlorpropham** Agriben, Condor
Stopper A **hexythiazox** Hoechst
Stop-Scald G **ethoxyquin** Hellafarm, Inagra, Monsanto, Pepro
Stopspot F **phenylmercury chloride**
Storaid Dust FG **tecnazene + thiabendazole** MSD AGVET
Storite F **thiabendazole** MSD AGVET
Storite SS FG **tecnazene + thiabendazole** MSD AGVET
Storm R **flocoumafen** Pliva, Sorex
Stratagem R **flocoumafen** Shell
Stratos H **cycloxydim** BASF
Strel H **carbaryl + propanil** Sandoz
streptomycin B [57-92-1] *O*-2-deoxy-2-(methylamino)-α-L-glucopyranosyl-(1→2)-*O*-5-deoxy-3-C-formyl-α-L-lyxofuranosyl-(1→4)-*N*,*N*'-bis(aminoiminomethyl)-D-streptamine, streptomycine, Agrept, Agrimycin, Agri-Strep, AS-50, Gerox, NSC 14083, Plantomycin
streptomycine B **streptomycin**
Streunex I **lindane** Shell
Strike Hormone Rooting Powder F **captan + 1-naphthylacetic acid** May & Baker
Strobane-T IRA **camphechlor** Agro-Quimicas de Guatemala
Stroller Kombi H **dicamba + MCPA** Ewos
Stroller Mossa H **chloroxuron** Ewos
Stroller Ograes H **benazolin + clopyralid** Kemi-Intressen
Stroller Ograes N H **clopyralid** Ewos
strychnine R [57-24-9] strychnidin-10-one, Certox, Gopher-Gitter, Hare-Rid, Kwik-Kil, Mole Death, Mous-Rid, Sanaseed, Strychnos, Taupicine
Strychnos R **strychnine**
Stuifzwavel F **sulphur** Aveve
Stulln FA **sulphur** Sapec
Stunt-Man G **maleic hydrazide**
Stutox R **zinc phosphide** JZD Pojihlavi
Subdue F **metalaxyl** Ciba-Geigy
Subitex HI **dinoseb** Hoechst
Sublimior F **sulphur** Sedagri
Substral Ameisen-Vernichter I **trichlorfon** Barnaengen
Substral Bladluisspray IA **piperonyl butoxide + pyrethrins** Verwet
Substral Bladluisspray N I **phenothrin + tetramethrin** Verwet

Substral bladlusspray I **piperonyl butoxide + pyrethrins** Barnaengen
Substral bladlusspray I **piperonyl butoxide + pyrethrins + rotenone** Barnaengen
Substral-Blattlausfrei IA **piperonyl butoxide + pyrethrins** Barnaengen
Substral Garten-Kalkstickstoff mit Unkrautstop FH **calcium cyanamide** Barnaengen
Substral Luseknekker I **pyrethrins** KeNord
Substral-Mehltaufrei F **lecithin** Barnaengen
Substral-Pflanzenschutz-Spray IA **piperonyl butoxide + pyrethrins + rotenone** Barnaengen
Substral Pflanzenschutzspray NEU I **dimethoate** Barnaengen
Substral plant pin I **butoxycarboxim** Barnaengen, Wacker
Substral ploentupinnar I **butoxycarboxim** Wacker
Substral Rasenduenger mit Moosvernichter H **iron sulphate** Barnaengen
Substral Rasenduenger mit Unkrautvernichter H **chlorflurecol-methyl + MCPA** Barnaengen
Substral Rosen-Spray FIA **piperonyl butoxide + pyrethrins + sulphur** Barnaengen
Substral Rosen-Spray neu FIA **lecithin + piperonyl butoxide + pyrethrins** Barnaengen
Substral Schnecken-Frei M **metaldehyde** Barnaengen
Substral Spray IA **methoxychlor + piperonyl butoxide + pyrethrins** Fincos
Substral Spray I **piperonyl butoxide + pyrethrins + rotenone** Barnaengen, Reckitt & Colman
Substral Unkraut-weg H **bromacil + diuron** Barnaengen
Sucker-Stuff G **maleic hydrazide** Drexel
Suelosana I **lindane** Agrocros
Sufenit I **fenitrothion** Agrocros
Suffa FA **sulphur** Drexel
Suffix H **benzoylprop-ethyl** Agrishell, Agroplant, Farm Protection, Kemira, Organika-Sarzyna, Shell, Szovetkezet, Zupa
Suffix BW H **flamprop-M-isopropyl** Shell
Suffix Plus H **flamprop-methyl** Shell
Sufralo FA **sulphur** Siegfried
Sufran FA **sulphur** Spiess, Urania
Sufrevil FA **sulphur** Inagra
Sugan R **warfarin** Neudorff
Suisect Jardim IF **diazinon + tetradifon + triforine** Permutadora
Sulcol AF **sulphur** Baslini
Suldan IA **endosulfan** Agronova
Sulerex H **metoxuron** Sandoz, Siapa
Sulf Muiabil F **sulphur** Tirnaveni
Sulf Pulbere F **sulphur** Tirnaveni, Zorka
Sulfacop LF **copper sulphate** Ingenieria Industrial
Sulfacube F **copper sulphate + maneb** Sipcam-Phyteurop
Sulfadan I **endosulfan** EniChem, Hellenic Chemical
sulfallate H [95-06-7] 2-chloroallyl diethyldithiocarbamate, thioallate, CDEC, Vegadex
Sulfamate H **ammonium sulphamate** Mitsui Toatsu
sulfamate d'ammonium H **ammonium sulphamate**
Sulfanebe F **copper oxychloride + copper sulphate + maneb** Agrishell
Sulfapron I **carbon disulphide + carbon tetrachloride** Probelte
Sulfastop F **copper sulphate + folpet** La Quinoleine
sulfat de cupru F **copper sulphate**
sulfate de cuivre F **copper sulphate**
Sulfater IA **endosulfan** Bimex
sulfato de cobre F **copper sulphate**
Sulfazul F **copper sulphate + zineb** Quimigal
Sulfex FA **sulphur** Excel
Sulfimix F **maneb + sulphur + zineb** Van Wesemael
Sulfitox F **sulphur** Van Wesemael
Sulflox FA **sulphur**
Sulfobar IAF **barium polysulphide** Luqsa
Sulfoben FA **sulphur** ERT
Sulfocoumar R **warfarin** Frere
Sulfocruz Coloidal S FA **sulphur** KenoGard
Sulfocruz Cuprico F **copper oxychloride + sulphur** KenoGard
Sulfocruz Micronizado F **sulphur** KenoGard
Sulfol VR F **sulphur** Du Pont
Sulfoma F **copper sulphate + maneb** La Cornubia
sulfometuron-methyl H [74222-97-2] methyl 2-[3-(4,6-dimethylpyrimidin-2-yl)ureidosulphonyl]benzoate, DPX 5648, Oust

Sulfonex I **endosulfan** Phytopharmaceutiki
Sulfor I **endosulfan** Sopepor
Sulforex F **sulphur** Protex
Sulforix F **sulphur** Du Pont
Sulfoshell Special F **sulphur** Shell
Sulfospor FA **sulphur** Farmers Crop Chemicals
Sulfosur F **sulphur** ERT
sulfotep IA [3689-24-5] *O,O,O',O'*-tetraethyl dithiopyrophosphate, sulfotepp, dithio, dithione, thiotepp, ENT 16273, ASP-47, Bayer E 393, Bladafum
sulfotepp IA **sulfotep**
Sulfotox R **sulphur** C.E.L.I.F.
Sulfovit F **sulphur** Eurofyto
Sulfox-Cide S **sulfoxide**
sulfoxide S [120-62-7] 1-methyl-2-(3,4-methylenedioxyphenyl)ethyl octyl sulphoxide, Sulfox-Cide, Sulfoxyl
Sulfoxyl S **sulfoxide**
Sulfozyl Microlux F **sulphur** Chimac-Agriphar
sulfur FA **sulphur**
sulglycapin H [51068-60-1] azepan-1-ylcarbonylmethyl methylsulphamate, BAS 46100H, BAS 461H
Sulifate Forte F **copper sulphate** Hoechst
Sulikol K F **sulphur** Spolana
Sulka NF **lime sulphur** Dimitrova
sulphaquinoxaline R [59-40-5] 4-amino-*N*-2-quinoxalinylbenzenesulfonamide
Sulphicol F **sulphur** Diana
Sulphohalcini F **copper oxychloride + sulphur + zineb** Gefex
sulphur FA [7704-34-9] sulphur, soufre, sulfur, Aquilite, Cosan, Elosal, Golden Dew, Kolodust, Kolofog, Kolospray, Kumulus, Magnetic 6, Solfa, Suffa, Sulfex, Sulfex, Sulflox, Sulfospor, Super Six, That, Thiolux, Thion, Thiovit, Zolvis
sulprofos I [35400-43-2] *O*-ethyl *O*-4-(methylthio)phenyl *S*-propyl phosphorodithioate, Bolstar, Helothion, NTN 9306
Sultox F **sulphur** Sedagri
sultropen F [963-22-4] 2,4-dinitrophenyl pentyl sulphone, sultropene, BTS 7901, RD 7901
sultropene F **sultropen**
Sulvite FA **sulphur** Sopepor
Sumagic G **uniconazole** Sumitomo
Sumbarit IAF **barium polysulphide** Zorka
Sumex H **diuron** Argos
Sumex-ATA H **amitrole + diuron** Argos
Sumi F **diniconazole** Agriben
Sumiagrex I **fenitrothion** Sadisa
Sumi-alfa I **esfenvalerate** Sumitomo
Sumi-alpha I **esfenvalerate** Sumitomo
Sumibac IA **fenvalerate** Sumitomo
Sumiboto F **procymidone** Agrocros, Sumitomo
Sumicidin I **fenvalerate** Agrishell, Agroplant, Felleskjoepet, Masso, Nordisk Alkali, Sapec, Shell, Siapa, Sumitomo, Zupa
Sumico F **carbendazim + diethofencarb** Sopra
Sumicombi I **fenitrothion + fenvalerate** Agrishell, Masso, Nordisk Alkali, Zupa
Sumidione F **diniconazole + iprodione** Rhodiagri
Sumifene I **fenitrothion** Condor, Ravit, Pepro
Sumifive I **fenvalerate** KenoGard
Sumifleece IA **fenvalerate** Sumitomo
Sumifly IA **fenvalerate** Sumitomo
Sumiherb H **bromobutide** Sumitomo
Sumilan I **fenitrothion** Aragonesas
Sumilex F **procymidone** Lucebni zavody, Mir, Nitrokemia, Sumitomo, Zupa
Sumimeton I **fenvalerate + oxydemeton-methyl** Agrishell
Sumimix I **fenitrothion + fenpropathrin** Masso
Sumipower I **fenitrothion** Sumitomo
Sumisclex F **procymidone** Agroplant, Bayer, ICI-Valagro, Masso, Schering, ScheringAAgrunol, Shell, Sopra, Sumitomo
Sumisclex MZ F **maneb + procymidone + zineb** Schering AAgrunol
Sumiseven G **uniconazole** Sumitomo
Sumistar F **carbendazim + diniconazole + iprodione** Agriben
Sumitan I **fenitrothion** Nordisk Alkali
Sumithion I **fenitrothion** Argos, Bjoernrud, Budapesti Vegyimuvek, Masso, Nordisk Alkali, Sumitomo
Sumithion forte I **fenitrothion + fenvalerate** Masso
Sumithrin I **phenothrin** Sumitomo
Sumitick IA **fenvalerate** Sumitomo
Sumition IA **fenitrothion** Kemira, Zupa
Sumitox IA **malathion** Rhone-Poulenc, Sedagri
Summit F **triadimenol** Bayer
Sumpor prah-F F **sulphur** Zorka Sabac
Sumporcin F **sulphur + zineb** Zorka Sabac

Sun 7 E Kevatruiskute IA **petroleum oils** Farmos
Sun Oil I **petroleum oils** Efthymiadis, Protex
Sun Superior Sprayoil IA **petroleum oils** SOPRO
Sun-Spray IA **petroleum oils** Protex
Suncide I **propoxur** Bayer
Sup'erb H **amitrole + bromacil + diuron** LHN
Sup'Operats R **bromadiolone** C.N.C.A.T.A.
Sup'Souryl R **bromadiolone** C.N.C.A.T.A.
Supasan "Lawn Sand" H **iron sulphate** Zaden van Engelen
Super Acarol IA **chlorobenzilate + parathion-methyl** Diana
Super Barnon H **flamprop-M-isopropyl** Shell, Spolana
Super bouillie Macclesfield FB **Bordeaux mixture** La Cornubia
Super Caid R **bromadiolone** Agriben, Lipha
Super-De-Sprout G **maleic hydrazide** Drexel
Super Fanox H **dinoseb** La Quinoleine, Maag
Super Granusol H **atrazine + dichlobenil** C.F.P.I.
Super Greenkeeper H **2,4-D + dicamba** Asef Fison, Moreels-Guano
Super Hedonal-Neste H **dichlorprop + MCPA** Berner
Super Helicide M **metaldehyde** Umupro
Super Herbicine H **amitrole + ammonium thiocyanate + simazine** Agriphyt
Super Herbogil H **dinoterb + mecoprop** Schering
Super Herboxy desherbant total H **amitrole + atrazine** Ciba-Geigy
Super Herboxy granule H **amitrole + atrazine + simazine** Ciba-Geigy
Super-Kabrol H **dinoseb** Pluess-Staufer
Super Laitom IA **azinphos-methyl** Lainco
Super Limaclor M **metaldehyde** SCAC-Fisons
Super Lovitox H **dinoseb** Chimac-Agriphar
Super Macclesfield F **copper sulphate + maneb + zineb** La Cornubia
Super Macclesfield F F **copper hydroxide + folpet** La Cornubia
Super Man F **maneb** Agriphyt
Super Monalox H **sethoxydim** Kwizda
Super mortivit R **difenacoum** Sovilo
Super Moss Killer and Lawn Fungicide H **dichlorophen** Murphy
Super Mosskil-A H **iron sulphate** Fisons
Super Mosstox HLFB **dichlorophen** May & Baker
Super Prodan I **sodium fluosilicate** Anorgachim, Tamogan
Super Raid I **piperonyl butoxide + pyrethrins + tetramethrin** Trans-Meri
Super Ramedit F **copper oxychloride + zineb** Agrochemiki
Super-Rasenduenger mit Moosvernichter H **iron sulphate** Wolf
Super remuron liquide H **amitrole + ammonium thiocyanate + diuron + simazine** Aseptan
Super-Rozol R **bromadiolone**
Super Sanzerb H **amitrole + atrazine + TCA** LHN
Super Sanzo H **atrazine + TCA** LHN
Super-Schachtox R **aluminium phosphide** Schacht
Super Schrumm I **dichlorvos + piperonyl butoxide + pyrethrins** Shell
Super Selectyl H **dichlorprop + MCPA + mecoprop** Agriphyt, Chimac-Agriphar
Super Sistan FIHN **metham-sodium** UCP
Super Six F **sulphur** Chiltern, Chimac-Agriphar
Super Slugran M **metaldehyde** Chimac-Agriphar
Super Sovitox R **difenacoum** Sovilo
Super-Stop Brot G **maleic hydrazide** Sadisa
Super Sucker-Stuff G **maleic hydrazide** Drexel
Super Syncuran H **chlorotoluron + chlorsulfuron** VCH
Super Ternet entretien H **amitrole + diuron + simazine** Ciba-Geigy
Super Tin F **fentin hydroxide** Chiltern, Chimac-Agriphar
Super Treflan H **trifluralin** Farmos
Super Trifloran H **prometryn + trifluralin** Diana
Super Verdone H **2,4-D + dicamba + ioxynil** ICI
Super Weedex H **amitrole + simazine** Murphy
Super X Macclesfield F **copper sulphate + maneb + zineb** La Cornubia
Super Z Macclesfield F **copper sulphate + zineb** La Cornubia
Superarsonate H **MSMA** Fermenta

Superaven H **difenzoquat methyl sulphate** Cyanamid
Superbax H **2,4-D** Sipcam-Phyteurop
Superbhex H **2,4-D + 2,4,5-T** Sipcam-Phyteurop
Superbix H **dicamba + MCPA** Sipcam-Phyteurop
Supercarb F **carbendazim** Pan Britannica
Superdust FAI **dicofol + dinocap + malathion + zineb** Protex
Superelgetol F **DNOC** Truffaut
Supergolfnet H **2,4-D + dicamba** Truffaut
Supergorsit H **sodium chlorate** Bayer
Supergreen and Weed H **2,4-D + mecoprop** May & Baker
Supergro-Extra H **2,4-D + dicamba** Fisons
Superhormona C H **2,4-D + MCPA** Condor
Supermix 74 H **2,4-D + mecoprop** Eurofyto
Supernox H **propanil** Cumberland, Crystal
Superormone concentre H **2,4-D + MCPA** Pepro
Supersinox HIA **DNOC** Bayer
Supertog H **atrazine + 2,4-D + diuron** Sipcam-Phyteurop
Supertox I **dichlorvos + methoxychlor** Bogena
Supertox H **2,4-D + mecoprop** Rhone-Poulenc
Supervelax I **lindane + parathion** Platecsa
Supervelax IA **azinphos-methyl** Platecsa
Supervelax F I **fenitrothion + lindane** Platecsa
Superzindacuivre F **copper oxychloride + zineb** Gaillard
Supona IA **chlorfenvinphos** Shell
Suprac IA **bromopropylate + methidathion** Ciba-Geigy
Supracid I **methidathion** Ciba-Geigy, Margesin, Sandoz
Supracol AF **sulphur** Tecniterra
Suprathion IA **methidathion** Makhteshim-Agan
Surcopur H **propanil** Bayer
Surecide I **cyanofenphos** Sumitomo
Surfassol H **dichlobenil** La Quinoleine
Surflan H **oryzalin** Elanco, ICI
Surpass H **vernolate** Stauffer
Surpass H **dichlormid + vernolate** OHIS, Serpiol
Surpur H **propanil** Bayer
Survan I **chlorfenvinphos + cypermethrin** Shell
Susokal N-Unkrautvernichter H **sodium chlorate** Schuster
Susvin IA **monocrotophos** Quimica Estrella
Sutan H **butylate** Avenarius, Du Pont, Efthymiadis, Ligtermoet, OHIS, Procida, Serpiol, Siapa, Stauffer, Urania
Sutan + H **butylate + dichlormid** Serpiol
Sutan A H **atrazine + butylate** Serpiol
Sutan Plus H **butylate + dichlormid** Stauffer
Sutazin H **atrazine + butylate** Serpiol, Siapa
Sutazin H **atrazine + butylate + dichlormid** ICI-Valagro
Sutazine H **atrazine + butylate** Stauffer
Sutene IA **endosulfan** Chimiberg
Sutiol I **fenitrothion** Afrasa
Suvamil I **carbaryl** Inagra
Suxon I **chlorpyrifos** SCEP botanica
Suzu F **fentin acetate** Hokko, Nihon Nohyaku
Suzu-H F **fentin hydroxide** Nihon Nohyaku
Svovelkalk F **lime sulphur** Berg, Bjoernrud, Langesaeter
Swebat I **temephos** Cyanamid
swep H [1918-18-9] methyl 3,4-dichlorocarbanilate, MCC, FMC 2995
Swiece Arrex R **zinc phosphide** Shell
Swiece Nortox F **sulphur** VIS
Swing IA **dichlorvos** Siapa
Swipe H **bromoxynil + ioxynil + mecoprop** Ciba-Geigy
Swirl S **petroleum oils** Shell
Sybol IA **pirimiphos-methyl** ICI
Sybol 2 Aerosol IA **pirimiphos-methyl + pyrethrins** ICI
Sydane IA **chlordane** Synchemicals
Sydex H **2,4-D + mecoprop** Synchemicals
Syford H **2,4-D** Synchemicals
Sygan F **captafol + cymoxanil + folpet** Du Pont
Syllit F **dodine** KenoGard, Kwizda, Ligtermoet, Organika-Azot
Sylvogam I **lindane** ICI
Symetox I **demeton-S-methyl** Agrotechnica
Synbetan D H **desmedipham** VCH
Synbetan Mix H **desmedipham + phenmedipham** VCH
Synbetan P H **phenmedipham** VCH
Synchemicals Couch and Grass Killer H **dalapon-sodium** Synchemicals

Synchemicals Mazide HG **maleic hydrazide** Synchemicals
Synchemicals Mazide Selective G **dicamba + maleic hydrazide + MCPA** Synchemicals
Synchemicals Medo FW **cresylic acid** Synchemicals
Syncuran H **chlorotoluron** VCH
Synergon 11 E S **petroleum oils** Kwizda
Synfloran H **trifluralin** VCH
Syngran H **simazine** Synchemicals
Syngran H **chlorotoluron + terbutryn** VCH
Synklor I **chlordane** Tamogan
Synleton H **chlorotoluron + mecoprop** VCH
Synlox H **asulam** VCH
Synox H **ioxynil + mecoprop** Synchemicals
Synpran 111 H **propanil + 2,4,5-T** Budapesti Vegyimuvek
Synpran 112 H **MCPA + propanil** Budapesti Vegyimuvek
Synpran N H **propanil** Chemolimpex, Budapesti Vegyimuvek
Synthrin I **resmethrin** Fairfield
Syphal F **captafol + copper oxychloride + cymoxanil** Du Pont
SYS 67 Actril C H **ioxynil + mecoprop** Schwarzheide
SYS 67 B H **2,4-DB** Bitterfeld, Schwarzheide
SYS 67 Buctril A H **bromoxynil + MCPA** Schwarzheide
SYS 67 Buctril DB H **bromoxynil + 2,4-DB** Schwarzheide
SYS 67 Buctril P H **bromoxynil + dichlorprop** Schwarzheide
SYS 67 Dambe H **dicamba + MCPA** Schwarzheide
SYS 67 Gebifan H **dichlorprop** Schwarzheide
SYS 67 MB H **MCPB** Schwarzheide
SYS 67 ME H **MCPA** Schwarzheide
SYS 67 MEB H **MCPA + MCPB** Schwarzheide
SYS 67 Mecmin H **mecoprop** Schwarzheide
SYS 67 Mprop H **mecoprop** Schwarzheide
SYS 67 Omnidel H **dalapon-sodium** Cequisa, Schwarzheide, Synthesewerk
SYS 67 Omnidel-Kombi H **dalapon-sodium + MCPA** Schwarzheide
SYS 67 Oxytril C H **bromoxynil + ioxynil + mecoprop** Schwarzheide
SYS 67 Prop H **dichlorprop** Schwarzheide
SYS 67 Prop Plus H **dichlorprop + MCPA** Schwarzheide
SYS 67 Ramex H **dichlorprop + MCPA** Schwarzheide
SYS 67 Wimex H **dicamba + dichlorprop** Schwarzheide
SYS Buratal H **2,4-DB** Schwarzheide
SYS Makasal H **MCPA** Schwarzheide
SYS Nadibut H **MCPB** Schwarzheide
SYS Omnidel H **dalapon-sodium** Schwarzheide
Systane C F **captan + myclobutanil** Pluess-Staufer
Systemic Fungicide FW **thiophanate-methyl** Murphy
Systemox IA **demeton** Bayer
Systemschutz D I **butocarboxim** Dow, Wacker
Systemschutz T F **thiabendazole** Dow
Systephos IA **mevinphos** Protex
Systhane F **myclobutanil** Ravit, Rohm & Haas
Systhane Combi F **myclobutanil + sulphur** Ravit
Systhane MZ F **mancozeb + myclobutanil** Ravit
Systhoate I **dimethoate** Procida
Systoate IA **dimethoate** Hoechst, Roussel Hoechst
Systosol I **disulfoton** Truffaut
Systox IA **demeton** Bayer
Sytam I **schradan**
Szklarniak I **dichlorvos** Organika-Azot
Szulfur F **sulphur** Budapesti Vegyimuvek

2,4,5-T H [93-76-5] (2,4,5-trichlorophenoxy)acetic acid, Brushwood Killer, Esteron, Weedone
T-Gas IF **ethylene oxide**
T.3 Spray I **pyrethrins** Sanders-Probel
Tabamex G **butralin** Etisa
Tabamor I **methamidophos** Agribus
Tabanex I **permethrin** AUV
Tabard Skordyrapenni I **cypermethrin** Temana
Tachigaren F **hymexazol** Agrishell, Kemira, Kwizda, Masso, Nordisk Alkali, Sankyo, SES, Sumitomo
Tachizolon I **parathion-methyl + phosalone** Hellafarm
Tackle H **acifluorfen-sodium** Radonja, Rhone-Poulenc

Taifen F **propineb** Visplant
Taifun H **fenthiaprop-ethyl**
Tairel F **benalaxyl** ICI
Tairel F F **benalaxyl + folpet** ICI
Tairel M F **benalaxyl + mancozeb** ICI
Tairel R F **benalaxyl + copper oxychloride** ICI
Taktic AI **amitraz** CAMCO, NOR-AM
Talan AF **dinobuton** Monteshell
Talcord I **permethrin** Pliva, Shell
Talent H **asulam + paraquat dichloride** Rhone-Poulenc
Talent S **tallow amine ethoxylate** Ravit
Talinex IN **carbofuran** Du Pont
Talis H **metazachlor + orbencarb** Siegfried
Talisman H **chlorotoluron** FCC
Talo-Sint F **bis(ethoxydihydroxydiethylamino) sulphate** Morera
Talon R **brodifacoum** ICI, JZD Pojihlavi
Talon I **chlorpyrifos** FCC
Talon PB F **chlorothalonil** Solfotecnica
Talon S Combi F **chlorothalonil + sulphur** Solfotecnica
Talonyl F **chlorothalonil** Protex
Talpan R **zinc phosphide** Marktredwitz
Talpan-Unkrautvernichtungsmittel H **sodium chlorate** Marktredwitz
Talpasep R **coumachlor** Sepran
Talpatox IV R **thallium sulphate** Chimac-Agriphar
Talpidin R **sulphaquinoxaline + warfarin** Industrialchimica
Talpkiller R **coumachlor** Bimex
Talstar I **bifenthrin** Agroslavonija, FMC, Gosa, Inagra, Pepro, Schering, Siegfried
Talunex R **strychnine** Luxan
Tam IA **methamidophos** Jin Hung
Tamanox IA **methamidophos** Crystal
Tamariz H **propanil + thiobencarb** Roussel Uclaf
Tamaron IA **methamidophos** Bayer
Tamex G **butralin** C.F.P.I., Pepro, Rhone-Poulenc, Union Carbide
Tanaflash grains souris R **crimidine** Zootherap
Tanaflash mulots R **chlorophacinone** Zootherap
Tanafourmi stilligoute I **sodium dimethylarsinate** Zootherap
Tanarat R **warfarin** Zootherap
Tanarat special Coumaspifene R **acetylsalicylic acid + warfarin** Zootherap
Tanazo-taupes R **chloralose** Zootherap
Tandem H **tridiphane** Dow
Tandex H **karbutilate** Ciba-Geigy
Tanex H **phenmedipham** Visplant
Tanone IA **phenthoate** Montedison
Tantizon H **isomethiozin** Bayer
Tanzene H **karbutilate + simazine** Ciba-Geigy
tar oils IHFX Carbo-Craven, Mortegg Emulsion, Sterilite
Taredan NI **S,S-di-sec-butyl O-ethyl phosphorodithioate** FMC
Tarexan H **metoxuron + trifluralin** Eli Lilly
Targa H **quizalofop-ethyl** Agriben, Maag, Nissan, Pepro, Radonja, Ravit
Target H **asulam + dalapon-sodium** Rhone-Poulenc
Tarsoden FA **binapacryl + tetradifon** Hoechst, Roussel Hoechst
Tarsol F **triforine** Schering
Tartan IA **cyanthoate**
Tartan H **asulam + diuron** Rhone-Poulenc
Tartan A **cyhexatin** Agrimont, ERT, Masso, Montedison
Tas F **fentin acetate** EniChem
Task IA **dichlorvos** Shell
Taterpex HG **chlorpropham** Mirfield
Tattoo I **bendiocarb** Schering
Taupicine R **strychnine** Agrarische Unie, Pradel
Taupinol R **strychnine** Gilbert
Taupizo R **chloralose** Zootherap
Taxene IA **azinphos-ethyl + dicofol** Pepro
Taxi-Zolone IA **parathion-methyl + phosalone** Condor
Taxylone IA **parathion-methyl + phosalone** Rhodiagri-Littorale
Taystuho I **diazinon + piperonyl butoxide + pyrethrins** Kemira
Tazalon H **atrazine** Hoechst, Procida, Roussel Hoechst
Tazastomp H **atrazine + pendimethalin** Cyanamid, Opopharma, Procida
2,3,6-TBA H [50-31-7] 2,3,6-trichlorobenzoic acid, acide trichlorobenzoique, TCBA, HC-1281
TBZ F **thiabendazole**
TCA H [650-51-1] sodium trichloroacetate, TCA-sodium, sodium-TCA, STCA, trichloroacetate de sodium, Konesta, NaTa, Sodium TCA, Tecane, Varitox
TCNB FG **tecnazene**
TCP F **fthalide**

TCTP H **chlorthal-dimethyl**
TD 82 H **amitrole + atrazine + diuron + simazine** Ciba-Geigy
TDE I [72-54-8] 1,1-dichloro-2,2-bis(4-chlorophenyl)ethane, DDD, OMS 1078, ENT 4225, Rhodane
Team H **benfluralin + trifluralin** Elanco
Team Four 80 S **tallow amine ethoxylate** Monsanto
Tebecap F **captan + thiabendazole** Agriplan
Tebefol F **folpet + thiabendazole** Agriplan
Tebepas G **chlormequat chloride + dichloroisobutyric acid+ ethephon** Bitterfeld
Tebulan F **dodine + fenarimol** Shell
tebutam H [35256-85-0] *N*-benzyl-*N*-isopropylpivalamide, tebutame, butam, Comodor, GPC-5544
tebutame H **tebutam**
tebuthiuron H [34014-18-1] 1-(5-*tert*-butyl-1,3,4-thiadiazol-2-yl)-1,3-dimethylurea, Bushwacker, EL-103, Graslan, Herbec, Perflan, Spike
Tebuzate F **thiabendazole** Prochimagro
Tebuzate corbeaux FX **anthraquinone + thiabendazole** Prochimagro
Tebuzate MG FIX **anthraquinone + endosulfan + lindane + thiabendazole** Prochimagro
Tebuzate TM F **thiabendazole + thiram** Prochimagro
Tebuzate TM corbeaux RFX **anthraquinone + thiabendazole + thiram** Prochimagro
Tebuzate triple FIX **anthraquinone + lindane + thiabendazole** Prochimagro
Tecane H **TCA** Schering
Techn'acid A **cyhexatin** Sipcam-Phyteurop
Techn'atral H **atrazine** Sipcam-Phyteurop
Techn'oate IA **dimethoate** Sipcam-Phyteurop
Techn'ocolor FHI **DNOC** Sipcam-Phyteurop
Techn'ufan I **endosulfan** Sipcam-Phyteurop
Techn'ur H **diuron** Sipcam-Phyteurop
Techolia I **petroleum oils** Dipon
Tecnasan F **tecnazene** Sanac
tecnazene FG [117-18-0] 1,2,4,5-tetrachloro-3-nitrobenzene, TCNB, Arena, Bygran, Easytec, Fusarex, Hickstor, Hystore, Hytec, Nebulin
Tecnicid I **carbaryl** Tecniterra
Tecnifos I **parathion** Tecniterra
Tecnifum FIHN **D-D** Tecniterra
Tecnil S **polyglycolic ethers** Serpiol
Tecnolio IA **petroleum oils** Tecniterra
Tecnoliofos I **parathion + petroleum oils** Tecniterra
Tectab F **thiabendazole** Merck Sharp & Dohme, MSD AGVET
Tecto F **thiabendazole** Agrolinz, Bayer, Ciba-Geigy, Collett, CTA, Decco-Roda, Deriva, Dow, Ewos, Margesin, Merck Sharp & Dohme, MSD AGVET, Pyrsos, Ravit, Scam
Tecto extra F **8-hydroxyquinoline sulphate + thiabendazole** MSD AGVET
Tecto plus F **imazalil + thiabendazole** Merck Sharp & Dohme
Tedane A **dicofol + tetradifon** Siapa
Tedane Combi FA **dicofol + dinocap + tetradifon** Siapa
Tedi-Kelt A **dicofol + tetradifon** Safor
Tediclor AI **chlorfenson + dicofol + tetradifon** Afrasa
Tedion V 18 A **tetradifon** Duphar, Galenika, Hortichem, KVK, Nordisk Alkali, Sandoz, Sodafabrik, Zootechniki
Tedion-Kelthane A **dicofol + tetradifon** Argos
Tedisol F **thiram** Fivat
teflubenzuron I [83121-18-0] 1-(3,5-dichloro-2,4-difluorophenyl)-3-(2,6-difluorobenzoyl)urea, tefluron, CME 134, Dart, Diaract, Nomolt
tefluron I **teflubenzuron**
tefluthrin I [79538-32-2] 2,3,5,6-tetrafluoro-4-methylbenzyl(1*R*,3*R*:1*S*,3*S*)-3-[(*Z*)-2-chloro-3,3,3-trifluoroprop-1-enyl]-2,2-dimethylcyclopropanecarboxylate, tefluthrine, Forca, Force, Forza, PP 993
tefluthrine I **tefluthrin**
Tekel A **dicofol + tetradifon** EniChem, Phytoprotect
Tekeldion A **dicofol + tetradifon** Masso
Tekeldion Ovicida A **tetradifon** Masso
Teknar I **Bacillus thuringiensis** Sandoz
Tekwaisa IA **parathion-methyl**
Telar H **chlorsulfuron** Du Pont
Teletox I **dimethoate** L.A.P.A., Protex
Telgor IA **dimethoate** Tecniterra
Telkar H **linuron** Montedison
Telodrin I **isobenzan** Shell
Telok H **norflurazon** Sandoz
Telone II N **1,3-dichloropropene** Condor, Dow, Kwizda, Ravit, Sanac, Schering
Telquel H **atrazine + 2,4-D + diuron** LHN

Telusol afrikansk insektpulver I **piperonyl butoxide + pyrethrins** Agro-Kemi
Telusol Afrikanskt-duft I **piperonyl butoxide + resmethrin** Kirk
Telusol blomsterspray I **piperonyl butoxide + pyrethrins** Kirk
Telusol Faneron H **bromofenoxim** Kirk
Telusol insecta-lac I **chlorpyrifos** Agro-Kemi
Telusol insektspray extra I **methoxychlor + piperonyl butoxide + pyrethrins** Kirk
Teman F **maneb** Tecniterra
temephos I [3383-96-8] *O,O,O',O'*-tetramethyl *O,O'*-thiodi-*p*-phenylene bis(phosphorothioate), OMS 786, ENT 27165, Abat, Abate, Abathion, AC 52160, Biothion, Lypor, Nimitex, Swebat
Temetid F **thiram** Zorka Sabac
Temik IAN **aldicarb** Agriben, Agrotec, BASF, Berner, Bjoernrud, Embetec, ICI-Valagro, Imex-Hulst, Nordisk Alkali, Pinus, Ravit, Rhone-Poulenc, Schering, Serpiol, Shell, Siapa, Union Carbide
Temik LD I **aldicarb + lindane** La Littorale, Union Carbide
Temik M IN **aldicarb + lindane** Rhodiagri-Littorale
Tempo H **linuron + terbutryn** Farm Protection
Tempo I **cyfluthrin** Bayer
Temus R **bromadiolone** Agro-Kemi, DLG, Nordisk Alkali
Tenac AS S **petroleum oils** Agrishell
Tender H **glyphosate** Monsanto
Teneran H **chloroxuron** Ciba-Geigy
tennecetin F **pimaricin**
Tenor F **prochloraz + triadimefon** Dow
Tenoran H **chloroxuron** Chromos, Ciba-Geigy, Ligtermoet, Liro, Schering, Shell
Tensiocid S **silicones** Ital-Agro
Tenso-Spray S **polyglycolic ethers** Chimac-Agriphar
Tensol S **polyglycolic ethers** Schering
Tentron H **fluroxypyr + isoproturon** Ligtermoet, Liro
Tentron F H **bromofenoxim + fluroxypyr + isoproturon** Liro
Tenysan A **tetradifon** Fahlberg-List
Tepeta F **folpet** Chimiberg
Tepeta Combi F **Bordeaux mixture + folpet** Chimiberg
Tephel H **trifluralin** Hellenic Chemical
TEPP AI [107-49-3] tetraethyl pyrophosphate, ethyl pyrophosphate, ENT 18771, Neotox, Tetron, Vapotone
Terabol FHIN **methyl bromide** Breymesser, Colkim, Degesch
terbacil H [5902-51-2] 3-*tert*-butyl-5-chloro-6-methyluracil, Du Pont Herbicide 732, Geonter, Sinbar
terbucarb H [1918-11-2] 2,6-di-*tert*-butyl-*p*-tolyl methylcarbamate, terbucarbe, terbutol, MBPMC, Azak, Hercules 9573
terbucarbe H **terbucarb**
terbuchlor H [4212-93-5] *N*-butoxymethyl-6'-*tert*-butyl-2-chloroacet-*o*-toluidide, terbuchlore, CP 46358, MON 0358
terbuchlore H **terbuchlor**
Terbuclor H **amitrole + terbutryn** Sadisa
terbuconazole F [107534-96-3] *α-tert*-butyl-*α*-(*p*-chlorophenylethyl)-1*H*-1,2,4-triazol-1-ethanol, Folicur, HWG 1608, Raxil
terbufos IN [13071-79-9] *S-tert*-butylthiomethyl *O,O*-diethyl phosphorodithioate, ENT 27920, AC 92100, Contraven, Counter
terbumeton H [33693-04-8] 2-*tert*-butylamino-4-ethylamino-6-methoxy-1,3,5-triazine, Caragard, GS 14259
Terbutex H **terbuthylazine + terbutryn** Protex
terbuthylazine H [5915-41-3] 2-*tert*-butylamino-4-chloro-6-ethylamino-1,3,5-triazine, Gardoprim, GS 13529, Primatol M
Terbutin I **permethrin + piperonyl butoxide + pyrethrins** Industrialchimica
terbutol H **terbucarb**
Terbutrex H **terbutryn** Aragonesas, COPCI, Edefi, Makhteshim-Agan
Terbutrex Combi H **simazine + terbutyn** Makhteshim-Agan
terbutryn H [886-50-0] 2-*tert*-butylamino-4-ethylamino-6-methylthio-1,3,5-triazine, terbutryne, Clarosan, GS 14260, Igran, Plantonit, Prebane, Terbutrex
terbutryne H **terbutryn**
Tercyl IG **carbaryl**
Teremec F **chloroneb** PBI Gordon
Terfit I **carbaryl** Caffaro
Terfos I **parathion** Terranalisi
Terfos Olio I **parathion + petroleum oils** Terranalisi
Terial I **chlorpyrifos** Shell
Teridox H **dimethachlor** Ciba-Geigy, Ligtermoet

Terlai FB **chloramphenicol + 8-hydroxyquinoline sulphate+ oxine-copper** Lainco
Terlate I **methomyl** Visplant
Termat Combi H **ethofumesate + lenacil** Schering
Termazina H **atrazine + simazine** ERT
Termi-Ded I **chlordane** Rigo
Termil H F **chlorothalonil** Ligtermoet
Ternet choc 2 H **amitrole + simazine + terbuthylazine** Ciba-Geigy
Ternet entretien H **atrazine + simazine** Ciba-Geigy
Ternet graminees H **amitrole + dalapon-sodium + thiazafluron** Ciba-Geigy
Ternet granule super H **amitrole + atrazine + diuron** Ciba-Geigy
Terodim I **dimethoate + trichlorfon** Inagra
Terpal G **ethephon + mepiquat chloride** Urania
Terpal C G **chlormequat chloride + ethephon** Urania
Terponel H **amitrole + atrazine + simazine + thiazafluron** Ciba-Geigy
Terprop T H **dichlobenil** La Quinoleine
Terr-O-Gas FHIN **chloropicrin + methyl bromide** Aporta, Great Lakes, Sobrom
Terra Coat F **etridiazole + quintozene** Efthymiadis
Terra Fume FHIN **metham-sodium** Dow
Terra-Systam IA **dimefox**
Terraclor F **quintozene** Efthymiadis, Uniroyal
Terraclor Super-X F **etridiazole + quintozene** Efthymiadis
Terraclor Super-X Plus FI **disulfoton + etridiazole + quintozene** Efthymiadis
Terracoat F **etridiazole + quintozene** Grima
Terracur NI **fensulfothion** Bayer
Terrafun F **quintozene** Organika-Azot
Terraklene H **paraquat dichloride + simazine** ICI, ICI-Zeltia, Sopra
Terraneb F **chloroneb** Kincaid
Terras- & gevelreiniger L **quaternary ammonium** Alabastine
Terrasan FHIN **metham-sodium** EniChem
terrasan Ameisentod I **lindane** Terrasan
terrasan-Moosentferner H **iron sulphate** Terrasan
terrasan Pflanzen-Spray IA **dichlorvos + dinocap + lindane** Terrasan
terrasan Rasen-Unkrautvernichter H **dicamba + MCPA** Terrasan
terrasan Rasenduenger mit Moosvernichter H **iron sulphate** Terrasan
terrasan Rasenduenger mit Unkrautvernichter H **2,4-D + dicamba** Terrasan
terrasan Rasenrein H **MCPA** Terrasan
terrasan Schnecken-Tod M **metaldehyde** Terrasan
Terrathion I **phorate** FCC
Terratin F **fentin hydroxide + propineb** UCB
Terratin M F **fentin hydroxide + maneb** Agriben, Bayer
Terrazim F **carbendazim + thiram** Hellenic Chemical
Terrazina H **atrazine** Terranalisi
Terrazole F **etridiazole** Efthymiadis, Grima, Siapa, Uniroyal
Terro-hyonteishavite IA **piperonyl butoxide + pyrethrins** Nordtend
Terro insektsspray I **piperonyl butoxide + pyrethrins** Nordtend
Tersan MI **fenitrothion + metaldehyde** Sepran
Tersan 1991 F **benomyl** Du Pont
Tersan LSR F **maneb** Du Pont
Tersan SP F **chloroneb**
Terset H **bromoxynil + ioxynil + isoproturon + mecoprop** Rhone-Poulenc
Tersiplene H **linuron + trifluralin** Sipcam-Phyteurop
Tertion I **disulfoton** Umupro
Terzyne H **amitrole + atrazine + simazine** Galpro
Tespi A **chloropropylate + tetradifon** Viopharm
Testor H **sodium chlorate** Martin
Tetan Rattenkoeder R **warfarin** Hawlik
Tetracap F **captan** Terranalisi
tetrachlorvinphos IA [22248-79-9] (*Z*)-2-chloro-1-(2,4,5-trichlorophenyl)vinyl dimethyl phosphate, CVMP, stirofos, OMS 595, ENT 25841, Appex, Debantic, Gardcide, Gardona, Rabon, Rabond, ROL, SD 8447
tetradifon A [116-29-0] 4-chlorophenyl 2,4,5-trichlorophenyl sulphone, tedion, ENT 23737, Tedion V18, V-18
tetradisul A **tetrasul**
Tetrafac IA **prothoate + tetradifon** Scam
Tetrafos I **parathion** Sipcam
Tetragil A **propargite + tetradifon** Scam

Tetralate I **resmethrin + tetramethrin** Fairfield
Tetram OM A **dicofol** Caffaro
tetramethrin I [7696-12-0] 3,4,5,6-tetrahydrophthalimidomethyl (±)-*cis,trans*-chrysanthemate, tetramethrine, phthalthrin, OMS 1011, FMC 9260, Neo-Pynamin, Py-Kill, SP 1103
tetramethrine I **tetramethrin**
Tetran A **cyhexatin** Visplant
Tetran Combi I **permethrin + piperonyl butoxide + tetramethrin** Sepran
Tetran K A **cyhexatin** Visplant
Tetranol A **tetradifon** Diana
Tetrapom FX **thiram** Visplant
O,O,O',O'-tetrapropyl dithiopyrophosphate I [3244-90-4] ASP-51, Aspon
Tetraram F **copper oxychloride** Terranalisi
Tetrasar F **thiram** EniChem
Tetrasol F **thiram** Terranalisi
tetrasul A [2227-13-6] 2,4,4',5-tetrachlorodiphenyl sulphide, diphenylsulphide, tetradisul, OMS 755, ENT 27115, Animert V101, V-101
Tetratox I **demeton-S-methyl** Diana
Tetron AI **TEPP** Chevron
Tevan-algmos L **quaternary ammonium** Teunis
TF-5 X **animal meal + animal oil + lanolin** Forst-Chemie
TF-35 H **2,4-D** Platecsa
TF-60 H **2,4-D** Platecsa
TF-1169 H **fluazifop-butyl** Ishihara Sangyo
TH 6040 I **diflubenzuron**
Th-Universal R **thallium sulphate**
thallium sulphate R [7446-18-6] thallium(I) sulfate, Denkarin Paste, Ratox, Th-Universal, Tharattin, Zelio
Thaneben FA **dinocap** ERT
Tharattin R **thallium sulphate** Hentschke & Sawatzki
That FA **sulphur** Stoller
Theiikos Halkos F **copper sulphate** Helinco
thiabendazole F [148-79-8] 2-(thiazol-4-yl)benzimidazole, TBZ, Apl-Luster, Arbotect, Comfuval, Mertect, Mycozol, Storite, Tecto, Thibenzole
thiadiazine F **milneb**
thiameturon-methyl H [79277-27-3] methyl 3-[[[[(4-methoxy-6-methyl-1,3,5-triazin-2-yl)amino]carbonyl]amino]sulfonyl]-2-thiophenecarboxylate, DPX M6316, Harmony
Thiamon AF **sulphur** Du Pont
Thianosan FX **thiram** Agriben, Agrotec, Anorgachim, Condor, Iberica, UCB
Thiasol FL **thiram** Van Wesemael
thiazafluron H [25366-23-8] 1,3-dimethyl-1-(5-trifluoromethyl-1,3,4-thiadiazol-2-yl)urea, thiazfluron, Erbotan, GS 29696
Thiazan F **thiram** Gen. Rep., Prochimagro
thiazfluron H **thiazafluron**
Thibenzole F **thiabendazole** MSD AGVET
Thicoper F **carbendazim** Bitterfeld
thidiazuron G [51707-55-2] 1-phenyl-3-(1,2,3-thiadiazol-5-yl)urea, Dropp, SN 49537
Thifor I **endosulfan** Rhone-Poulenc, SEGE
Thimenox IAN **phorate** Crystal
Thimet I **phorate** AVG, Cyanamid, Lapapharm, Zorka
Thimul IA **endosulfan** Condor
Thinsec IGE **carbaryl** ICI
thioallate H **sulfallate**
Thioate I **dimethoate + endosulfan** Hoechst
Thiobel I **cartap hydrochloride** Takeda
thiobencarb H [28249-77-6] *S*-4-chlorobenzyl diethyl(thiocarbamate), benthiocarb, thiobencarbe, B-3015, Bolero, Saturn, Saturno, Siacarb
thiobencarbe H **thiobencarb**
thiocarboxime IA [25171-63-5] 3-[1-(methylcarbamoyloxyimino) ethylthio]propionitrile, Talcord, WL 21959
Thiocide M FI **malathion + sulphur** R.S.R.
Thiocide M acaricide FIA **dicofol + malathion + sulphur** R.S.R.
Thiocron IA **amidithion**
Thiocuprazin F **copper oxychloride + sulphur + zineb** Solfotecnica
Thiocur F **myclobutanil + ziram** Schering
Thiocur F F **myclobutanil + sulphur** Rohm & Haas
Thiocur Flow F **myclobutanil** Schering
Thiocur S F **myclobutanil + sulphur** Schering
thiocyclam hydrogen oxalate I [31895-22-4] *N,N*-dimethyl-1,2,3-trithian-5-ylamine hydrogen oxalate, Evisect, Evisekt, SAN 155 I
Thiodan IA **endosulfan** Argos, Bjoernrud,

Budapesti Vegyimuvek, Chromos, FMC, Hoechst, ICI-Valagro, Kemira, Meoc, Pluess-Staufer, Procida, Promark, Quimigal, RousselHoechst, Spolana
Thiodan Extra I **endosulfan + parathion-methyl** Hoechst
Thiodan M.O. IA **endosulfan + petroleum oils** Hoechst
Thiodan-Combi IA **azinphos-methyl + endosulfan** Roussel Hoechst
Thiodane IA **endosulfan** Pluess-Staufer
thiodemeton IA **disulfoton**
Thiodex I **endosulfan** Gefex
thiodicarb I [59669-26-0] 3,7,9,13-tetramethyl-5,11-dioxa-2,8,14-trithia-4,7,9,12-tetra-azapentadeca-3,12-diene-6,10-dione, Larvin, Securex, Semevin, UC 51762
thiofanox IA [39196-18-4] 3,3-dimethyl-1-methylthiobutanone *O*-methylcarbamoyloxime, Dacamox, DS 15647
Thiofen I **fenitrothion** Ellagret
Thiofol F **folpet + sulphur** Marty et Parazols
Thiofrutt AF **sulphur** Margesin
Thiokar Combi FA **dinocap + sulphur** Solfotecnica
Thiolux FA **sulphur** Sandoz
Thiomat I **endosulfan** Agrofarm
Thiometilan IA **endosulfan + parathion-methyl** Roussel Hoechst
thiometon IA [640-15-3] *S*-2-ethylthioethyl *O,O*-dimethyl phosphorodithioate, dithiomethon, M-81, Bay 23129, Ekatin
Thiomex I **parathion** Protex
Thiomop IA **parathion-methyl** Marty et Parazols
Thion AF **sulphur** ATE, Aziende Agrarie Trento, Visplant
Thionam M F **ziram** Agriben
thionazin NI [297-97-2] *O,O*-diethyl *O*-pyrazin-2-yl phosphorothioate, Cynem, Nemafos, Nemaphos, Zinofos, Zinophos
Thione IA **azinphos-methyl** Chemia
Thionex IA **endosulfan** Alpha, Budapesti Vegyimuvek, Edefi, Makhteshim-Agan, Sapec
Thionic F **ziram** Agriben, UCB
Thionyl IA **parathion-methyl** Agriphyt
Thiophal F **folpet**
Thiophan FW **thiophanate-methyl** Jin Hung
thiophanate F [23564-06-9] diethyl 4,4'-(*o*-phenylene)bis(3-thioallophanate), thiophanate-ethyl, 3336-F, NF-35, Topsin E
thiophanate-ethyl F **thiophanate**
thiophanate-methyl FW [23564-05-8] dimethyl 4,4'-(*o*-phenylene)bis(3-thioallophanate), Cercobin, Cercobin M, Cycosin, Easout, Enovit M, Fungus Fighter, Mildothane, Neotopsin, NF-44, Pelt 44, Seal & Heal, Sigma, Thiophan, Topsin M
thiophos IA **parathion**
Thiopron FA **sulphur** BP
thioquinox AF [93-75-4] 1,3-dithiolo[4,5-*b*]quinoxaline-2-thione, quinothionate, chinothionat, Eradex, Eraditon, Erazidon
Thiormon G **gibberellic acid** Agriplan
Thiosulfan IA **endosulfan** Simplot
thiotepp IA **sulfotep**
Thiotox F **thiram** Diana, Sandoz
Thioval I **endosulfan** ICI-Valagro
Thiovit F **sulphur** Bjoernrud, Geopharm, Hanson & Moehring, Pan Britannica, Sandoz, Schering
thioxamyl IAN **oxamyl**
Thiozal F **sulphur** Hellenic Chemical
Thiradia F **thiram** Tradi-agri
thiram FX [137-26-8] tetramethylthiuram disulphide, thirame, thiuram, TMTD, ENT 987, AApirol, Cunitex, DRP 642532, Fernide, Hexathir, Hy-Vic, Nomersan, PolyramUltra, Pomarsol, Spotrete, Tetrapom, Thianosan, Thiotox, Thiramad, Thirasan, Tirampa, Trametan, Tripomol, Tuads, Vancide TM
Thiramad FX **thiram** Mallinckrodt
thirame FX **thiram**
Thiramvis F **thiram** Phytorgan
Thirasan FX **thiram** Condor, Pepro, Rhone-Poulenc
Thiratox F **thiram** Efthymiadis
Thirbane F **thiram** Agrishell
Thirodan F **thiram** Diana
Thirsol F **thiram** Scam
Thistrol H **MCPB** Union Carbide
thiuram FX **thiram**
Thripstick I **polybutene** Koppert
Thuricide I **Bacillus thuringiensis** Chromos, Geopahrm, Liro, Sandoz, Shell
Thuridan I **Bacillus thuringiensis** Polfa-Pabianice
Thylate FX **thiram** Du Pont, Zootechniki
TI-78 I **bensultap** Takeda
TI-1258 I **cartap hydrochloride** Takeda

TI-1671 I **bensultap** Takeda
Tian Wang H **butachlor + dymron**
Tiazin F **sulphur + zineb** Agrimont
tiazon NFHI **dazomet**
Tideran FX **thiram** Aragonesas
Tidon I **phosphamidon** Eli Lilly
Tiezene F **zineb** Agrimont, Montedison
Tifume F **thiabendazole** Merck Sharp & Dohme
Tigrex H **isoproturon** Pliva
Tiguvon I **fenthion** Bayer
Tijefon I **trichlorfon** Timosoara
Tiletole F **quintozene** Phytopharmaceutiki
Tilgin-Unkrautvertilgungsmittel H **sodium chlorate** Erlangen
Tillam H **pebulate** Geopharm, Serpiol, Stauffer
Tillantin Novo F **phenylmercury acetate** Bayer
Tillantox F **benquinox**
Tillermor G **chlormequat chloride + di-1-p-menthene** Mandops
Tilletia F **maneb + zineb** Mylonas
Tillox H **benazolin + bromoxynil + mecoprop** Schering, Schering AAgrunol
Tilozan F **TCMTB** Efthymiadis
Tilt F **propiconazole** Budapesti Vegyimuvek, Chromos, Ciba-Geigy, Fahlberg-List, Ligtermoet, Liro, VCH
Tilt 1 CB F **carbendazim + propiconazole** Ciba-Geigy
Tilt 2 Captafol F **captafol + propiconazole** Ciba-Geigy
Tilt 2 CT F **chlorothalonil + propiconazole** Ciba-Geigy
Tilt C F **carbendazim + propiconazole** Ciba-Geigy
Tilt CB F **carbendazim + propiconazole** Ciba-Geigy
Tilt CT F **carbendazim + propiconazole** Ciba-Geigy
Tilt Elite F **chlorothalonil + propiconazole + tridemorph** Ciba-Geigy
Tilt Excel F **carbendazim + chlorothalonil + propiconazole** Liro
Tilt SP F **carbendazim + chlorothalonil + propiconazole** Ciba-Geigy
Tilt top F **fenpropimorph + propiconazole** Ciba-Geigy
Tilt turbo F **propiconazole + tridemorph** Ciba-Geigy, Liro
Tilt Twin F **chlorothalonil + propiconazole** Liro
timbo IA **rotenone**
Timet I **phorate** Zorka Sabac
Tin Man F **fentin hydroxide + maneb** Chimac-Agriphar
Tinder Rat R **warfarin** Visplant
Tinestan FLM **fentin acetate** Lapapharm, Nihon Nohyaku
Tinhydroxyde F **fentin hydroxide** Bayer, Hoechst
Tioagrex AI **endosulfan** Sadisa
tiocarbazil H [36756-79-3] *S*-benzyl di-*sec*-butylthiocarbamate, Drepamon, M 3432
Tiocid IA **endosulfan** Zupa
Tiodal AF **sulphur** Fivat
Tiofolane F **folpet + sulphur** Chimiberg
Tioftal F **folpet** Agrimont
Tiogel AF **sulphur** Terranalisi
Tiolene AF **sulphur** Chimiberg
Tiolent H **cycloate** Zorka Sabac
Tiolerbane H **molinate** Chimiberg
Tiolerbane Combi H **molinate + perfluidone** Sivam
Tiomide IA **azinphos-methyl** Margesin
Tionazin IN **thionazin** Chemia
Tionazina IN **thionazin** Chemia
Tioneb F **sulphur + zineb** Tecniterra
Tionfos I **fenitrothion** Agriplan
Tioram F **copper oxychloride + sulphur** Mormino
Tiosol AF **sulphur** Sipcam
Tiospor AF **sulphur** Agrimont
Tiosur FX **thiram** ERT
Tiotox FX **thiram** Sandoz
Tiovit AF **sulphur** Sandoz
Tiowetting AF **sulphur** Scam
Tiozin F **copper oxychloride + zineb** Zorka, Zorka-Sabac
Tiozineb F **sulphur + zineb** Agronova
Tipoff G **1-naphthylacetic acid** ICI, Pepinieres Du Valois
Tipolan S **glycerol** Bayer
Tiptor F **cyproconazole + prochloraz** Sandoz
Tira-Hexalin IF **lindane + thiram** Borzesti
Tiradin F **thiram** Dudesti
Tiragrex FX **thiram** Sadisa
Tirahexa F **hexachlorobenzene + thiram** Borzesti
Tirama F **thiram** Kemira
Tirampa FX **thiram**
Tirep F **thiram** IPO-Warszawa
Tirex FX **thiram** Agriplan
Tison H **propyzamide + simazine** Visplant
Titan G **chlormequat chloride** Schering
Tiuram FX **thiram** Foret, Key

Tixit G **propham** Epro, Permutadora, Shell
Tixit C G **chlorpropham + propham** Sandoz
TMTD FX **thiram** Agronova, Chimiberg
TMZ 88 F **thiram + ziram** Chemia
Tobacol G **fatty alcohols** EniChem
Tobacron H **metobromuron + metolachlor** Viopharm
Tocsin F **thiophanate-methyl** Sandoz
Togastan I S **petroleum oils** Chemika
Tok H **nitrofen** Argos, Rohm & Haas
Tok Ultra H **linuron + nitrofen** Rohm & Haas
Tokkorn H **nitrofen** Rohm & Haas
Tokuthion I **prothiofos** Bayer
Tolban H **profluralin** Ciba-Geigy
tolclofos-methyl F [57018-04-9] *O*-2,6-dichloro-*p*-tolyl *O,O*-dimethyl phosphorothioate, Risolex, Rizolex, S-3349
Tolion H **linuron + nitrofen** Procida
Tolkan H **isoproturon** Agronorden, Agrotec, Ewos, Lucebni zavody, Ravit, Rhone-Poulenc
Tolkan A H **dinoterb + isoproturon** Agrotec, Rhone-Poulenc
Tolkan Fox H **bifenox + isoproturon** Agrotec
Tolkan S H **dinoterb + isoproturon** Agriben, Pepro, Rhone-Poulenc
Tolkan Super H **dinoterb + isoproturon** Agrotec
Tolouran H **chlorotoluron** Diana
Tolpel F **folpet** Top
Tolurane H **chlorotoluron** Chimiberg, Sivam
Tolureks H **chlorotoluron** Zupa
Tolurex H **chlorotoluron** Edefi, Makhteshim-Agan, Protex
Toluron H **chlorotoluron** Benagro, Edefi, Top
Tolux-torjunta-aine FIA **pine oil + potassium soap** To-Lu
tolylfluanid FA [731-27-1] *N*-dichlorofluoromethylthio-*N',N'*-dimethyl-*N-p* -tolylsulphamide, tolylfluanide, Bay 49854, Euparen M
tolylfluanide FA **tolylfluanid**
Tom Cat R **chlorophacinone** SEGE
Tom-Fix G **4-CPA** Phytopharmaceutiki
Tomacoop G **4-CPA** Ficoop
Tomador NT G **(2-naphthyloxy)acetic acid** Inagra
Tomadorane G **4-CPA** Chimiberg
Tomafix G **2,4-D + (2-naphthyloxy)acetic acid** Bayer
Tomapor G **naphthylmethyltolylphthalamic acid** Aporta
tomarin R **coumafuryl**
Tomathrel G **ethephon** Ciba-Geigy
Tomato Set G **chlorophenoxypropionic acid +(2-naphthyloxy)acetic acid** Shell
Tomatoben G **4-CPA** ERT
Tomatone G **4-CPA** C.F.P.I., Etisa, Pluess-Staufer
Tomatotone G **4-CPA** Luxan
Tombel I **quinalphos + thiometon** Hortichem, Sandoz
Tomorin R **coumachlor** Ciba-Geigy, Pliva
Top R **chlorophacinone** Sepran
Top 90 AF **sulphur** Schering
Top Albal H **hexazinone** Schering
Top antigerme G **chlorpropham** Top
TOP Borkenkaefermittel I **lindane + promecarb** Schering
Top Cop F **copper + sulphur** Stoller
Top Dendrocol X **copper naphthenate + natural resins** Huhtamaki, Schering
Top FZ R **zinc phosphide** Sepran
Top-Moosvernichter H **iron sulphate** Krupp
TOP Netzschwefel FA **sulphur** Schering
Top Oil S **petroleum oils** BASF
Top Special F **maneb + thiophanate-methyl** Lapapharm
Top Suc F **fentin hydroxide + sulphur** Hoechst
Topam G **propham** Ciba-Geigy
Topas F **penconazole** Ciba-Geigy
Topas C F **captan + penconazole** Ciba-Geigy
Topas Combi F **penconazole + sulphur** Ciba-Geigy
Topas Dac F **chlorothalonil + penconazole** Ciba-Geigy
Topas multivino F **copper oxychloride + folpet + penconazole** Ciba-Geigy
Topas MZ F **mancozeb + penconazole** Ciba-Geigy
Topaz F **penconazole** Ciba-Geigy, Ligtermoet, Liro
Topaz Fruit F **captan + penconazole** Liro
Topaz M Extra F **captan + maneb + penconazole** Ligtermoet
Topaz Speciaal F **captan + penconazole** Ligtermoet
Topaze F **penconazole** Ciba-Geigy

Topaze C F **captan + penconazole** Ciba-Geigy
Topaze multi F **folpet + penconazole** Ciba-Geigy
Topazol H H **ametryn + amitrole + 2,4-D** Ciba-Geigy
Topazol TL H **amitrole + ammonium thiocyanate + simazine** C.F.P.I., Ciba-Geigy
Topcide H **benzadox**
Topfane huileux H **dinoseb** Top
Topiclor I **chlordane**
Topidion R **bromadiolone** Formenti, Ravit, Siapa
Topin R **sulphaquinoxaline + warfarin** Industrialchimica
Topin B R **bromadiolone** Industrialchimica
Topirat R **coumatetralyl** Sici
Topitox R **chlorophacinone**
Toplan H **bromoxynil** C.F.P.I.
Toplawn H **2,4-D + dicamba** Pan Britannica
Topmaneb F **maneb** Top
Topmedifame H **phenmedipham** Top
Topmil F **fentin hydroxide + maneb** Hoechst
Topnebe F **mancozeb** Top
Topocid R **warfarin** Siapa
Topofin R **arsenic anhydride + sulphur** Agrocros
Topogard H **terbuthylazine + terbutryn** Ciba-Geigy, Ligtermoet, VCH, Viopharm
Toppel I **cypermethrin** Farm Protection
Topper H **ioxynil** Pluess-Staufer, Union Carbide
Topper 2 + 2 H **bromoxynil + ioxynil** Embetec
Toprazine H **atrazine** Top
Topsar F **maneb + thiophanate-methyl** Duphar, Schering AAgrunol
Topshot H **bentazone + cyanazine + 2,4-DB** Shell
Topsin combi F **mancozeb + thiophanate-methyl** KVK
Topsin E F **thiophanate** Cleary, KVK, Nippon Soda, Serpiol
Topsin M F **thiophanate-methyl** Bjoernrud, Duphar, Imex-Hulst, Kemira, KVK, Nippon Soda, Organika-Azot, Pennwalt, Schering, Schering AAgrunol
Topstop P R **warfarin** EniChem
Topsuc F **fentin hydroxide + sulphur** R.S.R.
Topthiram F **thiophanate-methyl + thiram** Duphar
Topusyn H **desmetryn** Bitterfeld
Topzineb F **zineb** Top
Torak IA **dialifos** Agrocros, Celemerck, Hellafarm, NOR-AM, Pliva, Schering, Shell, Siapa, Siegfried, Sopra
Toram H **bromacil + picloram** ICI
Torapron H **amitrole + atrazine + 2,4-D** BP
Torbin H **EPTC** Siapa
Torch H **bromoxynil** Union Carbide
Tordon H **picloram** Dow
Tordon 101 H **2,4-D + picloram** Condor, EniChem, Galenika, Kemira, Siapa
Tordon 22 H **picloram** Ciba-Geigy
Tordon 22 K H **picloram** Bianchedi, C.F.P.I., Chipman, Dow, Kwizda
Tordon 225 H **picloram + 2,4,5-T** Dow
Torero A **fluvalinate** Sandoz
Torgal H **2,4-D + picloram** Galenika
Tormona H **fenoprop** Shell
Tornade I **permethrin** Rhone-Poulenc
Tornado I **acephate**
Tornado E **carbaryl** ICI
Torpedo I **permethrin** ICI
Torpi 590 H **amitrole + ammonium thiocyanate + simazine** Aseptan
Torpi aqua H **amitrole + ammonium thiocyanate** Aseptan
Torpi TA H **amitrole + ammonium thiocyanate** Aseptan
Torque A **fenbutatin oxide** Agrishell, Agriplant, Grima, ICI, Kemira, Ravit, Shell
Torus I **fenoxycarb** Maag
Tosan H **TCA** Eli Lilly
Tota-Col H **diuron + paraquat dichloride** ICI, ICI-Zeltia, Sopra
Total H **glufosinate-ammonium** Hoechst
Total D H **dinitramine + linuron** Ravit
Total-Ex H **sodium chlorate** Neudorff
Total-Ex Super H **bromacil + diuron** Neudorff
Total Grima A **dicofol + tetradifon** Grima
Total-Unkrautvernichter Ektorex H **amitrole + diuron** Shell
Total Weed H **amitrole + ammonium thiocyanate + simazine** May & Baker
Total Weedkiller H **amitrole + simazine** Synchemicals
Totalcid H **amitrole + atrazine** Zorka Sabac
Totale RS H **TCA** Sipcam
Totalene IA **dimethoate + fenitrothion + trichlorfon** Chimiberg

Totazin H **atrazine + MCPA** Dimitrova
Totazina H **simazine** Chimiberg, Sepran
Toterbane H **diuron** Chimiberg
Totex Stroe H **atrazine + dichlobenil** Finnewos
Totril H **ioxynil** Agrolinz, Agronorden, Chromos, Condor, Ewos, Hortichem, Kemira, Plantevern-Kjemi, Rhodiagri-Littorale, Rhone-Poulenc
Tough H **pyridate** Gilmore
Touwen's Boombalsem FW **copper naphthenate** Touwen
Tox-Hid R **warfarin** Hopkins
Tox-Vetyl R **sulphaquinoxaline + warfarin** Vetyl-Chemie
Toxa Overdose R **difenacoum** Billen
Toxa Warfarin R **warfarin** Billen
Toxacar A **chlorobenzilate** Diana
Toxakil IRA **camphechlor** FMC
Toxal H **sodium chlorate** Chemia
Toxan plaene spray H **2,4-D + dichlorprop** Agro-Kemi
Toxan plaenerenser H **2,4-D + mecoprop** Agro-Kemi
Toxaphene IRA **camphechlor** Drexel
Toxation IA **azinphos-methyl** Visplant
Toxation Etil IA **azinphos-ethyl** Visplant
Toxation M IA **azinphos-methyl** Visplant
Toxatrin IA **azinphos-methyl** Enotria
Toxene I **trichlorfon** Scam
Toxer Canali H **dalapon-sodium** Visplant
Toxer Cipolle H **chlorthal-dimethyl** Visplant
Toxer EB H **2,4-D** Visplant
Toxer S H **2,4-D** Visplant
Toxer Total H **paraquat dichloride** Phytorgan, Visplant
Toxirat R **warfarin** Steininger
2,4,5-TP HG **fenoprop**
TPN F **chlorothalonil**
TPTA FLM **fentin acetate**
TPTH F **fentin hydroxide**
TPTOH F **fentin hydroxide**
TR16 I **Trichogramma maydis** UNCAA
Trabeal F **copper oxychloride** Platecsa
Tracapor I **malathion** Sopepor
Tracker H **dicamba** Burts & Harvey
Tracor I **malathion** Quimigal
Tradiacuivre FB **copper oxychloride** Tradi-agri
Tradiafume N **ethylene dibromide** Tradi-agri
Tradiagrane H **atrazine + 2,4-D + diuron** Tradi-agri
Tradiakill H **dinoseb** Tradi-agri
Tradianet D H **2,4-D** Tradi-agri
Tradianet debroussaillant H **2,4-D + 2,4,5-T** Tradi-agri
Tradianet G H **mecoprop** Tradi-agri
Tradianet M H **MCPA** Tradi-agri
Tradianet Total H **2,4-D + dalapon-sodium + diuron** Tradi-agri
Tradianol H **amitrole + atrazine** Tradi-agri
Tradiaphos I **dichlorvos** Tradi-agri
Tradiasim H **simazine** Tradi-agri
Tradiax H **TCA** Tradi-agri
Tradiazan Z F **zineb** Tradi-agri
Tradiazole TA H **amitrole + ammonium thiocyanate** Tradi-agri
Tradioate IA **dimethoate** Tradi-agri
Traduron. H **diuron** Tradi-agri
Trafos I **parathion** Aziende Agrarie Trento
Traitement coniferes F **fosetyl-aluminium** Umupro
Traitement d'hiver FI **anthracene oil + DNOC** Umupro
Trakephon H **buminafos** Bitterfeld
tralkoxydim H 2-[1-(ethoxyimino)propyl]-3-hydroxy-5-mesitylcyclohex-2-enone, Grasp, Grasp 604, PP 604, Splendor
tralomethrin I [66841-25-6] (*S*)-α-cyano-3-phenoxybenzyl (1*R*,3*S*)-2,2-dimethyl-3-[(*RS*)-1,2,2,2-tetrabromoethyl]cyclopropanecarboxylate, tralomethrine, HAG 107, RU 25474, Scout
tralomethrine I **tralomethrin**
Tramat H **ethofumesate** Diana, Fisons, Gullviks, Huhtamaki, Kwizda, Schering, Schering AAgrunol
Tramat Combi H **ethofumesate + lenacil** Schering
Tramazole H **amitrole + atrazine + simazine** Sipcam-Phyteurop
Trametan FX **thiram**
Tranid I **aldicarb + lindane** Ravit, Union Carbide
Trans-Vert H **MSMA** Union Carbide
Transplantone G **1-naphthylacetamide + 1-naphthylacetic acid** EniChem
Transplantone G **1-naphthylacetic acid** Agriben, C.F.P.I.
Transplantonic G **1-naphthylacetic acid** C.F.P.I., Sovilo
Trapan H **linuron + pendimethalin** Cyanamid, Opopharma, Sandoz
Trapex FHIN **methyl isothiocyanate** Schering

Trasan H **2,4-D + MCPA** Galenika
Trasplant H **2,4-D** Agrocros
Trasplant Forte H **2,4-D + MCPA** Agrocros
Traton H **alachlor + tebutam** Du Pont
Traubinol H **amitrole + MCPA + simazine** Protex
Travacid-Pulver/-poudre FW **8-hydroxyquinoline sulphate** Siegfried
Trazalex H **nitrofen + simazine** Berlin Chemie
Trazalex Extra H **nitrofen + simazine** Berlin Chemie
Tre Effe S F **etem + sulphur + zineb** Siapa
Tre-Hold G **1-naphthylacetic acid** C.F.P.I., Spiess, Union Carbide, Urania
Trebon I **ethofenprox** Mitsui Toatsu
Treflan H **trifluralin** Bjoernrud, Elanco, Eli Lilly, Grima, ICI-Valagro, Lilly Farma, Organika-Sarzyna, Radonja, Sintesa
Treplik H **neburon + pendimethalin** Cyanamid, Opopharma, Sopra
Tresex Gamma I **lindane** Nordisk Alkali
Trevespan H **ioxynil** Celamerck, Shell
Trevespan DP H **dichlorprop + ioxynil** Agrotec
Trevi A **hexythiazox** Siegfried
Tri VC 13 IN **dichlofenthion** Bourgeois, Pennwalt Holland
tri-allate H [2303-17-5] *S*-2,3,3-trichloroallyl di-isopropylthiocarbamate, triallate, Avadex BW, CP 23426, Far-Go
Tri-Ban R **pindone**
Tri-Chlor NIFRH **chloropicrin** Niklor
Tri-Farmon H **linuron + trifluralin** Farm Protection
Tri-Miltox F **copper carbonate (basic) + copper oxychloride + copper sulphate +mancozeb** Sandoz
Tri-Miltox B F **copper oxychloride + folpet** Sandoz
Tri-Miltox Forte F **copper carbonate (basic) + copper oxychloride + copper sulphate +mancozeb** Sandoz
Tri-Miltox Plus F **copper oxychloride + copper sulphate + cymoxanil + mancozeb** Sandoz
Triacetane F **fentin acetate** Chimiberg, Sivam
triadimefon F [43121-43-3] 1-(4-chlorophenoxy)-3,3-dimethyl-1-(1*H*-1,2,4-triazol-1-yl)butan one, triadimefone, Amiral, Bay MEB 6447, Bayleton
triadimefone F **triadimefon**
triadimenol F [55219-65-3] 1-(4-chlorophenoxy)-3,3-dimethyl-1-(1*H*-1,2,4-triazol-1-yl)butan-2-ol, Bay KWG 0519, Bayfidan, Baytan, Summit
Triagam I **lindane** Siegfried
Triagran H **bentazone + dichlorprop + MCPA** Collett
Trialat H **tri-allate** Sojuzchimexport
Trialex H **tri-allate** Protex, Top
triallate H **tri-allate**
triamiphos FIA [1031-47-6] *P*-5-amino-3-phenyl-1,2,4-triazol-1-yl-*N,N,N',N'*-tetramethylphosphonic diamide, Wepsyn 155
Triamyl H **tri-allate** Interphyto
Triangle LF **copper sulphate** Phelps-Dodge
Triaphyt H **atrazine + simazine** Sipcam-Phyteurop
Triarex extra fluide H **atrazine + simazine** Agrishell
triarimol F [26766-27-8] 2,4-dichloro-α-(pyrimidin-5-yl)benzhydryl alcohol, EL 273, Trimidal
Triasol H **atrazine + simazine** Aseptan
Triasol granule H **atrazine + 2,4-D + diuron** Aseptan
Triasyn F **anilazine** Bayer
Triatix AI **amitraz** Wellcome
Triatox AI **amitraz** Wellcome
triazine F **anilazine**
Triazol Super H **amitrole + atrazine + simazine** Ellagret
triazophos IAN [24017-47-8] *O,O*-diethyl *O*-1-phenyl-1*H*-1,2,4-triazol-3-yl phosphorothioate, Hoe 2960, Hostathion
triazotion IA **azinphos-ethyl**
Triban H **dicamba + dichlorprop + MCPA + mecoprop** KVK
Triblecar FIX **anthraquinone + lindane + mancozeb** La Littorale
Tribunil H **methabenzthiazuron** Agro-Kemi, Agroplant, Bayer, Berner, Hoechst, Imex-Hulst, Lucebnizavody, Pinus
Tribunil Combi H **dichlorprop + methabenzthiazuron** Bayer
Tribusan H **dinoseb** Bourgeois, Pennwalt France
Tribute H **dicamba + MCPA + mecoprop** Chipman
Tributon H **2,4-D + 2,4,5-T** Bayer
Tributyl H **2,4-D + 2,4,5-T** Agrofarm
S,S,S-tributyl phosphorotrithioate G [78-48-8] Chemagro B-1776, DEF, DEF Defoliant, De-Green, E-Z-off D, Fos-Fall A, Ortho Phosphate Defoliant

tricamba H [2307-49-5] 3,5,6-trichloro-*o*-anisic acid, Banvel T
Tricarbamix F **ferbam + maneb + zineb** Bourgeois, Cequisa, Chimac-Agriphar, Decco-Roda, Pennwalt Holland
Tricarbasul F **maneb + sulphur + zineb** Decco-Roda
Tricarnam I **carbaryl** Decco-Roda, Pennwalt Holland
trichlamide F [70193-21-4] (*RS*)-*N*-(1-butoxy-2,2,2-trichloroethyl)salicylamide, salclomide, Hataclean, NK 483
trichlophenidin I [53720-80-2] 1,3-bis-(3-chlorophenyl)-2-trichloromethylimidazolidine, CME 7002, UTH WP 50-7
Trichlorex I **trichlorfon** Gaillard, Procida
Trichlorfenson A **chlorfenson** Bourgeois
trichlorfon I [52-68-6] dimethyl 2,2,2-trichloro-1-hydroxyethylphosphonate, trichlorphon, DEP, metriphonate, chlorofos, dipterex, OMS 800, ENT 19763, Bay 15922, BayL13/59, Briten, Cekufon, Danex, Denkaphon, Dep, Dipterex, Ditrifon, Dylox, Nevugon, Proxol, Tugon
trichloroacetate de sodium H **TCA**
trichloronat I [327-98-0] *O*-ethyl *O*-2,4,5-trichlorophenyl ethylphosphonothioate, trichloronate, fenophosphon, OMS 412, OMS 578, ENT 25712, Agrisil, Agritox, Bay 37289, Phytosol
trichloronate I **trichloronat**
trichlorphon I **trichlorfon**
Trichoderma viride F Binab T
Tricimite A **dicofol + fenson** Fivat
triclopyr H [55335-06-3] 3,5,6-trichloro-2-pyridyloxyacetic acid, Crossbow, Dowco 233, Garlon, Turflon
Triclorkey I **trichlorfon** Key
Tricolan G **chlormequat chloride** Kwizda
Tricornox Special H **benazolin + dicamba + dichlorprop** Kemi-Intressen
Tricorta G **chlormequat chloride** KVK
Tricuper F **copper oxysulphate** SPE
Tricusan F **copper oxychloride + maneb + zineb** Aragonesas
tricyclazole F [41814-78-2] 5-methyl-1,2,4-triazolo[3,4-*b*]benzothiazole, Beam, Bim, Blascide, EL-291
tricyclohexyltin hydroxide A **cyhexatin**
Tridal F **nuarimol** Eli Lilly, Grima
Tridal Cap F **captan + nuarimol** Sandoz
Tridal MZ F **mancozeb + nuarimol** Eli Lilly, Radonja
Tridal S F **nuarimol + sulphur** Radonja
tridemorph F [81412-43-3] Reaction mixture of C_{11}-C_{14} 4-alkyl-2,6-dimethylmorpholine homologues containing 60-70% of 4-tridecylisomers, tridemorphe, Bardew, BASF 220, Calixin, Ringer
tridemorphe F **tridemorph**
Tridex F **mancozeb** Hoechst
Tridex H **cyanazine + trifluralin** Shell
Tridezol F **maneb + tridemorph** Shell
tridiphane H [58138-08-2] (*RS*)-2-(3,5-dichlorophenyl)-2-(2,2,2-trichloroethyl)oxirane, Dowco 356, Nelpon, Tandem
trietazine H [1912-26-1] 2-chloro-4-diethylamino-6-ethylamino-1,3,5-triazine, G 27901, Gesafloc, NC 1667
Trif H **trifluralin** Chemia
Trifanex H **DNOC** Bourgeois, Pennwalt France
Trifen F **fentin acetate** Pinus
Trifene F **fentin acetate** Sipcam
trifenmorph M [1420-06-0] 4-(triphenylmethyl)morpholine, Frescon
Trifenson A **fenson** Bourgeois
Trifina IAHF **DNOC** Pennwalt Holland
triflumizol F **triflumizole**
triflumizole F [68694-11-1] (*E*)-4-chloro-*α*,*α*,*α*-trifluoro-*N*-(1-imidazol-1-yl-2-propoxyethylidene)-*o*-toluidine, triflumizol, Duo Top, NF-114, Trifmine
Trifluragrex H **trifluralin** Sadisa
Triflural H **trifluralin** Interphyto
trifluralin H [1582-09-8] *α*,*α*,*α*-trifluoro-2,6-dinitro-*N*,*N*-dipropyl-*p*-toluidine, trifluraline, Digermin, Elancolan, Heritage, Ipersan, L-36352, Sinflouran, Treflan, Triflurex, Trigard, Tristar
Trifluralina H **trifluralin** Agrolac, CAG
trifluraline H **trifluralin**
Trifluran H **trifluralin** Diana, Terranalisi
Trifluran Kombi H **linuron + trifluralin** Diana
Trifluree H **linuron + trifluralin** Interphyto
Triflurex H **trifluralin** Anorgachim, Copci, Edefi, Makhteshim-Agan, Sinteza
Triflurex-Super H **prometryn + trifluralin** Ellagret
Triflurotox H **trifluralin** Organika-Sarzyna
Trifmine F **triflumizole** Nippon Soda
Trifocide IAHF **DNOC** Agronova, Chimac-Agriphar, Decco-Roda, Grima, KVK, Pennwalt France, Pennwalt Holland

Trifolex H **MCPB** Shell
Trifolex-tra H **MCPA + MCPB** Shell
Trifolin H **MCPB** Siegfried
Trifon plus I **fenitrothion + trichlorfon** Leu & Gygax
Trifoplex H **MCPA + MCPB** Burri
triforine F [26644-46-2] 1,1'-piperazine-1,4-diyldi-[*N*-(2,2,2-trichloroethyl)formamide], Cela W524, Denarin, Funginex, Saprol
Trifox H **bifenox + clopyralid + mecoprop** Eli Lilly
Trifrina IAF **DNOC** Du Pont, Elanco, Pennwalt France, Ravit, Sandoz
Trifulex H **trifluralin** Ellagret
Trifulon H **trifluralin** Hoechst
Trifungol F **ferbam** FMC, Pennwalt Holland
Trifusol F **quintozene** Bourgeois
Trigard I **cyromazine** Ciba-Geigy
Trigard H **trifluralin** Farmers Crop Chemicals
Trigger H **isoproturon + tri-allate** Rhone-Poulenc
Triherbide CIPC H **chlorpropham** Agronova, Decco-Roda, KVK, Pennwalt Holland
Triherbide IPC H **propham** Decco-Roda, Pennwalt Holland
Triherbide L H **chlorpropham** Pennwalt France
Triherbin H **trifluralin** Burri
Trihlorfon I **trichlorfon** Zorka Subotica
Trikepin H **trifluralin** Pinus
Triliane H **dicamba + linuron + trifluralin** Chimiberg
Trilin H **linuron + trifluralin** Protex
Trilixon H **chlorsulfuron + methabenzthiazuron** Bayer
Trilox H **alachlor + atrazine + pyridate** Agroplant
Trilox H **bromoxynil + dicamba + mecoprop** Agrolinz
Triluron H **linuron + trifluralin** EniChem
Trimangol F **maneb** ACA, Bos, Bourgeois, Decco-Roda, Desarrollo Quimico, Filocrop, Gullviks, Pennwalt, Pennwalt France, Pennwalt Holland
Trimanoc bleu F **mancozeb** Pennwalt
Trimanoc Neu F **maneb** Pennwalt Holland
Trimanoc Super F **maneb + zineb** Pennwalt Holland
Trimanzone F **ferbam + maneb + zineb** Bos, Bourgeois, Decco-Roda, Pennwalt, Pennwalt Holland, Shell
Trimaran H **chlorotoluron** La Quinoleine
Trimastan F **fentin acetate + maneb** Bos, Pennwalt, Pennwalt Holland
Trimaton FHIN **metham-sodium** Decco-Roda, Desarrollo Quimico, Pennwalt, Pennwalt Holland
Trimazol H **amitrole + simazine** Etisa
Trimec H **2,4-D + dicamba + mecoprop** PBI Gordon
Trimesol I **fenitrothion** Eli Lilly
trimethacarb IM [2686-99-9 (3,4,5-isomer); 3971-89-9 (2,3,5-isomer)] 3,4,5- and 2,3,5-trimethylphenyl methylcarbamate (ca. 4:1 mixture), Broot, Landrin
Trimethox I **diazinon + dimethoate + methoxychlor** Protex
Trimetion IA **dimethoate**
Trimeton I **carbophenothion + demeton-S-methyl + parathion-methyl** Hellenic Chemical
Trimidal F **nuarimol** Agrimont, Elanco, Eli Lilly, Radonja, Sandoz
Trimidal Beize F **imazalil + nuarimol** Elanco
Trimidan I **carbophenothion + phosmet** Efthymiadis, Serpiol
Trimifol F **copper oxychloride + folpet** Pliva, PVV
Trimilzan F **copper oxychloride + copper sulphate + cymoxanil** Aragonesas
Triminol F **nuarimol** Elanco
Trimisem FX **anthraquinone + maneb + nuarimol** Eli Lilly
Trimisem D F **imazalil + nuarimol** Eli Lilly
Trimisem SC F **imazalil + nuarimol** Eli Lilly
Trimisem total FIX **anthraquinone + lindane + maneb + nuarimol** Eli Lilly
Trinex I **permethrin** Protex
Trinoc IAF **DNOC** Chemia
Trinol H **DSMA + MCPA** Rhone-Poulenc
Trinol Super H **dicamba + MCPA + mecoprop** Rhodiagri-Littorale
Trinovin H **amitrole + atrazine + simazine** Efthymiadis
Trinulan H **linuron + trifluralin** Elanco, Eli Lilly, Maag, Radonja, Sadisa
Trio H **bromoxynil + ioxynil + pyridate** Leu & Gygax
Triocord I **cypermethrin** Sivam
Triona IA **petroleum oils** Shell

Trior H **dichlorprop + MCPA + mecoprop** L.A.P.A., Protex
Triormone D H **2,4-D + MCPA + mecoprop** Interphyto
Triormone DP H **dichlorprop + MCPA + mecoprop** Interphyto
Triotyl H **dichlorprop + MCPA + mecoprop** Agrishell
Trioxone H **2,4,5-T** ICI-Zeltia
Trioxone H **fenoprop** Prometheus
Tripart Arena F **tecnazene** Tripart
Tripart Beta H **phenmedipham** Tripart
Tripart Blue F **zineb** Tripart
Tripart Brevis G **chlormequat chloride** Tripart
Tripart Defensor F **carbendazim** Tripart
Tripart Faber F **chlorothalonil** Tripart
Tripart Gladiator H **chloridazon** Tripart
Tripart Imber F **sulphur** Tripart
Tripart Laurel H **terbutryn + trifluralin** Tripart
Tripart Legion F **carbendazim + maneb** Tripart
Tripart Ludorum H **chlorotoluron** Tripart
Tripart Merzo H **bentazone + mecoprop** Tripart
Tripart Mini Slug Pellets M **metaldehyde** Tripart
Tripart Obex F **maneb + zinc** Tripart
Tripart Sentinel H **propachlor** Tripart
Tripart Signum H **chloridazon + propachlor** Tripart
Tripart Systemic Insecticide IA **demeton-S-methyl** Tripart
Tripe G **chlorpropham + propham** Agronova, Decco-Rhoda
Tripece G **chlorpropham + propham** Pennwalt Holland
Tripenal H **2,4-D + dicamba + MCPA** Bayer
Triphenylacetate PH-50-11 F **fentin acetate** Duphar
Triphonil H **trifluralin** Filocrop
Tripion AZ H **MCPA-thioethyl + propanil** Inagra
Tripion CB H **MCPA-thioethyl** Sipcam
Tripion Extra H **MCPA-thioethyl + propanil + pyridate** Inagra
Triplam I **malathion** Ciba-Geigy
Triple Tin F **fentin hydroxide** Wesley
Triple XXX R **difenacoum** Certified Lab.
Triplen NT H **trifluralin** Sipcam
Tripomol F **thiram** Bos, Bourgeois, Decco-Roda, Pennwalt France, Pennwalt Holland
triprene I [40596-80-3] *S*-ethyl (*E*)-(*RS*)-11-methoxy-3,7,11-trimethyldodeca-2,4-dienethioate, Altorick, ZR 519
Triran A **cyhexatin** BP, Chemia
Trisanaat H **tri-allate** Sanac
Triscabol F **ziram** Pennwalt Holland
Trisila H **amitrole + 2,4-D + MCPA + TCA** Galpro
Trisol H **diuron + linuron + terbacil** Du Pont
Tristar H **trifluralin** Pan Britannica
Tristar H **alachlor + atrazine + pyridate** Agrolinz
Tritan F **fentin acetate** Terranalisi
Tritex H **alloxydim-sodium** Ewos
Trithac F **maneb + zineb** Dow
Trithion IA **carbophenothion** Serpiol, SPE, Stauffer
Trithion K AI **carbophenothion + dicofol** Serpiol
Trithion Kelthane AI **carbophenothion + dicofol** Serpiol
Trithion Oil IA **carbophenothion + petroleum oils** Serpiol, SPE
Triticol F **carbendazim** Spiess, Urania
Tritifen H **linuron + pendimethalin** Sandoz
Tritin A **cyhexatin** Scam
Tritisan F **quintozene** Hoechst
Tritoftorol F **zineb** Bos, Desarrollo Quimico, Pennwalt France, Pennwalt Holland
Triton CS 7 S **octylphenoxypolyethoxyethanol + sodium sulphosuccinate** Duphar, Rohm & Haas
Triton X 114 S **alkyl phenol ethoxylate** Rohm & Haas
Tritox H **dicamba + MCPA + mecoprop** Fisons
Tritox IA **hydrogen cyanide** Degesch
Triumph NI **isazofos** Ciba-Geigy
Triumph F **chlorothalonil + flusilazole** Du Pont
Trivax F **methfuroxam** Hellafarm
Trivax M F **maneb + methfuroxam** Hellafarm
Trizal H **linuron + trifluralin** Afrasa
Trizeb F **mancozeb + nuarimol** Eli Lilly
Trizilin H **nitrofen** Bitterfeld
Triziman F **maneb + zineb** Dow, Du Pont, Pennwalt, Pennwalt Holland
Trojan H **chloridazon** Schering
Trolene I **fenchlorphos**

Trolex H **amitrole + ammonium thiocyanate** Protex
Tromb H **bromoxynil + ioxynil + isoproturon** Ewos
Tropital S **piprotal**
Tropotone H **MCPB** Rhodiagri-Littorale
Tropotox H **MCPB** Agriben, Hortichem, Rhone-Poulenc
Tropotox Plus H **MCPA + MCPB** Embetec
Troysan FX **copper naphthenate** Troy Chemical
Truban F **etridiazole** Mallinckrodt
Trucidor IA **vamidothion** Rhone-Poulenc
Trump H **isoproturon + pendimethalin** Cyanamid
Trumpf Rasenduenger + Unkrautvernichter H **2,4-D + MCPA** Windhoevel
Trylone G **hydroxy-MCPA** Condor
Tschilla-Schneckenkorn M **metaldehyde** Marktredwitz
tserenox F **benquinox**
tsitrex F **dodine**
Tsumacide I **metolcarb** Gen. Rep., Mitsubishi
Tuads FX **thiram**
Tuba G **chlormequat chloride + diphenylurea** Mandops
Tubazole FB **organoiodine + thiabendazole** Wheatley
Tubergran F **quintozene** Wheatley
Tuberite G **propham** ICI-Valagro, ICI-Zeltia
Tuberite G **chlorpropham + propham** Pliva, Sopra
Tuberite Super G **chlorpropham + propham** ICI-Zeltia
Tubernet 2 G **chlorpropham** Sovilo
Tuberofen G **propham** Zupa
Tuberone G **4-indol-3-ylbutyric acid + 1-naphthylacetamide+ 1-naphthylacetic acid** Agriphyt
Tubosan F **cyprofuram** Schering AAgrunol
Tubostore FG **tecnazene** FCC
Tubotin F **fentin hydroxide** Rhone-Poulenc
Tudor-Corbo R **chloralose** Sovilo
Tudy AI **amitraz** Agrishell, Shell
Tue-Herbe H **amitrole + diuron + sodium thiocyanate** Truffaut
Tufane extra H **dinoseb** Agriphyt
Tufler H **butamifos**
Tugon I **trichlorfon** Agro-Kemi, Bayer, Berner
Tugon Perfekt-Stalspuitmiddel I **permethrin + plifenate** Bayer
Tula I **lindane** Aseptan
Tula rat R **chlorophacinone** Aseptan
Tula souris R **chlorophacinone** Aseptan
Tulipan F **quintozene** Zerpa
Tumar IA **chlorpyrifos-methyl** Dow, Du Pont
Tumaroben G **fenoprop** ERT
Tumbleblite F **propiconazole** Murphy
Tumbleleaf H **sodium chlorate** Wilbur-Ellis
Tumbleweed H **glyphosate** Murphy
Tumousse H **iron sulphate** Franco-Belge
Tunic H **methazole** Sandoz
Tupersan H **siduron** Du Pont
Turagil R **chlorophacinone** Rhone-Poulenc
Turagil Super B R **bromadiolone** Pepro
Turagil Super C R **chlorophacinone** Pepro
Turbair Acaricida A **dicofol + tetradifon** Vequer
Turbair BA F **dicloran + thiram** Vequer
Turbair DP F **dinocap** Vequer
Turbair Flydown I **piperonyl butoxide + pyrethrins**
Turbair fungicida de cobre F **copper oxychloride** Vequer
Turbair Grain Store Insecticide I **fenitrothion + permethrin + resmethrin** Pan Britannica
Turbair Insecticida General I **dimethoate** Vequer
Turbair-karpassumu IA **piperonyl butoxide + pyrethrins** Anticimex
Turbair KB H **propyzamide** Vequer
Turbair LD A **dicofol** Vequer
Turbair LO I **lindane** Vequer
Turbair MB F **maneb** Vequer
Turbair MN IA **malathion** Vequer
Turbair MZ F **mancozeb + zineb** Vequer
Turbair OT A **propargite** Vequer
Turbair RN I **resmethrin** Vequer
Turbair Rovral H **iprodione** Vequer
Turbair Systemic Insecticide IA **dimethoate** Pan Britannica
Turbair ZB F **zineb** Vequer
Turbo H **metolachlor + metribuzin** Mobay
Turbo TR F **propiconazole + tridemorph** Ciba-Geigy
Turboblanc D I **diazinon** Ciba-Geigy
Turbocuivre CS F **copper sulphate + maneb** Ciba-Geigy

Turbofal F **copper oxychloride + folpet** Ciba-Geigy
Turbofal PM F **copper oxychloride + copper sulphate + folpet** Ciba-Geigy
Turbosoufre F **sulphur** Ciba-Geigy
Turcam I **bendiocarb** Schering, NOR-AM
Turda-Cupral F **copper oxychloride** Turda
Turdavor F **copper oxychloride** Turda
Turf-Cal IH **calcium arsenate**
Turfcide F **quintozene** Uniroyal
Turfclear F **carbendazim** Fisons
Turfene H **dicamba + mecoprop** Chimiberg
Turflon H **triclopyr** Dow
Turflon D H **2,4-D + triclopyr** Dow
Turkan FHIN **metham-sodium** ERT
Turkenburg Mosbestrijder H **iron sulphate** Temana
Turkenburg onkruidbestrijder H **chlorthiamid** Temana
Turkenburg onkruidbestrijder voor gazons H **benazolin + dicamba + MCPA** Temana
Turkenburg Plantspray tegen insekten IA **dichlorvos** Intec
Turkenburg Tuinspray tegen insekten IA **dichlorvos** Intec
Tuta-RR H **2,4-D + mecoprop** Vogelmann
Tuta-Super-W-Unkrautvertilger H **dalapon-sodium + diuron + MCPA** Vogelmann
Tuta-SV-Schneckenvertilger M **metaldehyde** Vogelmann
Tuta-Total-Unkrautvertilger H **sodium chlorate** Vogelmann
Tutakorn-Streuunkrautvertilger H **borax + bromacil** Vogelmann
Tutakorn ZG H **diuron + methabenzthiazuron** Vogelmann
Tuttirat R **chlorophacinone + sulphaquinoxaline** Industrialchimica
Tuver acaricide IA **dicofol + ethion + parathion-methyl** R.S.R.
TwinSpan I **chlorpyrifos + disulfoton** Pan Britannica
Tyonal AF **sulphur** Chimiberg
Typholine gamma IA **lindane + parathion + petroleum oils** Procida
Typholine-C I **parathion + petroleum oils** EMTEA
Typholine-D I **parathion + petroleum oils** EMTEA
Typhon I **parathion** Procida
TZ-16 poudrage F **thiram + zineb** Bourgeois

U-5 I **dichlorvos + methoxychlor** Denka
U 46 Combi H **2,4-D + MCPA** BASF, Sapec
U 46 D H **2,4-D** BASF
U 46 DM H **2,4-D + MCPA** BASF
U 46 DP HG **dichlorprop** BASF, Zupa
U 46 KV H **mecoprop** BASF
U 46 KV Combi H **2,4-D + mecoprop** BASF
U 46 M H **MCPA** BASF, Ciba-Geigy, ICI, Intrachem, Wacker
U 46 MCPB H **MCPB** BASF
U 46 Super H **dichlorprop + MCPA + mecoprop** BASF
UC 70 F **copper sulphate + mancozeb** Rhodiagri-Littorale
UC 7744 IG **carbaryl** Union Carbide
UC 21149 IAN **aldicarb** Union Carbide
UC 21865 NIA **aldoxycarb** Union Carbide
UC 22463 H **dichlormate** Union Carbide
UC 51762 I **thiodicarb** Union Carbide
Ucegam I **lindane** Agriben
Ucesol G **chlormequat chloride** Agrotec
Ucetin F **fentin acetate** Agriben
Ugecap F **captan** Sipcam-Phyteurop
Ugecoil I **parathion + petroleum oils** Sipcam-Phyteurop
Ugecormone H **MCPA** Sipcam-Phyteurop
Ugecupric F **copper oxychloride** Sipcam-Phyteurop
Ugegamma I **lindane** Sipcam-Phyteurop
Ugress-Kverk-D H **2,4-D** Plantevern-Kjemi
Ujotin G **(2-naphthyloxy)acetic acid** Bitterfeld
Ukavau H **sodium chlorate** Geissler
Ukavau-Super H **bromacil + diuron** Geissler
Ukorzeniacz A G **benomyl + captan + indol-3-ylacetic acid** Chudzik
Ukorzeniacz B G **benomyl + captan + 1-naphthylacetic acid** Chudzik
Ultacron I **dicrotophos + methidathion** Ciba-Geigy
Ultima Plus H **bentazone + dichlorprop + MCPA** ICI
Ultima-DP H **dichlorprop + ioxynil** ICI
Ultima-MP H **ioxynil + mecoprop** ICI

Ultra Sofril F **sulphur** Agriben, Pepro, Rhone-Poulenc
Ultra Tephel H **prometryn + trifluralin** Hellenic Chemical
Ultracid I **methidathion** BASF, Budapesti Vegyimuvek, Chromos, Ciba-Geigy, Ligtermoet, Liro, Viopharm
Ultracide IA **methidathion** Ciba-Geigy
Ultranix F **sulphur** La Littorale
Ultrasofril F **sulphur** Condor
Ultrazolfo F **sulphur** Ergex
Ulvapron S **petroleum oils** BP, Megafarm
Umagam FI **lindane + thiram** Umupro
Umucortil I **lindane** Umupro
Umupro Cubetrine I **cypermethrin** Umupro
Umupro sol I **lindane** Umupro
Umuter D I **diazinon** Umupro
Umuxebe double FX **anthraquinone + maneb** Pepro
Undeen I **propoxur** Bayer
Unden I **propoxur** Bayer, Organika-Azot
Undene I **propoxur** Bayer, Pinus
Ungezieferkoeder Nexa-Lotte Spezial I **chlorpyrifos** Shell
Ungeziefer-Mittel Jacutin I **lindane** Shell
Ungeziefer-Mittel Jacutin F I **bromophos + pyrethrins** Shell
Ungeziefer-Puder Jacutin F I **bromophos** Shell
uniconazole G [76714-83-5] (*E*)-(*RS*)-1-(4-chlorophenyl)-4,4-dimethyl-2-(1*H*-1,2,4-triazol-1-yl)pent-1-en-3-ol, Ortho Prunit, Ortho Sumagic, Prunit, S-3307 D, Sumagic, Sumiseven, XE-1019
Unicrop Leatherjacket Pellets I **lindane** Universal Crop Protection
Unicrop Mini Slug Pellets M **metaldehyde** Universal Crop Protection
Unicrop Simatrole H **amitrole + simazine** Universal Crop Protection
Unidron H **diuron** Universal Crop Protection
Unifosz IA **dichlorvos** UVKSS
Unimeton IA **thiometon** Agriben
Uniran H **picloram** Ciba-Geigy
Unisan F **phenylmercury acetate** United Phosphorus
Unisol I **trichlorfon** UVKSS
Unitron I **trichlorfon** UVKSS
Universol I **DNOC + petroleum oils** Agrochemiki
Unkraut-Ende H **sodium chlorate** Schaefer
Unkraut-Ex H **sodium chlorate** Chemia, Stolte & Charlier
Unkraut-Ex 'frappant' 3 H **atrazine + diuron + simazine** Stolte & Charlier
Unkraut-frei H **bromacil + diuron** Vetyl-Chemie
Unkrautstab H **2,4-D + mecoprop** Shell
Unkraut-Tod H **sodium chlorate** Feit, Imhoff & Stahl, Overlack
Unkraut-Tod-Spezial H **amitrole + diuron** Overlack
Unkrautvernichter-Total H **sodium chlorate** Elsner
Unkrautvernichtungsmittel H **sodium chlorate** Fischar, Lopa
Unkrautvernichtungsmittel 313 T H **karbutilate** Urania
Unkrautvernichtungsmittel 371 H **amitrole + atrazine + 2,4-D + dichlorprop + terbuthylazine** Spiess, Urania
Unkrautvernichtungsmittel 371 DB H **amitrole + atrazine + 2,4-D + simazine** Urania
Unkrautvernichtungsmittel 371 M H **amitrole + atrazine + dichlorprop + diuron** Urania
Unkrautvernichtungsmittel 372 H **simazine** Spiess, Urania
Unkrautvernichtungsmittel 373 H **diuron + simazine** Urania
Unkrautvernichtungsmittel 374 H **diuron** Spiess, Urania
Unkrautvernichtungsmittel 447-68 DBS H **amitrole + atrazinc + 2,4-D + dichlorprop + terbuthylazine** Urania
Unkrautvertilger 4196 H **atrazine + picloram + simazine** Schering
Unkrautvertilger Tuta-Super H **bromacil + 2,4-D + diuron** Vogelmann
Unkrautvertilger UV H **sodium chlorate** Fischar
Unkrautvertilger W/Desherbant-Shell H **petroleum oils** Agroplant
Unkrautvertilger Waldschuetz H **sodium chlorate** Waldschuetz
Unkrautvertilgungsmittel H **sodium chlorate** Afalin
Unkrautvertilgungsmittel Vlinsora H **sodium chlorate** Orpil-Seifen
Unkrautweg Dom Samen H **bromacil + diuron** Barnaengen
Urab H **fenuron**
Uragan H **bromacil** Edefi, Makhteshim-Agan

Uragan D I **hydrogen cyanide** Dimitrova, Lucebni zavody, Mayr
Urame F **thiram** Quimigal
Ureon H **linuron** Scam
Uribest H **naproanilide** Mitsui Toatsu
Uroneb H **neburon** Aziende Agrarie Trento
Urox H **monuron trichloroacetate** Hopkins
Urox B H **bromacil** Hopkins
Urturanet H **sodium chlorate** Boucquillon
USB 3153 H **prodiamine**
USB 3584 H **dinitramine**
USR 604 FL **dichlone** Uniroyal
Ustaad I **cypermethrin** United Phosphorus
Ustilan H **ethidimuron** Bayer, C.F.P.I.
Ustilan GW H **dichlorprop + diuron + ethidimuron** Bayer
Ustilan NK H **amitrole + dichlorprop + ethidimuron** Bayer
Ustilan T H **amitrole + diuron + ethidimuron** Bayer
Ustinex H **amitrole + diuron** Bayer
Ustinex H **methabenzthiazuron + simazine** Bayer
Ustinex CN H **dichlobenil** Bayer
Ustinex F H **amitrole + bromacil + dichlorprop + diuron** Bayer
Ustinex GL H **amitrole + bromacil + 2,4-D + diuron** Bayer
Ustinex KR H **amitrole + MCPA + methabenzthiazuron** Bayer, Schering
Ustinex MS H **methabenzthiazuron + simazine** Agro-Kemi, Bayer
Ustinex NG H **atrazine + diuron** Bayer
Ustinex PA H **amitrole + diuron** Agro-Kemi, Bayer
Ustinex PD H **dalapon-sodium + diuron + MCPA** Bayer
Ustinex Speciaal H **amitrole + 2,4-D + diuron** Bayer
Ustinex Special H **amitrole + diuron + MCPA** Bayer, Pinus
Ustinex T H **amitrole + bromacil + diuron** Bayer
Ustinex TS H **methabenzthiazuron + simazine** Bayer
Ustinex-Unkrautfrei H **amitrole + diuron** Bayer
Ustinex W H **amitrole + diuron + MCPA** Bayer
Ustinex Z H **diuron + methabenzthiazuron** Bayer
Utacaffaro F **copper oxychloride** Caffaro
Utazin F **zineb** Caffaro
Utazin S F **sulphur + zineb** Caffaro
Utazolfo AF **sulphur** Caffaro
UTH I **trichlofenidine** Delicia
Utox CMPP H **mecoprop** Spiess, Urania
Utox DP H **dichlorprop** Spiess
Utox KV Combi H **2,4-D + mecoprop** Spiess, Urania
Utox M H **MCPA** Spiess, Urania
Utox-Super DPD H **2,4-D + dichlorprop** Spiess, Urania
Uvassa F **folpet** Sopepor
Uvon H **prometryn** Fahlberg-List
Uvon-Kombi H **prometryn + simazine** Fahlberg-List
UVS 99-Ex H **sodium chlorate** Biesterfeld

V 5 I **carbaryl** Sofital
V-18 A **tetradifon** Duphar
V-101 A **tetrasul** Duphar
V-Stalspuitmiddel I **permethrin + tetramethrin** Biopharm
Vacor R **pyrinuron**
Vaisox CS H **ammonium sulphamate** Aseptan
Vaisoxan H **2,4-D + 2,4,5-T** Aseptan
Vaisoxan AS H **dichlobenil** Aseptan
Vaisoxan G H **2,4-D + dichlorprop + MCPA** Aseptan
Vaisoxan limit G **mefluidide** Aseptan
Vaisoxan selectif H **dichlobenil** Aseptan
Vaisoxan suspension H **amitrole + atrazine** Aseptan
Vaisoxan TD H **2,4-D + triclopyr** Aseptan
Vaissol IA **petroleum oils** Ciba-Geigy
Valexon I **phoxim** Bayer
Valiant F **maneb** Berner
Valiant F **cymoxanil + folpet + fosetyl-aluminium** Pepro
Validacin F **validamycin A** Takeda
validamycin A F [37248-47-8] 1L-(1,3,4/2,6)-2,3-dihydroxy-6-hydroxymethyl-4-[(1*S*,4*R*,5*S*,6*S*)-4,5,6-trihydroxy-3-hydroxymethylcyclohex-2-enylamino]cyclohexyl β-D-glucopyranoside, Solacol, Validacin, Valimun
Valimun F **validamycin A** Takeda
Valinate H **chlorsulfuron + linuron** Du Pont
Valor H **linuron + simazine** Pluess-Staufer
Valsa Wax F **thiophanate-methyl** Nippon Soda

Vamidoate IA **vamidothion** Rhone-Poulenc
vamidothion IA [2275-23-2] *O,O*-dimethyl *S*-2-(1-methylcarbamoylethylthio)ethyl phosphorothioate, ENT 26613, 10465 RP, Kilval, NPH 83, Trucidor, Vamidoate
Vamin F **folpet + ofurace** Schering
Vamitrol H **amitrole + simazine** ICI-Valagro
Van Eenennaam dimoxol H **dichlorprop + flurecol-butyl + ioxynil + MCPA** Van Eenennaam
Van Rijn's Bladglans IA **petroleum oils** Van Rijn
Vancide TM FX **thiram** Vanderbilt
Vanfix H **lenacil + propham** Shell
Vangard H **phenmedipham** Farmers Crop Chemicals
Vangard F **etaconazole** Ciba-Geigy
VAP strip I **dichlorvos** Frowein
Vapagrex FHIN **metham-sodium** Sadisa
Vapam FHIN **metham-sodium** Agrotechnica, ICI-Valagro, Ligtermoet, Permutadora, Rohm & Haas, Serpiol, Shell, Sipcam, Stauffer
Vapasol FHIN **metham-sodium** Gen. Rep.
Vapazon FHIN **metham-sodium** Ellagret
Vapona IA **dichlorvos** Shell, Zorka
Vapona plantenspray IA **piperonyl butoxide + pyrethrins** Kortman
Vaponite IA **dichlorvos** Shell
Vapor Gard S **di-1-p-menthene** Agrichem, Chimiberg, Geopharm
Vaporthrin I **empenthrin** Sumitomo
Vapotone AI **TEPP** Chevron
Varikill I **fenoxycarb** Maag
Varitox H **TCA** Rhone-Poulenc
Vasgalic F **iron sulphate** Csepel
Vasteril FHIN **metham-sodium** Enotria
Vaztak I **alphacypermethrin** Spolana
Vazyl I **petroleum oils** C.C.L., Luxol, Oerter
VC 13 IN **dichlofenthion** Virginia-Carolina
VC9-104 NI **ethoprophos** Mobil
VCS-438 H **methazole**
VCS 506 I **leptophos**
Vectal H **atrazine** Efthymiadis, Schering, Schering AAgrunol
Vectobac I **Bacillus thuringiensis** Abbott
Veexcobre F **copper oxychloride** Agriplan
Vega H **bentazone + cyanazine + dichlorprop** BASF
Vegadex H **sulfallate**
Vegelux S **petroleum oils** C.C.L., Oerter
Vegemec H **2,4-D + prometon** PBI Gordon
Vegepron H **diuron + petroleum oils + simazine** UCAR
Vegestop H **amitrole + atrazine** MCK Diffusion
Vegestop D H **amitrole + diuron + simazine** MCK Diffusion
Vegestop D 900 H **amitrole + simazine** MCK Diffusion
Vegestop GR H **atrazine + 2,4-D + diuron** MCK Diffusion
Vegestop selectif H **simazine** MCK Diffusion
Vegeton I **dimethoate + fenvalerate**
Vegetox I **cartap hydrochloride** Takeda
Vegfru Foratox IAN **phorate** M/S Pesticides
Vegfru Fosmite AI **ethion** M/S Pesticides
Vegfru Malatox IA **malathion** M/S Pesticides
Vegoran H **bromofenoxim + terbuthylazine** Plantevern-Kjemi
Vegosol TD H **amitrole** Zootechniki
Vel-4207 H **cambendichlor**
Vel-5026 H **buthidazole**
Velan D FI **diazinon + lindane + thiram** Diana
Veliourini H **MSMA** Hellenic Chemical
Velpar H **hexazinone** Bayer, Du Pont, Farmos, Procida, Schering, Siapa, Spolana
Velpar K H **diuron + hexazinone** Du Pont
Velsicol 1068 I **chlordane** Sandoz
Velsicol 58-CS-11 H **dicamba**
Velvas F **iron sulphate** Pan Britannica
Velvetone Lawn Food and Weedkiller H **2,4-D + dicamba** Fisons
Vencedor LF **copper sulphate** Compania Quimica
Vendex A **fenbutatin oxide** Shell, Du Pont
Vengeance R **bromethalin** Eli Lilly
Venosal F **captan + carbendazim** ERT
Venotex Nouveau F **captan + dinocap** Protex
Ventiflor F **sulphur** Marty et Parazols
Ventifluid F **sulphur** Marty et Parazols
Ventilalt kenpor AF **sulphur** Kenorlo Uzem
Ventilene AF **sulphur** Mormino
Ventilene Acuprizzata F **sulphur + zineb** EniChem, Mormino
Ventilene Ramata F **copper oxychloride + sulphur** Mormino
Ventine F **ziram** Ciba-Geigy

Ventine MZ F **mancozeb** Ciba-Geigy
Vention FIA **parathion-methyl + sulphur** Marty et Parazols
Ventomic FA **sulphur** Gaillard
Ventox I **acrylonitrile** Degesch
Venturex D F **dodine** Terranalisi
Venturin F **thiram** Kwizda
Venturin-50 F **captan** Zupa
Venturol F **dodine** Celemerck, Geopharm, Margesin, Permutadora, Shell
Venzar H **lenacil** Agroplant, Bayer, Duphar, Du Pont, Farmos, La Qunioleine, OHIS, Organika-Zarow, Ravit, Rhone-Poulenc, Schering, Spiess, Urania
Veralin 1,5 IA **anthracene oil + DNOC** Maag
Veralin D IA **diazinon + petroleum oils** Maag
Veraline FI **anthracene oil + DNOC** Pepro, Rhone-Poulenc
Verapon VBM G **1-naphthylacetic acid hydrazide** Zerpa
Verazin F **ziram** Agronova
Verdane I **lindane** KenoGard
Verdane Mat IA **malathion** KenoGard
Verdecion AZ IA **azinphos-methyl** KenoGard
Verdecion Dia I **diazinon** KenoGard
Verdecion NA Super IA **naled** KenoGard
Verdecion Par I **parathion-methyl** KenoGard
Verdecion Para I **parathion-methyl** KenoGard
Verdecion SU I **fenitrothion** KenoGard
Verdecion TR I **trichlorfon** KenoGard
Verderame F **Bordeaux mixture** Sivam
Verderin E I **lindane** Caffaro
Verdet F **copper acetate** Duclos
Verdinal H **phenisopham** Schering
Verdone H **2,4-D + mecoprop** ICI
Verecar T IA **parathion-methyl + tetradifon** Du Pont
Vergasungsbrikett 'Rekord' R **phosphine** Schacht
Vergemaster H **MCPA + mecoprop** Burts & Harvey
Verigal H **bifenox + mecoprop** Agriben
Verindal Ultra I **lindane** Schering
Verisan F **iprodione** Agrotec, Rhone-Poulenc
Vermikill Duiven I **bioallethrin + piperonyl butoxide + pyrethrins** Van Vielandt
Vermortis I **naphthalene + sulphur** Joergensen
Vernam H **vernolate** Geopharm, Serpiol, Stauffer
vernolate H [1929-77-7] *S*-propyl dipropylthiocarbamate, R-1607, Reward, Saverit, Savirox, Vernam
Veromite IA **parathion-methyl + propargite** Uniroyal
Vertalec I **Verticillium lecanii** Bridet, Tate & Lyle, Microbial Resources
Verticillium lecanii I white halo fungus, Mycotal, Vertalec
Vertimec IA **abamectin** MSD AGVET
Vesakontuho DM H **2,4-D + MCPA** Kemira
Vesakontuho Tasku H **2,4-D** Kemira
Veto I **EPN + parathion-methyl** Drexel
Vetolan I **fenchlorphos** Land-Forst
Vetrazine I **cyromazine** Ciba-Geigy
Vibo I **dimethoate + fenitrothion** Zimmermann
Victenon I **bensultap** Takeda
Vidate IN **oxamyl** Du Pont
Vidden D N **D-D** Dow
Vigil F **diclobutrazol** ICI, Sopra
Vigil Combi F **diclobutrazol + sulphur** ICI, ICI-Valagro
Vigil K F **carbendazim + diclobutrazol** Kwizda
Vigil T F **captafol + diclobutrazol** Sopra
Vignor F **captafol + cymoxanil** R.S.R.
Vigor-Kill-Rat R **pindone + warfarin** Vigor
Viktor CI A **clofentezine + fenpropathrin** Schering
Viljavarasto-polyte IA **piperonyl butoxide + pyrethrins** Kemira
Vinagard H **amitrole + terbumeton + terbuthylazine** Ciba-Geigy
Vincit FX **anthraquinone + flutriafol + oxine-copper** Sopra
Vincit L F **flutriafol + thiabendazole** ICI
Vincit LU F **flutriafol + imazalil + thiabendazole** ICI
vinclozolin F [50471-44-8] (*RS*)-3-(3,5-dichlorophenyl)-5-methyl-5-vinyl-1,3-oxazolidine-2, 4-dione, vinclozoline, BAS 352F, Ornalin, Ronilan, Vorlan
vinclozoline F **vinclozolin**
Vinerba D H **amitrole + diuron** Hoechst
Vinhassa Ultra M F **copper oxychloride + maneb + zineb** Sopepor
Vinicoll F **folpet**
Vinicur F **cyprofuram** Schering
Vinipur F **copper oxychloride + zineb** Burri

Vinipur cupro extra FA **copper oxychloride + cyhexatin + sulphur + zineb** Burri
Vinipur extra FA **copper oxychloride + dicofol + sulphur + zineb** Burri
Vinipur Spezial F **copper oxychloride + folpet** Burri
Vinoril H **amitrole + atrazine** Hellenic Chemical
Vintox I **malathion** Geochem
Vinuran H **dichlobenil** Urania
Vinylphate IA **chlorfenvinphos**
Viozene I **fenchlorphos**
Viricobre F **copper oxychloride** Condor, Rhone-Poulenc
Viricuivre F **copper oxychloride** Pepro, Ravit, Rhone-Poulenc
Virifix F **copper oxychloride** Viopharm
Virimaag F **Bordeaux mixture** Maag
Viriman G **2,4-D** Platecsa
Virolex F **carbendazim** Protex
Virox I **Neodiprion sertifer NPV** Microbial Resources
Vis K S **hydroxyethyl cellulose + lauryl alcoholethoxylate** Visplant
Visbar FI **barium polysulphide** Visplant
Visene I **carbaryl** Quimigal
Visfenol F **fentin acetate** Visplant
Visfos I **parathion** Visplant
VIT-bejdse F **carboxin + imazalil + thiabendazole** Cillus
Vitacryl F **copper carbonate (basic) + copper sulphate + folpet** Vital
Vitaflo F **carboxin + thiram** Uniroyal
Vital R F **copper oxychloride + cymoxanil** Ravit
Vitam F **zineb** EniChem
vitamin D2 R **calciferol**
Vitamor H **sodium chlorate** Lambert
Vitan F **copper oxychloride + folpet** Agrocros
Vitan Extra F **Bordeaux mixture + copper oxychloride + folpet** Agrocros
Vitanebe C F **copper oxychloride + maneb + zineb** Hoechst
Vitaphos M IA **azinphos-methyl** Vital
Vitasan S F **folpet** Zupa
Vitavax F **carboxin** Berner, Bjoernrud, Maag, Mir, Uniroyal
Vitavax + TMTD + Lindan FI **carboxin + lindane + thiram** Uniroyal
Vitavax 2M F **carboxin + maneb** Hellafarm
Vitavax 20 D F **carboxin + thiram** Hellafarm
Vitavax 30 T F **carboxin + thiram** Du Pont, Ravit, Uniroyal
Vitavax 38/38 F **captan + carboxin** Ciba-Geigy
Vitavax 50 F **carboxin** Liro
Vitavax 75 F **carboxin** Condor, Hellafarm, Kemi-Intressen, Kwizda, Uniroyal
Vitavax 200 F **carboxin + thiram** Condor, Uniroyal
Vitavax 202 F **carboxin + imazalil + thiram** Uniroyal
Vitavax 300 F **captan + carboxin** Uniroyal
Vitavax 390 F **carboxin** Cillus, Ligtermoet, Liro
Vitavax C F **captan + carboxin + 8-hydroxyquinoline sulphate** Uniroyal
Vitavax Flo F **carboxin + thiram** Du Pont, Ravit
Vitavax GL F **carboxin** Kwizda
Vitavax K F **captan + carboxin** Hellafarm
Vitavax kombi F **carboxin + methoxyethylmercury silicate** Kwizda
Vitavax M F **carboxin + maneb** Uniroyal
Vitavax MC F **carboxin + 8-hydroxyquinoline sulphate + mancozeb** Uniroyal
Vitavax RS FI **carboxin + lindane + thiram** Cillus
Vitavax T F **carboxin + thiram** Berner, Condor, Uniroyal
Vitavid G **chlormequat chloride** Argos
Vitax Lawn Sand F **iron sulphate** Steetley
Vitazeb F **folpet + maneb** Vital
Vitene F **zineb + ziram** Sipcam
Vitex I **dimethoate** La Littorale
Vitex 8/68 F **cymoxanil + mancozeb** Siapa
Vitex Combi F **cymoxanil + mancozeb** Siapa
Vitex Marca Azzuro F **sulphur + zineb** Agrochemiki, Siapa
Vitex Marca Bianca F **zineb** Siapa
Vithiram F **thiram** Vital
Viti-Combi FA **copper compounds (unspecified) + fenbutatin oxide + folpet +sulphur** Agroplant
Viti-Folpet Pulver/poudre F **copper oxychloride + folpet** Agroplant
Viticarb IA **carbophenothion + parathion-methyl** Sipcam-Phyteurop
Viticol FA **sulphur** Afrasa
Vitifol F **captafol + folpet** Agrocros
Vitigran F **copper oxychloride** Bjoernrud, Hoechst, Pluess-Staufer, Roussel-Hoechst
Vitipec F **cymoxanil + folpet** Sapec

Vitipec C F **copper oxychloride + cymoxanil** Sapec
Vitiril F **oxadixyl + propineb** Bayer
Vitobel P7 I **pyrethrins** Lessennes
Vitol 96 IA **petroleum oils** Eli Lilly
Vitosan TD H **amitrole + diuron** Zootechniki
Vitrex IA **parathion**
Vivax G **chlormequat chloride + ethephon** Pepro
Vivithrin I **fenpirithrin**
Vizor H **lenacil** Farm Protection
VK-Karbolineumi HIA **anthracene oil + petroleum oils** Kemira
Vlido Huis en Tuin I **phenothrin + tetramethrin** Intradal
Volathion I **phoxim** Bayer
Volaton I **phoxim** Agro-Kemi, Bayer, Berner, Pinus
Volck AIHS **petroleum oils** Chevron
Volck invierno multiple IA **DNOC + petroleum oils** Agrocros
Volid R **brodifacoum** ICI
Vollernte M **metaldehyde** Lanz
Volnebe F **carbendazim + maneb + sulphur** Agrishell
Volparox F **fenitropan** EGIS
Volunteered HX **dalapon-sodium + terpenes** Mandops
Vondalhyd G **maleic hydrazide** Pennwalt, Pennwalt Holland
Vondathion F **fentin acetate** Decco-Roda
Vondcaptan F **captan** Pennwalt Holland
Vondeb F **zineb** Decco-Roda
Vondocarb F **carbendazim + mancozeb** Decco-Roda
Vondodine F **dodine** Pennwalt Holland
Vondofanex PDT H **dinoseb** Pennwalt France
Vondozeb F **maneb + zineb** Bourgeois, Burri, Candilidis, Grima, Kwizda, Leu & Gygax, PennwaltFrance, Pennwalt Holland
Vondozeb Plus F **mancozeb** Pennwalt Holland
Vondrax HG **chlorpropham + 2,4-D + maleic hydrazide** Bos
Vonduci H **chlorpropham + diuron** Decco-Roda
Vonduron H **diuron** Bourgeois, Pennwalt Holland
Voorjaar spuitmiddel Nexionol I **bromophos** Shell
Voraussaatherbizid Bi 3411 H **chloral hydrate** Bitterfeld
Vorex R **difenacoum** Ameco
Vorlan F **vinclozolin** Mallinckrodt
Vorlex FIN **D-D + methyl isothiocyanate** NOR-AM
Voronit F **fuberidazole** Bayer
Vorox (i) H **amitrole + atrazine + terbuthylazine** Urania
Vorox (i) 371 H **amitrole + atrazine + sebuthylazine** Urania
Vorox (s) Neu H **amitrole + atrazine + simazine** Spiess, Urania
Vorox 45-24 H **amitrole + ammonium thiocyanate + simazine** C.F.P.I.
Vorox entretien H **amitrole + simazine** C.F.P.I.
Vorox Plus H **amitrole + diuron + simazine** Spiess, Urania
Vorox TD H **amitrole + atrazine + simazine** Vieira
Vorox Unkrautvertilger H **amitrole + simazine** Urania
Vorpo F **Bordeaux mixture** Agrochemiki
Votromite A **chlorfenson + propargite** Ciba-Geigy, Hellafarm, Uniroyal
VP G **chlorpropham** Pluess-Staufer
VP 1940 FLM **fentin acetate** Hoechst
V.P. Fliegenspray I **piperonyl butoxide + pyrethrins** Sand
VPM FNHI **metham-sodium** Stauffer
V.P.M. Soil Fumigant FIHN **metham-sodium** Ravit
V.P. Poeder I **permethrin** Sandoz
V-R-100 IA **piperonyl butoxide + pyrethrins** Vetira
V/R 101 I **dichlorvos** Vetira
Vruchtboomcarbolineum HIA **anthracene oil** Asepta, Aveve, Sanac
VSM Algmos L **quaternary ammonium** Vosimex
Vulcan H **bromoxynil + clopyralid** Farm Protection
Vulkan-Feldmauspatrone R **potassium nitrate + sulphur** Laeubi
Vydate IN **oxamyl** Du Pont, Shell, Wacker

Wacker 83 F **copper oxychloride + sulphur** Wacker
Wacker DP H **dichlorprop** Dow
Wacker MP-D H **2,4-D + mecoprop** Dow
Wacker Netzschwefel FA **sulphur** Dow
Wacker S 14/10 IA **dimefox**
Wacker-Kombi H **2,4-D + MCPA** Dow

W.A.M. Wildafschrikmiddel X **animal oil** Schmidt
Warefog 25 G **chlorpropham** Wheatley
Warf R **warfarin**
Warfarat R **warfarin**
warfarin R [81-81-2] 4-hydroxy-3-(3-oxo-1-phenylbutyl)coumarin, warfarine, coumafene, zoocoumarin, Biotrol, Co-Rax, Cov-R-Tox, Cumarax, Dethmor, Dethnel, Kypfarin, Liqua-Tox, Mar-Frin, Mouse-Pak, Rat-Pak, Ratorex, RAX, Rodex, Rosex, Sakarat, Sewarin, Solfarin, Sorexa, Tox-Hid, Warf, WARF 42, Warfarat
warfarine R **warfarin**
Warrior H **linuron + trifluralin** ICI
Wartane F **dinocap** Caffaro
Wartox R **warfarin** Chimac-Agriphar, Druchema
Wasp Nest Destroyer I **resmethrin + tetramethrin** Sorex
Waspend I **pirimiphos-methyl + pyrethrins** ICI
Waterwax I F **imazalil** Fomesa
Waterwax P F **2-phenylphenol + thiabendazole** Fomesa
Waterwax TTT F **thiabendazole** Fomesa
Way Up H **pendimethalin** Cyanamid
WBA 8119 R **brodifacoum** Sorex
Weather Blok R **brodifacoum** ICI
Webo Slakkenkorrels M **metaldehyde** Van Wesemael
Weed Gun H **2,4-D + dicamba** ICI
Weed Out **alloxydim-sodium** May & Baker
Weed-E-Rad H **MSMA** Vineland
Weed-E-Rad H **DSMA** Vineland
Weed-Hoe H **MSMA** Vineland
Weed-Oil S **petroleum oils** Liro
Weed-Rhap H **MCPA** Vertac
Weedagro D H **2,4-D** Agronova
Weedagro M H **MCPA** Agronova
Weedar H **2,4-D** Bjoernrud, C.F.P.I.
Weedar ADS super H **amitrole + diuron + MCPA + terbacil** Avenarius
Weedar Ata-TL H **amitrole** Avenarius
Weedar MCP H **MCPA** Avenarius, EniChem
Weedar Riso H **dichlorprop** EniChem
Weedasept H **2,4-D** Asepta
Weedax H **dichlorprop + MCPA + mecoprop** Van Wesemael
Weedazine H **simazine** C.F.P.I.
Weedazol H **amitrole** Spiess, Union Carbide, Urania
Weedazol Super H **amitrole + diuron** Chimac-Agriphar
Weedazol Super H **amitrole + atrazine + diuron** Agriben
Weedazol TD H **amitrole** Union Carbide
Weedazol TL H **amitrole + ammonium thiocyanate** Agriben, C.F.P.I., Ciba-Geigy, Luxan
Weedazol TS H **amitrole + sodium thiocyanate** C.F.P.I.
Weedex H **glyphosate** Monsanto
Weedex H **simazine** Murphy
Weedex H **amitrole + 2,4-D + MCPA + TCA** C.F.P.I.
Weedex H **MCPA** Bjoernrud
Weedex A H **atrazine** Ciba-Geigy
Weedex KS H **diuron + terbacil** Gullviks
Weedex poudre H **amitrole + MCPA + TCA** C.F.P.I.
Weedex S H **simazine** Ciba-Geigy
Weedkiller LAPA H **dinoseb** L.A.P.A.
Weedmaster H **2,4-D + dicamba** Sandoz
Weedmaster H **chloridazon** Portman
Weedoclor H **amitrole + 2,4-D + sodium chlorate + TCA** C.F.P.I.
Weedol N **diquat dibromide + paraquat dichloride** ICI
Weedonac H **TCA** C.F.P.I.
Weedone 402 NV H **2,4-D** Avenarius, C.F.P.I.
Weedone D H **2,4-D** Avenarius
Weedone debroussaillant H **2,4-D + dichlorprop** C.F.P.I.
Weedone debroussaillant DP H **dichlorprop** C.F.P.I.
Weedone DP H **dichlorprop** Avenarius
Weedone Emulsamine H **2,4-D** EniChem
Weedone KV super H **2,4-D + mecoprop** Avenarius
Weedone LV H **2,4-D** C.F.P.I., EniChem
Weedone MCPP neu H **mecoprop** Avenarius
Weedoprol DP H **bromoxynil + dichlorprop** Urania
Weedopron H **dalapon-sodium** C.F.P.I.
Weedtrine-D H **diquat dibromide**
Weg-Rein H **sodium chlorate** Prokopp
Wege Unkraut-frei H **bromacil + diuron** Flora-Frey
Wegerein H **sodium chlorate** Bitterfeld
Wegerein K H **potassium chlorate** Bitterfeld
Wegit Unkrautvertilgungsmittel H **sodium chlorate** Kuhbier

Weguran H **dalapon-sodium + diuron + simazine** Urania
Weibulls Anti-Gnag X **thiram** Nordisk Alkali
Weibulls Biosekt I **malathion** Nordisk Alkali
Weibulls Nya Mossdoed H **chloroxuron** Nordisk Alkali
Weibulls Pyrex Insektsspray I **malathion + piperonyl butoxide + pyrethrins** Wikholm
Weibulls Taraxa H **2,4-D + dichlorprop** Nordisk Alkali
Wepsyn FIA **triamiphos**
Werrenkoerner Mioplant I **chlorpyrifos** Migros
Wetcol 3 Copper Fungicide F **copper hydroxide** Ford Smith
Whip H **fenoxaprop-ethyl** Hoechst
white halo fungus I **Verticillium lecanii**
white oils AIHS **petroleum oils**
Widol H **amitrole** Denka
Widolit-4 H **MCPA** Denka
Widoron HL **diuron** Denka
Wieder-D H **2,4-D + dicamba + MCPA** Asef Fison
Wilt-Pruf G **di-1-p-menthene** Synchemicals
Wiltz-65 FX **copper naphthenate**
Windhoevel-Superrasenduenger mit Unkrautvernichter H **2,4-D + MCPA** Windhoevel
Winner H **flurochloridone + neburon** Schering
Winter-Weissoel Promanal IA **petroleum oils** Neudorff
Wintol I **parathion-methyl + petroleum oils** Afrasa
Wintroil D.W. IA **petroleum oils** De Weerdt
Winylofos I **dichlorvos** Organika-Azot
Wipeout I **hydramethylnon** Cyanamid
Wireworm FS I **lindane** Dow
Witenol F **dinocap** Enotria
Witox H **EPTC** Chemolimpex, Sajobabony
WL 5792 H **chlorthiamid** Shell
WL 17731 H **benzoylprop-ethyl** Shell
WL 19805 H **cyanazine** Shell
WL 21959 IA **thiocarboxime**
WL 22361 F **fenfuram** Shell
WL 29761 H **flamprop-methyl** Shell
WL 43425 H **flamprop-M-isopropyl** Shell
WL 43467 I **cypermethrin** Shell
WL 43775 IA **fenvalerate** Shell
WL 63611 H **cyanatryn**
WL 85871 I **alphacypermethrin** Shell
WL 108366 R **flocoumafen** Shell
WL 115110 IA **flufenoxuron** Shell
Wofatox IA **parathion-methyl** Bitterfeld
Woldusin H **2,4-D** Bitterfeld
Wolf Geraete GMBH I **piperonyl butoxide + pyrethrins + rotenone** Graham
Wolf Unkrautvernichter mit Rasenduenger H **2,4-D + mecoprop** Wolf
Wondafdekmiddel W **triadimefon** Bayer
Wonuk H **atrazine** Bitterfeld, Fahlberg-List
Wotexit I **trichlorfon** Bitterfeld
Wuehl-Ex R **monochlorobenzene** Staehler
Wuehlmaus Ex Raeucherpatrone R **phosphine** Hawlik
Wuehlmaus Pille R **aluminium phosphide** Staehler
Wuehlmaus-Patrone Arrex Patrone R **phosphine** Shell
Wuehlmaus-raus R **phosphine** Flora-Frey
Wuehlmauskoeder Arrex R **zinc phosphide** Shell
Wuehlmauskoeder Wuelfel R **zinc phosphide** Wuelfel
Wuehlmaustod Arvicol R **zinc phosphide** Klepper
Wuehlmaustod Arvikol R **zinc phosphide** Klepper
Wurm-Killer IA **parathion** Bayer
Wurzelfix G **1-naphthylacetic acid** Urania
W.W.-Rasenduenger mit Moosvernichter H **iron sulphate** Windhoevel
Wypout H **barban** Uniroyal

X-52 H **chlomethoxyfen** Ishihara Sangyo
X-Pand H **isoxaben** Radonja
X-Spor FL **nabam** Campbell
Xantolio I **parathion + petroleum oils** Candilidis
XE-779L F **diniconazole** Chevron
XE-1019 G **uniconazole** Chevron
Xedafen S **alkyl phenol ethoxylate** Agromex
Xedamine G **diphenylamine** Xeda
Xedaquine G **ethoxyquin** Agromex, Xeda
Xedazole F **thiabendazole** Xeda
Xilin I **dimethoate + fenitrothion** Siegfried
XL H **benfluralin + oryzalin** Elanco
XL All Insecticide I **nicotine** Synchemicals

XMC I [2655-14-3] 3,5-xylyl methylcarbamate, Cosban, H-69, Macbal, Maqbal
XTR 4 G **ethephon** C.F.P.I.
xylachlor H [63114-77-2] 2-chloro-*N*-isopropylacet-2',3'-xylidide, AC 206784, Combat
Xyligen B F **furmecyclox** BASF
xylofop-ethyl H **quizalofop-ethyl**
xylylcarb I [2425-10-7] 3,4-xylyl methylcarbamate, MPMC, Meobal, S-1046

Yaltox IAN **carbofuran** Bayer
Yanock R **fluoroacetamide**
Yeh-Yan-Ku H **difenzoquat methyl sulphate** Cyanamid
Yellow Cuprocide F **cuprous oxide** Rohm & Haas
yellow oxide of mercury FW **mercuric oxide**
Yerbacid H **MCPA** Siegfried
Yerban H **diuron** Afrasa
Yerbaten H **chlorthal-dimethyl + methazole** Prometheus
Yerbatox H **mecoprop** Siegfried
Yerbatox-D H **2,4-D + mecoprop** Siegfried
Yerbatox Plus H **2,4-D + mecoprop** Siegfried
Yleisaerosoli IA **methoxychlor + piperonyl butoxide + pyrethrins** Bang
Yomesan M **niclosamide** Bayer
Yotol H **naproanilide + pretilachlor**
Yphos IA **parathion-methyl** Sipcam-Phyteurop
Yrodazin H **simazine** Bitterfeld
Yukamate H **dimepiperate** Inagra, Sipcam
Yukamate Combi H **dimepiperate + molinate** Inagra, Sipcam

Zalin H **linuron** Afrasa
Zamora 2 R **warfarin** Valmi
Zanovit-TD H **amitrole + atrazine** Zootechniki
Zantir F **copper oxychloride + sulphur + zineb** EniChem
Zaprawa Funaben T F **carbendazim + thiram** Organika-Sarzyna
Zaprawa Furadan IA **carbofuran** FMC
Zaprawa Nasienna GTS FI **lindane + thiram** Organika-Azot
Zaprawa Nasienna T F **thiram** Organika-Azot
Zaprawa Oxafun T F **carboxin + thiram** Organika-Azot
Zardex oleo AI **hexadecyl cyclopropanecarboxylate + petroleumoils** KenoGard
Zargon H **clopyralid + mecoprop** Prochimagro
Zeachlor H **alachlor** EniChem
Zealan H **linuron** Ravit
Zealin H **atrazine** Scam
Zeapos P H **atrazine + pyridate** Sajobabony
Zeapur H **atrazine** Elanco, Eli Lilly
Zeazin H **atrazine** Dimitrova, EniChem, Rumianca
Zeazin Mix H **atrazine + prometryn** Dimitrova
Zeazin Mix Extra H **atrazine + metolachlor + prometryn** Dimitrova
Zeazin S H **atrazine + simazine** EniChem
Zectran IAM **mexacarbate** Union Carbide
Zedesa I **methyl bromide** Desinsekta
Zedesa-Blausaeure I **hydrogen cyanide** Desinsekta
Zedesa-Pellets I **aluminium phosphide** Desinsekta
Zedesa-Tabletten I **aluminium phosphide** Desinsekta
zeidane I **DDT**
Zelan H **MCPA**
Zelas H **alachlor** Visplant
Zeldox A **hexythiazox** ICI-Zeltia
Zelio R **thallium sulphate** Bayer
Zellek H **haloxyfop-ethoxyethyl** Dow
Zeltacal I **calcium arsenate** ICI-Zeltia
Zeltamina G **2,4-D** ICI-Zeltia
Zeltia Sevin I **carbaryl** ICI-Zeltia
Zeltiafos IA **azinphos-methyl** ICI-Zeltia
Zelticel G **chlormequat chloride** ICI-Zeltia
Zeltion IA **dimethoate** ICI-Zeltia
Zeltivar I **trichlorfon** ICI-Zeltia
Zeltomate G **chlorophenoxypropionic acid + (2-naphthyloxy)acetic acid** ICI-Zeltia
Zeltoxone H **trifluralin** ICI-Zeltia
Zeltoxone doble H **linuron + trifluralin** ICI-Zeltia
Zemacon F **mancozeb** Visplant
Zennapron H **2,4-D + mecoprop** BP
Zephir H **terbutryn** Ciba-Geigy
ZERA-Gram H **trifluralin** Zera
Zergan H **triclopyr** Siapa
Zerlate F **ziram** Du Pont, Zootechniki

Zero One H **MCPA-thioethyl** Hokko
Zerox I **piperonyl butoxide + pyrethrins** Edialux
Zerpa S00 H **petroleum oils** Zerpa
Zerpa Slakkendood in Korrelvorm M **metaldehyde** Zerpa
Zertell IA **chlorpyrifos-methyl** Dow
Zes Doppio Blu F **sulphur + zineb** Mormino
Zetaram L F **copper oxychloride** Sipcam
Ziaran F **ziram** Aragonesas, Condor
Ziarnochron X **anthraquinone + lindane** Organika-Azot
Zicoluq F **copper oxychloride + zineb** Luqsa
Zicoluq 311 F **copper oxychloride + maneb + zineb** Luqsa
Zidil I **chlorpyrifos** Neudorff
Zierpflanzen-Spray Saprol F **triforine** Epro, Shell
Ziklon IAR **hydrogen cyanide** Heydt Bona
Ziman F **mancozeb** Zorka Sabac
Zimdal F **carbendazim** Grima
Zimmer-Pflanzenspray N IA **omethoate** Flora-Frey
Zimmerpflanzen-Spray Parexan IA **piperonyl butoxide + pyrethrins** Shell
Zimmerpflanzenspray Parexan F IA **piperonyl butoxide + pyrethrins + S421** Shell
Zin-Ram F **copper oxychloride + zineb** Chemia
Zinagran F **ziram** Inagra
Zinagrex F **zineb** Sadisa
Zinamix F **maneb + zineb** Bourgeois
zinc phosphide R [1314-84-7] trizinc diphosphide, phosphure de zinc, Denkarin Grains, Gopha-Rid, Phosvin, Pollux, Ratol, Ridall, Rodenticide AG, Zinc-Tox, ZP
Zinc-Tox R **zinc phosphide** All-India Medical
Zincofos R **zinc phosphide** Ital-Agro
Zincolor F **ziram** Platecsa
Zincosol F **sulphur + zineb** Eli Lilly
Zindan F **zineb** Diana
zineb F [12122-67-7] zinc ethylenebis(dithiocarbamate) (polymeric), zinebe, ENT 14874, Aspor, Dipher, Dithane Z 78, Ditiozin, Enozin, Hexathane, Kypzin, Lonacol, Phytox, Polyram-Z, Tiezene, Tritoftorol, Zinosan
Zinebane F **zineb** Agronova
zinebe F **zineb**
Zineber F **zineb** d'Oliveira
Zinebina F **zineb** Caffaro
Zinecor F **copper oxychloride + zineb** Ciba-Geigy
Zinecupryl F **copper sulphate + zineb** Visplant
Zinemil F **mancozeb** Permutadora
Zinene F **zineb** Chemia
Zineplan F **zineb** Agriplan
Zinevit F **sulphur + zineb** EniChem, Mormino
Zinfugan F **zineb** SPE
Zinkoneb F **zineb** Hellenic Chemical
Zinochlor F **anilazine** Bayer
Zinofos NI **thionazin** Cyanamid
Zinomag R **strychnine** Chimac-Agriphar
Zinomox super F **copper oxychloride + zineb** Marty et Parazols
Zinophos NI **thionazin** Cyanamid
Zinosan F **zineb** Agriben, Rhone-Poulenc
Zinotan F **zineb** Ellagret
Zinothio F **sulphur + zineb** Bredologos
Zinothion D F **sulphur + zineb** Ellagret
Zinothion WP F **maneb + zineb** Ellagret
Zinozan F **zineb** Rhone-Poulenc
Zintox H **paraquat dichloride** Agrochemiki
Zinugec F **zineb** Sipcam-Phyteurop
Ziracal F **ziram** Calliope
Ziragrex F **ziram** Sadisa
Ziraluq F **ziram** Luqsa
ziram FX [137-30-4] zinc dimethyldithiocarbamate, zirame, AAprotect, AAvolex, Carbazinc, Corozate, Cuman, Drupina, Fuclasin, Fungostop, Hexazir, Mezene, Pomarsol Z, Prodaram, Triscabol, Ziramvis, Zirasan, Zirberk, Zirex, Zitox
zirame FX **ziram**
Ziramex F **ziram** Bourgeois
Ziramine F **ziram** Hellenic Chemical
Ziramit F **ziram** Caffaro
Ziramon F **ziram** Du Pont
Ziramugec F **ziram** Sipcam-Phyteurop
Ziramvis F **ziram** Sepran, Phytorgan, Visplant
Zirane F **ziram** Chemia
Ziranol F **ziram** Diana
Zirargos F **ziram** Argos
Zirasan FX **ziram** Ravit, Rhone-Poulenc
Zirater F **ziram** Visplant
Zirberk FX **ziram**
Zircan F **ziram** Scam
Zirex FX **ziram** Agriplan, Rhone-Poulenc
Zirolam F **ziram** Ellagret
Zitan T **hydrolysed proteins** Zootechniki

Zithiol IA **malathion** Pepro, Rhone-Poulenc
Zitiol I **malathion** Ravit
Zitokor-D H **2,4-D** Zupa
Zitox FX **ziram**
Zizalon H **alloxydim-sodium** Lapapharm
Zizole H **amitrole + atrazine** LHN
Zlatica-fos I **phosmet** Radonja
Zlatica ofunak-P I **pyridafenthion** Radonja
Zlatica P I **quinalphos** Radonja
Zlatigal P I **carbaryl** Galenika
Z.M. 80 F **mancozeb** Scam
ZN-Omadini F **zinc omadine** Efthymiadis
Zocold AF **sulphur** Scam
Zofal D IA **petroleum oils** Siegfried
Zofarol IA **parathion + petroleum oils** Siegfried
zolaprofos I [63771-69-7] *O*-ethyl *S*-(3-methylisoxazol-5-ylmethyl) *S*-propyl phosphorodithioate, BAS 268I
Zolezine H **amitrole + atrazine** LHN
Zolfo AF **sulphur**
Zolfo Bagnabile AF **sulphur** Agrimont, Aziende Agrarie Trento, Bayer, Chimiberg, EniChem, Enotria, Mormino
Zolfo Colloidale AF **sulphur** Chimiberg, Aziende Agrarie Trento
Zolfo Flor PB AF **sulphur** Caffaro, Sofital
Zolfo Ramato F **Bordeaux mixture + sulphur** Caffaro, EniChem, Sepran, Siapa
Zolfo Ramato Raffinato Triventilato Scorrevole F **copper sulphate + sulphur** Caffaro
Zolfo scorrevole AF **sulphur** EniChem
Zolfo Spruzzabile AF **sulphur** Mormino
Zolfo Stella Ramato F **copper oxychloride + sulphur** Mormino
Zolfo Stella Scorrevole AF **sulphur** Mormino
Zolfo Stella Ventilato AF **sulphur** EniChem, Mormino
Zolfo Stella Ventilato Ramato F **copper oxychloride + sulphur** EniChem
Zolfo Stella Ventilato Scorrevole AF **sulphur** EniChem
Zolfo ventilato AF **sulphur** Sepran, Siapa
Zolon-Super I **dichlorvos + phosalone** Hellafarm
Zolone IA **phosalone** Agronorden, Chromos, Condor, Embetec, Hortichem, Maag, Organika-Azot, Ravit, Rhodiagri-Littorale, Rhone-Poulenc
Zolvis FA **sulphur** Visplant
Zomerolie "Elefant" IA **petroleum oils** Collumbien
zoocoumarin R **warfarin**
Zoofitar F **copper oxychloride + zineb** Platecsa
Zoramat H **ametryn + amitrole + atrazine** Zorka Sabac
Zorasan F **phenylmercury acetate + phenylmercury chloride** Zorka Sabac
Zorial H **norflurazon** Sandoz
Zorosan tecni F **phenylmercury acetate** Zorka Sabac
ZP R **zinc phosphide** Bell Labs
ZR F **copper oxychloride + zineb** Sipcam
ZR-512 I **hydroprene** Sandoz
ZR 515 I **methoprene**
ZR 519 I **triprene**
ZR Extra Blu F **copper oxychloride + zineb** Candilidis, Sipcam
Zufa-fosforbrinte R **aluminium phosphide** Zuschlag
Zufa K. mosegrisegift R **crimidine** Zuschlag
Zufa-stroepudder R **coumatetralyl** Zuschlag
Zufarin rottegift R **warfarin** Zuschlag
Zupazon H **bentazone** Zupa
Zupilan H **trifluralin** Zupa
Zurin R **potassium nitrate + sulphur** Ziegler
Zuvapin FHIN **metham-sodium** Zupa
Zwamdood F **copper oxychloride + sulphur** Chimac-Agriphar
Zwei-4-D-Dicopur H **2,4-D** Leu & Gygax
Zyban F **mancozeb + thiophanate-methyl** Mallinckrodt
Zyklon IR **hydrogen cyanide** Breymesser, Degesch, DGS, Geopharm, Heerdt-Lingler, Rentokil
Zynteb F **zineb** Scam
Zytron H **DMPA**
ZZ-Acaricida doble A **dicofol + tetradifon** ICI-Zeltia
ZZ-Aphox I **pirimicarb** ICI-Zeltia
ZZ-Bix F **ethirimol + maneb** ICI-Zeltia
ZZ-Cobre azul F **copper oxychloride + zineb** ICI-Zeltia
ZZ-Cobre triple azul F **copper oxychloride + maneb + zineb** ICI-Zeltia
ZZ-Cuprocol F **copper oxychloride** ICI-Zeltia
ZZ-Doricida I **bensultap** Takeda
ZZ-Fostion IA **malathion** ICI-Zeltia
ZZ-L I **lindane** ICI-Zeltia

ZZ-Mycodifol F **captafol + folpet** ICI-Zeltia
ZZ-Reforzado I **carbaryl + lindane** ICI-Zeltia
ZZ-Sulfocobre F **copper oxychloride + sulphur** ICI-Zeltia
ZZ-T Reforzado I **cartap hydrochloride** ICI-Zeltia

Sources of Information On Pesticides

Books

European Directory of Agrochemical Products (3rd Edition) (Vol 1 Fungicides; Vol 2 Herbicides; Vol 3 Insecticides, Acaricides etc.; Vol 4 Plant Growth Regulators, etc.) (RSC 1988)

The Agrochemicals Handbook (2nd Edition) (RSC 1987)

Pesticides 1988 (Pesticides Approved under the Control of Pesticides Regulations 1986) (MAFF, 1988)

Farm Chemicals Handbook 1988 (Meister Publishing Co., 1988)

Worthing, C. R. and Walker, B. (eds): **The Pesticide Manual (8th edition)** (BCPC, 1987)

Janes, N. F.: **Recent Advances in the Chemistry of Insect Control** (RSC, 1985)

Thomson, W. T.: **Agricultural Chemicals** (Vol 1 Insecticides; Vol 2 Herbicides; Vol 3 Fungicides; Vol 4 Fumigants, Growth Regulators, Repellents, and Rodenticides) (Thomson Publications)

Herbicide Handbook (Weed Science Society of America) (5th edition, 1983)

International Pesticide Directory (7th edition) (1987) (Supplement to International Pest Control)

1987 Pesticide Directory (Thomson Publications)

Japanese Chemicals Guide 1987 (Japan Plant Protection Association)

Crop Protection Chemicals Reference (3rd edition, 1987) (John Wiley)

Online Databases

In the pesticide area, the RSC has several online databases:

The Agrochemicals Handbook — gives scientific information on the active ingredients in pesticides, e.g. names, physical properties, analytical methods,uses,toxicological data, etc.

Available on Data-Star file AGRC and Dialog file 306

European Directory of Agrochemical Products — gives information on the pesticide products which are registered for use in Europe, including the UK.

Available on Data-Star file EDAP and Dialog file 316

Chemical Business NewsBase — gives business information on the activities of companies in the area of pesticides and fertilizers, as part of its total coverage of chemical -related industries

Available on Data-Star files CBNB and CBNX and on Dialog file 319

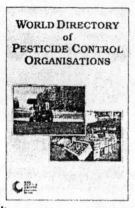

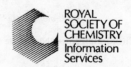

MARKETING COMPANY LIST

3M: *Sweden:* 3M Svenska AB, 191 89 SOLLENTUNA. Tel. 08/754 00 80.
- *UK:* 3M United Kingdom PLC, 3M House, PO Box 1, BRACKNELL, Berkshire RG12 1JU. Tel. (0344) 26726.
- *USA:* 3M Co., 3M Center, ST. PAUL MN 55101.

AAGrunol: *Netherlands:* AAGrunol BV, Oosterweg 127 9751 PE HAREN (PR).

Aako: *Netherlands:* Aako Agricultural Chemical B.V., Arnhemseweg 87, 3832 GK LEUSDEN. Tel. 033-948494.

Abbott: *Belgium:* Abbott S.A., Rue du Bosquet, 2, 1348 OTTIGNIES-LOUVAIN-LA-NEUVE. Tel. 010/41 87 60.
- *Switzerland:* Abbott Laboratoires SA, Rue Thalberg 2, 1200 GENEVE. Tel. 022 31 95 30.
- *USA:* Abbott Laboratories, Chemical & Agricultural Products Division, 14th Street & Sheriden Road, NORTH CHICAGO IL 60064. Tel. 1-800-323-9597.

Abelte: *Spain:* Abelte SA, Ctra Madrid Km, 384'6, Apartado 579, 30100 MURCIA. Tel. 968 83 14 50.

ABM Chemicals: *UK:* A.B.M. Chemicals Ltd., Unity Mills, Poleacre Lane, Woodley, STOCKPORT, Cheshire SK6 1PQ. Tel. (061 430) 4391.

ACF: *Netherlands:* ACF Chemiefarma N.V., Straatweg 2, Postbus 5, 3600 AA MAARSSEN. Tel. 030-452911.

ACP: *USSR:* Allunionsinstitut fuer chemische Pflanzenschutzmittel, MOSKAU.

Adam's: *Greece:* Adam's, 9 Anthimou Gazi St., GR-105 61 ATHENS. Tel. (01) 32.37.208; Telex 216665 IAM GR.

Aeropak: *Denmark:* Aeropak A/S, DAUGARD.

Aesculaap: *Belgium:* Aesculaap N.V., Groot Brittanielaan, 115, 9000 GENT. Tel. 091/25 01 03.
- *Netherlands:* Aesculaap B.V., Postbus 35, 5280 AA BOXTEL. Tel. 04116-75915.

Afepasa: *Spain:* Afepasa, c/ Real, 19, 43004 TARRAGONA. Tel. 977 21 22 50.

Afrasa: *Spain:* Industrias Afrasa c/ Ciudad de Sevilla, 53 Poligono Industrial Fuente del Jarro, 46988 PATERNA (Valencia). Tel. 96 132 17 00.

AgBioChem: *USA:* AgBioChem Inc., 3 Fleetwood Court, ORINDA CA 94563. Tel. (415) 254-0789.

Agil: *Belgium:* Agil L.T.D., Grimbertlaan, 23, 2120 SCHOTEN. Tel. 03/658 79 90.

Aglukon: *Netherlands:* Hoge Rijndijk 94a, 2313 KL LEIDEN. Tel. 07114-4053.

Agrarische Unie: *Netherlands:* Agrarische Unie-Vulcaan B.V., Pikeursbaan 15, Postbus 1000, 7400 BA DEVENTER. Tel. 05700-10955.

Agriben: *Belgium:* Agriben S.A., Galerie de la Porte Louise, 203 bte no 3., 1050 BRUXELLES. Tel. 02/513 41 40.
- *Netherlands:* Agriben Nederland B.V., Zeedijk 47, Postbus 209 ETTEN-LEUR. Tel. 01680-28920.

Agribus: *Greece:* Agribus G. & P. Christias, 10 Salaminas St., GR-546 25 THESSALONIKI. Tel. (031) 535.330, 543.444; Telex 418374.

Agrichem: *Netherlands:* Agrichem B.V., Koopvaardijweg 9, Postbus 295, 4900 AG OOSTERHOUT. Tel. 01620-31931.
- *Spain:* Agrichem, Sanchez Pacheco, 31 28022 MADRID. Tel. 91 415 83 90.
- *UK:* Agrichem Limited, Padholme Road, PETERBOROUGH PE1 5XL. Tel. (0733) 47881; Telex 32888.

Agricola Ucana: *Spain:* Agricola Ucana San Francisco, 1, 38300 LA ORATABA. 922 33 02 70.

Agricultural Chemicals: *Belgium:* Agricultural Chemicals N.V., Bredabaan, 672, 2170 WUUSTWEZEL. Tel. 031/669 77 47.

Agrikem (Borders): *UK:* Agrikem (Borders) Ltd., Sprouston Road, KELSO, Roxburghshire TD5 8EU. Tel. (0573) 24281.

Agrimont: *Italy:* Agrimont, Gruppo Montedison, Foro Bonaparto 31, 20121 MILANO.

Agrinet: *France:* Agrinet, B.P. 54, Le Patis - 10, rue Clement Ader, 785 12 RAMBOUILLET CEDEX. Tel. 30 41 03 03; Telex 698 3 18.

Agriphyt: *France:* Agriphyt S.A., B.P. 29, 229 rue Jean Jaures, 59970 FRESNES-SUR-ESCAUT. Tel. 27 26 00 17; Telex 110 750.

Agriplan: *Spain:* Reveex Agricola, SA (Agriplan), Raval de San Pere, 31, REUS (Tarragona). Tel. 977 34 02 11.

Agrishell: *France:* Agrishell, 14 rue du Professeur Deperet, 69 160 TASSIN-LA-DEMI-LUNE. Tel. 78 34 29 38; Telex 300 651 AGRISHEL.

Agro-Kanesho: *Japan:* Agro-Kanesho Co. Ltd., 2-4-1 Marunouchi, Chiyoda-ku, TOKYO 100. Tel. 03 216 5041. Telex 02223708 KGYCO J.

Agro-kemi: *Denmark:* Agro-kemi A/S, Postbox 80, 2605 BROENDBY. Tel. (02) 45 21 11.
- *Iceland:* Agro-kemi A/S, DANMOERKU,.

Agro-Quimicas de Guatemala: *Guatemala:* Agro-Quimicas de Guatemala S.A., Av. La Reforma 13-70 Zona 9, Edificio Real Reforma 3 Nivel, GUATEMALA CITY CA. Tel. 316 159 67962. Telex 5991 QUIGUA GU.

Agrochemiki: *Greece:* Agrochemiki-Isagogiki Ltd., 22 Favieru St., GR-104 38 ATHENS. Tel. (01) 52.43.362.

Agrocros: *Spain:* Agrocros, SA, c/ Recoletos, 22 Apartado 995, 28001 MADRID. Tel. 91435 40 60.

Agrofarm: *Greece:* Agrofarm, 32 Agias Paraskevis St., GR-121 32 Peristeri, ATTIKI. Tel. 57.22.014, 57.27.480.

Agrolinz: *Austria:* Agrolinz AG, Agrarchemikalein G.m.b.H., Postfach 296, St-Peter-Strasse 25, A-4021 LINZ. Tel. 0 73 2/591; Telex 02 1324.
- *Denmark:* Agrolinz AG, KOEBENHAVN S.
- *Hungary:* Agrolinz, Zorka Sabac Kepviselet, Dozsa Gyoergy ut 92/a BUDAPEST VI.

Agromex: *Greece:* Agromex, 100 Evrou St., GR-115 27 ATHENS.

Agronorden: *Denmark:* RP Agronorden A/S, Gladsaxevej 378, 2860 SOBORG. Tel. (01) 56 32 00.

Agronova: *Italy:* Agronova, Via Massarenti 221/6, 40138 BOLOGNA.

Agropharmaceutiki: *Greece:* Agropharmaceutiki, 21 Ol. Diamandi St., THESSALONIKI.

Agroplant: *Switzerland:* Agroplant AG, Postfach, 3052 ZOLLIKOFEN. Tel. 031 86 16 66.

Agrotec: *Germany (FRG):* Agrotec GmbH, Huettenstrasse, Postfach 1340, 5014 KERPEN-SINDORF. Tel. 0 22 73/50 76.
- *USA:* Agrotec Inc., Highway 35, North, P.O. Box 49, PENDLETON, NC 27862. Tel. (919) 585-1222.

Agrotechnica: *Greece:* Agrotechnica O.B.E.E., 1 Orphanidou St., GR-546 26 THESSALONIKI. Tel. (01) 32.17.104.

Agrotex: *Greece:* Agrotex Ltd., 100 Evrou St., GR-115 27 ATHENS.

Agway: *USA:* Agway Inc., P.O. Box 4933, SYRACUSE, NY 13221. Tel. (315) 477-7061.

Aifar: *Italy:* Aifar Agricola, Via Terpi 53/r, 16141 GENOVA.

Akzo: *Netherlands:* Akzo Zout Nederland Chemie B.V., James Wattstraat 100, Postbus 4080, 1009 AB AMSTERDAM. Tel. 020-5901911.
- *USA:* Akzo Chemie America, Armak Chemicals, 300 S Wacker Drive, CHICAGO, IL 60606. Tel. (312) 786-0400.

Alfasan: *Netherlands:* Alfasan B.V., Barwoutswaarder 13, 3449 HE WOERDEN. Tel. 03480-6945.

All-India Medical: *India:* All-India Medical Corp., 8th Road, Akhand Jyoti, Santacruz (East), BOMBAY 400055. Tel. 6124831. Telex 011 2476 NLIA.

Allied Chemical: *USA:* Allied Chemical, P.O. Box 2064R, MORRISTOWN, NJ 07960.

Alpha: *UK:* Alpha GB Ltd., 8/12 Brook Street, LONDON W1Y 1AA. Tel. (01 493) 9522-4.

MARKETING COMPANY LIST

Amvac: *USA:* Amvac Corp., 4100 E. Washington Boulevard, LOS ANGELES, CA 90023. Tel. (213) 264-3910. Telex 333589.

Anorgachim: *Greece:* Anorgachim Corp., 9 Mourouzi St., GR-106 74 ATHENS. Tel. (01) 72.36.676; Telex 215416 CHEM.

Anticimex: *Denmark:* Oy Anticimex Ab, Pulttitie 16, 00880 HELSINKI. Tel. 90 782 099.

- *Sweden:* Anticimex, Box 726, 101 30 STOCKHOLM. Tel. 08/23 15 60.

Aporta: *Spain:* Aporta, SA, Plaza Urquinaona, 6-9a, 08010 BARCELONA. Tel. 93 317 89 76.

Applied Biochemists: *USA:* Applied Biochemists Inc., 5300 West County Line Road, MEQUON, WI 53092. Tel. 1-800-558-5106.

Aprodas: *France:* Aprodas, St-Marcel, 13367 MARSEILLE CEDEX 4. Tel. 91 35 90 35; Telex 410 796 PROCIDA MARSL.

Aragonesas: *Spain:* Energia e Industrias Aragonesas, SA, P. Recoletos, 27, 28004 MADRID. Tel. 91 419 46 00.

Araujo: *Portugal:* J.M. Araujo, Lda., Rua de S. Joao 120-122, 4000 PORTO. Tel. 320407.

Argos: *Spain:* Industrias Quimicas Argos, SA, Plaza de Vicente Iborra, 4, 46003 VALENCIA. Tel. 96 331 44 00.

Arkovet: *France:* Arkovet, 2-4, rue Lionel Terray - B.P. 308, 92506 RUEIL-MALMAISON CEDEX. Tel. (1) 47 49 02 02; Telex CIGER A 203 012 F.

ASB-Gruenland: *Germany (FRG):* ASB-Gruenland Verkaufsgesellschaft, Porschestrasse 4, 7140 LUDWIGSBERG. Tel. 0 71 41/3 08-0.

Asef Fison: *Netherlands:* Asef Fison B.V., Pittelderstraat 20, 6942 GJ DIDAM. Tel. 08362-1041.

Asepta: *Belgium:* Asepta Belgium, Rozenlaan, 13, 2300 TURNHOUT. Tel. 014/41 52 92.

- *Netherlands:* Aseptafabriek B.V., Cyclotronweg 1, Postbus 33, 2600 AA DELFT. Tel. 015-569210.

Aseptan: *France:* Aseptan (Laboratories), Les Grandes-Raies - B.P. 15, 01320 CHALAMONT. Tel. 74 61 73 43; Telex ASEAGRO 375 832 F.

Atlansul: *Portugal:* Atlansul-Importacao e Exportacao, S.A.R.L. Rua Tomas Ribeiro 50-40, Apartado 1143, 1103 LISBOA CODEX.

Atlas-Interlates: *UK:* Atlas-Interlates Ltd., Fraser Road, ERITH, Kent DA8 1PN. Tel.(032 24) 32255; Telex 896176.

Aufra: *France:* Aufra (Laboratories), 5, rue d'Arsonval, 75015 PARIS. Tel. (1) 43 21 25 20.

Austria Metall: *Austria:* Austria Metall AG, Postfach 19, A-6230 BRIXLEGG. Tel. 0 53 37/25 51; Telex 5125126.

Avenarius: *Austria:* Avenarius, Chemische Fabrik Ges. m.b.H., Postfach 22, Burgring 1, A-1015 WIEN. Tel. 0 22 2/58 84 00; Telex 1-12581.

AVG: *Hungary:* Alkaloida Vegyeszeti Gyar, TISZAVASVARI.

Avitrol: *USA:* Avitrol Corp., 7644 East 46th Street, TULSA, OK 74145. Tel. (918) 622-7763. Telex 492 351 AVITROL TUL.

Avon Packers: *UK:* Avon Packers Ltd., Green Lane, FORDINGBRIDGE, Hants. SP6 1HT. Tel. (0425) 53354.

Aziende Agrarie Trento: *Italy:* Aziende Agrarie Trento Fitofarmaci, Via Allende, 4, 40139 BOLOGNA. Tel. (051) 54.10.09.

B.B.: *Belgium:* Belgische Boerenbond, (B.B.), Minderbroederstraat, 24, 3000 LEUVEN. Tel. 016/22 42 01.

Bakkers: *Netherlands:* Bakkers Zaadteelt en Zaadhandel B.V., Westerweg 354, 1852 PR HEILOO. Tel. 072-331315.

Bang: *Finland:* Bang & Co. Oy, PL 79, 00101 HELSINKI 10. Tel. 90 540 41.

Barenbrug: *Belgium:* Barenbrug Belgium Maes, Zoning industriel, 5800 GEMBLOUX. Tel. 081/61 29 31.

- *Netherlands:* Barenbrug Holland B.V., Stationsstraat 40, 6678 AC OOSTERHOUT (GLD). Tel. 08818-1545.

Barnaengen: *Austria:* Barnaengen Ges. m.b.H., St.-Veiter-Strasse 1, A-9560 FELDKIRCHEN/KAERNTEN. Tel. 0 42 76/27 01.

- *Denmark:* Barnaengens kemiske fabrik A/S, HERLEV.

- *Germany (FRG):* Barnaengen Deutschland GmbH, Alfred-Nobel-Strasse 1, 5020 FRECHEN. Tel. 0 22 34/1 09-1.
- *Sweden:* Barnaengen AB, Box 12080, 102 22 STOCKHOLM. Tel. 08/25 25 80.

BASF: *Austria:* BASF Oesterreich Ges. m.b.H., Postfach 1000, Hietzinger Hauptstrasse 119, A-1131 WIEN. Tel. 0 22 2/82 94 31-0, 82 94 41-0; Telex 13-4264.
- *Belgium:* BASF Chimie S.A., Avenue Hamoir, 14, 1180 BRUXELLES. Tel. 02/375 24 00.
- *Denmark:* BASF Danmark A/S, Ved Stadsgraven 15, Postbox 1734, 2300 KOEBENHAVN S. Tel. (01) 57 00 11.
- *Finland:* Suomen BASF Oy, PL. 5, 00241 HELSINKI. Tel. 90 159 81.
- *France:* BASF (Compagnie Franciase), 140, rue Jules-Guesde - B.P. 87, 92303 LEVALLOIS-PERRET CEDEX. Tel. (1) 47 30 55 00; Telex BASFC 620 445 F.
- *Germany (FRG):* BASF Aktiengesellschaft, Verkauf (AP/V) und Beratung (APL/B), Postfach 2 20, 6703 LIMBURGERHOF. Tel. 0 62 36/6 81.
- *Hungary:* BASF Magyarorszagi Iroda, Gervai u.4. BUDAPEST XIV.
- *India:* BASF India, Maybaker House, S.K. Ahire Marg., P.O. Box 19108, BOMBAY 400 025. Tel. 4930703.
- *Italy:* BASF-Agritalia, Via Matteo Bandello 6, 20123 MILANO. Tel. (02) 498.64.51.
- *Netherlands:* BASF Nederland B.V., Kadestraat 1, Postbus 1019, 6801 MC ARNHEM. Tel. 085-717171.
- *Norway:* BASF Norge AS, Postboks 131, 1312 SLEPENDEN.
- *Portugal:* BASF Portuguesa, S.A.R.L., Rua de Santa Barbara, 46 - 5, Apartado 1438, 1199 LISBOA CODEX. Tel. 56 25 11.
- *Spain:* BASF Espanola, SA, Paseo de Gracia, 99 Apartado 762, 08008 BARCELONA.
- *Sweden:* BASF Svenska AB, Lantbruksavd, Vretenvaegen 10, 171 54 SOLNA. Tel. 08/98 08 40.
- *Switzerland:* BASF (Schweiz) AG, Appital, Postfach, 8820 WAEDENSWIL/Au. Tel. 01 783 91 11.
- *UK:* BASF United Kingdom Ltd., Agrochemical Division, Lady Lane, HADLEIGH, Ipswich, Suffolk IP7 6BQ. Tel. (0473) 822531.
- *USA:* BASF Wyandotte Corp., P.O. Box 181, 100 Cherry Hill Road, PARSIPANNY, NJ 07054. Tel. (201) 263-3400.

Baslini: *Italy:* Ind. Chimiche Baslini, Via Gabrio Serbelloni 12, 20122 MILANO. Tel. (02) 77.82.

Battle, Hayward & Bower: *UK:* Battle, Hayward and Bower Ltd., Victoria Chemical Works, Crofton Drive, Allenby Road Industrial Estate, LINCOLN LN3 4NP. Tel. (0522) 29206-7.

Bauhaus: *Germany (FRG):* Bauhaus AG, Gutenbergstrasse 21, 6800 MANNHEIM 1. Tel. 06 21/3 90 51.

Bayer: *Austria:* Bayer Austria Ges. m.b.H., Geschaeftsbereich Pflanzenschutz, Postfach 109, Lerchenfelder Guertel 9-11, A-1164 WIEN. Tel. 0 22 2/92 55 21, 92 53 41; Telex 134655.
- *Belgium:* Bayer Belgium, Avenue Louise, 143, 1050 BRUXELLES. Tel. 02/537 13 75.
- *France:* Bayer France (Division Phytochim), 49-51 quai de Dion Bouton, 92815 PUTEAUX CEDEX. Tel. (1) 47 62 00 00; Telex BAYFR A 611 810.
- *Germany (FRG):* Bayer AG, PF-AT/Beratung-Deutschland, Pflanzenschutzzentrum Monheim, 5090 LEVERKUSEN BAYERWERK. Tel. 0 21 73/5 90 33 80.
- *Greece:* Bayer Epifa, 55-59 Deligiorgi St., GR-104 37 ATHENS. Tel. (01) 52.44.511; Telex 215144 BAY.
- *Hungary:* Bayer AG, Petoefi ter 2. BUDAPEST V.
- *Italy:* Bayer Italia, Divisione Agraria, Via Certosa, 126, 20156 MILANO. Tel. (02) 39.78.1.
- *Netherlands:* Bayer Nederland B.V., Divisie Bayer Farma, Nijverheidsweg 26, Postbus 80, 3640 AB MIJDRECHT. Tel. 02979-84151.

- *Netherlands:* Bayer Nederland B.V., Divisie Agro Chemie, Velperweg 28, Postbus 9217, 6800 HW ARNHEM. Tel. 085-681111.
- *Norway:* Bayer Norge A/S Postboks 311, 1324 LYSAKER.
- *Portugal:* Bayer Portugal, S.A.R.L., Rua Sociedade Farmaceutica, 3, Apartado 2365, 1109 LISBOA CODEX. Tel. 54 21 94.
- *Spain:* Bayer Hispania Comercial, SA, Pau Claris, 196, Apartado 1745, 08037 BARCELONA. Tel. 93 217 40 12.
- *Sweden:* Bayer (Sverige) AB, Div Agro-Kemi, Hemsoegatan 10 A, 211 24 MALMOE. Tel. 040/93 69 30.
- *UK:* Bayer Ltd., Agrochem Division, Eastern Way, BURY ST. EDMUNDS, Suffolk IP32 7AH. Tel. (0284) 63200.

Beaphar: *Netherlands:* Beaphar B.V., Linderteseweg 9, Postbus 7, 8100 AA RAALTE. Tel. 05720-51155.

Beecham: *Netherlands:* Beecham Veterinaire Produkten, Sportlaan 198, 1185 TH AMSTELVEEN. Tel. 020-459801.

Bell: *USA:* Bell Laboratories Inc., 3699 Kinsman Boulevard, MADISON, WI 53704. Tel. (608) 241-0202.

Bendien: *Netherlands:* H. P. Bendien B.V., Thierensweg 10-12, Postbus 17, 1411 EX NAARDEN. Tel. 02159-84151.

Berg: *Denmark:* Erik Berg A/S, KOEGE.

Berghoff: *Germany (FRG):* Caspar Berghoff KG, Moehnestrasse, 4788 WARSTEIN-ALLAGEN. Tel. 0 29 25/22 07/22 08.

Berlin Chemie: *Germany (GDR):* VEB Berlin Chemie, BERLIN.

Berner: *Finland:* Berner Oy, Kasvinsuojeluosasto, PL 15, 00131 HELSINKI. Tel. 90 176 521.

Bertram: *Germany (FRG):* Bertram GmbH, Dunzweiler Strasse 38, 6791 DITTWEILER. Tel. 0 63 86/2 71.

Billen: *Belgium:* Billen E., Rue de Stalle, 25, 1180 BRUXELLES. Tel. 02/356 79 20.

Bimex: *Italy:* Bimex, Via Cogolla 5, 36033 ISOLA VICENTINA (VI).

Bio-Innovation: *Sweden:* Bio-Innovation AB BINAB, Box 57, 545 00 TOEREBODA. Tel. 0506/12475.

Biochem Product: *Belgium:* Biochem Product, Rue de Prince Albert, 44, 1050 BRUXELLES. Tel. 02/516 61 11.

Biochem Products: *USA:* Biochem Products, P.O. Box 264, MONTCHANIN, DE 19710. Tel. (302) 654-0325. Telex 510 666 0470.

Biolchim: *Italy:* Biolchim, Via S. Carlo 2130, 40059 MEDICINA (BO). Tel. (051) 850.150.

Bitterfeld: *Germany (GDR):* VEB Chemiekombinat Bitterfeld, Zorbiger Strasse 1, 44 BITTERFELD.

Bjoernrud: *Norway:* A/S Edv. Bjoernrud, Postboks 98, Kaldbakken, 0902 OSLO 9.

Blitol: *Germany (FRG):* Blitol Marketing GmbH, Eiffestrasse 482-486, Postfach 26 05 26, 2000 HAMBURG 26. Tel. 0 40/21 15 41.

Bogena: *Netherlands:* Bogena B.V., Sluisweg 2, Postbus 150, 5145 AD WAALWIJK. Tel. 04160-36992.

Boots: *UK:* The Boots Company PLC, NOTTINGHAM, NG2 3AA. Tel. 0602 48522; Telex 37-7811.

Borax: *Germany (FRG):* Deutsche Borax GmbH, Reuterweg 14, 6000 FRANKFURT/MAIN. Tel. 0 69/15 90.

Borregaard: *Sweden:* Borregaard Trading Svenska AB, Box 14009, 400 20 GOETEBORG. Tel. 031/83 01 10.

Borzesti: *Romania:* Petrolchemisches Kombinat Borzesti, BORZESTI.

Bos: *UK:* Bos Chemicals Ltd., Paget Hall, Tydd St. Giles, WISBECH, Cambs PE13 5FL. Tel. (0945) 870014.

Bourgeois: *France:* Bourgeois (Ets), B.P. 7, 80380 VILLERS BRETONNEUX. Tel. 22 48 00 33; Telex 140 988.

BP: *Austria:* BP Austria AG, Postfach 207, Schwarzenbergplatz 13, A-1041 WIEN. Tel. 0 22 2/65 66 75; Telex 0-132163, 0-132159.

- *Belgium:* B.P. Belgium, Nieuwe Weg 1, 2730 ZWIJNDRECHT. Tel. 03/252 83 52.
- *Spain:* BP Espana, SA, P. Castellana, 91 28046 MADRID. Tel. 91 456 50 14.
- *UK:* BP Oil Ltd., Distributor Sales Division, BP House, Victoria Street, LONDON SW1E 5NJ. Tel. (01 821) 2000.

Braun: *Germany (FRG):* Braun GmbH, Heidenschestr. 14, 4920 LEMGO. Tel. 0 52 61/7 17 43.

Bredologos: *Greece:* Geoponiki Sif. Bredologos, Terma Therissou, GR-731 00 CHANIA, CRETE. Tel. (0821) 23.865.

Breymesser: *Austria:* Breymesser & Co. Ges. m.b.H., Rasumofskygasse 21, A-1030 WIEN. Tel. 0 22 2/72 14 65.

Brillocera: *Spain:* Brillocera-Retaro, SA, c/ En Proyecto, 8, Poligono Industrial Norte, 46469 BENIPARRELL (Valencia).

Brinkman: *Belgium:* Brinkman P.V.B.A., Lierseweg, 60, 2410 HERENTALS. Tel. 03/235 32 30.
- *Netherlands:* Brinkman B.V., Woutersweg 10, Postbus 2, 2690 AA 'S-GRAVENZANDE. Tel. 01784-13341.

Broeste: *Denmark:* P. Broeste, Lundtoftegardsvej 95, 2800 LYNGBY. Tel. (02) 93 33 33.

Brogdex: *Spain:* Brogdex, SA, Paseo Luis Belda, 18, GANDIA (Valencia).

Brummer: *Netherlands:* Firma J. Brummer, Aureliushof 123f, 6215 SP MAASTRICHT. Tel. 043-27705.

Buchrucker: *Austria:* Fr. Buchrucker, Laboratorium, Rudolfstrasse 36, A-4010 LINZ/DONAU. Tel. 0 73 2/23 20 55.

Buckman: *USA:* Buckman Laboratories Inc., 1256 N. McLean Boulevard, MEMPHIS, TN 38108. Tel. (901) 278-0330.

Budapesti Vegyimuvek: *Hungary:* Budapesti Vegyimuvek, Ken u. 5 BUDAPEST IX.

Burri: *Switzerland:* Andre Burri, Agricide, 2514 LIGERZ BE. Tel. 032 85 16 86.

Burts & Harvey: *UK:* Burts & Harvey, Crabtree Manorway North, BELVEDERE, Kent DA17 6BQ. Tel. (01 311) 7000.

C.F.P.I.: *France:* C.F.P.I. (Compagnie Francaise des Produits Industriels), 28, boulevard Camelinat - B.P. 75, 92233 GENNEVILLIERS CEDEX. Tel. (1) 47 99 99 55; Telex 610 626 CFPI GENVL.

C.G.I.: *France:* C.G.I. (Compagnie Generale des Insecticides), Rue Lavoisier - Z.I. d'Epluches, 95310 SAINT-OUEN-L'AUMONE, B.P. 12 - 95311 CERGY-PONTOISE CEDEX. Tel. (1) 34 64 11 73 & 30 37 15 90; Telex 696 240F.

C.N.C.A.T.A.: *France:* C.N.C.A.T.A., 83, avenue de la Grande-Armee, 75782 PARIS CEDEX 16. Tel. (1) 45 01 54 15; Telex 612 192.

Caffaro: *Italy:* Caffaro, Via Privata Vasto 1, 20121 MILANO. Tel. (02) 62.01.1.

CAG: *Spain:* Coop Agropecuaria de Guissona, 25210 GUISSONA (Lerida).

Calliope: *France:* Calliope, 1, villa Aublet, 75017 PARIS. Tel. (1) 42 67 95 49; Telex 643 370 F CALLIOP.

Campbell: *UK:* J. D. Campbell & Sons Ltd. and J. D. Campbell (Sales) Ltd., 18, Liverpool Road, Great Sankey, WARRINGTON, Cheshire WA5 1QR. Tel. (0925) 33232-3.

Candilidis: *Greece:* M. Candilidis & Co., 310 Sygrou Ave., GR-176 73 KALLITHEA. Tel. (01) 95.23.755; Telex 225258 MCAN.

CARAL: *France:* CARAL, Usine de la Chapelle, REANVILLE, 27950 SAINT-MARCEL.
- *France:* CARAL, B.P. 452, 27204 VERNON CEDEX. Tel. 32 52 42 60.

Cebeco: *Netherlands:* Cebeco-Handelsraad, Blaak 31, Postbus 182, 3000 AD ROTTERDAM. Tel. 010-4544911.

Cequisa: *Spain:* Cequisa, c/ Muntaner, 322, 08021 BARCELONA. Tel. 93 200 03 22.

Certified Laboratories: *France:* Certified Laboratories, quartier Champbenoist, 77160 PROVINS. Tel. (1) 64 00 12 20; Telex 692 889.

Chafer: *UK:* J. W. Chafer Ltd., Chafer House, 19, Thorne Road, DONCASTER, S. Yorks DN1 2HQ. Tel. (0302) 67371.

Chefam: *Netherlands:* Chemische Fabriek en Handelsonderneming Chefam, Abennestraat 19 hs, AMSTERDAM.

MARKETING COMPANY LIST

Chemia: *Austria:* Chemia, Ges. m.b.H., Sparte Landwirtschaft, Lerchenfelder Guertel 9-11, A-1160 WIEN. Tel. 0 22 2/92 55 21, 92 53 41; Telex 134655.
- *Italy:* Chemia, Casella Postale 7, Via Statale 327, 40040 DOSSO (Fe). Tel. (0532) 84.43.5.

Chemie Linz: *UK:* Chemie Linz UK Ltd., 12 The Green, RICHMOND, Surrey TW9 1PX. Tel. (01 948) 6966.

Chemolimpex: *Hungary:* Chemolimpex (Hungarian Trading Company for Chemicals), P.O. Box 121, BUDAPEST 1805. Tel. 36 1 183 976. Telex 22 4351.

Chempar: *USA:* Chempar, Division of Lipha Chemicals Inc., 3101 W Custer Avenue, MILWAUKEE, WI 53209. Tel. (414) 462-7600.

Chevita: *Netherlands:* Chevita Holland B.V., De Del 34, 6991 CW RHEDEN. Tel. 08309-2587.

Chevron: *USA:* Chevron Chemical Company, Agricultural Chemicals Division, P.O. Box 7144, SAN FRANCISCO CA 94120. Tel. (415) 894-2378; Telex 34410.

Chiltern: *UK:* Chiltern Farm Chemicals Ltd., 11 High Street, THORNBOROUGH, Buckingham MK18 2DF. Tel. (0280) 817099.

Chimac-Agriphar: *Belgium:* Chimac-Agriphar, Rue de l'Etuve, 52, 1000 BRUXELLES. Tel. 02/513 68 00.

Chimiberg: *Italy:* Chimiberg, Via Tonale 15, 24061 ALBANO S. ALESSANDRO (Bg). Tel. (035) 58.11.20.

Chimique: *France:* Chimique (la) de Paris, 16, avenue Foch, 75116 PARIS. Tel. (1) 45 00 57 29.

Chinoin: *Hungary:* Chinoin Gyogyszer- es Vegyeszeti Termekek Gyara, To u. 1-5 BUDAPEST IV.

Chipman: *UK:* Chipman Ltd., The Goods Yard, HORSHAM, Sussex RH12 2NR. Tel. (0403) 60341-5; Telex 887723 CHIPKO G.
- *USA:* Chipman Chemicals Inc., 800 Marion, P.O. Box 718, RIVER ROUGE, MI 48218.

Chudzik: *Poland:* Spoldzielnia Rzemieslnicza Chemitechnika, Wytwornia Chemiczna Zwiazkow Organicznych R. Chudzik, ul. Traktorowa 17 B, 91-116 LODZ.

Chugai: *Japan:* Chugai Pharmaceutical Co. Ltd., Meiho Building, Nishi-Shinjuku, 1-21-1 Shinjuku-ku TOKYO 160.

Cia-Shen: *Taiwan:* Cia-Shen Co. Ltd., 67-672 TAIPEI.

Ciba-Geigy: *Austria:* Ciba-Geigy Ges. m.b.H., Division Agro, Breitenfurter Strasse 251, A-1231 WIEN. Tel. 0 22 2/84 26 11; Telex 13-1923.
- *Belgium:* Ciba-Geigy, Noordkustlaan, 18, 1720 DILBEEK. Tel. 02/465 74 00.
- *Denmark:* Ciba-Geigy A/S, Lyngbyvej 172, 2100 KOEBENHAVN OE. Tel. (01) 29 14 22.
- *Finland:* Ciba-Geigy Oy, Pasilanraitio 5, 00240 HELSINKI. Tel. 90 145 322.
- *France:* Ciba-Geigy S.A., 2-4, rue Lionel-Terray - B.P. 308, 92506 RUEIL-MALMAISON CEDEX. Tel. (1) 47 49 02 02; Telex CIGER A 203 012 F.
- *Germany (FRG):* Ciba-Geigy GmbH, Division Agrarchemie, Liebigstrasse 51-53, Postfach 11 03 53, 6000 FRANKFURT/MAIN. Tel. 0 69/71 55-0.
- *Greece:* Ciba-Geigy Hellas, Anthousis Ave., GR-100 110 Anthousa, ATHENS. Tel. (01) 66.66.612-3; Telex 215536 GYAT.
- *Hungary:* Ciba-Geigy Kepviselet, Belgrad rkp. 25, BUDAPEST V.
- *Italy:* Ciba-Geigy, Casella Postale 88, 20147 SARONNO (Va). Tel. (02) 96.54.1.
- *Poland:* Ciba-Geigy, ul. Lektykarska 9/15, 01-687 WARSZAWA. Tel. 33 58 71, 33 96 77; Telex 814880.
- *Portugal:* Ciba-Geigy, Lda., Av. 5 de Outubro, 48, Apartado 1006, 1001 LISBOA CODEX. Tel. 53 68 61.
- *Spain:* Ciba-Geigy, SA, Paseo de Carlos I, 206, Apartado 1628, 08013 BARCELONA.
- *Sweden:* Ciba-Geigy, Division Agro, Stora Nygatan 55, 211 37 MALMOE. Tel. 040/10 17 55.
- *Switzerland:* Ciba-Geigy AG, Division Agro., Verkauf Schweiz, 4002 BASEL. Tel. 061 37 11 11.

- *UK:* Ciba-Geigy Agrochemicals, Whittlesford, CAMBRIDGE CB2 4QT. Tel. (0223) 833621-7.
- *USA:* Ciba-Geigy Corp., P.O. Box 18300, GREENSBORO, NC 27419. Tel. (919) 292-7100.

CIFO: *Italy:* CIFO, Via Oradour 6, 40016 S. GIORGIO DI PIANO.

Cillus: *Denmark:* Cillus A/S, Kobbervej 8, 2730 HERLEV. Tel. (02) 91 60 55.

Cimelak: *France:* Cimelak, 63, quai Joseph Gillet, 69004 LYON. Tel. 78 29 98 22.

Cit: *Austria:* Cit, Fabrik chem.-techn. Produkte, A-8051 GRAZ-GOESTING. Tel. 0 31 6/62 4 69, 53 4 23.

Cleanacres: *UK:* Cleanacres Ltd., Andoversford, CHELTENHAM, Glos. GL54 4LZ. Tel. (0242) 820481-4.

Cleanacres (Northern): *UK:* Cleanacres (Northern) Ltd., Kirton-in-Lindsey, GAINSBOROUGH, Lincs. DN21 4BE. Tel. (0652) 648461.

Cleary: *USA:* W. A. Cleary Chemical Corp., 1049 Somerset Street, P.O. Box 10, SOMERSET, NJ 08873. Tel. (201) 247-8000.

Clifton: *UK:* Clifton Chemicals Ltd., 49 Queen Square, BRISTOL BS1 4LW. Tel. (0272) 290677.

Colkim: *Italy:* Colkim, Via Piemonte 26, 40064 OZZANO EMILIA (Bologna). Tel. (051) 79.94.45.

Collett: *Norway:* Collett Kjemi A/S, Postboks 205, 1371 ASKER.

Colvoo: *Belgium:* Colvoo Belgium, Opitterkiezel, 81, 3690 BREE. Tel. 011/46 33 13.

Comac: *UK:* Comac Agrochemicals, PO Box 8, ALDRIDGE, West Midlands WS9 8DS.

Comlets: *Taiwan:* Comlets Chemical Industrial Co. Ltd., 61 Shinping Road, Taiping Hsiang, TAICHUNG HSIEN 406. Tel. 04 2702121.

Commercia: *France:* Commercia, 23, boulevard Albert I, PRINCIPAUTE DE MONACO. Tel. 93 30 12 67.

Compania Quimica: *Argentina:* Compania Quimica SA, Sarmiento 329, 1041 BUENOS AIRES.

Compo: *Germany (FRG):* Compo GmbH, Produktions- und Vertriebsgesellschaft, Postfach 2107, 4400 MUENSTER/WESTF. Tel. 02 51/39 71.

Condor: *Spain:* Insecticidas Condor, SA, c/ Villanueva, 13, 28001 MADRID. Tel. 91 435 43 75.

Cooper: *France:* Cooper France, 22-24, rue du Chateau, 92200 NEUILLY-SUR-SEINE. Tel. (1) 47 47 11 71; Telex COOFRA 630 966 F.

Coopers: *Denmark:* Coopers-Veterinaire S.A. 36, avenue de l'Epinette B.P. 142, 77100 MEAUX. Tel. (1) 64 33 17 32; Telex: 990 449.

- *Netherlands:* Coopers Agrovet B.V., IJsselmeerlaan 2, 1380 AC WEESP. Tel. 02940-804022.

COPCI: *France:* COPCI, 44, rue La Boetie, 75008 PARIS. Tel. (1) 45 63 11 22; Telex 280 837.

Copper Chemical: *USA:* Copper Chemical Corp., 29 Dunnell Lane, P.O. Box 634, PAWTUCKET, RI 02860.

Copyr: *Italy:* Copyr SpA, Via dei Giovi 6, 20032 CORMANO (Mi). Tel. (02) 619.65.33.

Coswig: *Germany (GDR):* VEB Chemiewerk Coswig, COSWIG/ANHALT.

Covagri: *France:* Covagri, 5, rue Francois 1, 75383 PARIS CEDEX 08. Tel. (1) 42 56 66 11; Telex 280 193.

CP Chemicals: *USA:* CP Chemicals Inc., Arbor Street, P.O. Box 158, SEWAREN, NJ 07007. Tel. (201) 636-4300. Telex 235002.

Craven: *UK:* The Craven Chemical Co. Ltd., Church Street, EVESHAM, Worcestershire.

Crystal: *USA:* Crystal Chemical Inter-America, 1525 North Post Oak Road, HOUSTON, TX 77055. Tel. (713) 682-1221. Telex 795 168.

CTA: *Switzerland:* Chemisch-Technische Agrarprodukte AG, 4657 DULLIKEN. Tel. 062 35 44 22.

Cumberland: *USA:* Cumberland International Corp., 1523 North Post Oak Road, HOUSTON, TX 77055. Tel. (713) 682-1221. Telex 795 168.

MARKETING COMPANY LIST

Cyanamid: *Austria:* Cyanamid Ges. m.b.H., Tendlergasse 13, A-1090 WIEN. Tel. 0 22 2/42 44 33 and 42 54 30.
- *Belgium:* Cyanamid Benelux S.A., Rue du Bosquet, 1348 MONT-ST-GUIBERT. Tel. 010/41 51 50.
- *France:* Cyanamid S.A., 74, rue d'Arcueil, immeuble lena Silic 275, 94578 RUNGIS CEDEX. Tel. (1) 46 87 23 19; Telex 270 571 F.
- *Germany (FRG):* Cyanamid GmbH, Pfaffenrieder Strasse 7, 8190 WOLFRATSHAUSEN/OBB. Tel. 0 81 71/2 21.
- *Italy:* Cyanamid Italia, Div. Agricolo-Veterinaria, Via Delle Sette Chiese 233, 00147 ROMA. Tel. (06) 51.84.70.
- *Netherlands:* Cyanamid Benelux N.V. P/A Lederle Ned B.V., Stationsplein 23, 4872 XL ETTEN-LEUR. Tel. 01680-37800.
- *Spain:* Cyanamid Iberica, SA, Carretara, Madrid-Irun, km 23, Apartado 471, San Sebastian Reyes, 28700 MADRID. Tel. 91 431 13 43.
- *Sweden:* Cyanamid Nordiska AB, Archimedesvaegen 4, 161 70 BROMMA. Tel. 08/98 83 80.
- *UK:* Cyanamid of Great Britain Ltd., Agricultural Division, Fareham Road, GOSPORT, Hants PO13 0AS. Tel. (0329) 224000.
- *USA:* American Cyanamid Co., Agricultural Division, 1 Cyanamid Plaza, WAYNE, NJ 07470. Tel. (201) 831-2000.

Dansoll: *Denmark:* Dansoll-Sandager, OESTER-ULSLEV.

De Ceuster: *Belgium:* De Ceuster N.V., Lintsesteenweg, 132, 2570 DUFFEL. Tel. 015/31 22 57.

De Rauw: *Belgium:* De Rauw K., Meerskant, 34, 9371 DENDERBELLE. Tel. 052/21 29 66.

De Weerdt: *Belgium:* De Weerdt P.V.B.A., Welleweg, 54 E, 9440 AALST. Tel. 053/21 99 69.

Debauche: *Belgium:* Debauche et Fils, Rue des Chaufours, 7600 PERUWELZ. Tel. 069/73 13 61.

Debrella: *Sweden:* Debrella AB, Box 7002, 720 07 VAESTERAS.

Decco-Roda: *Italy:* Decco-Roda, Via Consolare 2952, 47032 BERTINORO (FO).

Defensa: *Brazil:* Defensa, Industria de Defensivos Agricolas S.A., Rua Gen. Andrade Neves, 106 Caixa Postal 2679 90.000 PORTO ALEGRE. Tel. 0512 25 40 22. Telex 051 1331.

Degesch: *France:* Degesch France, 97, rue Jean-Jaures, 92300 LEVALLOIS-PERRET. Tel. (1) 42 70 66 82; Telex 630 883 IDIS.
- *Germany (FRG):* Degesch GmbH, Weismuellerstrasse 32-40, 6000 FRANKFURT/MAIN. Tel. 0 69/41 00 21.
- *Netherlands:* Handelsonderneming Degesch-Benelux, Dr. Hub. Van Doorneweg 99H, 5026 RB TILBURG. Tel. 013-635470.
- *USA:* Degesch America Inc., P.O. Box 116, WEYERS CAVE, VA 24486. Tel. (703) 234-9281. Telex 829323.

Degussa: *Tel. (201) 288-6500. Telex 134445:* Degussa Corp., Route 46 at Hollister Road, P.O. Box 2004, TETERBORO, NJ 07608.

Dehner: *Germany (FRG):* Dehner GmbH & Co. KG, Postfach 11 60, 8852 RAIN AM LECH. Tel. 0 90 02/7 70.

Delicia: *Germany (GDR):* VEB Delicia, DELITZSCH.

Delis: *Greece:* D. A. Delis SA, 5 P. Benizelou St., GR-105 56 ATHENS. Tel. (01) 32.50.302.

Delitia: *Germany (FRG):* Dr. Werner Freyberg Chemische Fabrik Delitia Nachf., Postfach 9, 6941 LAUDENBACH/BERGSTRASSE. Tel. 0 62 01/70 80.

Denka: *Netherlands:* Denka Chemie B.V., Tolnegenweg 61, Postbus 1, 2250 AA VOORTHUIZEN. Tel. 03429-2424.

Desinfection Integrale: *Belgium:* Desinfection Integrale, Parc Industriel, 4822 PETIT-RECHAIN. Tel. 087/33 12 51.

Desinsectisation: *France:* Desinsectisation Moderne France, 34, rue du Contrat Social, 76000 ROUEN. Tel. 35 70 14 42 & 35 89 46 49; Telex 180 911 Permaro.

Desinsekta: *Germany (FRG):* Desinsekta GmbH, Steinbacher Hohl 42, 6000 FRANKFURT 90. Tel. 0 69/76 30 40.

Desur: *Spain:* Desur Almda Capuchinos, 50, 29014 MALAGA. Tel. 952 25 80 08.

Detia Freyberg: *Germany (FRG):* Detia Freyberg GmbH, Postfach 9, 6941 LAUDENBACH/BERGSTRASSE. Tel. 0 62 01/76 76.

Devel: *Netherlands:* Devel Chemical Division, Laurentiusdijk 21, 4651 TJ STEENBERGEN. Tel. 01660-2912.

DGS: *Germany (FRG):* Deutsche Gesellschaft fuer Schaedlingsbekaempfung mbH, Packersweide 19, 2000 HAMBURG 28. Tel. 0 40/7 54 00 31.

- *Germany (FRG):* Deutsche Gesellschaft fuer Schaedlingsbekaempfung mbH, Oederweg 52-54, 6000 FRANKFURT/MAIN. Tel. 0 69/55 04 17.

Diamond Shamrock: *Belgium:* Diamond Shamrock Europe, Chaussee de la Hulpe, 185, 1170 BRUXELLES. Tel. 02/377 20 50.

- *Greece:* Diamond Shamrock, 24 Viziinou, N. Smirni, ATHENS.
- *USA:* Diamond Shamrock Corp., 1100 Superior Avenue, CLEVELAND, OH 44114.

Diana: *Denmark:* Diana Skovtjaere v/Tage Hansen, NOERRE-ALSLEV.

- *Greece:* Diana, G. Servos & Co. Ltd., 19 Tsimiski St., GR-546 24 THESSALONIKI. Tel. (031) 261.225-8; Telex 412526.

Dillen: *Netherlands:* H.J.B. van Dillen, Van der Veldstraat 58, 2161 ZG LISSE. Tel. 02521-13557.

Dimitrova: *Czechoslovakia:* Chemicke zavody 'J Dimitrova', np, BRATISLAVA.

Dirlexco: *Netherlands:* Dirlexco N.V., Groot Hertoginnelaan 156, 2517 EM DEN HAAG. Tel. 070-469312.

Diversey: *Austria:* Diversey Ges. m.b.H., Palmgasse 3, Postfach 36, A-1150 WIEN. Tel. 0 22 2/83 15 72.

- *Germany (FRG):* Diversey GmbH, Grosse Friedberger Strasse 23-27, D-6, FRANKFURT/MAIN.
- *Netherlands:* Diversey B.V., Kuipersweg 18, 3449 JA WOERDEN. Tel. 03480-18744.

Doff-Portland: *UK:* Doff-Portland Ltd., Bolsover Street, Hucknall, NOTTINGHAM NG15 7TY. Tel. (0602) 632842.

Dom-Samen-Fehlemann: *Germany (FRG):* Dom-Samen-Fehlemann KG, Postfach 55, 4178 KEVELAER 1. Tel. 0 28 32/44 01.

Dow: *Austria:* Dow Chemical Austria Ges. m.b.H., Wohllebengasse 6, A-1040 WIEN. Tel. 0 22 2/65 89 21; Telex 13-4678.

- *Belgium:* Dow Chemical Belgium S.A., Industrie Park, 3980 TESSENDERLO. Tel. 013/66 29 71.
- *Denmark:* Dow Chemicals A/S, Strandvejen 171, 2900, HELLERUP. Tel. (01) 62 79 99.
- *Germany (FRG):* Dow Pflanzenschutz GmbH, Berg-am-Laim-Strasse 47, 8000 MUENCHEN 80. Tel. 0 89/63 80 50.
- *Greece:* Dow Chemical (Hellas) SA, 178 Kifissia Ave., GR-154 10 Psychico, ATHENS. Tel. (01) 67.26.080-7; Telex 215526, 214769.
- *Hungary:* Dow Chemical Iroda, Vaci u. 19-21, BUDAPEST V.
- *Italy:* Dow Chemical Italia, Viale del Ghisallo 20 20151 MILANO. Tel. (02) 30021; Telex 311589.
- *Netherlands:* Dow Chemical (Nederland) B.V., Aert van Nesstraat 45, Postbus 3010, 3000 BH ROTTERDAM. Tel. 010-4174280.
- *Poland:* Dow, ul. Smolenskiego 1/11, 01-698 WARSZAWA. Tel. 33 22 22; Telex 816446.
- *Spain:* Dow Chemical Iberica, SA, Avda de Burgos, 109. Apartado 36.208, 28050 MADRID.
- *Sweden:* Dow Chemical AB, Karlavaegen 53, 114 49 STOCKHOLM. Tel. 08/24 66 80.
- *UK:* Dow Agriculture, Latchmore Court, Brand Street, HITCHIN, Herts SG5 1HZ. Tel. (0462) 57272; Telex 82414.
- *USA:* Dow Chemical U.S.A., Agri Products Department, P.O. Box 1706, MIDLAND, MI 48640. Tel. (517) 636-1000; Telex 227455.

MARKETING COMPANY LIST

Drexel: *USA:* Drexel Chemical Co., 2487 Pennsylvania Street, P.O. Box 9306, MEMPHIS, TN 38109. Tel. (901) 774-4370.

Dreyfus: *France:* Ste Louis Dreyfus, Herschtel & Co., 3, avenue du Coq, 75009 PARIS. Tel. (1) 48 74 07 45 & 42 80 61 66.

Drogenhansa: *Austria:* Drogenhansa Drogerie- und Reformwaren, Michelbeuerngasse 9a, A-1090 WIEN. Tel. 0 22 2/48 03/560; Telex 135289.

Drugtrade: *Finland:* Oy Drugtrade Ab, Ruosilantie 14, 00390 HELSINKI. Tel. 90 546 244.

Du Pont: *Belgium:* Du Pont de Nemours, Rue de la Fusee, 100, 1130 BRUXELLES. Tel. 02/722 06 25.

- *Denmark:* Du Pont de Nemours Nordiska AB, SKOVLUNDE.
- *France:* Du Pont de Nemours France S.A., (Departement Agrochimie), 137 rue de Universite 75007 PARIS. Tel. (1) 45 50 65 50; Telex 280 866.
- *Germany (FRG):* DuPont de Nemours (Deutschland) GmbH, Postfach 11 29, Bleichstrasse 29, 6120 ERBACH. Tel. 0 60 62/20 18.
- *Greece:* Du Pont de Nemours Dev. s.a.r.l., 238 Sygrou St., GR-176 72 Kallithea, ATHENS. Tel. (ol) 95.68.287, 95.68.298; Telex 219252 DUPATHS.
- *Hungary:* Du Pont Chemical Company, Rajk L. u. 11. BUDAPEST XIII.
- *Italy:* Du Pont Conid S.p.A., Amonn Fitochimica Division, Via Piave 2, 39100 BOLZANO. Tel. (0471) 998111.
- *Netherlands:* Du Pont de Nemours (Nederland) B.V., Helftheuvelweg 1, Postbus 2060, 5202 CB 'S-HERTOGENBOSCH. Tel. 073-206911.
- *Norway:* Du Pont de Nemours Nordiska AB, Filial Norge, Postboks 156, Oekern, 0509 OSLO 5.
- *Poland:* Du Pont, Intraco Center, ul. Stawki 2, 00-193 WARSZAWA. Tel. 39 58 86; Telex 81 23 76.
- *Spain:* Du Pont Iberica, SA, Tuset, 23, 3, 08006 BARCELONA MADRID-34. Tel. 93 200 73 11.
- *Sweden:* Du Pont de Nemours Nordiska AB, Djaeknegatan 4, 211 35 MALMOE. Tel. 040/718 45.
- *Switzerland:* Du Pont de Nemours International S.A., Biochemicals Departement, 78-82, route des Acacias, 1221 GENEVE 24. Tel. 022 42 16 00.
- *UK:* Du Pont UK Ltd., Agricultural Chemicals Department, Wedgwood Way, STEVENAGE, Herts. SG1 4QN. Tel. (0438) 734457.
- *USA:* E.I. du Pont de Nemours & Co., Inc., 1007 Market Street, WILMINGTON, DE 19898. Tel. (302) 774-2421.

Duclos: *France:* Duclos S.A., Departement Agricole - B.P. 205, 34400 LUNEL. Tel. 67 71 12 72; Telex 490 937 I.HUMUS.

Duphar: *Belgium:* Duphar, Boulevard Emile Bockstael 122, 1020 BRUXELLES. Tel. 02/428 00 50.

- *Netherlands:* Duphar B.V., Drentestraat 11, Postbus 7133, 1007 JC AMSTERDAM. Tel. 020-440911.

Duratox: *Belgium:* Duratox S.A., Chaussee de Louvain 270, 1130 BRUXELLES. Tel. 02/354 26 40.

Duslo: *Czechoslovakia:* Duslo, Narodny Podnik, SALA.

Duthoit: *France:* Duthoit, Avenue Industrielle, 59930 LA CHAPELLE-D'ARMENTIERES. Tel. 20 35 85 13; Telex 110 199.

Dyrup: *Denmark:* S. Dyrup & Co. A/S, SOEBORG.

Edefi: *Spain:* Espanola de Desarrollo Financiero, SA, c/ Sagasta, 30 28004 MADRID. Tel. 91 447 74 54.

Edialux: *Belgium:* Edialux, Grote Dries, 36-38, 2520 EDEGEM. Tel. 03/457 75 35.

Efthymiadis: *Greece:* K. & N. Efthymiadis, Dodekanisou 24, P.O. Box 107 32 THESSALONIKI. Tel. (031) 798.403, 527.430; Telex 410254 KNE, 412021 SIND.

Egesa: *Germany (FRG):* Egesa Einkaufsgenossenschaft der Samenhaendler AG, Karl-Benz-Strasse 9, Postfach 11 06 20, 6300 GIESSEN. Tel. 06 41/7 20 89.

EGIS: *Hungary:* EGIS Gyogyszervegyeszeti Gyar, Keresztúri ut 30-38 BUDAPEST X.

EGYT: *Hungary:* EGYT Pharmaceutical Works, P.O. Box 100, BUDAPEST.

Eimermacher: *Germany (FRG):* Ferdinand Eimermacher oHG, Chemische Fabrik, Eimermacherweg 22-26, 4400 MUENSTER. Tel. 02 51/21 41 84.

Elanco: *Austria:* Eli Lilly & Elanco GmbH, Huetteldorfer Strasse 65, A-1150 WIEN. Tel 0 22 2/95 15 55; Telex 13-6918.
- *Denmark:* Elanco, Toemmerup Stationsvej 10, 2770 KASTRUP. Tel. (01) 50 77 00.
- *Germany (FRG):* Elanco Abt. der Eli Lilly GmbH, Postfach 14 41, Saalburgstrasse 153, 6380 BAD HOMBURG. Tel. 0 61 72/27 30.
- *Greece:* Elanco Hellas SA, 335 Mesogion St., GR-152 31 Halandri, ATHENS. Tel. (01) 67.26.380; Telex 223384.
- *Hungary:* Elanco Kepviselet, Petoefi ter 2. BUDAPEST V.
- *Poland:* Elanco Intraco Center, ul. Stawki 2, 00-193 WARSZAWA. Tel. 39 80 17, 39 80 24; Telex 815640.
- *Spain:* Elanco Quimica, SA, Paseo de la Industria, s/n (Zona Industrial-Alcobendas), Apartado 585, 28080 MADRID. Tel. 91 653 45 00.
- *UK:* Elanco Products Ltd., Kingsclere Road, BASINGSTOKE, Hants. RG21 2XA. Tel. (0256) 53131.

Eli Lilly: *Belgium:* Eli Lilly Benelux, Rue de l'Etuve, 53, 1000 BRUXELLES. Tel. 02/512 51 50.
- *France:* Lilly France (Departement ELANCO), 203, bureaux de la Colline - Bat. D, 92213 SAINT-CLOUD. Tel. (1) 49 11 34 34; Telex LILLYFRA 270 227.
- *Italy:* Eli Lilly Italia, Via Tolara di sotto 85, 40064 OZZANO EMILIA (BO).
- *Netherlands:* Eli Lilly Nederland, Stationsplein 97, 3511 ED UTRECHT. Tel. 030-316364.
- *Sweden:* Eli Lilly Sweden AB, Box 30037, 104 25 STOCKHOLM. Tel. 08/54 66 11.
- *USA:* Elanco Products Co., Div. of Eli Lilly & Co., 740 South Alabama Street, INDIANAPOLIS, IN 46285. Tel. (317) 261-3000.

Ellagret: *Greece:* Ellagret SA, 38 Aristotelous St., GR-104 33 ATHENS. Tel. (01) 82.25.501-5; Telex 219468 ELLA.

Elsner: *Germany (FRG):* Pflanzenschutz Elsner, Postfach 64, 2841 WAGENFELD. Tel. 0 54 44/2 32.

EM Industries: *USA:* EM Industries Inc., Plant Protection Division, 5 Skyline Drive, HAWTHORNE, NY 10532.

Embetec: *UK:* Embetec Crop Protection Ltd. Springfield House, Kings Road, HARROGATE HG1 5JJ. Tel. (0423) 509731-5; Telex 57486.

EMTEA: *Greece:* EMTEA, 1 Mitropoleos St., GR-105 57 ATHENS.

EniChem: *Italy:* EniChem Agricoltura SpA, Via Medici del Vascello 26, 20138 MILANO. Tel. 02/5201; Telex 310246.
- *USA:* EniChem Americas, Agip USA Inc., 1221 Avenue of the Americas, NEW YORK, NY 10020. Tel. (212) 382-6300. Telex 1 2288.

Enna: *Netherlands:* Enna Aerosols B.V., Holwerderweg 11, 9101 PA DOKKUM. Tel. 05190-6215.

Enotria: *Italy:* Enotria, S.S. 193 Km. 8, 96010 MELILLI (Siracusa). Tel. (095) 37.33.23.

Epple: *Germany (FRG):* Mineraloelwerk Epple GmbH, Laemmleshalde 67, 7000 STUTTGART 50. Tel. 07 11/54 12 81 & 54 12 84.

Epro: *Austria:* Epro Ges. m.b.H., Postfach 95, Belghofergasse 17, A-1121 WIEN. Tel. 0 22 2/84 36 01-600; Telex 13-2430.

Equitable Trading: *Taiwan:* Equitable Trading Co. Ltd., P.O. Box 70-251, TAIPEI. Tel. 02 711 2513. Telex 23234 EQUITE.

Ergex: *Greece:* Ergex Ltd., 35 Deligiorgi St., GR-104 37 ATHENS. Tel. (01) 52.35.946.

ERT: *Spain:* Union Explosivos Rio Tinto SA, (ERT), Paseo de la Castellana, 20. Apartado 66, 28080 MADRID. Tel. 91 431 30 40.

Esbjerg: *Denmark:* Esbjerg Kemikaliefabrik A/S, Madevej 80, 6705 ESBJERG OE. Tel. (05) 12 70 00.

Esoform: *Italy:* Esoform, Viale del Lavoro 10, 45100 ROVIGO.

Etisa: *Spain:* Especialidades Tecnico Industriales, SA, (Etisa), Avda Meridiana, 133, 3, 08026 BARCELONA. Tel. 93 245 87 24.
Eurofill: *Netherlands:* Eurofill B.V., Grote Tocht 99, 1507 CE ZAANDAM. Tel. 075-175651.
Eurofyto: *Belgium:* Eurofyto Industrielaan 6, 8900 IEPER. Tel. 057/20 53 01.
Europlant: *Netherlands:* Europlant B.V., Vaart N.Z. 2, Postbus 2, 8426 AN APPELSCHA. Tel. 05162-1216.
Eurozolfi: *Italy:* Eurozolfi, Zona industriale 14 strada, angolo 3 strada, 95100 CATANIA.
Ewos: *Denmark:* Ewos A/S, SILKEBORG.
- *Sweden:* Ewos AB, Box 618, 151 27 SOEDERTAELJE. Tel. 0755/340 80.

Excel: *India:* Excel Industries Ltd., 184-87 Swami Vivekanand Road, Jogeshwari West, BOMBAY 400 102. Tel. 571431 2 3. Telex 011 71307 EXCEL IN.
Fabel-Maxenri: *Belgium:* Fabel-Maxenri F.A., Rue Jos. Claes, 50, 1060 BRUXELLES. Tel. 02/537 50 52.
Fagaras: *Romania:* Chemisches Kombinat Fagaras, FAGARAS.
Fahlberg-List: *Germany (GDR):* VEB Fahlberg-List, MAGDEBURG.
Fair Products: *USA:* Fair Products Inc., P.O. Box 386, 806 Reedy Creek Road, CARY, NC 27511. Tel. (919) 467-8352.
Fairfield: *USA:* Fairfield American Corp., 238 Wilson Avenue, NEWARK, NJ 07104. Tel. (201) 589-0263. Telex 47 54111.
Fairmount: *USA:* Fairmount Chemical Co. Inc., 2317 Versailles Road, LEXINGTON, KY 40504.
FALI: *Germany (FRG):* FALI Landwirtschaftliche Service GmbH, Liebigstrasse 51-53, 6000 FRANKFURT/MAIN 11. Tel. 0 69/71 55-0.
Farm Protection: *UK:* Farm Protection Ltd., Glaston Park, Glaston, OAKHAM, Leicestershire LE15 9BX. Tel. (0572) 822561.
Farmer: *Italy:* Farmer Italiana, Localita Piazzano 32, 66030 PIAZZANO d'ATESSA (CH).
Farmipalvelu: *Finland:* Suomen Farmipalvelu Oy, PL 43, 21531 PAIMIO. Tel. 921 734 666.
Farmix: *Netherlands:* Farmix B.V., Nijverheidsweg 2, 3881 LA PUTTEN. Tel. 03418-52224.
Farmon Agrovia: *Greece:* Farmon Agrovia, 20 Kalogera St., GR-113 61 Kypseli, ATHENS.
Farmos: *Finland:* Farmos-Yhtyma Oy, Maatalousryhma, PL 425, 20101 TURKU 10. Tel. 921 662 211.
Fattinger: *Austria:* Fattinger Agrarchemie, Erzeugung und Vertrieb chem.-techn. Produkte Ges. m.b.H., Liebenauer Hauptstrasse 89, A-8041 GRAZ. Tel. 0 31 6/42 0 81; Telex 03-1175.
FBC-Schering: *Belgium:* FBC-Schering, Mommaertslaan 14, 1920 MACHELEN. Tel. 02/720 49 10.
FCC: *UK:* Farmers Crop Chemicals Ltd., County Mills, WORCESTER WR1 3NU. Tel. (0905) 27733.
Feed Farm: *Netherlands:* Feed Farm B.V., Lissenveld 7, 4941 VK RAAMSDONKSVEER. Tel. 01621-4550.
Felleskjoepet: *Norway:* Felleskjoepet, Rosenkrantzgt. 8, 0159 OSLO 1.
Fermenta: *USA:* Fermenta Plant Protection Co., 7528 Auburn Road, P.O. Box 348, PAINESVILLE, OH 44077. Tel. (216) 357-3361. Telex 196191 FPPC UT.
Ferrosan: *Denmark:* Ferrosan A/S, SOEBORG.
- *Sweden:* AB Ferrosan, P.O. Box 201, 10 MALMOE.

Fettchemie: *Germany (GDR):* VEB Fettchemie, KARL-MARX-STADT.
Ficoop: *Spain:* Ficoop, SA, Paseo de la Castellana, 83-85, 28046 MADRID. Tel. 91 456 60 02.
Filocrop: *Greece:* Filocrop SA, 26 Carolou St., GR-104 37 ATHENS. Tel. (01) 52.34.404; Telex 219129.
Fina: *France:* Fina France, 19 rue du General Foy, 75361 PARIS CEDEX 08. Tel. (1) 45 22 90 10; Telex 650 468.
Finnewos: *Finland:* Finnewos Oy, PL 21, 01641 VANTAA 64. Tel. 90 840 200.
Fischar: *Germany (FRG):* Otto Fischar GmbH & Co. KG, Kaiserstrasse 221, 6601 SAARBRUECKEN-SCHEIDT. Tel. 06 81/81 40 31.

Fischer: *Germany (FRG):* G. & A. Fischer, Oelkerstr. 4, Postfach 18 01 29, 4800 BIELEFELD 18 (Hillegossen). Tel. 05 21/20 00 07.

Fisons: *Germany (FRG):* Fisons plc. Horticulture Division, Technisches Buero Dr. E. G. Lange, Karl-Peters-Strasse 41, 6200 WIESBADEN. Tel. 0 61 21/76 18 77.
- *Switzerland:* Fisons AG, Baarerstrasse 10, 6300 ZUG. Tel. 042 21 10 55.
- *UK:* Fisons PLC, Horticulture Division, Paper Mill Lane, Bramford, IPSWICH, Suffolk IP8 4BZ. Tel. (0473) 830492; Telex 98168 FIBRAM.

Fivat: *Italy:* Fivat Industria Chimica, Via Castiglione 6 bis, 10132 TORINO. Tel. (011) 83.02.45.

Flora-Frey: *Germany (FRG):* Flora-Frey GmbH & Co. KG, Forcher Strasse 30-34 Postfach 16 01 47, 5650 SOLINGEN 16. Tel. 0 21 22/3 83-1.

Floralis: *Germany (FRG):* Floralis GmbH, Postfach 14 01 23, 7530 PFORZHEIM 14. Tel. 0 72 31/5 30 55.

Fluegel: *Germany (FRG):* Hans Fluegel, Forstschutzmittel, Westerhoefer Strasse 45, 3360 OSTERODE 23. Tel. 0 55 22/8 23 60.

FMC: *Austria:* FMC Chemikalien Handelsgesellschaft m.b.H. Paulanergasse 13/3, A-1040 WIEN. Tel. 0 22 2/57 95 35; Telex 136511.
- *Belgium:* F.M.C. Europe S.A., Avenue Louise 523, Bte 1, 1050 BRUXELLES. Tel. 02/640 50 00.
- *Greece:* FMC Hellas Ltd., 124 Kifisias St., GR-115 26 ATHENS. Tel. (01) 69.26.920, 69.11.070; Telex 214961.
- *Switzerland:* FMC International AG, Agric. Chem. Division, Postfach 246, 6301 ZUG. Tel. 042 21 56 22.
- *USA:* FMC Agricultural Chemical Group, 2000 Market Street, PHILADELPHIA, PA 19103. Tel. (215) 299-6565.

Fomesa: *Spain:* Fomesa, c/ Jesus Morante Borras, 24, 46012 VALENCIA. Tel. 96 323 69 10.

Ford Smith: *UK:* Ford Smith & Co., Lyndean Industrial Estate, Felixstowe Road, Abbey Wood, LONDON SE2 9SG. Tel. (01 310) 8127.

Foret: *Spain:* Foret, SA, c/ Corcega, 293. Apartado 582, 08008 BARCELONA. Tel. 93 218 79 00.

Formulex: *Belgium:* Formulex P.V.B.A., Grote Dries, 36-38, 2520 EDEGEM. Tel. 03/457 20 10.

Forst-Chemie: *Germany (FRG):* Forst-Chemie Ettenheim GmbH, Postfach 2 70, 7637 ETTENHEIM/BADEN. Tel. 0 78 22/50 37.

Franco-Belge: *Belgium:* Comptoir Franco-Belge, Avenue Reine-Astrid, 242, 4880 SPA. Tel. 087/77 10 69.

Franken-Chemie: *Germany (FRG):* Franken-Chemie, Elisabethstrasse 55, 4932 LAGE (Lippe). Tel. 0 52 32/35 38.

Freiberg: *Germany (FRG):* Wolfgang Freiberg, Untere Bliesstrasse 63, 6680 NEUNKIRCHEN 5. Tel. 0 68 21/4 79 45.

Frere: *Belgium:* Frere et Cie S.A., Avenue des Noisetiers, 7, 1170 BRUXELLES. Tel. 02/672 23 30.

Frowein: *Germany (FRG):* 808 Apparate und Praeparate Walter Frowein GmbH + Co. KG, Am Reislebach 83, 7470 ALBSTADT 2 (Tailfingen). Tel. 0 74 32/33 40-30 48.

Frunol Chemie: *Germany (FRG):* Frunol Chemie, Bahnhofstrasse 34, 4750 UNNA. Tel. 0 23 03/1 26 50.

GABI-Biochemie: *Germany (FRG):* GABI-Biochemie Huendersen, 4902 BAD SALZUFLEN. Tel. 0 52 22/2 10 05.

GAF: *USA:* GAF Corp., 1361 Alps Road, WAYNE, NJ 07470. Tel. (201) 628-3000. Telex 130374.

Gaillard: *Switzerland:* Philippe Gaillard, Produits agricoles, 1907 SAXON VS. Tel. 026 6 30 40.

Gartenpracht: *Germany (FRG):* Gartenpracht GmbH, Oskar-Hoffmann-Strasse 61, 4630 BOCHUM 1.

Gefex: *Greece:* Gefex Ltd., 10 Piraeus & Zinovos St., GR-104 31 ATHENS. Tel. (01) 52.36.737; Telex 218565 AGRI.

Geissler: *Germany (FRG):* Geissler GmbH, Drachenfelsstrasse 54-56, 6700 LUDWIGSHAFEN 15. Tel. 06 21/57 34 66.

Geistler: *Germany (FRG):* August Geistler, chem.-pharm. Fabrik GmbH & Co. KG, Moselstrasse 12a, 4040 NEUSS/RHEIN. Tel. 0 21 01/4 21 32.

Gen. Rep.: *Greece:* General Representations, 38 Aristotelous St., GR-104 33 ATHENS. Tel. (01) 82.33.028.

GENP: *Taiwan:* GENP International Corp., P.O. Box 68-1561, TAIPEI. Tel. 02 521 4789. Telex 23022 GENPINTL.

Geochem: *Greece:* Geochem, 1 Piraeus St., GR-105 52 ATHENS. Tel. (01) 32.47.881-3.

Geopharm: *Greece:* Geopharm SA, 57 Panepistimiou St., GR-105 64 ATHENS. Tel. (01) 32.18.733-5; Telex 219156.

Geopharmaceutiki: *Greece:* Geopharmaceutiki, 4 Peristeriou St., GR-546 25 THESSALONIKI.

Geophyt: *Greece:* Geophyt, Aphoi Phrendzou & Co. Ltd., 15 Lada Chr., GR-121 32 ATHENS. Tel. (01) 57.47.315-6.

Gerhardtz: *Germany (FRG):* Hugo Gerhardtz jr. Duengemittelfabrik, Postfach 13 01 26, 5650 SOLINGEN-MERSCHEID. Tel. 0 21 22/33 20 31-32.

Gilbert: *France:* Gilbert (Laboratoires), Route de Lion-sur-Mer, 14200 HEROUVILLE-SAINT-CLAIR. Tel. 31 93 17 94; Telex LABOGIL 171 904 F.

Gilmore: *USA:* Gilmore Inc., 5501 Murray Road, MEMPHIS, TN 38119. Tel. (901) 761-5870. Telex 682 8022.

Gist-Brocades: *Netherlands:* Gist-Brocades N.V., Wateringseweg 1, Postbus 1, 2600 MA DELFT. Tel. 015-799111.

Globol-Werk: *Germany (FRG):* Globol-Werk GmbH, Postfach 25, 8858 NEUBURG/DONAU. Tel. 0 84 31/22 65.

Glowacki: *Poland:* Spoldzielnia Rzemieslnicza Wytworczosci Roznej, Wytwornia Chemiczna Zwiazkow Organicznych A. Glowacki, S. Ignasiak, ul. Krakowska 13, 85-950 BYDGOSZCZ. Tel. 41 18 26.

Gobbi: *Italy:* L. Gobbi, Via G. Murtola 55 r, 16157 GENOVA-PALMARO.

Goncalves: *Portugal:* Rolao Goncalves, Rua de Marvila, 1-7, 1900 LISBOA. Tel. 38 39 15.

Gorivaerk: *Denmark:* Gorivaerk A/S, KOLDING.

Gosa: *Spain:* Gosa Rambla Pulido, 36, S/C DE TENERIFE. Tel. 06 351 34 26.

Graham: *Belgium:* Graham F.A., Villabaan, 13, 1610 SINT-PIETERS-LEEUW. Tel. 02/376 20 20.

Graines Loras: *France:* Graines Loras, B.P. 4, 69811 TASSIN-LA-DEMI-LUNE CEDEX. Tel. 78 34 10 75; Telex 310 917 Public CL Lyon F pour 0065.

Great Lakes: *USA:* Great Lakes Chemical Corp., P.O. Box 2200, WEST LAFAYETTE, IN 47906. Tel. (317) 463-2511. Telex 27 9428.

Greenwood: *USA:* Greenwood Chemical Co., P.O. Box 26, State Highway 690, GREENWOOD, VA 22943. Tel. (703) 456-6832. Telex 710 9560 187.

Griffin: *USA:* Griffin Ag Products Co. Inc., P.O. Box 1847, Rocky Ford Road, VALDOSTA, GA 31601. Tel. (912) 242-8635. Telex 804639.

Grima: *Portugal:* Sociedade Portuguesa de Desinfeccao Grima, Lda., Rua da Manutencao, 23-3., Esq., 1900 LISBOA. Tel. 38 27 20.

– *Spain:* Grima Quimica, SA, Avda Industria s/n, Alcobendas, 28100 MADRID. Tel. 91 653 45 00.

Gripen: *Sweden:* Tekn fabriken Gripen, Box 1213, 581 11 LINKOEPING. Tel. 013/12 99 50.

Grondmix: *Netherlands:* Grondmix B.V., Duinstraat 51, 9494 RN YDE. Tel. 05906-1279.

Groves: *UK:* W. H. Groves, Lenelby Road, SURBITON, Surrey.

Grower: *Greece:* Grower, 17 Bizaniou St., GR-156 69 Papagou, ATHENS.

Guenther: *Germany (FRG):* Paul Guenther Cornufera GmbH, Weinstrasse 19, 8520 ERLANGEN 2. Tel. 0 91 31/6 00 33.

Gulf Oil: *USA:* Gulf Oil Co., P.O. Box 3766, HOUSTON, TX 77001.

Gullviks: *Sweden:* Gullviks Fabriks AB, Box 50132, 202 11 MALMOE. Tel. 040/18 11 20.

Gustafson: *USA:* Gustafson Inc., 17400 Dallas North Parkway DALLAS, TX 75252.
Guth: *USA:* Guth Corp., 352 S. Center Street, HILLSIDE, IL 60162. Tel. (312) 547-7030.
Haco: *Belgium:* Haco, Maaseikersteenweg 98, 3760 LANAKEN. Tel. 011/71 45 99.
Halldor Jonsson: *Iceland:* Halldor Jonsson hf., Dugguvogi 10, REYKJAVIK.
Hankkija: *Finland:* Keskusosuusliike Hankkija, PL 80, 00101 HELSINKI 10. Tel. 90 7291.
Hanson & Moehring: *Sweden:* AB Hanson & Moehring, Box 104, 401 21 GOETEBORG. Tel. 031/17 35 40.
Hartsikemia: *Finland:* Hartsikemia Erkki Karhio & Co., Pohjoiskaari 15, 00200 HELSINKI 20. Tel. 90 6926 122.
Hawlik: *Germany (FRG):* W. Hawlik, Ludwigstrasse 9, 8901 STETTENHOFEN. Tel. 08 21/49 14 49.
Hecquet: *France:* Hecquet (Ets), Rue Calquieres-Basses - B.P. 5, 34120 PEZENAS. Tel. 67 98 10 71.
Helena: *USA:* Helena Chemical Co., 5100 Poplar, Suite 3200, MEMPHIS, TN 38137.
Helinco: *Greece:* Helinco Ltd., 10 Amerikis Ave., GR-106 72 ATHENS. Tel. (01) 36.24.524; Telex 225204.
Hellafarm: *Greece:* Hellafarm, 30 Zenonos St., ATHENS.
Hellenic Chemical: *Greece:* Hellenic Chemical Products & Fertilizer Co. Ltd., 20 Amalias Ave., GR-105 57 ATHENS. Tel. (01) 32.36.091-9; Telex 215160 OXEA GR.
Helm: *Germany (FRG):* Karl O. Helm, HAMBURG.
Hema: *Netherlands:* Hema B.V., Frankemaheerd 2, 1102 AN AMSTERDAM. Tel. 020-5904323.
Henkel: *Austria:* Henkel Austria Ges. m.b.H., Erdbergstrasse 29, A-1031 WIEN. Tel. 0 22 2/75 04-0; Telex 133101.
- *Denmark:* Skandinavisk Henkel A/S, VALBY.
- *Netherlands:* Henkel Nederland B.V., Brugwal 11, 3432 NZ NIEUWEGEIN. Tel. 03402-73911.

Hentschke & Sawatzki: *Germany (FRG):* Hentschke und Sawatzki, Kampstrasse 85, 2350 NEUMUENSTER-GADELAND. Tel. 0 43 21/78 11.
Hercon: *USA:* Hercon Laboratories Corp., 200 B Corporate Court, Midlesex Business Center, SOUTH PLAINFIELD, NJ 07080.
Hermoo: *Belgium:* Hermoo Belgium, Zepperenweg, 99, 3800 SINT-TRUIDEN. Tel. 011/67 55 34.
Heydt Bona: *Spain:* Juan Luis Heydt Bona, Av Pio XII, 35, 28016 MADRID. Tel. 91 250 55 32.
Hickson & Welch: *UK:* Hickson & Welch Ltd., Ings Lane, CASTLEFORD, W. Yorks WF 10 2JT. Tel. (0977) 556565.
Hico: *India:* Hico Products Ltd., P.B. 16483, 771 Pt. Satavlekar Marg., Mahim, BOMBAY 400 016. Tel. 45 72 31. Telex 011 71032.
Hightex: *Spain:* Hightex SA, Capuchinos 60, Igualada, 08700 BARCELONA.
Hindustan Insecticides: *India:* Hindustan Insecticides Ltd., Hans Bhawan, Wing 1, Bahadur Shah Zafar Marg., NEW DELHI 110002. Tel. 331 2397 or 331 9472. Telex 31 65928.
Hinsberg: *Germany (FRG):* Otto Hinsberg GmbH, Mainzer Strasse 130, 6506 NACKENHEIM/RHEIN. Tel. 0 61 35/23 15.
Hodogaya: *Japan:* Hodogaya Chemical Co. Ltd., 4-2 Toranomon 1-Chome, Minato-ku, TOKYO 105. Telex 0222 2237 HDCHEM J.
Hoechst: *Austria:* Hoechst Austria AG, Altmannsdorfer Strasse 104, A-1121 WIEN. Tel. 0 22 2/85 05; Telex 133701.
- *Belgium:* Hoechst Belgium, Chaussee de Charleroi, 111-113, 1060 BRUXELLES. Tel. 02/536 41 11.
- *Denmark:* Hoechst Danmark A/S, Islevdalvej 110, 2610 ROEDOVRE. Tel. (02) 91 26 22.
- *Finland:* Oy Hoechst Fennica Ab, PL 237, 00101 HELSINKI 10. Tel. 90 820 022.
- *Germany (FRG):* Hoechst Landwirtschaftliche Entwicklungsabteilung, Hessendamm 1-3, 6234 HATTERSHEIM 1. Tel. 0 61 90/8 03-1.

- *Germany (FRG):* Hoechst Aktiengesellschaft Verkauf Landwirtschaft, Postfach 80 03 20, 6320 FRANKFURT/MAIN 80. Tel. 0 69/3 05-1.
- *Greece:* Hoechst Hellas AG, Tatoiou St., GR-102 40 N. Erythraea, ATHENS. Tel. (01) 80.10.811-19; Telex 215890.
- *Hungary:* Hoechst AG Kepviselet, Bartfai u. 54. BUDAPEST XI.
- *Netherlands:* Hoechst Holland B.V., Hogehilweg 21, 1101 CB AMSTERDAM (Z.O.). Tel. 020-5908911.
- *Norway:* Norske Hoechst A/S, Postboks 177, Oekern, 0509 OSLO 5.
- *Poland:* Hoechst Intraco Center, ul. Stawki 2, 00-950 WARSZAWA, P.O. Box 15. Tel. 39 75 76, 39 75 62, 39 97 18; Telex 812298.
- *Portugal:* Hoechst Portuguesa S.A.R.L., Apartado 6, 2726 MEM MARTINS Codex. Tel. 291 21 60.
- *Spain:* Hoechst Iberica, SA, Travesera de Gracia, 47-49, 08021 BARCELONA. Tel. 93 209 31 11.
- *Sweden:* Svenska Hoechst AB, Box 502 31, 202 12 MALMOE. Tel. 040/93 53 50.
- *UK:* Hoechst Ltd., Agriculture Division, East Winch Hall, East Winch, KING'S LYNN, Norfolk PE32 1HN. Tel. (0553) 841581.
- *USA:* Hoechst-Roussel Agri-Vet Co., Rt. 202-206, NORTH SOMERVILLE, NJ 08876. Tel. (201) 231-2908.

Hoek's: *Netherlands:* NV WA Hoek's Machine en Zuurstoffabriek, Havenstraat 1, 3115 HC SCHIEDAM. Tel. 010-731122.

Hokko: *Japan:* Hokko Chemical Industry Co. Ltd., Mitsui Building No. 2, 4-4-20, Nihonbashi Hongoku-cho, Chuo-ku, TOKYO 103.

Hokochemie: *Switzerland:* Hokochemie AG, Postfach 230, 4900 LANGENTHAL. Tel. 063 43 18 43.

Holler: *Germany (FRG):* Carl Holler, Kalkofenstrasse 52, 6602 SAARBRUECKEN-DUDWEILER.

Hooker: *USA:* Hooker Chemicals and Plastics Corp., Industrial Chemicals Group, 360 Rainbow Boulevard, SOUTH NIAGARA FALLS, NY 14303. Tel. (716) 286-3000.

Hopkins: *USA:* Hopkins Agricultural Chemical Co., P.O. Box 7532, MADISON, WI 53707. Tel. (608) 221-6200.

HORA: *Germany (FRG):* HORA Landwirtschaftliche Betriebsmittel GmbH, Liebigstrasse 51-53, Postfach 11 03 53, 6000 FRANKFURT/MAIN. Tel. 0 69/71 55-1.

Hortichem: *UK:* Hortichem, 14 Edison Road, Churchfields Industrial Estate, SALISBURY, Wiltshire SP2 7NU. Tel. (0722) 20133.

Huebecker: *Germany (FRG):* Paul Huebecker, Rosenstr. 77, 4154 TOENISVORST 1. Tel. 0 21 51/79 00 33-34.

Huhtamaki: *Finland:* Huhtamaki Oy, Laaketehdas Leiras, Kasvinsuojeluosasto, PL 415, 20101 TURKU 10. Tel. 921 623 111.

Hui Kwang: *Taiwan:* Hui Kwang Chemical Co. Ltd. P.O. Box 33, 17-10 Ling Tzyy Lin, Matou 22109, TAINAN HSIEN.

Hunt: *UK:* Octavius Hunt Ltd., 5, Dove Lane, Redfield, BRISTOL BS5 9NQ. Tel. (0272) 556107.

Hygaea: *Denmark:* A/S Hygaea, AALBORG.

Hygan: *Germany (FRG):* Hygan Chemie & Service GmbH & Co KG, Justus-von-Liebig-Strasse 3, 2200 ELMSHORN. Tel. 0 41 21/8 10 37-8.

Iberica: *Spain:* Quimica Iberica, SA, Plaza Marques de Salamanca, 11, 28006 MADRID. Tel. 91 435 43 60.

ICI: *Australia:* ICI Australia Ltd., ICI House, 1 Nicholson Street, P.O. Box 4311, MELBOURNE, Victoria 3001.

- *Austria:* ICI Oesterreich Ges. m.b.H., Schwarzenbergplatz 7, A-1030 WIEN. Tel. 0 22 2/72 66 16; Telex 131446.
- *Belgium:* ICI Belgium, Everslaan, 45, 3078 EVERBERG. Tel. 02/758 92 11.
- *Denmark:* I.C.I. Danmark A/S, Islands Brygge 41, 2300 KOEBENHAVN S. Tel. (01) 57 62 64.

- *Germany (FRG):* Deutsche ICI GmbH, Agrar-Abteilung, Lyoner Strasse 36, Postfach 71 02 55 6000 FRANKFURT/MAIN 71. Tel. 0 69/6 60 01.
- *Greece:* ICI Hellas SA, 183 Syngrou St., GR-171 21 N. Smyrni, ATHENS. Tel. (01) 93.58.368; Telex 215922 ICI.
- *Hungary:* ICI Iroda, Rozsa F.U. 72. BUDAPEST VI.
- *Italy:* ICI - Solplant, Via S. Sofia 21, 20122 MILANO. Tel. (02) 546.68.51.
- *Netherlands:* ICI Holland B.V., Wijnhaven 107, Postbus 551, 3000 AN ROTTERDAM. Tel. 010-4171911.
- *Poland:* ICI, ul. Nieklanska 23, 03-924 WARSZAWA, P.O. Box 73. Tel. 17 62 42, 17 06 58; Telex 815362.
- *Portugal:* ICI-Valagro, S.A.R.L., Av. D. Carlos I, 42-3., 1200 LISBOA. Tel. 67 01 74/75/76.
- *UK:* ICI Garden Products, Woolmead House East, Woolmead Walk, FARNHAM, Surrey, GU9 7UB. Tel. 0252 724525; Telex 858270.
- *UK:* ICI Agrochemicals, Imperial Chemical Industries PLC, Fernhurst, HASLEMERE, Surrey GU 27 3JE. Tel. 0428 4061.
- *UK:* ICI Professional Products, Imperial Chemical Industries PLC, Fernhurst, HASLEMERE, Surrey GU27 3JE. Tel. 0428 4061.
- *USA:* ICI Americas Inc., Agricultural Chemicals Division, WILMINGTON, DE 19897. Tel. (302) 575-3000.

Imex-Hulst: *Netherlands:* Imex-Hulst B.V., Zoutestraat 109, 4561 TB HULST. Tel. 01140-14853.

Inagra: *Spain:* Investigaciones Agricolas SA, c/ Salvador Giner, 14, 46003 VALENCIA. Tel. 96-331 95 01, 96-331 86 08; Telex 64469.

Inca: *Spain:* Inca, Islas Canarias, SA, c/ Anselmo Benitez, 5, SANTA CRUZ DE TENERIFE (Tenerife).

Industrialchimica: *Italy:* Industrialchimica, Via Lion 9, 35020 MASERA (PD).

Ingenieria: *Mexico:* Ingenieria Industrial, S.A. de C.V., Av. Coyoacan No. 1878-403, COL DEL VALLE 03100.

Ingvars: *Sweden:* Kjell Ingvars Foersaeljinings AB, Skraevlingevaegen S6 B' 212 31 MALMOE. Tel. 040/495520.

Intec: *Netherlands:* Intec B.V., Laan Der Techniek 22, 3903 AT VEENENDAAL. Tel. 08385-61933.

Interfertil: *Switzerland:* Interfertil SA, Bd. de la Foret 47, 1012 PULLY VD. Tel. 021 28 44 78.

Interphyto: *France:* Interphyto, 14, route de Montesson, 78110 LE VESINET. Tel. (1) 30 71 45 78; Telex 695 286 (INPHYTO).

Intolas: *Netherlands:* Intolas, Rijksstraatweg 78, 7383 AT VOORST. Tel. 05758-544.

Intrachem: *Greece:* Intrachem Hellas Ltd., 31 Kephisias Ave., GR-115 23 ATHENS. Tel. (01) 64.60.700/800; Telex 216152.

Intracrop: *UK:* Intracrop Ltd., The Crop Centre, Waterstock, OXFORD OX9 1LJ. Tel. (08447) 377/267.

Intradal: *Netherlands:* Intradal Nederland B.V., Brabantsestraat 17, 3812 PJ AMERSFOORT. Tel. 033-618822.

IPO-Sarzyna: *Poland:* IPO Warszawa, Zaklady Doswiadczalny Organika-Sarzyna, 37-310 NOWA SARZYNA. Tel. Rzeszow 340 47; Telex 814844.

Ishihara: *Japan:* Ishihara Sangyo Kaisha Ltd., 10-30 Fujimi 2-chome, Chiyoda-ku, TOKYO 102. Tel. 03 230 8650. Telex 2324306 ISK J.

Ital-Agro: *Italy:* Ital-Agro, Via Cravero 110, 10095 GRUGLIASCO (To). Tel. (011) 78.18.75.

Janssen: *Greece:* Janssen Pharmaceutica AEBE, 282 Kifissias Ave., GR-152 32 Halandri, ATHENS.

- *Netherlands:* Janssen Pharmaceutica N.V., Turnhoutseweg 30, 2340 BB BEERSE,. Tel. 014-602965.
- *USA:* Janssen Pharmaceutica, Box 344, Bear Tavern Road, WASHINGTON CROSSING, NJ 08560. Tel. (609) 737-3700.

Jewnin-Joffe: *Israel:* Jewnin-Joffe Chemicals Ltd., P.O. Box 29511, TEL AVIV 61294. Tel. 03 650034. Telex 37 1453 RIMI IL.

Jin Hung: *Korea:* Jin Hung Fine Chemicals Co. Ltd., 543-6 Kajwa-Dong, Buk-ku, INCHEON.

JSB: *France:* JSB (John et Stephen B.) 38, avenue Hoche, 75008 PARIS. Tel. (1) 42 89 14 34; Telex 642 469 AGRIJET.

Kaken: *Japan:* Kaken Pharmaceutical Co. Ltd., 4-10 Nihonbashi-Honcho, 3-Chome, Chuo-ku, TOKYO 103. Tel. 03 270 4351. Telex 2222353 KAKNP J.

Kalo: *USA:* Kalo Inc., 4550 West 109 Street, Suite 222, OVERLAND PARK, KS 66211.

Kanesho: *Japan:* Kanesho Co. Ltd., Room 33, Marunouchi Building, Maunouchi, Chiyoda-ku, TOKYO.

Katzilakis: *Greece:* A. Katzilakis, 46 Cornarou Sq., GR-712 01 IRAKLIO, CRETE.

KemaNord: *Sweden:* KemaNord AB, Box 111 20, 100 61 STOCKHOLM. Tel. 08/743 40 00.

Kemi-Intressen: *Sweden:* Kemi-Intressen AB, Box 6018, 172 06 SUNDBYBERG. Tel. 08/98 12 25.

Kemichrom: *Spain:* Kemichrom, SA, c/ Caracas, 15, BARCELONA-30.

Kemira: *Finland:* Kemira Oy, PL 330, 00101 HELSINKI 10. Tel. 90 13 211.
- *Sweden:* Kemira Svenska AB, Grynbodgatan 3 2111 33, MALMOE. Tel. 040/119900.

KenoGard: *Denmark:* KenoGard ApS, Smedeland 26, 2600 GLOSTRUP. Tel. (02) 45 11 66.
- *Norway:* KenoGard A/S, Verpet Industrioraade, 1540 VESTBY.
- *Spain:* KenoGard, SA, Diputacion 279, 08007 BARCELONA. Tel. 93 301 61 12.
- *Sweden:* KenoGard AB, Box 110 33, 100 61 STOCKHOLM. Tel. 08/743 40 00.

Kerr-McGee: *USA:* Kerr-McGee Chemical Corp., Kerr-McGee Center, OKLAHOMA CITY, OK 73125. Tel. (405) 270-1313. Telex 747 128.

Key: *Spain:* Industrial Quimica Key, SA, Av de Barcelona, km. 511,600 TARREGA (Lerida).

Killgerm: *UK:* Killgerm Chemicals Ltd., PO Box 2, Wakefield Road, Flushdyke, OSSETT, West Yorkshire WF5 9BW.

Kiltin: *Denmark:* Kiltin, KOEBENHAVN NV.

Kincaid: *USA:* Kincaid Enterprises Inc., P.O. Box 671, NITRO, WV 25143. Tel. (304) 755-3377.

Kinkel & Gebel: *Germany (FRG):* Kinkel & Gebel, Am Dornbusch 4, 3167 BURGDORF/HAN. Tel. 0 51 36/12 04.

Kirk: *Denmark:* Kirk Chemicals A/S, Skelstedet 16, Troeroed, 2950 VEDBAEK. Tel. (02) 89 12 33.

Klasmann: *Netherlands:* Klasmann Benelux B.V., Nieuwemeerdijk 87, 1171 NE BADHOEVEDORP. Tel. 02968-7256.

Klepper: *Germany (FRG):* Rudolf Klepper, Chemische Fabrik, Postfach 1 48, 3508 MELSUNGEN. Tel. 0 56 61/22 39.

Kobanyai: *Hungary:* Kobanyai Gyogyszerarugyar, Gyomroi ut 19-21, BUDAPEST X.

Koch & Reis: *Belgium:* Koch et Reis, Ijzerlaan, 3, 2000 ANTWERPEN. Tel. 03/233 87 21.

Kocide: *USA:* Kocide Chemical Corp., 12701 Almeda Road, HOUSTON, TX 77045.

Kommer: *Netherlands:* P. Kommer B.V., Industrieterrein Oosterzij 14, 1851 NT HEILOO. Tel. 072-332836.

Koppert: *Netherlands:* Koppert B.V., Veilingweg 8A, 2651 BE BERKEL EN RODENRIJS. Tel. 01891-4044.

Korleti: *Greece:* I. Korleti, Beroias 10, THESSALONIKI.

Kornitol: *Germany (FRG):* Kornitol Chemie-Plastik Produktions- und Vertriebsgesellschaft mbH, 8100 GARMISCH-PARTENKIRCHEN. Tel. 0 88 21/27 00.

Kortman: *Netherlands:* Kortman Nederland B.V., Laan der Techniek 22, 3903 AT VEENENDAAL. Tel. 08385-61911.

Kraft: *Germany (FRG):* Wolfgang Kraft, Garten Kraft, Frielick 8-13, 4700 HAMM-HEESEN. Tel. 0 23 81/6 02 73.

Krishi Rasayan: *India:* Krishi Rasayan (Bihar), World Trade Centre, 14/1B Ezra Street, CALCUTTA 700 001. Tel. 26 9727/26-7710. Telex GITA IN.

Kumiai: *Japan:* Kumiai Chemical Industry Co. Ltd., 4-26 Ikenohata 1-chome, Taitoh-ku, TOKYO 110. Tel. 03 823 1701. Telex 265 5526 KUMIKA J.

Kureha: *Japan:* Kureha Chemical Industry Co. Ltd., 1-9-11 Nihonbashi Horidome-cho, Chuo-ku, TOKYO 103. Tel. 03 662 9611. Telex 2522737 KUREHA J.

KVK: *Denmark:* Kemisk Vaerk Koege A/S, Gl. Lyngvej 2, 4600 KOEGE. Tel. (03) 65 75 85.
– *Sweden:* Kemisk Verk Koege AB, Karbingatan 15, 252 55 HELSINGBORG. Tel. 042/16 20 90.

Kwizda: *Austria:* F. Joh. Kwizda Ges. m.b.H., Dr.-Karl-Lueger-Ring 6, A-1011 WIEN. Tel. 0 22 2/63 46 01; Telex 11-2294.

L.A.P.A.: *France:* L.A.P.A. (Laboratoire d'Achat pour l'Agriculture), 08450 RAUCOURT-OMICOURT. Tel. 24 36 40 01.

L.C.B.: *France:* L.C.B. (Laboratoire de Chimie et de Biologie), La Salle, 71260 LUGNY. Tel. 85 37 50 10; Telex 800 774.

L.F.P.C.: *France:* L.F.P.C. (Laboratoire Francais de Produits Chimiques), 40, avenue de la Republique, 06300 NICE. Tel. 93 89 58 18 & 93 30 12 67.

La Cornubia: *France:* La Cornubia, 85, quai de Brazza - B.P. 55, 33016 BORDEAUX CEDEX. Tel. 56 86 45 00; Telex 560 961 CORNBIA.

La Quinoleine: *France:* La Quinoleine et ses Derives, 43, rue de Liege, 75008 PARIS. Tel. (1) 45 22 17 49; Telex Paris 660 166 F.

Lainco: *Spain:* Lainco, SA, Av Bizet, 8-12. Apartado 73, 08191 RUBI (Barcelona). Tel. 93 699 17 00.

Lambert: *Belgium:* Lambert G. S.P.R.L., Champ de Pihot, 3, 4510 SAIVE. Tel. 041/62 74 56.
– *France:* Lambert et Fils, 306, Chemin des 4-Chemins, 06600 ANTIBES. Tel. 93 33 38 50; Telex 462 128.

Land-Forst: *Austria:* Land-Forst Betriebsmittel-Ges. m.b.H., Postfach 231, Oppolzergasse 4, A-1011 WIEN. Tel. 0 22 2/63 46 01; Telex 112294.

Lantmaennen: *Sweden:* Svenska Lantmaennens Riksfoerbund, Box 122 38, 102 26 STOCKHOLM. Tel. 08/13 15 00.

Lapapharm: *Greece:* Lapapharm, 73 Menandrou St., GR-104 37 ATHENS. Tel. (01) 52.46.011-15; Telex 215158 LACA.

Lassen & Wedel: *Denmark:* Lassen & Wedel A/S, HUMLEBAEK.

Lauff: *Germany (FRG):* J. M. Lauff Chemische Produkte GmbH, Hertzstrasse 2a, 5000 KOELN 40 (LOEVENICH). Tel. 02 21/7 71 00.

Ledona: *Switzerland:* Ledona AG, Zentralstrasse 30, 6030 EBIKON. Tel. 041 33 10 01.

Lefki: *Greece:* P. Lefki, 8 Bouboulinas St., GR 106 82 ATHENS.

Leu & Gygax: *Switzerland:* Leu & Gygax AG, Agrarhilfsstoffe, Fellstrasse 1, 5413 BIRMENSTORF AG. Tel. 056 85 15 15.

Leuna-Werke: *Germany (GDR):* VEB Leuna-Werke 'Walter Ulbricht', LEUNA.

Leyffer & Nellen: *Germany (FRG):* Leyffer & Nellen, Holzmuehlenstrasse 12, 2000 HAMBURG 70. Tel. 0 40/6 56 00 61.

LHN: *France:* LHN (Laboratoires Hygiena), 10, rue A. Becquerel - B.P. 107, 21302 CHENOVE CEDEX. Tel. 80 52 17 41; Telex 350 447.

Ligtermoet: *Netherlands:* Ligtermoet Chemie B.V., Stepvelden 10, Postbus 1048, 4700 BA ROOSENDAAL. Tel. 01650-32912.

Lilly Farma: *Portugal:* Lilly Farma, Produtos Farmaceuticos, Lda., Rua Eugenio Castro Rodrigues, 9A, 1700 LISBOA. Tel. 52 53 02/20/32.

Lipha: *France:* Lipha, 34, rue Saint-Romain - B.P. 8481, 69359 LYON CEDEX 08. Tel. 78 75 44 24; Telex LIPHA LY 340 495 F.
– *Iceland:* Lipha Lyon, Frakklandi, REYKJAVIK.

Liro: *Belgium:* Liro Belgium, Steenweg naar Asse, 34, 1702 ASSE. Tel. 02/452 87 96.

Lobel: *USA:* Lobel Chemical Corp., 100 Church Street, NEW YORK, NY 10007. Tel. (212) 267-4265.

Lodi: *France:* Lodi, 12, rue de Rouen, 95450 LE BORD'HAUT DE VIGNY. Tel. (1) 30 39 28 44; Telex LODI 697 402 F.

Logissain: *France:* Logissain (Laboratories), Z.I. Argeisans, 90800 BAVILLIERS. Tel. 84 21 15 27.

Lonza: *Switzerland:* Lonza AG, Muenchensteinerstrasse 38, 4002 BASEL. Tel. 061 50 88 50.

- *USA:* Lonza Inc. 22-10 Route 208, FAIRLAWN, NJ 07410. Tel. (201) 794-2400.

Lonza-Werke: *Germany (FRG):* Lonza-Werke GmbH, Postfach 2 49, 7890 WALDSHUT-TIENGEN 1. Tel. 0 77 51/8 23 73.

Los Angeles: *USA:* Los Angeles Chemical Co., 4545 Ardine Street, SOUTH GATE, CA 90280. Tel. (213) 583-4761.

Luqsa: *Spain:* Lerida Union Quimica, SA (Luqsa), c/ Afueras, s/n, 25173 SUDANELL (Lerida). Tel. 973 72 02 56.

Luxan: *Belgium:* Luxan Belgie, Wondelgemkaai, 4, 9000 GENT. Tel. 091/53 11 32.

- *Netherlands:* Luxan B.V., Chem. Pharmaceutische Industrie, Industrieweg 2, Postbus 9, 6660 AA ELST (Gld.),. Tel. 08819-72144.

Maag: *Switzerland:* Dr. R. Maag AG, Chemische Fabrik, 8157 DIELSDORF. Tel. 01 853 12 55.

- *USA:* Maag Agrochemicals Inc., 5699 N Kings Highway, P.O. Box 6430, VERO BEACH, FL 32961. Tel. (305) 567-7506.

Maasmond: *Netherlands:* Cooperatieve Land- en Tuinbouwvereniging Maasmond G.A., Jan van de Houtweg 8, Postbus 7, 2678 ZG DE LIER. Tel. 01745-13591.

Maataloupalvelu: *Finland:* Maatalouspalvelu Oy, Lapinlahdenkatu 21 B 5 krs, 00180 HELSINKI 10. Tel. 90 694 7633.

Mahle Duenger: *Germany (FRG):* Mahle Duenger GmbH, Salzstrasse 178-180, 7100 HEILBRONN. Tel. 0 71 31/1 08 68.

Makhteshim: *Hungary:* Makhteshim Chemical Works Kepviselet, Rajk L. u. 11. BUDAPEST XIII.

Makhteshim-Agan: *Israel:* Makhteshim-Agan, P.O. Box 60, BEER-SHEVA 84100. Tel. 057656; Telex 5276 5312.

Maldoy: *Belgium:* Maldoy-Donck, Mechelsesteenweg, 5, 2580 SINT-KATELIJNE-WAVER. Tel. 015/31 16 06.

Mallinckrodt: *USA:* Mallinckrodt Inc., P.O. Box 5439, ST. LOUIS, MO 63147. Tel. (314) 982-5043. Telex 216 496.

Mandops: *UK:* Mandops (Agrochemical Specialists) Ltd., Tower Industrial Estate, Chickenhall Lane, EASTLEIGH, Hants. SO5 5NZ. Tel. (0703) 641826.

Manica: *Italy:* Manica, Via all'Adige, 4, Borgosacco, 38068 ROVERETO (Tn). Tel. (0464) 3.37.05/3.44.10.

Mansson: *Sweden:* Lennart Mansson Bekaempningsmedel, Box 700, 251 07 HELSINGBORG. Tel. 042/12 00 05.

Marba: *Spain:* Insecticidas Marba, SL, Dr Vila Barbera, 11, 46007 VALENCIA. Tel. 96 341 14 73.

Margesin: *Italy:* Margesin, Via S. Floriano 3, 39011 LANA D'ADIGE (Bz). Tel. (0473) 5.11.71.

Marks: *UK:* A. H. Marks & Co. Ltd., Wyke Lane, Wyke, BRADFORD, W. Yorks. BD12 9EJ. Tel. (0274) 675231.

Marktredwitz: *Germany (FRG):* Chem. Fabrik Marktredwitz AG, Postfach 1 65, 8590 MARKTREDWITZ/Bay. Tel. 0 92 31/40 04-05.

Marty et Parazols: *France:* Marty et Parazols, 50, avenue Pierre Semard, 11012 CARCASSONNE. Tel. 68 47 84 72; Telex 500 442 CALIMOP.

Masso: *Spain:* Comercial Quimica Masso, Viladomat, 321, 5, 08029 BARCELONA. Tel. 93 321 83 00.

May & Baker: : see Rhone-Poulenc.

MCK Diffusion: *France:* MCK Diffusion, Les Essarts cidex 481, 69400 POUILLY-LE-MONIAL. Tel. 74 03 81 23.

McLaughlin Gormley King: *USA:* McLaughlin Gormley King Co., 8810 Tenth Avenue, NORTH MINNEAPOLIS, MN 55427.

MCU: *USSR:* Ministerium fuer chemische Industrie der UdSSR MOSKAU.

Medley: *Sweden:* Snada Medley Jr., Katarina Bangata 71, 116 39 STOCKHOLM. Tel. 08/402029.

Megafarm: *Greece:* Megafarm, 13 Mitropoleos St., GR-546 24 THESSALONIKI. Tel. (031) 276.414; Telex 412072.

Meiji Seika Kaisha: *Japan:* Meiji Seika Kaisha Ltd., 4-16, Kyobashi 2-chome, Chuo-ku, TOKYO 104.

Meoc: *Switzerland:* Meoc S.A., Manufacture d'engrais organiques, 1906 CHARRAT VS. Tel. 026 5 36 39.

Merck Sharp & Dohme: *Denmark:* Merck Sharp & Dohme, Marielundvej 46 C, 2730 HERLEV. Tel. (02) 91 77 66.
- *Germany (FRG):* Therapogen-Werk Zweigniederlassung der MSD Sharp & Dohme GmbH, Abt. Pflanzenschutz, Toelzer Str. 1, 8022 GRUENWALD. Tel. 0 89/64 18 71.
- *Netherlands:* Merck Sharp & Dohme B.V., Waarderweg 39, Postbus 581, 2031 BN HAARLEM. Tel. 023-319330.
- *Portugal:* Merck Sharp & Dohme, Lda, Rua Barata Salgueiro, 37, 1., 1200 LISBOA. Tel. 54 34 89.
- *Sweden:* Merck Sharp & Dohme (Sweden) AB, 161 20 BROMMA. Tel. 08/98 07 80.

Meriel: *France:* Meriel (Laboratoires), 32, route de Soissons - B.P. 3, 02370 VAILLY-SUR-AISNE. Tel. 23 54 70 77; Telex 150 724 F.

Meristem: *Spain:* Quimicas Meristem, SL, Ctra Moncada-Naquera Km, 1'700, Apartado 30, 46113 MONCADA (Valencia). Tel. 96 139 45 11.

Meyer: *Germany (FRG):* Lucas Meyer, Ausschlaeger Elbdeich 62, 2000 HAMBURG 28. Tel. 0 40/78 17 01.

Microbial Resources: *USA:* Microbial Resources Inc., 1303-2 Cynwyd Club Drive, WILMINGTON, DE 19808. Tel. (302) 998-2320.

Midkem: *UK:* Midkem Ltd., 20 Rothersthorpe Avenue NORTHAMPTON NN4 9JH. Tel. (0604) 64027.

Midol: *Denmark:* A/S Midol, Industrivangen 12, 2635 ISHOEJ. Tel. (02) 73 16 77.

Migros: *Switzerland:* Migros-Genossenschaftsbund, Marketing Food-Agrarprodukte, Blumen und Pflanzen, Limmatstrasse 152, 8031 ZUERICH. Tel. 01 44 44 11.

Mikasa: *Japan:* Mikasa Chemical Ind. Co. Ltd., 9-1, 4-chome, Tenjin, FUKUOKA 810.

Miller: *USA:* Miller Chemical & Fertilizer Corp., P.O. Box 333, HANOVER, PA 17331. Tel. (717) 632-8921; Telex 840448 MILLER CH HNVR.

Mirfield: *UK:* Mirfield Sales Services Ltd., Moorend House, Moorend Lane, DEWSBURY, W. Yorks. WF13 4QQ. Tel. (0924) 409782 or 408571.

Mitchell Cotts: *UK:* Mitchell Cotts Chemicals Ltd., P.O. Box No. 6, Stearnard Lane, MIRFIELD, W. Yorks. WF14 8QB. Tel. (0924) 493861-6.

Mitsubishi: *Japan:* Mitsubishi Chemical Industries Ltd., 5-2, Marunouchi 2-chome, Chiyoda-ku, TOKYO 100. Tel. 03 283 6995. Telex 24901.

Mitsui: *Japan:* Mitsui & Co. Ltd., Fine Chemicals Division, 1-2-1 Ohtemachi, Chiyoda-ku, TOKYO.

Mitsui Toatsu: *Japan:* Mitsui Toatsu Chemicals Inc., 2-5, kasumigaseki 3-chome, Chiyoda-ku, TOKYO 100.

Mobay: *USA:* Mobay Chemical Corp., Agricultural Chemicals Division, P.O. Box 4913, 8400 Hawthorn Road, KANSAS CITY, MO 64120. Tel. (816) 242-2000. Telex 426116.

Mobil: *USA:* Mobil Chemical Co., P.O. Box 26683, RICHMOND, VA 23261.

Mommersteeg: *Netherlands:* Mommersteeg International B.V., Wolput 72, Postbus 1, 5251 AA VLIJMEN. Tel. 04108-9116.

Monsanto: *Austria:* Monsanto Ges. m.b.H., Am Stadtpark (Hilton Center), A-1030 WIEN. Tel. 0 22 2/72 46 24; Telex 135601.
- *Belgium:* Monsanto Europe S.A., Avenue de Tervuren, 270-272, 1150 BRUXELLES. Tel. 02/762 11 12.
- *Denmark:* Monsanto A/S, Smedeland 6, 2600, GLOSTRUP. Tel. (02) 45 77 99.
- *Finland:* Monsanto Oy, Vainamoisenkatu 19 A 6, 00100 HELSINKI 10. Tel. 90 409 122.
- *France:* Monsanto S.A. (Division Agriculture), 4, allee de Lausanne - B.P. 52, St-Quentin-Fallavier, 38290 VA VERPILLIERE. Tel. 74 95 55 55; Telex MOFAGRI 900 494.
- *Germany (FRG):* Monsanto (Deutschland) GmbH, Immermannstrasse 3-5, 4000 DUESSELDORF. Tel. 02 11/3 67 51.

- *Italy:* Monsanto Italiana, Via Melchiorre Gioia 8, 20124 MILANO. Tel. (02) 62.73.
- *Netherlands:* Monsanto B.V., Jan van Nassaulaan 53-55, 2596 BP 'S-GRAVENHAGE. Tel. 070-245015.
- *Poland:* Monsanto, Intraco Center, ul. Stawki 2, 00-950 WARSZAWA, P.O. Box 344. Tel. 39 81 62; Telex 812747.
- *Portugal:* Sociedade Portuguesa de Desenvolvimento Quimico de Monsanto, Lda., Rua D. Joao V, 25-1. Esq, 1200 LISBOA. Tel. 65 11 31/65 67 65.
- *Spain:* Monsanto Espana, SA, c/ Orense, 4; 9 B, 28020 MADRID. Tel. 91 456 62 50.
- *Sweden:* Monsanto Scandinavia AB, Norra Neptunigatan 3, 211 20 MALMOE. Tel. 040/736 25.
- *Switzerland:* Monsanto S.A., Tiefenhoehe 10, 8022 ZUERICH. Tel. 01 211 54 07.
- *UK:* Monsanto PLC, Agricultural Division, Thames Tower, Burleys Way, LEICESTER LE1 3TP. Tel. (0533) 20864.
- *USA:* Monsanto Agricultural Products Co., 800 N. Lindbergh Boulevard, ST.LOUIS, MO 63167. Tel. (314) 694-1000.

Montedison: *Austria:* Montedison Handelsgesellschaft m.b.H., Karlsplatz 1, A-1010 WIEN. Tel. 0 22 2/65 46 51-55; Telex 132260, 133695.

- *Belgium:* Montedison Belgio S.A., Avenue Louise 326, Bte 500, 1050 BRUXELLES. Tel. 02/649 60 70.
- *France:* Montedison France, Tour Franklin - Cedex 11, 92081 PARIS-LA DEFENSE. Tel. (1) 47 76 41 16; Telex 620 232 MONTED & 620 574 EDIMONT.
- *Germany (FRG):* Montedison Deutschland GmbH, Postfach 56 26, Frankfurter Strasse 33-35, 6236 ESCHBORN BEI FRANKFURT. Tel. 0 61 96/49 21.
- *Greece:* Montedison Hellas SA, 46 Thiseos St., GR-176 76 Kallithea, ATHENS. Tel. (01) 95.22.091-7; Telex 215212 MONT.
- *Hungary:* Montedison Kepviselet, Vaci u. 19-21, BUDAPEST V.
- *Spain:* Montedison Iberica, SA, Plaza de Espana, 18. 28008 MADRID. Tel. 91 247 64 06.
- *Switzerland:* Montedison (Suisse) SA, St. Jakobstr. 25, 4000 BASEL. Tel. 061 22 54 77.
- *UK:* Montedison UK Ltd., Agrochemicals Dept. P.O. Box 8, High Street, NEWMARKET, Suffolk CB8 6JY. Tel. (0638) 660011.

Moreels-Guano: *Belgium:* Moreels-Guano, Koopvaardijlaan, 180-182, 9000 GENT. Tel. 091/51 13 55.

Morera: *Spain:* Jose Morera, SL, Plaza de Almansa, 1, 46001 VALENCIA. Tel. 96 331 59 44.

Mormino: *Italy:* P. Mormino e Figlio Mormino, Via Lungomolo 16, 90018 TERMINI IMERESE (Pa). Tel. (091) 8141004 & 8141512.

Mortalin: *Denmark:* Mortalin Brabyvej 74-76 4690 HASLEV.

Motomco: *USA:* Motomco Ltd., 29 N Fort Harrison Avenue, CLEARWATER, FL 33515. Tel. (813) 447-3417. Telex 810 866 0349 RODENT CWRT.

MSD AGVET: *France:* M.S.D. Agvet (Merck, Sharp et Dohme), 3, avenue Hoche, 75008 PARIS. Tel. (1) 47 55 97 22; Telex AGVET FR 642 266 F.

- *Italy:* Divisione della Merck Sharp & Dohme (Italia) S.p.A., Via Milano 141, 20021 BARANZATE DI BULLATE (MI).
- *Spain:* MSD Agvet (Merck, Sharp and Dohme) c/ Josefa Valcarcel, 38, 28027 MADRID. Tel. 91 742 60 12.
- *Switzerland:* MSD AGVET AG, Chamerstr. 67, 6300 ZUG. Tel. 042 21 91 88.
- *UK:* MSD AGVET, Division of Merck, Sharp & Dohme Ltd., Hertford Road, HODDESDON, Herts. EN11 9BU. Tel. (0992) 467272.
- *USA:* MSD AGVET, Division of Merck & Co. Inc., P.O. Box 2000, RAHWAY, NJ 07065. Tel. (201) 750-8605.

Mylonas: *Greece:* Sim. I. Mylonas, 20 Paparigopoulou St., GR-546 30 THESSALONIKI. Tel. (031) 528.198.

Nagel: *Austria:* Albert Nagel, Fabrikation chem. Praeparate und Pflanzenschutzmittel, Drogerie, Postfach 43, A-6973 HOECHST/Vorarlberg. Tel. 0 55 78/52 18.

National Chemsearch: *Belgium:* National Chemsearch, Verbrandebrugsteenweg 58, 1850 GRIMBERGEN. Tel. 02/251 43 32.
– *France:* Zone Industrielle B.P. 102, 77160 PROVINS. Tel. (1) 6400 1223; Telex 600 232.
Nenninger: *Germany (FRG):* Ludwig Nenninger, Bueltenweg 48, 3300 BRAUNSCHWEIG. Tel. 05 31/33 33 71.
Neudorff: *Germany (FRG):* W. Neudorff GmbH KG Chemische Fabrik, An der Muehle 3, Postfach 12 09, 3254 EMMERTHAL 1. Tel. 0 51 55/6 32 50.
Nicholas-Mepros: *Netherlands:* Nicholas-Mepros B.V., Industrieweg 1, Postbus 23, 5530 AA BLADEL. Tel. 04977-1833.
Nickerson: *UK:* Nickerson Seed Specialists Ltd., Stow Bardolph, KING'S LYNN, Norfolk PE34 3JA. Tel. (0366) 382426.
Nielsen: *Denmark:* A/S Albert Nielsen, Kem. Fabrik, ARHUS.
Nihon Nohyaku: *Japan:* Nihon Nohyaku Co. Ltd. Eitaro Building, 2-5 Nihonbashi 1-chome, Chuo-ku, TOKYO 103.
Nihon Tokushu Noyaku Seizo: *Japan:* Nihon Tokushu Noyaku Seizo K.K., 7-1, 2-chome, Nihonbashi Honcho, Chuo-ku, TOKYO 103.
Niklor: *Japan:* Niklor Chemical Co. Inc., 2060 E. 220th Street, LONG BEACH, CA 90810. Tel. (213) 830-2253.
Nippon Kayaku: *Japan:* Nippon Kayaku Co. Ltd., Tokyo Kaijo Building 2-1, Marunouchi 1-chome, Chiyoda-ku, TOKYO 100.
Nippon Soda: *Japan:* Nippon Soda Co. Ltd., Agropharm Div., 2-2-1 Ohtemachi, Chiyoda-ku, TOKYO 100. Tel. 03 245 6190.
Nissan: *Japan:* Nissan Chemical Industries Ltd., Agricultural Chemicals Division, Kowa-Hitosubashi Building, 7-1, 3-chome, Kanda-Nishiki-cho, Chiyoda-ku, TOKYO 101. tel. 03 296 8235. Telex 222 2379.
Nitor: *Sweden:* Tekniska Fabriken Nitor AB, Box 90, 763 03 SKEBOBRUK. Tel. 0175/402 80.
Nitrofarm: *Greece:* Nitrofarm, P.A. Kirgidis & Co., 13 Fragon St., THESSALONIKI. Tel. (031) 514.440; Telex 410074 NITR GR.
Nitrokemia: *Hungary:* Nitrokemia Ipartelepek FUZFOGYARTELEP.
Nitto Kasei: *Japan:* Nitto Kasei Co. Ltd., 17-14 Nishiawaji 3-chome, Higashiyodogawa-ku, OSAKA 533. Tel. 06 322 4351. Telex 523 3467 NTKOSA J.
Nobel: *Sweden:* Nobel Consumer Goods AB, Box 12080 102 22 STOCKHOLM. Tel. 08/804520.
NOC: *Netherlands:* North Sea Breeding Company B.V., Van Harenstraat 49, 9076 ZN ST ANNAPAROCHIE. Tel. 05180-2425.
NOR-AM: *USA:* NOR-AM Chemical Co., 3509 Silverside Road, P.O. Box 7495, WILMINGTON, DE 19803. Tel. (302) 575-2000.
Nordisk Alkali: *Denmark:* Nordisk Alkali Biokemi A/S, Islands Brygge 91, 2300 KOEBENHAVN S. Tel. (01) 57 61 00.
– *Sweden:* Nordisk Alkali A/S, Sturkoegatan 10, 211 24 MALMOE. Tel. 040/18 11 10.
Nordox: *Norway:* Nordox A/S, Ostensjoveien 13, 06661 OSLO 6.
Nordtend: *Finland:* Oy Nordtend Ab, PL 19, 01301 VANTAA 30. Tel. 90 833 671.
– *Sweden:* Nordtend AB, Box 32025, 126 11 STOCKHOLM.
Norsk Pyrethrum: *Norway:* A/S Norsk Pyrethrum, Aslakveien 20, 0753 OSLO 7.
Novotrade: *Denmark:* Novotrade ApS, KOEBENHAVN.
O.P.: *Austria:* Oesterreichische Pflanzenschutz- und Saatgut-Gesellschaft m.b.H., Postfach 109, Lerchenfelder Guertel 9-11, A-1164 WIEN. Tel. 0 22 2/92 61 02; Telex 134655.
Occidental: *USA:* Occidental Chemical Co., P.O. Box 1185, HOUSTON, TX 77001.
Olin: *France:* Olin Europe S.A., 108-110 Boulevard Haussmann, 75008 PARIS.
– *USA:* Olin Corp., P.O. Box 991, LITTLE ROCK, AR 72203.
Opopharma: *Switzerland:* Opopharma AG, Kirchgasse 42, 8001 ZUERICH. Tel. 01 47 65 00.
Organika-Azot: *Poland:* Zaklady Chemiczne Organika-Azot, ul. Szopena 94, 32-510 JAWORZNO. Tel. 644 42; Telex 0312621.

Organika-Fregata: *Poland:* Gdanskie Zaklady Chemiczne Organika-Fregata, ul. Grunwaldzka 497, 90-309 GDANSK-OLIWA. Tel. 52 00 27; Telex 0512177.

Organika-Rokita: *Poland:* Nadodrzanskie Zaklady Przemyslu Organicznego Organika-Rokita, ul. Sienkiewicza 4, 56-120 BRZEG DOLNY. Tel. Wroclaw 44 32 11; Telex 0715606, 0715452.

Organika-Sarzyna: *Poland:* Zaklady Chemiczne Organika-Sarzyna, 37-310 NOWA SARZYNA. Tel. Rzeszow 386 21; Telex 0632302, 0632303.

Organika-Zarow: *Poland:* Dolnoslaskie Zaklady Chemiczne Organika, 58-130 ZAROW, ul. Armii Czerwonej 59. Tel. 451; Telex 0742451.

Ortho: *Germany (FRG):* Deutsche Ortho GmbH, Marienbader Platz 20, 6380 BAD HOMBURG v.d.H. Tel. 0 61 72/2 30 47-48.

Otsuka: *Japan:* Otsuka Chemical Co. Ltd., 10 Bungo-machi, Higashi-ku, OSAKA 540.

Overlack: *Germany (FRG):* Gebr. Overlack, Chemische Fabrik, Aachener Strasse 258, 4050 MOENCHENGLADBACH. Tel. 0 21 61/3 21 71.

Oxon: *Hungary:* Oxon Kepviselet, Lajos u. 11-15, BUDAPEST III.

Pamol: *Israel:* Pamol Ltd., Luxembourg Chemicals, P.O. Box 13, TEL-AVIV 61000. Tel. 03 370566. Telex 341255.

Pan Britannica: *UK:* Pan Britannica Industries Ltd., Britannica House, WALTHAM CROSS, Herts. EN8 7DY. Tel. (0992) 23691; Telex 23957.

Parco: *Germany (FRG):* Parco GmbH Hamburg, Guetersloher Strasse 127, 4837 VERL 1. Tel. 0 52 46/21 01-3.

Parimco: *Netherlands:* Parimco B.V., Blaak 31, Postbus 182, 3000 AD ROTTERDAM. Tel. 010-4544351.

PBI Gordon: *USA:* PBI Gordon Corp., 1217 West 12th Street, KANSAS CITY, MO 64101. Tel. (816) 421-4070.

Pelton: *France:* Pelton, Service Commercial, 45200 MONTARGIS.

Penick: *USA:* Penick-Bio UCLAF Corp., 1050 Wall Street West, LYNDHURST, NJ 07071. Tel. (201) 935-9090. Telex 420 839.

Pennwalt: *Spain:* Pennwalt Iberica, SA, c/ Villa de Madrid, 50, Poligono Industrial Fuente del Jarro, 46980 PATERNA (Valencia). Tel. 96 132 08 00.

- *UK:* Pennwalt Chemicals Ltd., Dorman Road, CAMBERLEY, Surrey. Tel. (0276) 61212.
- *USA:* Pennwalt Corp., Ag. Chemical Division, 3 Parkway, PHILADELPHIA, PA 19102. Tel. (215) 587-7219. Telex 845 234.

Pennwalt France: *France:* Pennwalt France (Departement Agchem Decco), B.P. 65 - 1, rue des Freres-Lumiere, 78372 PLAISIR CEDEX. Tel. (1) 30 55 80 45; Telex 696 023.

Pennwalt Holland: *Netherlands:* Pennwalt Holland B.V., Tankhoofd 10, Postbus 7120, 3000 HC ROTTERDAM. 010-381200.

Pepinieres Du Valois: *France:* Pepinieres Du Valois, 02600 VILLERS-COTTERETS. Tel. 23 96 19 12.

Peppas: *Greece:* G. Peppas-E. Sotriiadou & Co., 1 Eolou St., GR-105 55 ATHENS.

Pepro: *France:* Pepro, B.P. 5 - 69131 ECULLY Cedex L'Oree d'Ecully - Bat. C - Chemin de la Forestiere, 69130 ECULLY. Tel. 78 33 66 77; Telex PEPRO LYON 310 001.

Perycut: *Switzerland:* Perycut-Chemie AG, Wehrenbachhalde 54, 8053 ZUERICH. Tel. 01 55 53 69.

Pestcon: *USA:* Pestcon Systems Inc., P.O. Box 469, 221 Poplar Boulevard, ALHAMBRA, CA 91802. Tel. (213) 283-2761. Telex 698635.

Pfizer: *Finland:* Pfizer Oy, PL 26, 02101 ESPOO 10. Tel. 90 460 600.

- *Greece:* Pfizer Hellas, 5 Alecetou St., GR-116 33 ATHENS. Tel. (01) 75.17.981; Telex 215024.

Phelps Dodge: *USA:* Phelps Dodge Refining Corp., 300 Park Avenue, NEW YORK, NY 10022.

Phyto: *Germany (FRG):* Phyto Research Pflanzenschutz GmbH, Im Niederfeld, 5014 KERPEN. Tel. 0 22 37/25 88.

Phytopharmaceutiki: *Greece:* Phytopharmaceutiki, 56 Arapaki St., GR-176 76 Kallithea, ATHENS. Tel. (01) 34.61.938; Telex 219970.

Phytoprotect: *Greece:* Phytoprotect, 9 Chatzigianni Mexi St., GR-115 28 ATHENS.

Phytorgan: *Greece:* Phytorgan, 4-6 Philippidou St., GR-104 34 ATHENS. Tel. (01) 88.24.180.
Phytosan: *France:* Phytosan S.E., 28, rue du Pont Hardy - B.P. 71, 77400 LAGNY. Tel. 64 30 21 07 & 64 30 68 16; Telex 600 888 F.
Pillar: *Taiwan:* Pillar International Co., P.O. Box 70-111, TAIPEI. Tel. 02 704 0001. Telex 23623 PINTER.
Plant Products: *Canada:* Plant Products Co. Ltd., BRAMALEA.
Planters: *Philippines:* Planters Products Inc., PPI Building, Esteban Street, Legaspi Village, MAKATI MANILA. Tel. 818 21 19. Telex 66408 PPI PN.
Plantevern-Kjemi: *Norway:* A/S Plantevern-Kjemi, Huggenes Gard, 1580 RYGGE.
Platecsa: *Spain:* Platecsa, Menendez Y Pelayo, 16, 1, 08012 BARCELONA. Tel. 93 207 41 34.
PLK: *Denmark:* Plantekemi Odense A/S, Ove Gjeddes Vej 16, 5220 ODENSE SOE. Tel. (09) 15 90 91.
Pluess-Staufer: *Switzerland:* Pluess-Staufer AG, 4665 OFTRINGEN. Tel. 062 43 11 11.
Polfa-Kutno: *Poland:* Kutnowskie Zaklady Farmaceutyczne Polfa, ul. Sienkiewicza 25, 99-300 KUTNO. Tel. 33 125, 33 129; Telex 83379, 83594, 83622.
Polfa-Pabianice: *Poland:* Pabianickie Zaklady Farmaceutyczne Polfa, ul. Zymierskiego 5, 95-200 PABIANICE. Tel. Pabianice 21 11, Lodz 424 89.
Polisenio: *Italy:* Polisenio, Via S. Andrea 10, 48022 LUGO (Ravenna). Tel. 24.56.0.
Portman: *UK:* Portman Agrochemicals Ltd., 67-69, George Street, LONDON W1H 5PJ. Tel. (01 486) 8114.
PPG: *USA:* PPG Industries Inc., One PPG Place, PITTSBURGH, PA 15272. Tel. (412) 434-2252.
Praestrud & Kjeldsmark: *Denmark:* Praestrud & Kjeldsmark A/S, KOEBENHAVN.
Prentiss: *USA:* Prentiss Drug and Chemical Co. Inc., 2100 Vernon Street, C.B. 2000, Floral park, NEW YORK, NY 11001. Tel. (516) 326-1919.
Probelte: *Spain:* Probelte, SA, Carretera de Madrid, km, 384'6., Apartado 579, 30100 MURCIA.
Prochimagro: *France:* Prochimagro, (Division de DOW CHEMICAL France), Route des Cretes, Parc de Sophia Antipolis B.P. 31, 06561 VALBONNE CEDEX. Tel. 93 33 91 02; Telex DOWCHEM 970 923.
Procida: *France:* Procida, Saint-Marcel 13367 MARSEILLE CEDEX 11. Tel. 91 35 90 35; Telex 410 796 PROCIDA MARSL.
- *France:* Procida, 163, avenue Gambetta 75020 PARIS. Tel. (1) 43 60 01 65; Telex AGROV 212 167 F.
- *Spain:* Procida Iberica, SA, San Rafael, 3, 28100, ALCOBENDAS (Madrid).

Productos: *Argentina:* Productos OSA, Ave. De Mayo 1161, 1085 BUENOS AIRES. Tel. 38 0645. Telex 17340 OSA AR.
Productos Life: *Spain:* Guadalquivir, 58, 46026 VALENCIA. Tel. 96 334 32 63.
Proefstation: *Netherlands:* Proefstation voor de Tuinbouw onder glas, Zuidweg 38, 2671 MN NAALDWIJK,. Tel. 01740-26541.
Progallus: *France:* Progallus, 16, avenue Foch, 75116 PARIS. Tel. (1) 45 00 28 52.
Promark: *UK:* Promark, Hoechst UK Ltd., Agriculture Division, East Winch Hall, East Winch, KING'S LYNN, Norfolk PE32 1HN. Tel. (0553) 841581.
Prometheus: *Greece:* Prometheus F.T.C., 4b Thessalonikis St., GR-182 33 Ag. I. Rentis, Piraeus, ATHENS. Tel. (01) 49.30.802; Telex 241136.
Propfe: *Germany (FRG):* Heinrich Propfe Chemische Fabrik GmbH, Duesseldorfer Strasse 9-11, Postfach 81 05 11, 6800 MANNHEIM 81. Tel. 06 21/85 10 12-13.
Protekta: *Netherlands:* Protekta B.V., Mathenessestraat 27-29, 4834 EA BREDA. Tel. 076-659223.
Proval: *France:* Proval, 27, rue de la Gare de Reuilly, 75012 PARIS. Tel. (1) 43 46 96 79.
Puteaux: *France:* Puteaux (Etablissements), 8-10, place de la Loi, 78150 LE CHESNAY. Tel. 39 54 53 59; Telex CCVER 698 958 F (Puteaux).
PVV: *Hungary:* Peremartoni Vegyipari Vallalat, PEREMARTON-GYARTELEP.
Pyrsos: *Greece:* Pyrsos Ltd., 38 Aristotelous, ATHENS.

Quimagro: *Portugal:* Quimica Agricola Industriel, S.A.R.L., Vala do Carregado, 2580 ALENQUER. Tel. 9 14 06.

Quimica Estrella: *Argentina:* Quimica Estrella, Agrochemical Department, Av. Constituyentes 2995, 1427 BUENOS AIRES. Tel. 541 52 7847. Telex 17754.

Quimicas del Valles: *Spain:* Industrias Quimicas del Valles SA c/ Rafael de Casanova, 73, MOLLET DEL VALLES (Barcelona).

Quimicas Oro: *Spain:* Quimicas Oro, SA, Ctra Valencia-Ademuz, Km, 13'100, 46184 SAN. ANT. BENAGEBER (Valencia). Tel. 96 132 04 50.

Quimigal: *Portugal:* Quimigal - Quimica de Portugal, E.P., Divisao de produtos para a agricultura Marketing Pesticidas, Rua dos Navegantes, 48-53, 1200 LISBOA. Tel. 60 10 51.

- *Portugal:* Quimigal - Centro de Desenvolvimento Agricola, Quinta dos Almosteis, 2685 SACAVEM. Tel. 251 31 81.

Quiminor: *Spain:* Quiminor, SA, c/ Pintor Sorolla, 5, VALENCIA-2.

R.S.R.: *France:* R.S.R. (Raffineries de Soufre Reunies), l, place du General-de-Gaulle - B.P. 1818 13221 MARSEILLE CEDEX 01. Tel. 91 54 90 60; Telex 430 581 SOUFRES MARSEILLE.

RACROC: *Switzerland:* RACROC AG, Import-Export, Postfach 81, 2557 STUDEN. Tel. 032 53 25 44.

Radermecker: *Belgium:* Radermecker P. Interchimie S.A., Rue de Steppes, 103, 4000 LIEGE. Tel. 041/27 53 18.

Radix: *Switzerland:* Radix AG, Chemische Fabrik, 9314 STEINEBRUNN TG. Tel. 071 66 11 12.

Rallis India: *India:* Rallis India Ltd., Fertilizers & Pesticides Division, Ralli House 21, Damodardas Sukhadvala Marg., BOMBAY 400 001. Tel. 2048221. Telex 011 2357 RILB IN.

Rasmussen: *Denmark:* Kaj Rasmussen & Son Arslev ApS, Kirstinebjergvej 1-3, 5792 ARSLEV. Tel. (09) 99 13 31.

Ravit: *Italy:* Ravit, Viale degli Ammiragli 91, 00136 ROMA. Tel. (06) 637.90.51.

Reckhaus: *Germany (FRG):* Reckhaus GmbH & Co. KG, Industriestrasse 53, 4800 BIELEFELD 11 SENNESTADT. Tel. 0 52 05/40 53-54.

Reckitt: *Germany (FRG):* Reckitt GmbH, Suderwichstrasse 100 Postfach 10 14 63, 4350 RECKLINGSHAUSEN.

Reckitt & Colman: *Belgium:* Reckitt et Colman, Rue de la Bienvenue, 9, 1070 BRUXELLES. Tel. 02/376 20 66.

- *Switzerland:* Reckitt & Colman AG, Postfach, 4002 BASEL. Tel. 061 26 90 88.

Reis: *Portugal:* Sociedades Reunidas Reis, Lda, Rossio, 102-108-1., 1100 LISBOA. Tel. 36 25 21.

Rentokil: *Belgium:* Rentokil S.A., Avenue Rogier, 385, 1030 BRUXELLES. Tel. 02/733 98 70.

- *Denmark:* A/S Rentokil, HERLEV.
- *Finland:* Oy Rentokil Ab, Niittykummuntie 4, 02200 ESPOO 20. Tel. 90 423 033.
- *Netherlands:* BV Rentokil Chemie, Strijkviertel 27, 3454 PH DE MEERN. Tel. 03406-64704.
- *UK:* Rentokil Ltd., Felcourt, EAST GRINSTEAD, West Sussex RH19 2JY.

Rhodiagri-Littorale: *France:* Rhodiagri-Littorale, Parc du Millenaire B 13, B.P. 9622, Route de Mauguio, 34036 MONTPELLIER CEDEX. Tel. 6769 7600; Telex 490 712 F.

Rhodic: *France:* Rhodic Espaces Verts, (Rhone-Poulenc Agrochemie), 42, Chemin du Moulin-Carron, 69130 ECULLY. Tel. 7229 48 48; Telex 370 232.

Rhone-Poulenc: *Austria:* Rhone-Poulenc Handelsgesellschaft m.b.H., Grosse Neugasse 8, A-1040 WIEN. Tel. 0 22 2/56 36 26; Telex 135 336.

- *France:* Rhone-Poulenc Chimie de Base, 47, rue de Villiers - B.P. 122, 92527 NEUILLY-SUR-SEINE CEDEX. Tel. (1) 47 30 60 60.
- *Germany (FRG):* Rhone-Poulenc GmbH, Delegation Agrochimie, Staedelstr. 10, Postfach 70 10 29, 6000 FRANKFURT/MAIN 70. Tel. O 69/6 09 31.

- *Greece:* Rhone-Poulenc Hellas, 308 Mesoghion Ave., 2 Arkadiou St., GR-155 62 Holargos, ATHENS. Tel. (01) 65.32.492-7; Telex 216148.
- *Hungary:* Rhone-Poulenc S.A. Budapesti Iroda, Vaci utca 16, BUDAPEST V.
- *Poland:* Rhone-Poulenc, ul. Ursynowska 34, 02-605 WARSZAWA. Tel. 44 95 20, 44 77 47; Telex 812715.
- *Portugal:* Rhone Poulenc AGROP - Produtos Quimicos Lda, Rua Antonio Enes, 25-2., Dt., 1000 LISBOA. Tel. 54 41 80.
- *Sweden:* Rhone Poulenc Sverige AB, Box 4189, 102 62 STOCKHOLM. Tel. 08/23 74 80.
- *Switzerland:* Rhone-Poulenc (Suisse) S.A., 63, rue de Lausanne, 1211 GENEVE 21. Tel. 022 31 59 50.
- *UK:* Rhone-Poulenc Agricultural Division, Regent House, Hubert Road, BRENTWOOD, Essex CM14 4TZ. Tel. (0277) 230522.
- *USA:* Rhone-Poulenc Inc., Agrochemical Division, P.O. Box 125, Black Horse Lane, MONMOUTH JUNCTION, NJ 08852. Tel. (201) 297 0100.

Riem: *Belgium:* Riem S.P.R.L., Chaussee de Charleroi, 19, 6338 SOMBREFFE. Tel. 071/81 34 34.

Rigo: *USA:* Rigo Corp., P.O. Box 89, BUCKNER, KY 40010. Tel. (502) 222-1456.

Riwa: *Netherlands:* Riwa B.V., Mathenessestraat 27-29, 4834 EA BREDA. Tel. 076-654650.

Roberts: *Netherlands:* Roberts Holland B.V., Parallelweg, Postbus 64, 3360 AB SLIEDRECHT. Tel. 01840-14266.

Rodent Control: *UK:* Rodent Control Ltd., 70/78 Queen's Road, READING, Berks RG1 4BZ.

Rohm & Haas: *Austria:* Rohm and Haas Ges. m.b.H. Austria, Kelsenstrasse 2, A-1030 WIEN. Tel. 0 22 2/78 66 22-0; Telex 133199.
- *France:* Rohm and Haas France S.A., La Tour de Lyon - 185, rue de Bercy, 75579 PARIS CEDEX 12. Tel. (1) 40 02 50 00; Telex 220 819.
- *Greece:* Rohm & Haas Greece, 8 Adrianou St., GR-154 51 N. Psychico, ATHENS. Tel. (01) 67.17.876 & 64.72.185.
- *Hungary:* Rohm & Haas, Gyarmat u. s/a. BUDAPEST XIV.
- *Italy:* Rohm & Haas Italia, Via Vittor Pisani 26, 20124 MILANO. Tel. (02) 62.56.1.
- *Portugal:* Rohm and Haas (Portugal), Av. Columbano Bordalo Pinheiro, 74-7., Dto, 1000 LISBOA. Tel. 72 31 02/72 36 69.
- *Spain:* Rohm & Haas Espana, SA, c/ Provenza, 216, 08036 BARCELONA. Tel. 93 323 20 66.
- *Sweden:* Rohm and Haas Nordiska AB, Box 11045, 161 11 BROMMA. Tel. 08/98 07 70.
- *UK:* Rohm & Haas Ltd., Lennig House, 2, Mason's Avenue, CROYDON, Surrey CR9 3NB. Tel. (01 686) 8844.
- *USA:* Rohm & Haas Co., Independence Mall West, PHILADELPHIA, PA 19105. Tel. (215) 592-3000; Telex 845 247.

Rotox: *Denmark:* Rotox, NYKOEBING F.

Roussel: *UK:* Roussel Laboratories Ltd., Broadwater Park, North Orbital Road, UXBRIDGE, Middlesex UB9 5HP. Tel. (0895) 834343.

Roussel Hoechst: *Italy:* Roussel Hoechst Agrovet S.p.A., Piazza Turr 5, 20149 MILANO. Tel. (02) 31.07.1.

Rowi: *Netherlands:* Rowi B.V., Dodewaardlaan 8, 4006 EA TIEL. Tel. 03440-14150.

Sadisa: *Spain:* Servicios Agricolas Diversos, SA, (Sadisa), c/ Joaquin Costa, 61, 28002 MADRID. Tel. 91 261 49 53.

Sadolin: *Denmark:* Sadolin Malervarer A/S, KOEBENHAVN S.

Sajobabony: *Hungary:* Eszakmagyarorszagi Vegyimuvek, SAJOBABONY.

Salomez: *France:* Salomez (Establissements), Route de Avallon, Z.I. Pature des Chaumes, 89130 TOUCY. Tel. 86 44 21 75.

Samen-Maier: *Germany (FRG):* Samen-Maier, 8311 BODENKIRCHEN/BAYERN. Tel. 0 87 45/10 11.

Sanac: *Belgium:* Sanac Belgium S.A., Menenstraat, 66A, 8660 GELUWE. Tel. 056/51 24 52.

Sandoz: *Austria:* Sandoz Ges. m.b.H., Agrar-Vertrieb, Brunner Strasse 59, A-1235 WIEN. Tel. 0 22 2/86 45 46; Telex 134720.
- *Belgium:* Sandoz S.A., Chaussee de Haecht, 226, 1030 BRUXELLES. Tel. 02/216 80 95.
- *France:* Sandoz (Produits), 14, boulevard Richelieu - B.P. 318, 92506 RUEIL-MALMAISON CEDEX. Tel. (1) 47 32 75 11; Telex SANDOZ 203 203 F.
- *Germany (FRG):* Sandoz AG, Techn. Buero fuer die Bundesrepublik, Seebachweg 1, 7121 MUNDELSHEIM. Tel. 0 71 43/57 34.
- *Hungary:* Sandoz Informacios Koezpont, Csalogany u. 6-10, BUDAPEST I.
- *Italy:* Sandoz Prodotti Chimici, Divisione Agrochimica, Via C. Arconati 1, 20135 MILANO. Tel. (02) 57.95.
- *Netherlands:* Sandoz B.V., Loopkantstraat 25, Postbus 91, 5400 AB UDEN. Tel. 04132-65911.
- *Poland:* Sandoz, ul. Smolenskiego 4/13, 01-698 WARSZAWA. Tel. 33 12 88; Telex 815562.
- *Portugal:* Produtos Sandoz, Lda, Rua de S. Caetano, 4, 1200 LISBOA. Tel. 67 50 11.
- *Spain:* Sandoz, SAE, Gran Via de les Corts Catalanes, 764, Apartado 708, 08013 BARCELONA.
- *Switzerland:* Sandoz AG, Agro-Filiale Schweiz, Postfach, 4002 BASEL. Tel. 061 24 11 11.
- *UK:* Sandoz Products Ltd., Agrochemicals, Norwich Union House, 16/18 Princes Street, IPSWICH, Suffolk IP1 1QT. Tel. (0473) 55972.
- *USA:* Sandoz Crop Protection 341 East Ohio Street, CHICAGO, IL 60611. Tel. (312) 670-4500.

Sanerings: *Sweden:* Svenska Sanerings AB, Box 56, 182 71 STOCKSUND. Tel. 08/85 30 02,-04, -20.

Sanigene: *France:* Sanigene (Laboratoires), 5, rue de l'Industrie, PRINCIPAUTE DE MONACO. Tel. 93 30 12 67; Telex 469 271 MC.

Sankyo: *Japan:* Sankyo Co. Ltd., 7-12, Ginza 2-chome, Chuo-ku, TOKYO 104.

Sapec: *Portugal:* Sapec, Rua Vitor Cordon, 19, Apartado 2349, 1108 LISBOA CODEX. Tel. 36 07 15.

Sarabhai: *India:* Sarabhai M. Chemicals, P.O. Box 80, BARODA 390001. Tel. 82821. Telex 0175 473.

SCAC-Fisons: *France:* SCAC-Fisons S.A., La Galboisiere - B.P. 338, 37705 SAINT-PIERRE-DES-CORPS CEDEX. Tel. 47 44 54 32; Telex FISCAC 750 054 F.

Scam: *Italy:* Scam, Via Bellaria 164, 41050 S. MARIA DI MUGNANO (Mo). Tel. (0462) 30.91.31.

Schacht: *Germany (FRG):* F. Schacht GmbH + Co. KG, Bueltenweg 48, Postfach 48 23, 3300 BRAUNSCHWEIG. Tel. 05 31/33 33 71.

Schaufler: *Austria:* Schaufler, Import-Export, Wiesengasse 8, A-3441 JUDENAU. Tel. 0 22 74/445, 0 22 74/512.

Scheidler: *Germany (FRG):* Wilhelm Scheidler KG, Kutenhauser Strasse 13, Postfach 31 45, 4950 MINDEN/WESTF. Tel. 05 71/4 15 66.

Schering: *Austria:* Schering Wien Ges. m.b.H., Scheringgasse 2, A-1147 WIEN. Tel. 0 22 2/97 15 36; Telex 131 192.
- *Denmark:* Schering A/S Agro, Fjeldhammervej 8, 2610 ROEDOVRE. Tel. (01) 70 55 55.
- *France:* Schering, 5, rue Le Corbusier - Silic 237, 94528 RUNGIS CEDEX. Tel. (1) 46 87 23 45; Telex SCHER 204 494 F.
- *Germany (FRG):* Schering Aktiengesellschaft, Pflanzenschutz Deutschland, Werftstrasse 37, Postfach 12 21, 4000 DUESSELDORF 1. Tel. 02 11/5 00 60.
- *Hungary:* Schering, Petoefi ter 2. BUDAPEST V.
- *Iceland:* Schering AG Agrochemical Division, THYSKALANDI.
- *Italy:* Schering, Divisione Agricoltura, Via L. Mancinelli 7/11, 20131 MILANO.

- *Spain:* Schering Espana, SA, Pol. Ind. El. Pla. Parcela, 17, 46290 ALCACER (Valencia). Tel. 96 123 24 12.
- *Switzerland:* Schering Zuerich AG, Hermetschloostrasse 75, 8048 ZUERICH. Tel. 01 62 57 76.
- *UK:* Schering Agriculture, UK Agriculture Head Office, Nottingham Road, Stapleford, NOTTINGHAM NG9 8AJ. Tel. (0602) 390202; Telefax (0602) 390202.

Schering AAgrunol: *Netherlands:* Schering AAgrunol B.V., Oosterweg 127, Postbus 100, 9750 PE HAREN (Gr.). Tel. 050-333911.

Schippers: *Netherlands:* M. Schippers, Bleijenhoek 17, 5531 BK BLADEL. Tel. 04977-2017.

Schomaker: *Germany (FRG):* Rudolf Schomaker KG, 4471 LAHN/EMSLAND. Tel. 0 59 51/5 63.

Schwarzheide: *Germany (GDR):* VEB Synthesewerk Schwarzheide, SCHWARZHEIDE.

Schwebda: *Germany (FRG):* Feinchemie Schewbda GmbH, Bahnhof 2, 3446 MEINHARD-SCHEWBDA. Tel. 0 56 51/6 00 38.

Schwefelveredlung: *Germany (GDR):* VEB Schwefelveredlung, LEIPZIG.

Schweizer-Effax: *Netherlands:* Schweizer-Effax, P.C., Hooftstraat 5, 1071 BL AMSTERDAM. Tel. 020-641311.

SDS Biotech: *Belgium:* SDS Biotech Europe Corporation, Bd Bisschoffsheim 39 - Bte 15, B-1000 BRUXELLES.

- *Japan:* SDS Biotech K.K., 2-12-7, Higashi Shinbashi 2-chome, Minato-ku, TOKYO 105. Tel. 03 436 3811. Telex 23221 J.
- *UK:* SDS Biotech UK Ltd., Bayheath House, 4, Fairway, Petts Wood, ORPINGTON, Kent BR5 1EG. Tel. (0689) 74011.

Sectolin: *Netherlands:* Sectolin B.V., Parkweg 109-111, 7545 MV ENSCHEDE. Tel. 053-312160.

Security: *USA:* Security Lawn & Garden Products Co., P.O. Box 938, FORT VALLEY, GA 31030. Tel. (912) 825-5511.

Sedagri: *France:* Sedagri, Departement Rhone-Poulenc Agrochimie, Immeuble Norly, 42, Chemin du Moulin Carron. 69130 ECULLY. 78 33 39 09.

SEGE: *Greece:* SEGE, 14 Victoros Ougo, ATHENS.

Sema Vinyl: *Belgium:* Sema Vinyl S.A., Wavreumont, 11, 4970 STAVELOT.

Seppic: *Spain:* Isisa-Seppic, SA, Poniente, 38, 28036 MADRID. Tel. 91 202 57 01.

Sepran: *Italy:* Sepran Agrochimici, Via Cogolla 5/B, 36033 ISOLA VICENTINA (VI). Tel. (051) 78.10.90.

Serpiol: *Spain:* Serpiol, SA, Juan de Austria, 46002 VALENCIA. Tel. 96 351 60 51.

Serpis: *Spain:* Electroquimica del Serpis, SA, Partida Almansa, s/n, 46721 POTRIES (Valencia). Tel. 96 280 04 75.

Servigeco: *France:* Servigeco (Service d'Hygiene Ecologique) Cheminde Mennecy, 9 1840 SOISY-SUR-ECOLE. Tel. 64 98 02 91.

Shell: *Austria:* Shell Austria AG, Postfach 174, A-1011 WIEN. Tel. 0 22 2/78 03-0; Telex 133241.

- *Belgium:* Belgian Shell, Kantersteen, 47, 1000 BRUXELLES. Tel. 02/511 65 76.
- *Denmark:* Shell Kemi A/S, Kampmannsgade 2, 1604 KOEBENHAVN V. Tel. (01) 93 53 40.
- *Germany (FRG):* Deutsche Shell Chemie GmbH, Nibelungenplatz 3, 6000 FRANKFURT/MAIN. Tel. 0 69/1 56 51.
- *Germany (FRG):* Shell Agrar GmbH & Co., KG, P.O. Box 200, 6507 INGELHEIM/RHEIN. Tel. 0 61 32/789-0.
- *Greece:* Shell Chemicals (Hellas) Ltd., 2 El. Venizelou St., GR-176 76 Kallithea, ATHENS. Tel. (01) 92.32.222; Telex 215162.
- *Hungary:* Shell International Petroleum Company, Rajk L. u. 11. BUDAPEST XIII.
- *Italy:* Shell Agricoltura, Via F. Londonio 2, 20154 MILANO. Tel. (02) 38.85.
- *Netherlands:* Shell Nederland Chemie B.V., Hofplein 19, Postbus 2960, 3000 CZ ROTTERDAM. Tel. 010-4696915.
- *Norway:* A/S Norske Shell, Postboks 1154, Sentrum, 0107 OSLO 1.

- *Portugal:* Shell Portuguesa, S.A.R.L., Av. da Liberdade, 249, Apartado no. 2008, 1200 LISBOA. Tel. 53 41 41.
- *Spain:* Sociedad Petrolifera Espanola Shell, SA, c/ Barquillo, 17. Apartado 652, 28004 MADRID. Tel. 91 521 47 41.
- *Sweden:* AB Svenska Shell, Fack, 171 79 SOLNA. Tel. 08/730 80 00.
- *UK:* Shell Chemicals Ltd., Agricultural Division, 39-41 St. Mary's Street, ELY, Cambs. CB7 4HG. Tel. (0353) 3671.
- *UK:* Shell Chemicals Ltd., Liquid Seed Treatment Department, The Pines, Fordham Road, Exning, NEWMARKET. Tel. (0638) 720926.
- *UK:* Shell International Chemical Co. Ltd., LONDON.
- *USA:* Shell Chemical Co., P.O. Box 3871, HOUSTON, TX 77253.

Shen Hong: *Taiwan:* Shen Hong Agricultural Chemical Co., 5F.206 Nanking East Road, Sec 2 Taipei, P.O. Box 46-209, TAIPEI. Tel. 02 5223034. Telex 085 27075 SHACCO.

Shionogi: *Japan:* Shionogi & Co. Ltd., 3-12 Doshomachi, Higashi-ku, OSAKA 541.

Siapa: *Greece:* Siapa, 22 Phavierou St., GR-104 38 ATHENS.
- *Italy:* Siapa, Via Yser 16, 00198 ROMA. Tel. (06) 84.34.5.

Sici: *Italy:* Sici, Via Torino 41, 00184 MILANO.

Sico: *France:* Sico, B.P. 59, 38120 SAINT-EGREVE. Tel. 76 75 30 45; Telex 320 094 F.

Sideco-Chemie: *Germany (FRG):* Sideco-Chemie Dr. Schirm GmbH, Am Schlutuper Markt 4, Postfach 16 01 06, 2400 LUEBECK 16. Tel. 04 51/6 95 73.

Siegfried: *Switzerland:* Siegfried Aktiengesellschaft, Abt. Agrochemikalien, 4800 ZOFINGEN. Tel. 062 50 22 93.

Sierra: *USA:* Sierra Chemical Co., 1001 Yoshemite Drive, MILPITAS, CA 95035. Tel. (408) 263-8080. Telex 910 338 0565.

SIMAR: *Italy:* S.I.M.A.R., Via Appia Nuova, 1220/1253, 00178 ROMA. Tel. 06/7995272-7994210.

Simonis: *Netherlands:* B.V. Ind-en Handelsondern, Simonis, Groothandelsgebouw D6, 3000 AA ROTTERDAM. Tel. 010-4113200.

Simplot: *USA:* Simplot, J. R., Co., Minerals & Chemical Division, Western Region, P.O. Box 198, LATHROP, CA 95330. Tel. (209) 858-2511. Telex 510 762 6631.

Sintesul: *Brazil:* Sintesul, Rua Joao Thomaz Munhoz 218, Caixa Postal No. 263, Pelotas, TIO GRANDE DO SUL. Tel. (0532) 25 86 66.

Sipcam: *France:* Sipcam-Phyteurop, Siege social: 5, avenue des Chaussers, 75017 PARIS. Tel. (1) 47 66 03 01; Telex 660 574.
- *France:* Sipcam-Phyteurop, Departement Technique, 16, rue Mederic, 75017 PARIS. Tel. (1) 47 66 51 50; Telex 643 265 SIPCAM.
- *Italy:* Sipcam, Viale Gian Galeazzo 3, 20136 MILANO. Tel. (02) 837.48.41/832.12.41.
- *Sweden:* Sipcam Sverige AB, c/o Translaw AB, Box 7269 103 89 STOCKHOLM. Tel. 08/249 450.

Sivam: *Italy:* Sivam, Via Scarlatti 30, 20124 MILANO. Tel. (02) 27.04.51.

Skadedyr: *Denmark:* A/S Skadedyrcentralen, ESBJERG.

SKW Trostberg: *Germany (FRG):* SKW Trostberg AG, Marktbereich Landwirtschaft, Postfach 11 50/11 60, 8223 TROSTBERG/OBB. Tel. 0 86 21/8 61.

Sme-Lang: *Netherlands:* Sme-Lang, Sinnerstraat 36, 7461 TT RIJSSEN,. Tel. 5480-16605.

Sodafabrik: *Switzerland:* Schweizerische Sodafabrik, Baerengasse 29, 8022 ZUERICH. Tel. 01 211 03 30.

Sofital: *Italy:* Sofital, 14 Strada Zona Industriale, 95030 CATANIA.

Sojuzchimexport: *USSR:* Sojuzchimexport, MOSKAU.

Solfotecnica: *Italy:* Solfotecnica Italiana, Via G. Matteotti 16, 48100 RAVENNA. Tel. (0544) 3.80.92.

Solvay Svenska: *Sweden:* Solvay Svenska AB, Strandvaegen 5A 114 51, STOCKHOLM. Tel. 08/231025.

Sopepor: *Portugal:* SOPEPOR - Sociedade Comercial de Pesticidas Portugueses, S.A.R.L., Rua Condessa da Junqueira 182, 2080 ALMEIRIM. Tel. 5 20 23.

Sopra: *France:* Sopra, 1, avenue Newton - B.P. 208, 92142 CLAMART CEDEX. Tel. (1) 45 37 51 11; Telex ICIFRAN 200 909.

SOPRO: *Switzerland:* SOPRO AG, Bahnhofstrasse 44, 3000 BERN 5. Tel. 031 25 28 45.

Sorex: *UK:* Sorex Ltd., Trading Estate, St. Michaels Road, WIDNES, Cheshire WA8 8TJ.

Source Technology Biologicals: *USA:* Source Technology Biologicals Inc., 2850 Metro Drive, Suite 309, BLOOMINGTON, MN 55420. Tel. (612) 854-9436.

Sovilo: *France:* Sovilo, Rue Andre-Huet - B.P. 406, 51064 REIMS CEDEX. Tel. 26 87 06 11; Telex 830 728 SOVILO.

SPE: *Greece:* SPE SA, 27-29 Politechniou St., GR-546 26 THESSALONIKI. Tel. (031) 517.456.

Spedro: *Switzerland:* Spedro AG, Laengfeldweg 119, 2501 BIEL. Tel. 032 42 31 21.

Spezialchemie Leipzig: *Germany (GDR):* VEB Spezialchemie Leipzig, LEIPZIG.

Spiess: *Germany (FRG):* C. F. Spiess und Sohn GmbH & Co., 6719 KLEINKARLBACH UEBER GRUENSTADT/PFALZ. Tel. 0 63 59/8 01-1.

Spolana: *Czechoslovakia:* Spolana np, NERATOVICE.

Sporex: *Greece:* Sporex-El. Grigoriadou & Co., 11 3rd Septembriou St., ATHENS.

Spyridakis: *Greece:* C. Spyridakis, 19 Evans St., GR-712 01 Iraklion, CRETE. Tel. (081) 224.761.

Staehler: *Austria:* Staehler Agrochemie GmbH & Co. KG, Postfach 2047, Stader Elbstrasse, 2160 STADE. Tel. 0 41 41/20 16.

Stahl: *Germany (FRG):* Dr. Stahl & Sohn GmbH & Co. KG, Postfach 12 27, 7770 UEBERLINGEN. Tel. 0 75 51/46 03.

Stauffer: *Austria:* Stauffer Chemical GmbH, Schwindgasse 20, A-1040 WIEN. Tel. 0 22 2/65 76 56; Telex 135010.

- *Belgium:* Stauffer Chemical Belgium, Parc Industriel Tyberchamps, 6198 SENEFFE. Tel. 064/55 53 51.
- *Netherlands:* Stauffer Chemical B.V., Het Witte Huis, Wijnhaven 3A, 3011 WG ROTTERDAM. Tel. 010-4135542.
- *Switzerland:* Stauffer Chemicals S.A., 18 rue de Lancy 1227 CAROUGE-GENEVE. Tel. 022 42 27 00.
- *USA:* Stauffer Chemical Co., Agricultural Chemical Div., WESTPORT, CT 06881. Tel. (203) 222-3000.

Stefes: *Germany (FRG):* Herbert Stefes Agrarchemikalien, Im Niederfeld, 5014 KERPEN. Tel. 0 22 37/25 88.

Steininger: *France:* Steininger Pest Control S.A., 15, avenue de la Resistance - RN 4 - B.P. 1094, 54523 LAXOU CHAMP LE BOEUF CEDEX. Tel. 83 96 43 42; Telex 960 423 F NANCY.

Sterling: *France:* Sterling, 96, boulevard Victor-Hugo, 92115 CLICHY CEDEX. Tel. (1) 42 70 83 70; Telex 620 680 F.

Stingel-Chemie: *Netherlands:* Stingel-Chemie, Heinrich-Otto-Str. 42, 7317 DD WENDLINGEN (W.Dtsl.). Tel. 07024-7419.

Stinnes: *Germany (FRG):* Stinnes Agrarchemie GmbH, Grosse Bleichen 19, Postfach 30 52 26, 2000 HAMBURG 36. Tel. 0 40/35 09 29.

Stodiek: *Germany (FRG):* Wilhelm Stodiek & Co. KG, Quellental 9, 4972 LOEHNE/WESTF. Tel. 0 57 32/30 61.

Stoller: *USA:* Stoller Chemical Co. Inc., 8582 Katy Freeway, Suite 200, HOUSTON, TX 77024. Tel. (713) 461-2910. Telex 79 1212 AGRIMEN.

Stolte & Charlier: *Germany (FRG):* Stolte & Charlier GmbH, Gertrudenstrasse 14, 2000 HAMBURG 1. Tel. 040/33 73 83.

Stroem: *Sweden:* AB Karl H. Stroem, Verktygsvaegen 8 552 71 JOENKOEPING. Tel. 036/188460.

Stuifbergens: *Sweden:* Stuifbergens Blomsterloeks - inport AB, Box 9062 121 09 JOHANNESHOV. Tel. 08/918140.

Sumitomo: *Germany (FRG):* Sumitomo Deutschland GmbH, Georg-Glock-Strasse 4, 4000 DUESSELDORF. Tel. 02 11/4 57 01.

- *Japan:* Sumitomo Chemical Co. Ltd., 15 5-chome, Kitahama Higashi-ku, OSAKA. Tel. 06 220 3745.
- *Netherlands:* Sumitomo Europa B.V., Weena 108, 3012 CP ROTTERDAM. Tel. 010-4331133.
- *UK:* Sumitomo Ltd., 107 Cheapside, LONDON EC2 6DQ. Tel. 01 726 6262.
- *USA:* Sumitomo Chemical America Inc., 345 Park Avenue, NEW YORK, NY 10154. Tel. (212) 207-0600.

Sundat: *Malaysia:* Sundat (S) Pte. Ltd., P.O. Box 434, Jurang Town Post Office, 26 Gul Crescent, SINGAPORE 2262. Tel. 8612460. Telex 35186 HHL.

Sunko: *Taiwan:* Sunko Chemical Co. Ltd., 12 Lane 42, Jen-Hua Road, Ta-Li Hsiang, TAICHUNG HSIEN. Tel. 04 3302126-7. Telex 56318.

Superfos: *Denmark:* Superfos Kemi A/S, Frydelundsvej 30 Postbox 39 2950 VEDBAEK.

Svaloef: *Sweden:* Svaloef AB/Hammenhoegs, 270 50 HAMMENHOEG. Tel. 0414/40400.

Synchemicals: *UK:* Synchemicals Ltd., 44, Grange Walk, LONDON SE1 3EN. Tel. (01 232) 1225; Telex 886387.

Syntex: *Netherlands:* Syntex Agribusiness Nederland B.V., Wattstraat 44, 2723 RC ZOETERMEER. Tel. 079-410144.

Szovetkezet: *Hungary:* Agrokemia Szovetkezet, SELLYE.

T & D Mideast: *Cyprus:* T & D Mideast Chemicals Ltd., P.O. Box 1644, LIMASSOL. Tel. 051 52546. Telex 52546.

Taiwan Tainan Giant: *Taiwan:* Taiwan Tainan Giant Industrial Co. Ltd., 191, Sec. Ming Sheng Road, TAINAN. Tel. 06 2221221. Telex 71592 YONYU.

Takeda: *Japan:* Takeda Chemical Industries Ltd., 12-10, Hihonbashi, 2-chome, Chuo-ku, TOKYO 103. Tel. 03278 2567. Telex 24252 TAKEDA J.

Tamogan: *Israel:* Tamogan Chemicals Ltd., 61 Jabotinsky Street, Petach-Tikva, P.O. Box 2438, TEL-AVIV 61024. Tel. 03 9265272. Telex 342507 OSEM IL.

Tanaco: *Denmark:* Tanaco-Partners ApS, ESBJERG.

Tarsoulis: *Greece:* D. Tarsoulis, 1 Piraeus St., GR-105 52 ATHENS.

Tate & Lyle: *UK:* Tate & Lyle Group Research & Development, Philip Lyle Memorial Research Laboratory, White Knights, READING, Berks.

Tecniterra: *Italy:* Tecniterra, Via Nino Bixio 34, 20129 MILANO. Tel. (02) 20.01.25.

Teknosan: *Sweden:* Teknosan, Fack, 201 10 MALMOE. Tel. 040/93 66 00.

Temana: *Netherlands:* Temana Nederland B.V., Duitslandweg 7, Postbus 65, 2410 AB BODEGRAVEN. Tel. 01726-19021.
- *UK:* Temana Bees Ltd., Sealand, CHESTER CH1 6BA.

Tenneco: *UK:* Tenneco Organics Ltd., Rockingham Works, AVONMOUTH, Bristol BS11 0YT. Tel. (0272) 823611.

Tennessee: *USA:* Tennessee Chemical Co., 3475 Lenox Road N.E., Suite 670, ATLANTA, GA 30326. Tel. (404) 233-6811. Telex 54 2347.

Terminator: *Sweden:* Terminator AB, Box 20089 104 60 STOCKHOLM. Tel. 08/7728900.

Terranalisi: *Italy:* Terranalisi, Via G. Donizetti 2/A, 44042 CENTO (Fe). Tel. (051) 90.62.07.

Terrasan: *Germany (FRG):* Terrasan-Chemie GmbH, Hindenburgstrasse 20a, 8070 INGOLSTADT. Tel. 08 41/8 10 47-48.

Thompson: *Germany (FRG):* Thompson-Siegel GmbH, Erkratherstrasse 230, 4000 DUESSELDORF 1. Tel. 02 11/73 52-0.
- *Germany (FRG):* Thompson-Siegel GmbH, Henkelstrasse 67, Postfach 11 26, 4000 DUESSELDORF.

Thompson-Hayward: *USA:* Thompson-Hayward Chemical Co., P.O. Box 2383, KANSAS CITY, KS 66106. Tel. (913) 321-3131.

Tifa: *USA:* Tifa Ltd., Tifa Square, 50 Division Ave., MILLINGTON, NJ 07946. Tel. (201) 647-4570. Telex 136366.

Top: *France:* Top S.A., Place du 14-Juillet, 80380 VILLERS-BRETONNEUX. Tel. 22 48 00 33 & 22 48 01 35; Telex 140 988.

Tradi-Agri: *France:* Tradi-Agri, 38 avenue Hoche, 75008 PARIS. Tel. (1) 42 89 14 34; Telex 642 469 AGRIJET.

Traital: *France:* Traital S.A., Z. A. Les Pierrelets, 45610 CHAINGY. Tel. 38 88 02 41; Telex 780 859 F.

Trans-Meri: *Finland:* Oy Trans-Meri Ab, PL 228 00101 HELSINKI. Tel. 90 520 211.

Trimp: *Netherlands:* Trimp Handelsonderneming, Schieweg 85, 3038 AH ROTTERDAM. Tel. 010-4657653.

Tripart: *UK:* Tripart Farm Chemicals Ltd., Swan House, 17, Beulah Street, Gaywood, KING'S LYNN, Norfolk PE30 4DN. Tel. (0553) 674303.

Tripmacker: *Germany (FRG):* Johann Tripmacker, Wether Strasse 28, 2168 DROCHTERSEN-ASSEL. Tel. 0 41 48/12 07.

Troy: *USA:* Troy Chemical Corp., One Avenue L, NEWARK, NJ 07105. Tel. (201) 589-2500. Telex 13 8930.

Truchem: *UK:* Truchem Ltd., Brook House, 30 Larwood Grove, Sherwood, NOTTINGHAM NG5 3JD. Tel. (0602) 260762.

Truffaut: *France:* Truffaut, 21, rue des Pepinieres Les Noels B.P. 9, 41350 VINEUIL. Tel. 54 42 64 62; Telex TRUFOVI 750 192.

Tschupp: *Switzerland:* Tschupp & Co. AG, Fabrik chem.-techn. Produkte, 6275 BALLWIL LU. Tel. 041 89 13 13.

Tsilis: *Greece:* Chr. Tsilis, 7-9 Korai St., IOANNINA.

Tuhamij: *Netherlands:* Tuhamij, Woutersweg 10, Postbus 95, 2690 AB 'S-GRAVENZANDE. Tel. 01748-15198.

Turba Projar: *Spain:* Turba Projar, SA, Apartado 526, 39080 SANTANDER. Tel. 942 25 38 50.

UCAR: *Switzerland:* Union des Cooperatives Agricoles Romandes, Av. des Jordils 3, 1006 LAUSANNE. Tel. 021 27 51 51.

UCB: *USA:* UCB Chemicals Corp., 801 Water Street, PORTSMOUTH, VA 23704. Tel. (804) 393-3210.

Umupro: *France:* Umupro (Division de la C.M.P.A.), 216, rue des Escarceliers - B.P. 1203, 34010 MONTPELLIER CEDEX. Tel. 67 40 43 43; Telex CMPA 490 712.

Union Carbide: *Austria:* Union Carbide Austria Ges. m.b.H., Storchengasse 1, A-1150 WIEN. Tel. 0 22 2/83 45 36, 83 06 01.

- *Germany (FRG):* Union Carbide Deutschland GmbH, Techn. Buero Dr. Uhl, Bismarckstr. 29, 6360 FRIEDBERG/HESSEN. Tel. 0 60 31/21 91.
- *Greece:* Union Carbide Hellas SA, P.O. Box 3100, GR-102 10 ATHENS. Tel. (01) 68.22.400; Telex 216136.
- *Italy:* Union Carbide Italia, Via Durini 28, 20122 MILANO.
- *Netherlands:* Union Carbide Benelux N.V., Zonnebaan 18, Postbus 1508, 3600 BM MAARSSEN. Tel. 030-437946.
- *Switzerland:* Union Carbide Europe SA, 15 chemin Louis-Dunant, 1211 GENEVE 20. Tel. 022 39 61 11.
- *UK:* Union Carbide Ltd., Agricultural Chemicals Division, Springfield House, King's Road, HARROGATE, N. Yorks. HG1 5JJ. Tel. (0423) 509731-3.
- *USA:* Union Carbide Agricultural Products Co., Inc., P.O. Box 12014, T.W. Alexander Dr., RESEARCH TRIANGLE PARK, NC 27709. Tel. (919) 549-2000; Telex 420542.

Uniroyal: *France:* Uniroyal S.A.R.L., Department Agrochimique, 13 avenue du General-Coronat 83000 TOULON. Tel. 94 41 23 84.

- *UK:* Uniroyal Ltd., Agrochemical Research Station, Brooklands Farm, Cheltenham Road, EVESHAM WR11 6LW. (0386) 47251.
- *USA:* Uniroyal Chemical, Div. of Uniroyal, Inc., Elm Street, NAUGATUCK, CT 06770.

United Agri-Products: *USA:* United Agri-Products Inc., P.O. Box 1286, GREELEY, CO 80632. Tel. (303) 356-4400. Telex 450363.

United Phosphorus: *India:* United Phosphorus Ltd., 167 Dr. Annie Besant Road, Worli, BOMBAY 400 018. Tel. 22 493 5666. Telex 1175074 UFOS IN.

Universal Crop Protection: *UK:* Universal Crop Protection Ltd., Park House, Maidenhead Road, COOKHAM, Berks. SL6 9DS. Tel. (06285) 26083.

Urania: *Germany (FRG):* Pflanzenschutz Urania GmbH, Alsterufer 20, Postfach 30 40 31, 2000 HAMBURG 36. Tel. 0 40/44 19 61.

UVKSS: *Hungary:* Universal Vegyi, Kulturcikk es Szolgaltato Szovetkezet, SZEGED.

Vadheim: *Norway:* A/S Vadheim Elektrochemiske Fabriker, Postboks 1814, 5011 NORDNES.

Valimex: *Spain:* Valimex, SL, Pelleter, 2, 46008 VALENCIA. Tel. 96 326 53 52.

Valmi: *France:* Valmi (Laboratoires), 119, rue Henri Barbusse, 59155 FACHES-THUMESNIL. Tel. 20 96 05 80.

Van Eenennaam: *Netherlands:* Van Eenennaam & Zn. B.V., Jannewekken 2, 4301 HH ZIERIKZEE. Tel. 01110-3851.

Van Loosen: *Germany (FRG):* Hans-Joachim van Loosen, Bismarckstrasse 160, 4270 DORSTEN 2/WESTF. Tel. 0 23 62/6 26 67.

Vanderbilt: *USA:* Vanderbilt, R. T., Co. Inc., 30 Winfield Street, NORWALK, CT 06855. Tel. (203) 853-1400. Telex 710 468 2940.

Vass: *UK:* L. W. Vass (Agricultural) Ltd., Station Road, AMPTHILL, Bedfordshire. Tel. (0525) 403041.

Vequer: *Spain:* Comercial Industrial Vequer, SA, Disputacion, 239, 2, 4a, 08007 BARCELONA. Tel. 93318 51 48.

Vernooy: *Netherlands:* Harry Vernooy Import/Export B.V., Nassaustraat 2, 2161 RK LISSE. Tel. 02521-16352.

Vertac: *USA:* Vertac Chemical Corp., 5100 Poplar Avenue, Suite 2414, MEMPHIS, TN 38137. Tel. (901) 767-6851.

Verwet: *Netherlands:* Verwet B.V., Laan van Westroyen 6, 4003 AX TIEL. Tel. 03440-18644.

Vetira: *Netherlands:* Vetira Romance Aerosols B.V., Sluispolderweg 9, 1505 HJ ZAANDAM. Tel. 075-156755.

Vetyl-Chemie: *Germany (FRG):* Vetyl-Chemie L. P. Braun, Gewerbestrasse 12-14, 6688 ILLINGEN. Tel. 0 68 25/4 40 71.

Vieira: *Portugal:* J. L. Vieira - Comercio de Importacao e Exportacao, Lda, Av. Joao Crisostomo, 79-2., Esq., Apartado 1375, 1011 LISBOA CODEX. Tel. 53 43 34.

Vineland: *USA:* Vineland Chemical Co. Inc., 1611 W. Wheat Road, VINELAND, NJ 08360. Tel. (609) 691-3535. Telex 510 687 8949.

Viopharm: *Greece:* Viopharm SA, 111 Avlonos St., GR-104 43 ATHENS. Tel. (01) 51.24.111-13; Telex 215754 VPHM.

Visplant: *Italy:* Visplant-Chimiren, Via Curiel 27, 40013 CASTELMAGGIORE (Bo). Tel. (051) 71.30.80.

Vital: *France:* Vital (Manufacture des engrais), 84320 ENTRAIGUES. Tel. 90 83 17 33.

Vitapharm: *Netherlands:* Vitapharm Nederland B.V., De Steiger 187, 1351 AT ALMERE-HAVEN. Tel. 03240-17279.

Vogelmann: *Germany (FRG):* Bruno Vogelmann Chemische Fabrik, Postfach 4 40, 7180 CRAILSHEIM/WUERTT. Tel. 0 79 51/2 10 45.

Volrho: *India:* Volrho Ltd., Patencheru 502319, Medak District, ANDRA PRADESH. Tel. 822730. Telex 0155 580.

Voorbraak: *Netherlands:* Chem. Fabr. Brabant J.W Voorbraak B.V., Koopvaardijweg 9, Postbus 295, 4900 AG OOSTERHOUT. Tel. 01620-31931.

Vorratsschutz: *Germany (FRG):* Vorratsschutz GmbH, Postfach 10, 6941 LAUDENBACH/BERGSTRASSE. Tel. 0 62 01/76 78.

Vosimex: *Netherlands:* Vosimex Int., Graafschap Hornelaan 182, 6004 HT WEERT. Tel. 04950-38257.

Wacker: *Denmark:* Wacker-Chemie A/S, KOEBENHAVN K.

- *Germany (FRG):* Wacker Chemie GmbH, Abteilung Pflanzenschutz, Prinzregentenstrasse 22, 8000 MUENCHEN 22. Tel. 0 89/2 10 91.
- *Netherlands:* Wacker-Chemie Nederland B.V., Zaanweg 63, Postbus 88, 1520 AB WORMERVEER. Tel. 075-288511.

Wallco: *Sweden:* Wallco AB, Box 30, 127 21 SKAERHOLMEN. Tel. 08/7405440.

Wellcome: *Belgium:* Wellcome, Industriezone, 3, 9440 AALST. Tel. 053/77 88 66.

- *France:* Wellcome S.A. (Laboratoires), 159, rue Nationale, 75640 PARIS CEDEX 13. Tel. (1)45 85 04 60 & 45 84 60 10; Telex Tabloid 204 276.
- *Netherlands:* Wellcome Nederland B.V., Ijsselmeerlaan 10, 1382 JT WEESP. Tel. 02940-18383.
- *Poland:* Wellcome, ul. Smolenskiego 4 n. 1, 01-698 WARSZAWA. Tel. 33 09 49; Telex 816929.
- *UK:* Wellcome Foundation Ltd., Ravens Lane, BERKHAMSTED, Herts.

Wesemael: *Netherlands:* R. van Wesemael B.V., Zoutestraat 109, 4561 TB HULST. Tel. 01140-14853.

Wesley: *USA:* Wesley Industries Inc., P.O. Box 490, MONTROSE, AL 36559. Tel. (205) 626-2040. Telex 467039.

Wheatley: *UK:* Wheatley Chemical Co. Ltd., Langthwaite Grange Industrial Estate, South Kirkby, PONTEFRACT, W. Yorks. WF9 3AP. Tel. (0977) 47578 or 47771.

Wider: *Germany (FRG):* F. A. Wider Chemische Fabrik, Postfach 4 05, 7057 WINNENDEN. Tel. 0 71 95/20 74.

Wikholm: *Sweden:* Knut Wikholm & Co., Nordenflychtsvaegen 64, 112 51 STOCKHOLM. Tel. 08/53 94 01-02.

Wilbur-Ellis: *USA:* Wilbur-Ellis Co., P.O. Box 1286, FRESNO, CA 93715. Tel. (209) 442-1220.

Wilson: *UK:* Kenneth Wilson (Holdings) Ltd., Morwick Hall, York Road, LEEDS LS15 4ND. Tel. (0532) 737373.

Windhoevel: *Germany (FRG):* W. Windhoevel GmbH & Co. KG, Kaisstrasse 1-3, 4000 DUESSELDORF 1. Tel. 02 11/30 56 71 & 30 51 11.

Wolf: *Germany (FRG):* Wolf-Geraete GmbH, Postfach 8 60 & 8 80, 5240 BETZDORF/SIEG. Tel. 0 27 41/28 10.
- *Netherlands:* Wolf Nederland, Reehorsterweg 50, Postbus 8, 6710 BA EDE. Tel. 08380-33455.

Woolfolk: *USA:* Woolfolk Chemical Works Inc., P.O. Box 938, FORT VALLEY, GA 31030.

Wuelfel: *Germany (FRG):* Chemische Fabrik Wuelfel, Just & Dittmar GmbH & Co., Hildesheimer Strasse 305, Postfach 89 01 09, 3000 HANNOVER 89. Tel. 05 11/83 13 41-42.

Wyss Samen & Pflanzen: *Switzerland:* Wyss Samen und Pflanzen AG, Postfach, 4500 SOLOTHURN. Tel. 065 26 21 21.

Xeda: *France:* Xeda International S.A., 58, rue Pottier, 78150 LE CHESNAY. Tel. (1) 39 54 74 45; Telex 695 612 F.

Young Il: *South Korea:* Young Il Chemical Co. Ltd., Woo-Jin Building, 212-2 Seocho-Dong, Gangnam-ku, SEOUL-135.

Zera: *Germany (FRG):* Zera Vertriebsges. fuer Agrarchemikalien mbH, Oldesloer Strasse 8, 2060 TRAVENBRUECK-TRALAU. Tel. 0 45 31/57 24.

Zerpa: *Netherlands:* Handelsonderneming Zerpa B.V., Pr. Marijkelaan 52, Postbus 16, 4020 BA MAURIK. Tel. 03449-1579.

Ziegler: *Switzerland:* A. Ziegler AG, Desinfektionsmittelfabrik, Gutstrasse 73, 8055 ZUERICH. Tel. 01 33 27 50.

Zoecon: *USA:* Zoecon Corp., Division of Sandoz Crop Protection, Professional Pest Management Division, 12200 Denton Drive, DALLAS, TX 75234. Tel. (214) 243-2321.

Zootechniki: *Greece:* Zootechniki SA, 38 Aristoteleos St., GR-104 33 ATHENS. Tel. (01) 88.31.814; Telex 215935.

Zootherap: *France:* Zootherap (Laboratoires), 4, boulevard Diderot, 75012 PARIS. Tel. (1) 46 28 28 05.
- *France:* Zootherap (Laboratoires), Le Pont Roch B.P. 11, Audrieu, 14250 TILLY-SUR-SEULLES. Tel. 31 80 20 46.

Zucchet: *Italy:* Zucchet S.p.A., Vicolo Pian Due Torri 52, 00146 ROMA. Tel. (06) 5270841.

Zuschlag: *Denmark:* A/S Arnold Zuschlags Laboratorium, KOEBENHAVN F.

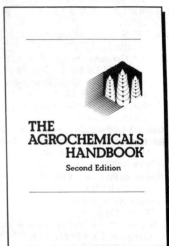

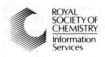

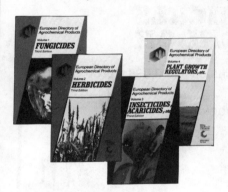

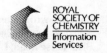